ELECTRONICS AND ELECTRICAL ENGINEERING

PROCEEDINGS OF THE 2014 ASIA-PACIFIC CONFERENCE ON ELECTRONICS AND ELECTRICAL ENGINEERING (EEEC 2014, SHANGHAI, CHINA, 27–28 DECEMBER 2014)

Electronics and Electrical Engineering

Editor

Alan Zhao
Shanghai Jiao Tong University, China

CRC Press is an imprint of the
Taylor & Francis Group, an **informa** business

A BALKEMA BOOK

CRC Press/Balkema is an imprint of the Taylor & Francis Group, an informa business

© 2015 Taylor & Francis Group, London, UK

Typeset by MPS Limited, Chennai, India

All rights reserved. No part of this publication or the information contained herein may be reproduced, stored in a retrieval system, or transmitted in any form or by any means, electronic, mechanical, by photocopying, recording or otherwise, without written prior permission from the publishers.

Although all care is taken to ensure integrity and the quality of this publication and the information herein, no responsibility is assumed by the publishers nor the author for any damage to the property or persons as a result of operation or use of this publication and/or the information contained herein.

Published by: CRC Press/Balkema
P.O. Box 11320, 2301 EH Leiden, The Netherlands
e-mail: Pub.NL@taylorandfrancis.com
www.crcpress.com – www.taylorandfrancis.com

ISBN: 978-1-138-02809-8 (Hardback)
ISBN: 978-1-315-68532-8 (eBook PDF)

Table of contents

Foreword	IX
Organizing Committee	XI

Fuzzy based direct torque and flux control of induction motor drives 1
C. Ning

Coordinated frequency control of thermal units with DC system in a wind-thermal-bundled
system transmitted by High Voltage Direct Current (HVDC) line 7
J. Dang, Y. Tang, M.C. Song, J. Ning & X. Fu

A biomedical system combined fuzzy algorithm for telemedicine applications 13
P.L. Peng, P.Z. Chen, C.Y. Pan, G.J. Jong & B.H. Lin

A preventive control method for overload in a multi-source grid 17
Z.W. Zhang, F. Xue, Y. Zhou, X.F. Song & L. Zhou

Effects of thermal annealing on the tungsten/lanthanum oxide interface 23
H. Wong, J.Q. Zhang, K. Kakushima, H. Iwai, J. Zhang & H. Jin

A study on the capacity optimization and allocation of wind/solar/diesel and energy
storage hybrid micro-grid systems 27
J.G. Zhang, P.Y. Liu & H. Zhang

A neural network Proportional-Integral-Differential (PID) control based on a genetic
algorithm for a coupled-tank system 33
Y.S. Li & H.X. Li

Research into the reactive power compensation of a new dual buck non-isolated grid inverter 39
P. Sun, Y. Xie, Y. Fang, L.J. Huang & Y. Yao

Modelling condition monitoring inspection intervals 45
A. Raza & V. Ulansky

A coordinated voltage control strategy for a Doubly-Fed Induction Generator (DFIG) wind farm system 53
J.J. Zhao, X.G. Hu, X. Lv & X.H. Zhang

SOM-based intrusion detection for SCADA systems 57
H. Wei, H. Chen, Y.G. Guo, G. Jing & J.H. Tao

Nonlinear and adaptive backstepping speed tracking control of a permanent magnet
synchronous motor despite all parameter uncertainties and load torque disturbance variation 63
H.I. Eskikurt & M. Karabacak

The convolution theorem associated with fractional wavelet transform 71
Y.Y. Lu, B.Z. Li & Y.H. Chen

Robustness testing method for intelligent electronic devices 75
H.T. Jiang, Y. Yang, W. Huang & Y.J. Guo

Design and analysis of quasi-optics for a millimeter wave imaging system 83
N.N. Wang, J.H. Qiu, Y. Zhang, Y. Zhang, P.Y. Zhang, H. Zong & L.Y. Xiao

A simply fabricated hybrid type metal based electrode for application in supercapacitors 87
S.C. Lee, U.M. Patil & S.C. Jun

Joint scheduling based on users' correlation in MU-CoMP 89
Y.F. Wang & D.L. Wang

A monitoring system of harmonic additional loss from distribution transformers 95
Z. Liu, Y. Liu, Q.Z. Cao & Z.L. Zhang

A performance evaluation of higher order modulation considering Error Vector Magnitude (EVM) in a Long Term Evolution (LTE) advanced downlink *X.S. Liu & Z.G. Wen*	101
Research on the control method of the driving system for the in-wheel driven range-extended electric vehicle *S.T. Wang & X. Zhang*	107
Research on TSP based two phases path planning method for sweep coverage of a mobile wireless sensor network *Z.Y. Zhang, W.L. Wang, Q.S. Fang & H.M. Cheng*	113
A monitor method of a distribution transformer's harmonic wave compatible to its loss *D. Yu, Y. Zhao, Y. Zhang, Y. Tan, J.M. Zhang & Z.L. Zhang*	117
Identification of a gas-solid two-phase flow regime based on an electrostatic sensor and Hilbert–Huang Transform (HHT) *J.X. Hu, X.H. Yao & T. Yan*	123
A fast multilevel reconstruction method of depth maps based on Block Compressive Sensing *T. Fan & G.Z. Wang*	129
Dynamic modelling with validation for PEM fuel cell systems *Y. Chen, H. Wang, B. Huang & Y. Zhou*	135
A highly sensitive new label-free bio-sensing platform using radio wave signal analysis, assisted by magnetic beads *J.H. Ji, K.S. Shin, Y.K. Ji & S.C. Jun*	143
Noise analysis and suppression for an infrared focal plane array CMOS readout circuits *P.Y. Liu, J.L. Jiang & C.F. Wang*	147
Speaker recognition performance improvement by enhanced feature extraction of vocal source signals *J. Kang, Y. Kim & S. Jeong*	151
Online detection and disturbance source location of low frequency oscillation *J. Luo, F.Z. Wang, C.W. Zhou & B.J. Wen*	155
A soft-start Pulse Frequency Modulation-controlled boost converter for low-power applications *M.C. Lee, M.C. Hsieh & T.I. Tsai*	161
Thermal analysis of phase change processes in aquifer soils *D. Enescu, H.G. Coanda, O. Nedelcu, C.I. Salisteanu & E.O. Virjoghe*	167
The development of a slotted waveguide array antenna and a pulse generator for air surveillance radar *M. Wahab, D. Ruhiyat, I. Wijaya, F. Darwis & Y.P. Saputera*	177
A harmonic model of an orthogonal core controllable reactor by magnetic circuit method *W.S. Gu & H. Wang*	181
A risk assessment model of power system cascading failure, considering the impact of ambient temperature *B.R. Zhou, R.R. Li, L.F. Cheng, P.Y. Di, L. Guan, S. Wang & X.C. Chen*	185
Target speech detection using Gaussian mixture modeling of frequency bandwise power ratio for GSC-based beamforming *J. Lim, H. Jang, S. Jeong & Y. Kim*	191
A compressive sampling method for signals with unknown spectral supports *E. Yang, X. Yan, K.Y. Qin, F. Li & B. Chen*	195
Design and analysis of SSDC (Subsynchronous Damping Controller) for the Hulun Buir coal base plant transmission system *G.S. Li, S.M. Han, X.D. Yu, S.W. Xiao & X.H. Xian*	201
Stage division and damage degree of cascading failure *X.Q. Yan, F. Xue, Y. Zhou & X.F. Song*	205
The design of a highly reliable management algorithm for a space-borne solid state recorder *S. Li, Q. Song, Y. Zhu & J.S. An*	211

A patrol scheme improvement for disconnectors based on a logistic regression analysis *J.S. Li, Y.H. Zhu & Z.Q. Zhao*	215
Progress on an energy storage system based on a modular multilevel converter *B. Ren, C. Liu, Y.H. Xu, C. Yuan, S.Y. Li & T. Wu*	219
Robust fall detection based on particle flow for intelligent surveillance *C.Q. Zhang & Y.P. Guan*	225
An IEC 61850 based coordinated control architecture for a PV-storage microgrid *H.Y. Huang, F.J. Peng, X.Y. Huang, A.D. Xu, J.Y. Lei, L. Yu & Z. Shen*	231
The design of an IED for a high voltage switch operating mechanism based on IEC 61850 *Z.Q. Liu & X.R. Li*	237
A profile of charging/discharging loads on the grid due to electric vehicles under different price mechanisms *M.Y. Li & B. Zou*	241
Algorithm design of the routing and spectrum allocation in OFDM-based software defined optical networks *S. Liu, X.M. Li & D.Z. Zhao*	247
The impact of Negative Bias Temperature Instability (NBTI) effect on D flip-flop *J.L. Yan, X.J. Li & Y.L. Shi*	253
The electrical property of a three dimensional graphene composite for sensor applications *M.S. Nam, I. Shakery, J.H. Ji & C.J. Seong*	259
A method of automatically generating power flow data files of BPA software for a transmission expansion planning project *B. Zhou, T. Wang, L. Guan, Q. Zhao, Y.T. Lv & L.F. Cheng*	261
The analysis of training schemes for new staff members from substation operation and maintenance departments *Y.T. Jiang, Y.B. Ren, X.H. Zhou, L. Mu, Y. Jiang & H.K. Liu*	267
Research of source-grid-load coordinated operation and its evaluation indexes in ADN *W. Liu, M.X. Zhao, H. Hui, C. Ye & S.H. Miao*	271
Progress on the applications of cascaded H-bridges with energy storage systems and wind power integrated into the grid *S.Y. Li, T. Wu, Y.S. Han, W. Cao & Y.H. Xu*	275
Aluminium alloy plate flaw sizing by multipath detection *D.B. Guo, X.Z. Shen & L. Wang*	281
An equilibrium algorithm clarity for the network coverage and power control in wireless sensor networks *L. Zhu, C.X. Fan, Z.G. Wen, Y. Li & Z.Y. Zhai*	285
Three-layer architecture based urban photovoltaic (PV) monitoring system for high-density, multipoint, and distributed PV generation *H. Gang, P. Qiu & D.C. He*	289
Modelling and estimation of harmonic emissions for Distributed Generation (DG) *L.F. Li, N.H. Yu, J. Hu & X.P. Zhang*	293
Mechanism and inhibiting methods for cogging torque ripples in Permanent Magnet Motors *H. Zhang, G.Y. Li & Z. Geng*	299
Content-weighted and temporal pooling video quality assessment *F. Pan, C.F. Li, X.J. Wu & Y.W. Ju*	305
A filter structure designing for an EAS system *M. Lin & J.L. Jiang*	311
A mini-system design based on MSP430F249 *M.M. Yang, Y.M. Tian & H.W. Wang*	317
A control system for the speed position of a DC motor *V.V. Ciucur*	321

Intelligent wireless image transmission system *M.M. Zhang, J.Y. Li & M.F. Wang*	325
Model predictive control for a class of nonlinear systems via parameter adaptation *C.X. Zhang, W. Zhang & D.W. Zhang*	329
Control strategy of BESS for wind power dispatch based on variable control interval *T. Lei, W.L. Chai, W.Y. Chen & X. Cai*	333
A study of the maintenance of transformer using a cost-effectiveness optimization model *L.J. Guo & S.M. Tao*	339
A new decision support system for a power grid enterprise overseas investment *L. Tuo & Z. Yi*	343
A coal mine video surveillance system based on the Nios II soft-core processor *P.J. Wei & L.L. Shi*	347
Research on a mechanism for measuring force in material moulding and the features of its measuring circuits *H.P. An, Z.Y. Rui & R.F. Wang*	351
Author index	355

Foreword

The purpose of EEEC-14 is promoting the creativity of the Chinese nation in the scope of Electronics and Electrical Engineering. Electronics, as well as Electrical Engineering are always the companion of Electrical and Electronics.

When we were collecting the papers for the Conference on Electronics and Electrical Engineering, the interesting things were that a number of authors were quite keen to look for the same chance to make public the theoretical works of intellectual creativity or formal results of scientific research or practice, written by those who are their former class mates, campus fellows, friends, relatives, colleagues and cooperators. It is really a chance for us to organize a chance for our smart intellectuals to expose, exchange and confidently approve each other and the value of their arduous work.

Electronics is one of the most important matters of our life and with us as ever we have existed, unlike computer or information technology but unfortunately we have never known all Electronics we have used around us. Nevertheless, we have been working very hard to discover new sources of energy we are in need of and striving non-stop for new synthetic stuffs or man-made matters. High demand has pushed our scholars, experts and professionals to continue the mission, not only for the materials themselves but non-perilous to the life and environment as well, causing the least hazard to the world. We appreciate our authors consciously to involve into that mission.

Asia-Pacific Electronics and Electrical Engineering Conference was scheduled to hold in Shanghai from December 27–28, 2014; experts and scholars, a group of the authors and other related people have attended the conference with apparent interest; we are expecting a full success of the conference.

We appreciate those who responded to our proposal and submitted their papers, especially those whose papers have been selected for the conference EEEC-14, the sponsors who have provided their valuable and professional suggestions and instructions and the scholars and professors who have spent their efforts as peer reviewers.

Thank you!
Samson Yu
December 1, 2014

Organizing Committee

General Chairs

Prof. Soloman Kim, *Shanghai Jiao Tong University, China*

Technical Program Committee

Dr. Yuh-Huei Shyu, *Tamkang University, Taiwan, China*
Dr. Yudi Gondokaryono, *Institute of Teknologi Bandung, Indonesia*
Dr. Mohd Nazri Ismail, *Universiti Kebangsaan, Malaysia*
Dr. Yongfeng Fang, *Xidian University, China*
Dr. Funian Li, *Wuhan University of Science and Technology, China*
Dr. Gabriel Alungbe, *University of Cincinnati, USA*
Dr. Mingsi Su, *Lanzhou University, China*
Dr. V. L. Manekar, *S. V. National Institute of Technology, India*
Prof. Xinge You, *Huazhong University of Science and Technology, China*
Prof. Yuwen Chen, *Shenyang Pharmaceutical University, China*
Prof. Jikang Jian, *Xinjiang University, China*
Prof. Ling Song, *Guangxi University, China*
Prof. Shaohua Teng, *Nanjing University, China*
Prof. Jinyuan Jia, *Tongji University, China*
Prof. Huailin Shu, *Guangzhou University, China*
Prof. Yibin He, *Wuhan Institute of Technology, China*
Prof. Qiuling Tang, *Guangxi University, China*
Prof. Qingfeng Chen, *Guangxi University, China*
Prof. Lianming Wang, *Northeast Normal University, China*
Prof. Lei Guo, *Beihang University, China*
Prof. Zongtian Liu, *Shanghai University, China*
Prof. Yimin Chen, *Shanghai University, China*
Prof. Xiaoping Wei, *China University of Mining and Technology, China*
Prof. Xiaodong Wang, *North China Electric Power University, China*
Prof. Jianning Ding, *Jiangsu University, China*
Prof. Xiaodong Jiang, *School of Electrical Engineering & Automation of Tianjin University, China*
Prof. Jinan Gu, *Jiangsu University, China*
Prof. Xueping Zhang, *Henan University of Technology, China*
Prof. Yingkui Gu, *Jiangxi University of Science and Technology, China*
Prof. Shengyong Chen, *Zhejiang University of Technology, China*
Prof. Qinghua You, *Shanghai Maritime University, China*
Prof. Bintuan Wang, *Beijing Jiaotong University, China*
Prof. Pengjian Shang, *Beijing Jiaotong University, China*
Prof. Yiquan Wu, *Nanjing University of Aeronautics and Astronautics, China*
Prof. Hongyong Zhao, *Nanjing University of Aeronautics and Astronautics, China*
Prof. Gang Ren, *Southeast University, China*
Prof. Jianning Ding, *Nanjing Normal University, China*
Prof. Chen Peng, *Nanjing Normal University, China*
Prof. Huajie Yin, *South China University of Technology, China*
Prof. Yuhui Zhou, *Bejing Jiaotong University, China*
Prof. Zhongjun Wang, *Wuhan University of Technology, China*
Prof. Zhongqiang Wu, *Yanshan University, China*
Prof. Wenguang Jiang, *Yanshan University, China*
Prof. Fuchun Liu, *Guangdong University of Technology, China*
Prof. Kangshun Li, *South China Agricultural University, China*

Prof. Jie Ling, *Guangdong University of Technology, China*
Prof. Lin Yang, *Shanghai Jiaotong University, China*
Prof. Xinfan Feng, *Jinan University, China*
Prof. Zongfu Hu, *Tongji University, China*
Prof. Wanyang Dai, *Nanjing University, China*

Fuzzy based direct torque and flux control of induction motor drives

Chuang Ning
Professional Engineering Electrical Design, Perth, Australia

ABSTRACT: This paper investigates direct torque and flux control of an induction motor drive based on the Fuzzy Logic (FL) control technique. Direct torque and flux control has become a widely acceptable alternative to field oriented control The hysteresis-band controller for the stator flux and the electro-magnetic torque was designed using a Fuzzy Logic System. (FLS) in MATLAB. Simulation results show that the direct torque and flux control using an FL approach performs very fast dynamic response and has a simple structure which makes it to be more popularly used in the industry.

1 INTRODUCTION

Induction Motor (IM) drives may be classified into two main control strategies. Scalar control, of the IM voltage magnitude and frequency, is one of ac drives which produces good steady-state performance but poor dynamic response. There is an inherent coupling effect using the scalar control because both torque and flux are the functions of voltage or current and frequency. This results in sluggish response and is prone to instability because of 5th order harmonics. However, vector control (VC) decouples these effects.

A second IM drive method can be either the field oriented control (FOC) or the direct torque and flux control (DTFC or DTC). The principle used in these drive methods is to asymptotically decouple the rotor speed from the rotor flux in vector controlled ac motor drives. The two most commonly used methods of the vector control are direct vector control and indirect vector control. Control with field orientation may either refer to the rotor field, or to the stator field, where each method has own merits [1].

Direct torque and flux control is also denoted as direct self-control (DSC) which is introduced for voltage-fed pulse-width-modulation (PWM) inverter drives. This technique was claimed to have nearly comparable performance with vector-controlled drives [2]. Another significance mentioned in [3] is that DTFC does not rely on machine parameters, such as the inductances and permanent-magnet flux linkage. Consequently, a number of research approaches have been proposed for a wide range of industrial applications where [3] is proposed for direct-drive PMSG wind turbines.

If an IM is being operated under its steady state, the three-phase drive can be easily presented as just one-phase because all the variables on the IM are sinusoidal, balanced and symmetrical. However, if the operation requires dealing with dynamics of motor speeds or varying torque demands in a sudden change, the motor voltages and currents are no longer in a sinusoidal waveform. Hence, the IM drive scheme using vector control or direct torque and field control is able to provide faster transient responses due to these dynamics.

This paper is arranged as follows. Section 2 lists the notations used in this paper. Section 3 explains the induction motor dynamics. Section 4 describes the direct torque and flux control. Section 5 presents the design of the fuzzy logic controller. Section 6 illustrates the performance of the controller in the simulation. Specific conclusions are drawn in Section 7.

2 NOMENCLATURE

d^e-q^e synchronous reference frame direct, quadrature axes
d^s-q^s stationary reference frame direct, quadrature axes
u_{sd}; u_{sq} stator voltages
i_{sd}; i_{sq} stator currents
u_{rd}; u_{rq} rotor voltages
i_{rd}; i_{rq} rotor currents
ψ_s; ψ_r stator, rotor flux vector
R_s; R_r stator, rotor resistance
L_s; L_r stator, rotor self-inductance
L_m magnetizing inductance
σ resultant leakage constant
ω_e; ω_r synchronous, rotor speed
P number of motor pole pairs
J Total inertia
T_e; T_L electromagnetic, load torque

3 INDUCTION MOTOR DYNAMICS

3.1 Machine model in arbitrary reference frames

In a three-phase ac machine, there are three main reference frames of motion, which could be used to model

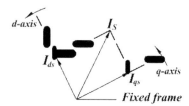

Figure 1. Reference frame of motion in an ac machine.

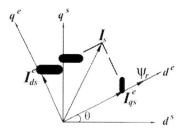

Figure 2. Decoupling between rotor flux and torque.

its three main regions of operation. These are the stationary reference frame for startup, the synchronous reference frame for equilibrium motion, and the rotor reference frame for changing speeds by acceleration or deceleration. The two commonly employed coordinate transformations with induction machines are the stationary and the synchronous reference frames as shown in Fig. 1.

In special reference frames, the expression for the electromagnetic torque of the smooth-air-gap machine is similar to the expression for the torque of a separately excited dc machine.

These mathematical transformations of rotor ABC variables to rotor d-q variables, which are known as Park Transformation [4], can facilitate understanding of the variation of the mutual inductance between the stator and the rotor under differing rotation conditions.

All the transformation equations from ds – qs frame to de – qe frame, and vice verse, remain the same as in an induction motor. The complex stator current space vectors sq and isd are defined as:

$$i_{qs}^s = \frac{2}{3}i_{as} - \frac{1}{3}i_{bs} - \frac{1}{3}i_{cs};$$
$$i_{ds}^s = \frac{1}{\sqrt{3}}(i_{cs} - i_{bs}). \quad (1)$$

A dc motor-like electromechanical model can be derived for an ideal vector-controlled drive in the d-q co-ordinate. One of the advantages of the separately excited dc motor of being able to decouple the flux control and the torque control that is thereby opened up. Fig. 2 shows this concept that the rotating vectors are orthogonal, but decoupled.

3.2 Modeling of an induction motor in d–q co-ordinate

The mathematical model [5] of an induction motor in d–q reference frame can be written as the stator voltage differential equations (2):

$$u_{sd} = R_s i_{sd} + \frac{d}{dt}\psi_{sd} - \omega_e \psi_{sq};$$
$$u_{sq} = R_s i_{sq} + \frac{d}{dt}\psi_{sq} + \omega_e \psi_{sd}. \quad (2)$$

And the rotor voltage differential equations (3):

$$u_{rd} = 0 = R_r i_{rd} + \frac{d}{dt}\psi_{rd} - (\omega_e - \omega_r)\psi_{rq};$$
$$u_{rq} = 0 = R_r i_{rq} + \frac{d}{dt}\psi_{rq} + (\omega_e - \omega_r)\psi_{rd}. \quad (3)$$

These linked fluxes of stator and rotor are as follows:

$$\psi_{sd} = L_s i_{sd} + L_m i_{rd};$$
$$\psi_{sq} = L_s i_{sq} + L_m i_{rq};$$
$$\psi_{rd} = L_r i_{rd} + L_m i_{sd};$$
$$\psi_{rq} = L_r i_{rq} + L_m i_{sq}. \quad (4)$$

Since the direction in d-axis is aligned with the rotor flux this means that the q-axis component of the rotor flux space vector is always zero.

$$\psi_{rq} = 0 \quad \text{and} \quad \frac{d}{dt}\psi_{rq} = 0. \quad (5)$$

The slip frequency ω_{sl} can be calculated from the reference values of stator current components are defined in the rotor flux oriented reference frame as follows:

$$\omega_{sl} = \omega_e - \omega_r \quad (6)$$

We can calculate the rotor speed from the relation $\omega_r = \omega_e - \omega_{sl}$, since ω_{sl} and ω_e can be determined as [6]:

$$\omega_{sl} = \frac{(1+\sigma\tau_r s)L_s i_{qs}}{\tau_r(\psi_{ds} - \sigma L_s i_{ds})};$$
$$\omega_e = \frac{(V_{qs}^s - i_{qs}^s R_s)\psi_{ds}^s - (V_{ds}^s - i_{ds}^s R_s)\psi_{qs}^s}{\psi_s^2}. \quad (7)$$

Here τ_r is the rotor time constant denoted as $\tau_r = \frac{L_r}{R_r}$, and σ is a resultant leakage constant defined as:

$$\sigma = 1 - \frac{L_m^2}{L_s L_r}$$

The state equations of a linear model for the induction motor dynamics can be obtained as:

$$\dot{x} = \begin{bmatrix} \frac{(L_r^2 R_s + L_m^2 R_r)}{\sigma L_s L_r^2} & 0 & \frac{L_m R_r}{\sigma L_s L_r^2} & \frac{L_m \omega_r}{\sigma L_s L_r} \\ 0 & \frac{(L_r^2 R_s + L_m^2 R_r)}{\sigma L_s L_r^2} & \frac{L_m \omega_r}{\sigma L_s L_r} & \frac{L_m R_r}{\sigma L_s L_r^2} \\ \frac{L_m R_r}{L_r} & 0 & -\frac{R_r}{L_r} & -\omega_r \\ 0 & \frac{L_m R_r}{L_r} & \omega_r & \frac{R_r}{L_r} \end{bmatrix} x + \begin{bmatrix} \frac{1}{\sigma L_s} & 0 \\ 0 & \frac{1}{\sigma L_s} \\ 0 & 0 \\ 0 & 0 \end{bmatrix} \begin{bmatrix} V_{sd} \\ V_{sq} \end{bmatrix} \quad (8)$$

$$x = [i_{sd}\ i_{sq}\ \psi_{rd}\ \psi_{rq}]^T$$

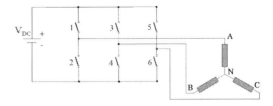

Figure 3. Basic scheme of 3-phase inverter connected to an AC motor.

However, induction motors belong to a class of multi-variable nonlinear systems which could lead to the control task to be rather complicated due to to unknown disturbances (load torque) and changes in values of parameters during its operation. Such a challenge have been stated in hybrid vector control or DTFC induction motor drives [6]–[10].

The electromagnetic torque equation can be written as:

$$\bar{T}_e = \frac{3}{2}(\frac{P}{2})\bar{\psi}_s \times \bar{I}_s$$
$$= \frac{3}{2}(\frac{P}{2})\frac{L_m}{L_r L_s}\bar{\psi}_r \times \bar{\psi}_s \quad (9)$$

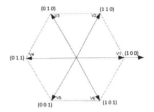

Figure 4. SVPWM voltage vectors

Thus, the magnitude of the torque is

$$T_e = \frac{3}{2}(\frac{P}{2})\frac{L_m}{L_r L_s} \mid \psi_r \mid \mid \psi_s \mid \sin \theta_{sr} \quad (10)$$

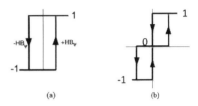

Figure 5. (a) Two-level stator flux hysteresis-band. (b) Three-level stator torque hysteresis-band.

where $\theta_{sr} = \theta_s - \theta_r$ is the angle between the stator flux and the rotor flux.

If the rotor flux remains constant and the stator flux is incrementally changed by the stator voltage $V^-s \times \Delta t$ with the corresponding angle change of $\Delta \theta_{sr}$, the incremental torque is then obtained as:

$$\Delta T_e = \frac{3}{2}(\frac{P}{2})\frac{L_m}{L_r L_s} \mid \psi_r \mid \mid (\bar{\psi}_s + \Delta \bar{\psi}_s) \mid \sin \Delta \theta_{sr} \quad (11)$$

The mechanical speed of the rotor in terms of the load constants can be computed as:

$$T_e = J\frac{d\omega_r}{dt} + B\omega_r + T_L \quad (12)$$

4 STUDY ON DIRECT TORQUE AND FLUX CONTROL

4.1 Space vector modulation of 3-phase voltage source in-verter with DTFC

During the IM drive operation, a control strategy for the voltage-fed space vector pulse-width-modulation (SVPWM) may be required with direct torque and field control. The structure of a typical 3-phase power inverter is shown in Fig. 3, where V_A, V_B and V_C are the voltages applied to the star-connected motor windings, and where V_{DC} is the continuous inverter input voltage.

The SVPWM method of generating the pulsed signals fits the above requirements and minimizes the harmonic contents. The inverter voltage vectors provide eight possible combinations for the switch commands. These eight switch combinations determine eight phase voltage configurations. SVPWM supplies the ac motor with the desired phase voltages. A diagram in Fig. 4 depicts these combinations.

4.2 Control strategy of DTFC

The magnitudes of command stator flux $\hat{\psi}_s^*$ and T_e^* are compared with the respective estimated values, and the errors are processed through the hysteresis-band (HB) controllers, as shown in Fig. 5.

The circular trajectory of the command flux vector $\hat{\psi}^*$ with the hysteresis-band rotates in an anti-clockwise direction. The flux hysteresis-band controller has two levels of digital output according to the following relations [2]:

$$\begin{aligned} H_\psi &= 1 \quad \text{for } E_\psi > +HB_\psi \\ H_\psi &= -1 \quad \text{for } E_\psi < -HB_\psi \end{aligned} \quad (13)$$

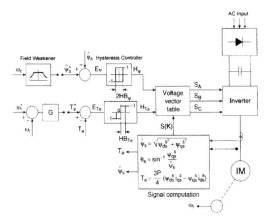

Figure 6. Block diagram of direct torque and field control.

Table 2. Flux and torque variations due to applied voltage vectors.

Table 2 shows the effect of the voltage vectors on the stator flux and the electromagnetic torque, which the arrows indicate the magnitudes and directions [2].

Table 1. Switching table of inverter voltage vectors.

H_ψ	H_{Te}	S_1	S_2	S_3	S_4	S_5	S_6
1	1	V_2	V_3	V_4	V_5	V_6	V_1
	0	V_0	V_7	V_0	V_7	V_0	V_7
	−1	V_6	V_1	V_2	V_3	V_4	V_5
−1	1	V_3	V_4	V_5	V_6	V_1	V_2
	0	V_7	V_0	V_7	V_0	V_7	V_0
	−1	V_5	V_6	V_1	V_2	V_3	V_4

Also, a torque controller proportional gain is chosen by G = 1.5. The overall control of the DTFC is shown in Fig. 6.

When the inverter voltage sequence $V_1 - V_6$ is properly selected as shown in Fig. 4, the stator flux rotates at the desired synchronous speed within the specified band. As the stator resistance is small enough to be neglected, we may consider that the stator flux monotonically follows the stator voltage at each step time Δt.

Thus, changing the stator flux space vector can be accomplished by changing the stator voltage during a desired period of time which can be expressed as follows:

$$\bar{V}_s = \frac{d}{dt}(\bar{\psi}_s) \Rightarrow d\bar{\psi}_s = \bar{V}_s dt; \quad (14)$$
$$\text{Thus,} \quad \Delta\bar{\psi}_s = \bar{V}_s \Delta t$$

Depending on the sector that the voltage reference in Fig. 4, two adjacent vectors are chosen. The binary representations of two adjacent basic vectors differ in only one bit from 000 to 111. This means the switching pattern moves from one vector to the adjacent one. The two vectors are time weighted in a sample period T to produce the desired output voltage to the inverter.

Table 1 applies the selected voltage vectors, which essentially affects both the flux and torque simultaneously.

5 DESIGN OF THE FUZZY LOGIC CONTROLLER

5.1 Structure of fuzzy control in DTFC

Fuzzy logic has been widely applied in power electronic systems. Recently developed approaches in DTFC [11]–[13] have been proven to be more robust and improved performance for dynamic responses and static disturbance rejections using fuzzy logic control. [11] also designs a hysteresis-band controller in DTFC using a fuzzy logic method, but their fuzzy controller appears not precise enough because of the less membership functions being chosen.

Now a fuzzy logic controller is considered for the stator flux $\hat{\psi}^*$ and torque T^* of the DTFC induction motor drive. The fuzzy inference system (FIS) onsists of a formulation of the mapping from a given input set of E and CE to an output set using FL Mamdany type method in this study.

According to the switching table of the inverter voltage vectors, the triangular membership functions (MF) of the seven linguistic terms for each of the two inputs $e(pu)$ and $ce(pu)$ are defined in per unit values. $du(pu)$ is the output of the fuzzy inference system. Here, $e(pu)$ is selected as the flux error $\hat{\psi}^* - \hat{\psi}_s$ for the difference between the command stator flux and the actual stator flux ψ_s, and $ce(pu)$ is the rate of change of $\frac{d}{dt}\Delta\hat{\psi}_s$. The represented linguistic variables in the fuzzy rule matrix are:

NB = negative big   NM = negative medium
NS = negative small   Z = zero
PS = positive small   PM = positive medium
PB = positive big

The general considerations in the design of the proposed fuzzy logic controller for this DTFC are:

1) If both $e(pu)$ and $ce(pu)$ are zero, then maintain the present control setting $du(pu) = 0$ (V_0/V_7).
2) If $e(pu)$ is not zero but is approaching to this value at a satisfactory rate, then maintain the present control setting of the voltage vector.

Table 3. Fuzzy logic rule matrix.

ce(pu) \ du(pu), e(pu)	NB	NM	NS	Z	PS	PM	PB
NB	V_1	V_1	V_2	V_2	V_6	V_6	V_0
NM	V_1	V_2	V_2	V_6	V_6	V_7	V_3
NS	V_2	V_2	V_6	V_6	V_0	V_3	V_3
Z	V_2	V_6	V_6	V_7	V_3	V_3	V_5
PS	V_6	V_6	V_0	V_3	V_3	V_5	V_5
PM	V_6	V_7	V_3	V_3	V_5	V_5	V_4
PB	V_0	V_3	V_3	V_5	V_5	V_4	V_4

3) If $e(pu)$ is growing, then change the control signal $du(pu)$ depending on the magnitude and sign of $e(pu)$ and $ce(pu)$ to force $e(pu)$ towards zero.

$$\begin{aligned} E_\psi(k) &= \hat{\psi}_s^* - \hat{\psi}_s; \\ CE_\psi(k) &= E_\psi(k) - E_\psi(k-1) \\ e_\psi(pu) &= \frac{E_\psi(k)}{GE} \end{aligned} \quad (15)$$

Here, GE is a respective scale factor to convert the fuzzy input and output variables into as per unit.

The suitable fuzzy rules are selected for the DTFC, such that the variations of the fuzzy output $du(pu)$ depend on the required inverter voltage vectors as listed in Table 2. This yields a 49-rule structure as shown in Table 3.

5.2 Fuzzy logic control implementation in MATLAB

The Fuzzy Logic Toolbox in MATLAB provides a very comprehensive user friendly environment to handle high-level engineering issues. Fig. 7 illustrates the MATLAB FIS editor for implementing this DTFC system.

The variation of the developed electromagnetic torque can also be obtained from the equations (10), (11) and (12). Once obtained the flux error, the torque error and the angle θ_{sr} can be achieved in MATLAB/Simulink as follows in Fig. 8.

6 SIMULATION RESULTS

6.1 Performance of the DTFC drive

Once the fuzzy algorithm had been developed, the performance test of the DTFC drive was carried out on a fairly large 150 kW induction motor in simulation. The rotor speed was set at 500 rpm and the stator flux was set to 0:8 Wb. The motor was 80% loaded by 960 Nm at 1:5 s after running.

Fig. 9 shows the stable flux locus and the stator current responding to the load. Fig. 10 illustrates the

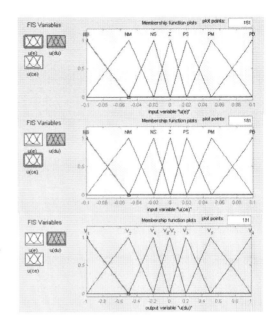

Figure 7. Membership functions of the fuzzy logic controller (a) Input u(e), (b) Input u(ce), (c) Output u(du).

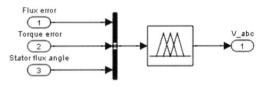

Figure 8. Fuzzy hysteresis-band controller in MATLAB/Simulink.

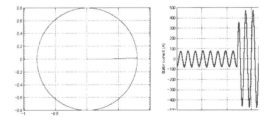

Figure 9. (a) Stator d-q axis flux. (b) Stator current response to the load change.

deviation responses from the rotor speed and the electromagnetic torque respectively due to the load change at t = 1:5 s.

It can be observed that the motor reaches the desired speed in 0:6 s. The rotor speed deviation affected by the load is 28 rpm to yield a speed error about 5:6% instantaneously. The torque response following the load is considerably fast with less ripples. The steady-state three-phase current of the motor is plotted in Fig. 11, which shows that it is smoothly balanced with less distortion.

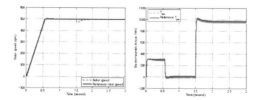

Figure 10. (a) Rotor speed deviation vs the reference rotor speed. (b) Electromagnetic torque vs the reference torque.

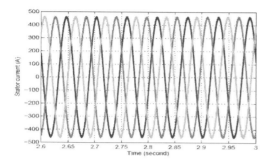

Figure 11. The steady-state three-phase current of the motor.

Table 4. The induction motor parameter values.

R_s	14.85×10^{-3} Ω	P	2 poles
R_r	9.295×10^{-3} Ω	V_{rated}	415 V
L_s	0.3027×10^{-3} H	f	50 Hz
L_r	0.3027×10^{-3} H	J	3.1 kg m^2
L_m	10.46×10^{-3} H	B	0.08 N m s

The simulated induction motor parameters are given in Table 4.

We may note that one significant advantage of the DTFC drive is reasonably easy to be implemented numerically and conceptually less complicated in design, because of the absence of the vector transformations from the d-q synchronous frame to the d-q stationary frame such as the FOC method. No feedback current control is required for the DTFC drive, but the motor seems to draw the higher current.

7 CONCLUSION

DTFC offers superior induction motor drive performances by effectively combining the accurate and fast flux and torque control. The fuzzy logic based direct torque and flux control of induction motor drives appears quite simple and robust. The power inverter operational control plays an important key role in modern power electronics motor drives and the DTFC drive has made its cost effective due to its simple control structure.

The DTFC technique combined with the FLS technique has demonstrated a fast electromagnetic torque response when the motor is 80% loaded and the stator magnetic flux can be kept upon the desired flux-band during the operation. Also, a good tracking of the motor speed under the load change has been verified.

REFERENCES

[1] J. Holtz S nsorl sscontrol of induction motor drives. Proceedings of the IEEE., Vol. 90, pp. 1359–1394, 2002.

[2] B. K. Bos Modem power electronics and AC driv s. Prentice Hall 2002.

[3] Z. Zhang, Y. Zhao, V. Qiao and L. Qu A discrete-tim direct-torU and flux control for direct-drive PMSG wind turbin. IEEE Industry Applications Conference, 1–8, 2013.

[4] R.J. Lee, P. Pillay and R.G. Harley. D. Q ref rence frames for the simulation of induction motors. Electric Power Systems Research, Vol. 8, pp. 15–26, 1984/1985.

[5] J. Holtz. On the spatial propagation of transient magnetic fields in AC machin s. IEEE Transactions on Industry Applications, Vol. 32(4), pp. 927–937, 1996.

[6] B. K. Bose, M. G. SimB D. R.Cr hus K. R ash kara and R.Martin. Speed sensorl ss hybrid vector controlled induction motor driv Industry Applications Coference, Vol. 1, pp. 137–143, 1995.

[7] P. Ttin n and M. Surandra. The next generation motor control method. DTC direct torqu tontrolIEEE International Coernce on Power Electronics. Drives and Energy Systems for Industrial Growth, Vol. 1, pp. 37–43, 1996.

[8] A.M rabet M. Ouhrouch and R. T. Bui Nonlin prediutiv control with disturbance obs rver for induction motor drive. IEEE International Symposium on Industrial Electronics, Vol. 1, pp. 86–91, 2006.

[9] A. Merabet. H. Arioui and M. Ouhrouche Cascaded predictive controll :rd ign for sp dtontrol and load toU r tion of indution motor. American Control Conference, Vol.1, pp. 1139–1144, 2008.

[10] Y.S. Lai V.K. Wang and Y.C. Ch n Novel swi hing t chniqu for reducing the speed ripple of AC drives with direct torque control. IEEE Transactions on Industrial Electronics, Vol. 51(4), pp. 768–775, 2004.

[11] A.Lokriti Y. Zidani and S. Doubabi Fuzzy logic control contribution to the dirωttor u and fluxtontrol of an induction machin Interna – tional Conference on Multimedia Co uting and Systems (ICMCS), pp. 1–6, 2011.

[12] T. RameshA. K. Panda and S. S. Kumar 1)'pe-1 and type-2 fuzzy logi't p dtontroll r based high performanc dirωt torqu and flux controlled induction motor drive. Annual IEEE India Conference (INDICONp) p. 1–6, 2013.

[13] T. Ramesh A. K. Panda and S. S. Kumar Slidingmode and fuzzy logtucontrol based MRAS sp d stimators for s nsorl ss dirωt torU and flux control of an induction motor drive. Annual IEEE India Conference (INDICON), pp. 1–6, 2013.

Coordinated frequency control of thermal units with DC system in a wind-thermal-bundled system transmitted by High Voltage Direct Current (HVDC) line

Jie Dang
Central China Power Dispatching and Communication Center, Wuhan, China

Yi Tang, Mengchen Song & Jia Ning
School of Electrical Engineering, Southeast University, Nanjing, China

Xiangyun Fu
State Grid Jiangsu Electric Power Company, Lianyungang Power Supply Company, Lianyungang, China

ABSTRACT: The wind-thermal-bundled power system transmitted by High Voltage Direct Current (HVDC) line has been an important development mode for large-scale wind power bases in northwest of China nowadays. In terms of the frequency stability problem of sending-end system caused by wind power fluctuation and some faults, the necessity for HVDC participating in frequency regulation is put forward. In this paper, the coordinated frequency control strategies of thermal units and HVDC link are proposed. Two cases are considered: the fluctuation of wind power and sending-end system faults. Simulations are carried out based on an equivalent wind-thermal-bundled power system to verify the effectiveness of the control strategies.

1 INTRODUCTION

Large amounts of wind power need to be transmitted to load centres by long transmission lines, because ten million kilowatt level wind bases are distributed in the northwest, the north, and the northeast of China which are far from the load centres. However, it is almost impossible to transmit wind power alone to the load centres over a long distance due to the fluctuation of wind power. Nowadays, the developmental pattern of wind-thermal-bundled power systems transmitted by HVDC transmission lines has been proposed, thus making full use of wind resources and mitigating the effects of fluctuations in wind power on receiver systems (Guo et al. 2012).

Large-scale wind power integration has a series of impacts on power system stability (Doherty et al. 2010, Chen et al. 2011). Research into the respect of frequency control focuses primarily on two points. One point is about wind turbine generation and wind power base research into wind turbine generators' participation in the system's frequency regulation, for examples of the analysis of frequency regulation capability of wind turbine (Zhang et al. 2012, Conroy et al. 2008), and inertia control of wind turbine (Keung et al. 2009, Miao et al. 2010). The other point is about power grid control research into wind power's impacts on frequency regulation and reserve (Wu et al. 2011), active power control strategies (Li et al. 2013) and an optimal generating unit tripping scheme during transmission (Chen et al. 2013).

An HVDC transmission system has multiple operation patterns and is usually applied to improving the stability of the AC-DC system. HVDC modulation participating in system frequency regulation is always applied in island operating mode of the system whose sending terminals are thermal power units (Chen et al. 2013). A variable structure DC power modulation and an optimal coordinated control strategy is proposed for restraining AC system frequency oscillation caused by a sudden change of load (Zhu & Luo, 2012). Considering the fluctuation of wind power, researchers (Zhu et al. 2013) put forward a coordinated control strategy on the basis of a DC system tracking the fluctuation in wind power and this strategy reduced the switching times of the HVDCs' tap-changers and AC filters by a DC step control.

As a new development mode of power transmission, the wind-thermal-bundled power system with AC/DC transmission lines is different from either the distributed development pattern in Europe or the existing development pattern of wind power in China. It is a typical structure of a large power supply, long transmission lines, and a weak power grid and its frequency control strategy remains to be further researched. In this paper, the importance of DC system participating in the frequency regulation of a sending-end system is discussed. The coordinated frequency control strategy of thermal generator units and DC system is proposed for solving wind power fluctuation problems and sending-end systems' failure problems. At the end of this paper, the proposed strategy is applied in

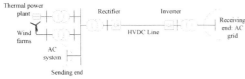

a. A simplified system of the wind-thermal-bundled islanded system with HVDC link

b. A simplified system of the wind-thermal-bundled interconnected system with HVDC link

Figure 1. Wind-thermal-bundled system with HVDC link.

a simulation case based on the equivalent northwest system with wind-thermal-bundled generation and the results show the validity of the proposed strategy.

2 INTRODUCTION OF THE STUDIED SYSTEM AND ITS FREQUENCY REGULATION METHODS

2.1 The system studied

The power system studied is shown in Figure 1. The wind-thermal-bundled power system transmitted by a DC line has two typical topologies: a wind-thermal-bundle islanded system with HVDC link (see Figure 1a) and a wind-thermal-bundle interconnected system with HVDC link (see Figure 1b). In an islanded system, the sending end consists of only wind farms and a thermal power plant. While in an interconnected system, the power source should be connected with the AC system.

In this paper, an equivalent system of WTB shown in Figure 1 is established. In this system, the total installation capacity of WTB system is 3 GW and the actual transmission power is 2 GW. The wind power base and the thermal generation base are connected to a converter station by 750 kV transmission lines. The converter station transmits 2 GW output with a ±800 kV HVDC transmission system and the type of transmission line is a double loop. When working in the interconnected mode, the sending-end is connected to the AC system by 750 kV double loop AC transmission lines. A thermal generator is modelled on a 6 order synchronous generator model, with an excitation system and a speed control system in this case. The wind farm consists of GE1.5MW type doubly-fed induction generators (DFIGs).

2.2 Existing problem

The active power output of wind turbine generators varies with wind energy. Nowadays wind turbine generators do not participate in frequency regulation in China. Thus, when the system's frequency changes, wind turbine generators could not response to it, which provides no help to stabilize the system's frequency. Moreover, due to the small inertia of the wind turbine, the whole system's inertia decreases with some synchronous generators replace by wind turbine generators. The frequency regulation capability significantly declined because the ratio of the installed capacity of the wind turbine generators is continuously increasing. This situation aggravates the frequency regulating burden of the sending-end system. In addition, relying solely on thermal power units, it is not possible to response to fast and large frequency changes. Therefore, it is necessary to study other frequency regulation strategies.

Frequency fluctuation is severe in a wind-thermal-bundled islanded system because in this pattern load frequency regulation effect does not exist. In a wind-thermal-bundled interconnected system, if the transmission power is large, the short-circuit current of the sending system is small. In this situation, the transmission system is recognized as a weak AC system which easily suffers from the influence of wind power fluctuation, wind farm failure, and disturbances of the AC system.

2.3 Frequency regulation methods in sending-end system

2.3.1 Primary Frequency Regulation (PFR) of thermal units

Primary frequency regulation means the function that the generator control system struggles to make the active power reach a new equilibrium and keeps the system's frequency in an acceptable range by raising or reducing the generating power automatically according to changes in the system's frequency.

Several parameters like the dead band ε and the limited range will affect the primary frequency regulation ability of thermal units.

The governor dead band ε must be set up in a reasonable range. If the dead band is too small, even slight frequency deviations will cause the governor to react. If the dead band is too large, it will affect the effeteness of primary frequency regulation because the governor does not response to large power shortages.

The limited range of PFR means the maximum controllable frequency range that generator units can achieve and this parameter determines the regulating quantity of governor.

According to the guidelines for the power system's operation, the dead band of the thermal generator units, based on electro hydraulic the turbine control system, is usually within ±0.033 Hz (±2r/min) and the limited range is usually at 6%.

2.3.2 DC modulation

A notable advantage of DC transmission systems over those of AC transmission systems is their fast controllability. Therefore, the active power of a DC system can be regulated during disturbances, in order to improve the stability of the system.

In this paper, DC frequency modulation and DC emergency power control are considered.

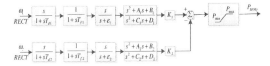

Figure 2. Schematic diagram of the DC frequency modulation controller.

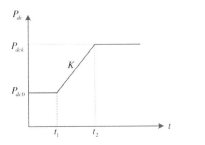

Figure 3. DC active power curve of DC power modulation.

The configuration and parameters of a DC frequency modulation controller is shown in Figure 2. The input signal is the frequency deviation derived from an AC line, through a differential link, filtering link, pre-guided compensation link, notch filtering link, amplification link and limiter, getting the output signal P_{MOD}. Where ω is the frequency deviation of the AC system; T_d is the time constant in a differential link; T_f is the time constant in the filter; ε is the guide compensation factor; $A\backslash B\backslash C\backslash D$ are parameters of notch filter; K is the gain of the controller; P_{max} and P_{min} are the upper limit and lower limit of the controller respectively.

DC emergency power control means that the output of the DC power is artificially changed according to the rules shown in Figure 3. In Figure 3, t_1 represents the start time of the raising/reducing power, and t_2 represents the end time of the raising/reducing power, K represents the modulation rate and the formula relationship of DC power is:

$$P'_{dc} = P_{dc0} + K(t_2 - t_1) \qquad (1)$$

3 COORDINATED FREQUENCY CONTROL STRATEGY WITH WIND POWER FLUCTUATION

3.1 Coordinated frequency control philosophy and strategy

In consideration of wind speed fluctuation, the key of frequency regulation strategy is the coordination and cooperation between PFR and DC frequency modulation. Coordinated control strategy can be divided into two modes according to regulated quantity and action sequence:

Mode 1: the DC system works as an auxiliary of the thermal generator unit when regulating frequencies.

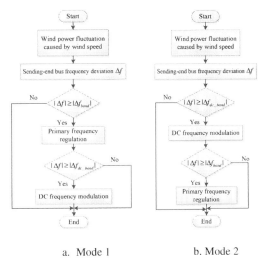

a. Mode 1   b. Mode 2

Figure 4. Coordinated frequency control strategies under normal conditions.

Table 1. Parameters of three control strategies

Control Strategies	Thermal units		DC modulation	
	dead band	limitation	Td/s	limitation
Strategy 1	0.033 Hz	6%	20	10%
Strategy 2	0.1 Hz	5%	10	20%
Strategy 3	0.033 Hz	6%	–	–

Mode 2: the thermal generator unit works as an auxiliary of the DC system when regulating frequencies.

Flow charts of two modes are shown in Figure 4.

Considering wind speed fluctuation, DC frequency modulation controller is used to participate in frequency regulation. The two control modes above can be realized by setting different thermal generator units' governor dead band ε and derivative time constant T_d of DC frequency modulation. For example, if governor dead band ε is small and T_d is large, it can realize the control objective of Mode 1. Similarly, if the governor dead band ε is large and T_d is small, it can realize the control objective of Mode 2. A simulation case is set up to compare the control effect of two modes for a wind-thermal-bundled islanded system and a wind-thermal-bundled interconnected system separately.

3.2 Wind-thermal-bundled islanded system

Strategies are tested in WTB islanded system based on the system shown in Figure 1a. In order to compare and analyse, three control strategies are tested in the simulation case: Mode 1, Mode 2, and PFR only. Parameters of these three strategies are shown in Table 1.

It assumes that a gust wind whose max speed is 1 m/s happens at 10 s and lasts 10 s. Figure 5 shows

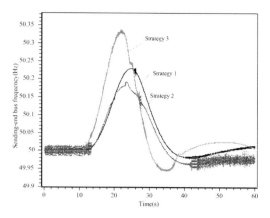

Figure 5. Comparison of frequency curve of sending-end bus.

Table 2. The results of three control strategies.

Control strategies	Maximum frequency deviation/Hz	Steady state
Strategy 1	0.23	better
Strategy 2	0.193	oscillation
Strategy 3	0.331	better

Table 3. The results of three control strategies.

Control strategies	Maximum frequency deviation/Hz	Steady state value/Hz
Strategy 1	0.23	50.01
Strategy 2	0.26	49.96
Strategy 3	0.34	50.02

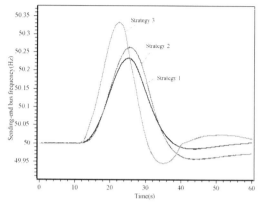

Figure 6. Comparison of sending-end bus frequency curve.

the changes of sending-end bus frequency with three control strategies.

When only PFR worked, the maximum deviation of system frequency reached 0.34 Hz. Compared with Mode 1, Mode 2 helped lower the maximum deviation of system frequency effectively but its stabilized frequency value was worse and frequency fluctuation happened in the pattern of island operation condition. In all, both the resulting curve of frequency regulation and the new stabilized frequency value was better when Mode 1 was applied in the simulation system.

Table 2 shows the results of simulation case with three different strategies. Overall, it is reasonable to choose Mode 1 as frequency control strategy in the wind-thermal-bundled islanded pattern with HVDC transmission lines.

3.3 Wind-thermal-bundled interconnected system

Wind-thermal-bundled interconnected system is established based on Figure 1b. The proposed strategies are separately tested in this simulation case. Parameters are the same with Table 1 and Figure 6 show the simulation results.

Table 3 shows the results of the simulation case with three control strategies.

When only relying on PFR, the deviation of system frequency is the largest and the frequency restored well. Apparently the result of Mode 1 is better than that of Mode 2 owing to a lower system frequency deviation and a higher frequency restoration value. In all, Mode 1 is very suitable for wind-thermal-bundled interconnected pattern with HVDC transmission lines.

To sum up, in the condition of wind power fluctuation, Mode 1 is the most competitive control strategy for both wind-thermal-bundled islanded pattern and interconnected pattern.

4 COORDINATED FREQUENCY CONTROL STRATEGY UNDER FAULT CONDITIONS

4.1 Coordinated frequency control philosophy and strategy

When power disturbance occurs in the power grid, such as the input/excision of load, cut off generators, wind farms getting off, and HVDC monopole block, etc., the sudden change in power will lead to imbalance between generation and load. If the DC system operates at constant power mode or constant current mode, the thermal generators will increase or decrease its electromagnetic power according to its frequency regulation characteristic in order to achieve a dynamic balance. If the response speed of the regulation system is slow or the generator has reached to its upper limit, the system will appear as a power persistent disequilibrium, leading to the collapse of the whole system eventually.

The coordinated control strategy under fault conditions is as follows. Under normal circumstances, DC transmission lines operate at a fixed power mode. When a fault occurs in the sending-end AC system, the fault signal is detected by the fault detecting device and the DC emergency power modulation is started. The active power changed can be obtained by off-line calculation and online match, as well as dispatch

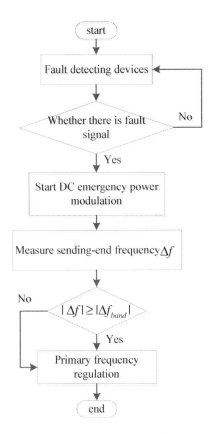

Figure 7. Coordinated frequency control strategy under fault conditions.

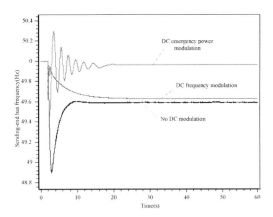

Figure 8. Sending-end bus frequency of wind-thermal-bundled islanded system.

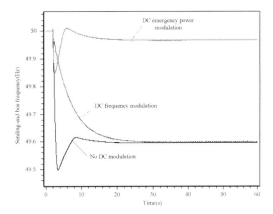

Figure 9. Sending-end bus frequency of wind-thermal-bundled interconnected system.

instructions by dispatchers. At the same time, thermal units assist frequency regulation. The specific control block diagram is shown in Figure 7.

4.2 Simulation

Simulations are carried out on wind-thermal-bundled islanded and interconnected systems based on Figure 1 to verify the control strategies. At 20 second, parts of wind farms in sending system are out of operation, with a loss of power of 327 MW. Comparison is made among the three methods: 1) No DC modulation; 2) DC frequency modulation; 3) DC emergency power modulation. The results are shown in Figures 8 and 9.

In wind-thermal-bundled islanded system, due to the serious power imbalance in the DC rectifier caused by a fault, sending-end bus frequency decreased fast. Primary frequency regulation by thermal units cannot satisfy the frequency regulation requirements. DC frequency modulation can effectively avoid frequency declining when faults occurred, while the stable value cannot meet the standards. DC emergency power modulation can effectively regulate the system's frequency and make the frequency return to the required value. In addition, the former two ways are differential regulations, which need AGC or dispatcher to adjust the transmission power in order to keep the frequency at the specified value.

In wind-thermal-bundled interconnected systems, without a DC modulation, the sending-end bus frequency exceeds the prescribed value at fault moments and DC frequency modulation can effectively alleviate the frequency drop at fault times. However, the controller can only react to changed signals, so the frequency values can reach to specified values after the faults; when DC emergency power modulation controller is adopted, it can quickly adjust the DC active power to alleviate the imbalance. Therefore, the recovery of system frequency is at its optimum when using DC emergency power modulation.

To sum up, under sending-end system faults, DC emergency power modulation should be adopted in wind-thermal-bundled islanded systems or wind-thermal-bundled interconnected systems, taking thermal units as an auxiliary frequency regulation.

5 CONCLUSIONS

This paper proposed coordinated frequency control strategies between thermal units and HVDC system to cope with frequency stability problems in wind-thermal-bundled sending-end system. The coordinated strategies are suitable for two cases: wind power fluctuation and sending-end system faults. The conclusions are as follows:

1) For wind power fluctuation, primary frequency regulation is the main regulation method, taking HVDC system as an auxiliary frequency regulation mode.
2) For sending-end system faults, HVDC active power modulation is first started, taking thermal generators as an auxiliary frequency regulation mode.

This method is more suitable for the receiving-end system is strong while the sending-end system is relatively weak, such as in the three northern areas of China. Furthermore, the specific parameters of the strategy should be set according to the actual grid structure.

ACKNOWLEDGEMENT

This study was supported by State Key Laboratory of Alternate Electrical Power System with Renewable Energy Sources (Grant No. LAPS14017).

REFERENCES

[1] X. Guo, S. Ma, and H. Shen, et al, "HVDC grid connection schemes and system stability control strategies for large-scale wind power," Automatic of Electric Power Systems, vol. 36, no. 22, pp. 107–115, 2012. (in Chinese).
[2] R. Doherty, A. Mullane, G. Lalor, D. J. Burke, A. Bryson, and M. O'Malley, "An assessment of the impact of wind generation on system frequency control," IEEE Trans. Power Syst., vol. 25, no. I, pp. 452–460, Feb. 2010.
[3] Z. Chen, Y. Chen, and Z. Xing, et al, "A control strategy of active power intelligent control system for large cluster of wind farms part two: Coordination control for shared transmission of wind power and thermal power," Automation of Electric Power Systems, vol. 35, no. 21, pp. 12–15, 2011. (in Chinese).
[4] Z. S. Zhang, Y. Z. Sun, J. Lin, and G. J. Li, "Coordinated frequency regulation by doubly fed induction generator-based wind power plants," IET Renew. Power Gen., vol. 6, no. 1, pp. 38–47, Jan. 2012.
[5] J. F. Conroy, and R. Watson, "Frequency response capabil-ity of full converter wind turbine generators in comparison to conventional generation," IEEE Trans. Power Systems, vol. 23, no. 2, pp. 649–656, May. 2008.
[6] P. K. Keung, P. Li, H. Banakar, and B. T. Ooi, "Kinetic energy of wind-turbine generators for system frequency support," IEEE Trans. Energy Systems, vol. 24, no. 1, pp. 270–287, Feb. 2009.
[7] Z. Miao, L. Fan and D. Osborn, et al. Wind farms with HVDC delivery in inertial response and primary frequency-cy control [J]. Energy Conversion, IEEE Transactions on, 2010, 25(4): 1171–1178.
[8] C. Wu, "Research on the influence of wind power on power balancing and reserve capacity," East China Electric Power, vol. 39, no. 6, pp. 993–996, 2011 (in Chinese).
[9] Q. Li, T. Liu, and X. Li, "A new optimized dispatch method for power grid connected with large-scale wind farms," Power System Technology, vol. 37, no. 3, pp. 733–739, 2013. (in Chinese).
[10] S. Chen, H. Chen and X. Tang et al, "Generator Tripping Control to Uphold Transient Stability of Power Grid Out-wards Transmitting Thermal-Generated Power Bundled With Wind Power," Power System Technology, vol. 37, no. 2, pp. 515–519, 2013. (in Chinese).
[11] Y. Chen, Z. Cheng, and K. Zhang, et al, "Frequency regulation strategy for islanding operation of HVDC," Proceedings of the CSEE, vol. 33, no. 4, pp. 96–102, 2013. (in Chinese)
[12] H. Zhu, and L. Luo, "Improving frequency stability of parallel AC-DC hybrid systems by power modulation strategy of HVDC link," Proceedings of the CSEE, vol. 32, no. 16, pp. 36–43, 2012 (in Chinese).
[13] Y. Zhu, P. Dong and G. Xie, et al, "Real-Time Simulation of UHVDC Cooperative Control Suitable to Large-Scale Wind Farms," Power System Technology, vol. 37, no.7, pp. 1814–1819, 2013. (in Chinese).

A biomedical system combined fuzzy algorithm for telemedicine applications

Peng-Liang Peng, Pin-Zhang Chen, Chien-Yuan Pan, Gwo-Jia Jong & Bow-Han Lin
National Kaohsiung University of Applied Sciences, Kaohsiung, Taiwan, R.O.C

ABSTRACT: In order to reduce the number of patients in hospitals for health examination as those suffering from chronic diseases among an aging population gradually increases, we hope that doctors can still do health detection by a long-distance home care service. We have proposed a biomedical information network platform which integrates Wi-Fi and Radio Frequency Identification (RFID) systems in this paper. With the medical instruments of Wireless Sensor Network (WSN) chips, the technology of Zigbee, and a medical decision-making system, we established a low-noise region to collect data. After constructing an interconnection between cloud servers and medical systems, users can scan these historical medical records through the website platform and readily grasp the physical condition of patients and save medical resources.

1 INTRODUCTION

With the improvement of living standards and health literacy, people increase their emphasis on physical health. With the condition of medical staff shortages, we transmit the measured physiological data to the cloud server by way of wireless transmission and create a database. In order to reduce the waste of medical resources and occupational time, medical personnel can monitor the patients' condition with a web platform in order that patients who really need medical care can have proper care. Therefore, it is worth discussing and developing important issues to construct a biomedical health system that is adapted for home adoption.

This paper proposes a method of conducting records and monitoring data through the integration of biomedical devices, wireless sensor networks and cloud servers [1]. a) It obtains the users' physiological information by using wireless transmission technology. The use of RFID distinguishes each individual and receives their physiological information. b) Next, it transmits this information to cloud databases, in order to build a web platform where patient and physicians can observe their condition in the pipeline of information. When it shows abnormal physiological condition of patients, clinics will inform the patient back to the hospital through the network platform with preliminary diagnosis and the physician's professional assessment. Historical physiological data is recorded in the web platform to facilitate the physician in a system of long-term follow up history and complete the goal of long-term care for the chronically ill and tracking them.

2 METHODLOGY

As shown in Figure 1, physicians keep controlling patients' physiological information through automatic monitoring systems management platforms. The patients measured biomedical information is sent over the internet to cloud servers for data preservation. All physiological information previously stored measurements will always be in the medical record database for monitoring and information [2] and for setting the standard values for the measured values (such as diastolic, systolic, mean arterial pressure). If the measured data is too low or too high, the system is going to automatically highlight. Achieving the goal of effective prevention is better than treatment.

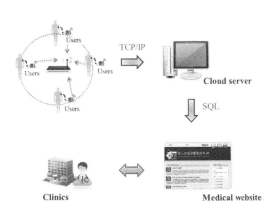

Figure 1. Schematic platform.

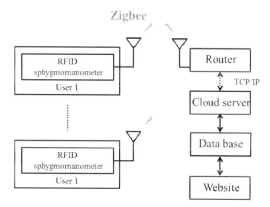

Figure 2. Blood pressure chart in a cloud network platform.

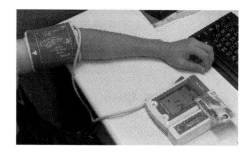

Figure 3. The sphygmomanometer's actual measurement chart.

Figure 4. The medical records databases schema diagram.

3 SYSTEM ARCHITECTURE

Figure 2 is the block diagram of the system's architecture. It can mainly be divided into the transmitter end, the transmission medium, and the receiving end of the three blocks. The following three parts were made for this introduction.

3.1 Transmitter end

There is a RFID sensor in each sphygmomanometer. By way of a wireless mesh network physiological information is sent to the cloud servers via Zigbee modules. With the use of the massive storage space in the cloud, recording and saving masses of physiological information becomes viable and provides important information as a reference for doctors during diagnosis.

3.2 Wireless sphygmomanometer

Using the RFID's recognition function, it is possible to lower tremendously the chances of having a mistake in recording a patients' history. The user will place the identification card on the reader, and it will take the physiological data and save them in the cloud server in an according IP. Therefore, it will be convenient for doctors to monitor the patients. As shown in Figure 3, the time displayed on the instrument will also be saved in the cloud server to prevent any medical conflict related issues in the future.

3.3 Wireless sensor network

WSN combined wireless internet technology, sensors, data recording instrument and information technology will be useful in health and medical attention care. It can also be used in ecological monitoring, business, home automation, and in different more special fields.

This paper is using Zigbee module [4], [5] to show the great advantages of transmitting data wirelessly. The great advantages are the low cost of energy, its light weight, high sensitivity, and system compatibility. Compared to cable transmission, wireless transmission lowers labor costs and improves the convenience for users and enjoys energy saving.

3.4 Cloud servers

Taking a large amount of data transmission into consideration, we are planning to use the cloud server to handle enormous physiological data usage. Using the high speed internet, we are able to connect servers around the world to form a highly efficient data storage system, the "Cloud Main Frame." More than one server will be operating at the same time, therefore, even when disconnection occurs during transmission, webpages lagging, due to the system crashing, can still be avoided.

3.5 Received end

The definition of a receiver in this paper is a cases database. Its construction is shown in Figure 4. Analyzing and comparing based on a users' physiological data to determine patients' state of health, doctors can understand patients' health condition and make instantaneous diagnostic and treatment decisions. Not to mention that the cases database is growing day by day, which means the reliability for this comparison increases relatively.

Medical decision-making system

The cases database contains all kinds of patients' physiological information, personal information, and history of blood pressure, etc.

As shown in Figure 5, we build a medical decision-making system combined with a fuzzy algorithm. Data analysis is done according to data storage and fuzzy control rules.

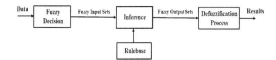

Figure 5. Fuzzy medical decision-making system's architectural diagram.

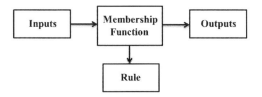

Figure 6. The block diagram of a fuzzy decision.

As shown in Figure 6, we used fuzzyfication to measure physiological information about patients. Next, we compare the reference which are defined by fuzzy rules and use defuzzification to get the value of the users' mean arterial pressures.

The output is a mean arterial pressure that uses systolic and diastolic blood pressure in medical decision-making.

$$\text{MAP} = \text{DBP} + \frac{1}{3} \times (\text{SBP} - \text{DBP}) \quad (1)$$

The following table is a fuzzy defined rule base. There are ten samples in Table 1. After increasing samples to more than one hundred, the output of the interval can be done in greater detail with the enhancement of the number of samples. Relatively, the complexity will also increase.

In order to get exact output values, we added the algorithm of defuzzification at the system's end. The steps are as following [6], [7], [8], [9]:

Order reduction of the Type-2 membership values by computing upper MF y_r by equation (2):

$$y_r = \frac{\sum_{i=1}^{N} f^i y_r^i}{\sum_{i=1}^{N} f^i} \quad (2)$$

y_r upper MF of output; i is a variable value changed from 1 to N; N is number of non-zero membership function values created by firing the rules; f^i is the fuzzy value of input of i; y_r^i is the mean value of the output upper MF.

Then, compute lower MF y_l by equation (3):

$$y_l = \frac{\sum_{i=1}^{N} f^i y_l^i}{\sum_{i=1}^{N} f^i} \quad (3)$$

y_l is the lower MF of output; y_l^i is the mean value of the output lower MF

Table 1. Defined rule bases in fuzzy algorithm.

Number	Age	Sex	Weight (kg)	SBP (mmHg)	DBP (mmHg)
1	19	Female	55	105.00	72.33
2	23	Female	55	120.33	78.67
3	24	Female	60	79.33	50.00
4	28	Male	68	114.00	71.33
5	34	Female	65	108.00	71.33
6	38	Female	77	122.67	74.00
7	48	Male	80	124.00	85.67
8	51	Female	70	110.00	78.67
9	59	Female	75	156.33	89.33
10	64	Female	92	121.33	84.67

Figure 7. The home page of web platform.

Compute the output of IT2FS by using centre of sums as a defuzzification method as in equation (4):

$$y = \frac{y_r + y_l}{2} \quad (4)$$

y is the output of the fuzzy system.

4 EXPERIMENTAL RESULTS

The experimental results are presented by a web platform. As shown in Figure 7, the home pages include the newest medical information and the connection with each hospital. The processes are fuzzification, an inference system, a database, a rule base and defuzzification, as shown in Figure 1.

Next, users enter their personal account and passwords to view historical information stored in the cloud server which stored patients' medical information. As shown in Figure 8. We choose line charts to present our experimental results, as shown in Figure 9. Among the advantages are the fact that doctors and patients can scan these physiological information conveniently. We could notice at any time if one of the data has anomalies. Besides, we added a function to prevent any written modification in order to protect the rights of both sides and avoid having medical malpractice disputes in the future. Neither patients nor clinics are entitled to tamper with any relevant information in

Figure 8. The user interface of supervision.

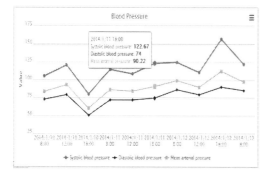

Figure 9. Blood pressure measurement line charts.

the biomedical platform. If having access to the network, users could check their own health information. This significantly enhances the users' convenience and flexibility.

5 CONCLUSIONS

We have proposed WSD and RFID technology applications in this paper. Under the Zigbee transmission, the inputs are diastolic and systolic. The output is called mean arterial pressure. In accordance with the definition of normal blood pressure to determine the current physical condition of patients, doctors can evaluate and save diagnostic timing. Through the cloud concept, patients can view their own medical records. The aging phenomonon is more severe, relatively, the probability of getting chronics is higher. The medical resources are required urgently. In order to reduce medical resources consumption, the medical web platform can achieve effective solutions. In the future, the system will increase the number of users in the process. We would like to realize this public convenience and avoid waste of medical resources.

REFERENCES

[1] Janghorbani, A., Arasteh, Abdollah, Moradi, M.H., "Application of local linear neuro-fuzzy model in prediction of mean arterial blood pressure time series", Biomedical Engineering (ICBME), 2010 17th Iranian Conference of, pp. 1–4, Nov. 2010.
[2] L. Constantinescu, Jinman Kim, D.D. Feng, "SparkMed: A Framework for Dynamic Integration of Multimedia Medical Data into Distributed m-Health Systems", Information Technology in Biomedicine, IEEE Transactions on, pp. 40–52, Jan. 2012.
[3] Po Yang, Wenyan Wu, Moniri, M., Chibelushi, C.C., "Efficient Object Localization Using Sparsely Distributed Passive RFID Tags", Industrial Electronics, IEEE Transactions on, pp. 5914–5924, Dec. 2013.
[4] Dhaka, H., Jain, A., Verma, K. "Impact of Coordinator Mobility on the throughput in a Zigbee Mesh Networks". Advance Computing Conference (IACC), IEEE 2nd International, pp. 279–284, Feb. 2010.
[5] Zhou Yiming, Yang Xianglong, Guo Xishan, Zhou Mingang, Wang Liren, "A Design of Greenhouse Monitoring & Control System Based on ZigBee Wireless Sensor Network", Wireless Communications, Networking and Mobile Computing. WiCom. International Conference, pp. 2563–2567, Sept. 2007.
[6] Al-Jaafreh, M.O., Al-Jumaily, A.A. "Type-2 Fuzzy System Based Blood Pressure Parameters Estimation", Modeling & Simulation, 2008. AICMS 08. Second Asia International Conference on, pp. 953–958, May 2008.
[7] Morsi, I., Abd El Gawad, Y.Z., "Fuzzy logic in heart rate and blood pressure measureing system", Sensors Applications Symposium (SAS), 2013 IEEE, pp. 19–21, Feb. 2013.
[8] Janghorbani, A., Arasteh, Abdollah, Moradi, M.H., "Application of local linear neuro-fuzzy model in prediction of mean arterial blood pressure time series", Biomedical Engineering (ICBME), 2010 17th Iranian Conference of, pp. 3–4, Nov. 2010.
[9] Chin-Teng Lin, Shing-Hong Liu, Jia-Jung Wang, Zu-Chi Wen, "Reduction of interference in oscillometric arterial blood pressure measurement using fuzzy logic", Biomedical Engineering, IEEE Transactions on, pp. 432–441, April 2003.

ABSTRACT: The power system structure has gradually become complicated due to the integration of renewable energy sources, which gives preventive control measures more alternatives. The problem becomes how to choose the optimal preventive control strategy and this has become a focus of research. This paper proposes a preventive control method under overload conditions in a multi-source grid. The proposed method formulates a control performance index, based on control sensitivity, and also takes the characteristics of renewable energy sources and the control cost of different types of units into consideration. The objective function is to minimize the control cost and post-control overload risk. In the end, the viability of the method is demonstrated through the simulation of the IEEE 39-bus test system.

A preventive control method for overload in a multi-source grid

Zaiwei Zhang
Hohai University, Nanjing, China

Feng Xue, Ye Zhou, Xiaofang Song & Ling Zhou
Nari Technology Development Limited Company, Nanjing, China

1 INTRODUCTION

With rapidly developing generational technologies of renewable energy (wind, photovoltaic, etc.), power systems face a high level of penetration from renewable energy sources and have gradually formed an interactive complex multi-source grid, such as is found in most areas in northwest China (Gansu, Qinghai, etc.). However, the intermittence of wind and photovoltaic energy brings many uncertainties. Hence, we are faced with how to integrate the maximum amount of intermittent energy and effectively control various types of power sources at the same time. This has become an urgent issue to resolve.

Taking Gansu power grid as an example, there coexists many types of power generation, such as wind power, photovoltaic (PV) power, hydroelectric power, thermal power, and gas turbine generation. From the aspect of safety, if there occurs a N-1 or N-2 fault in the tie line between the power sending and the receiving end, and the power flow in it is relatively large, the remaining transmission lines or transformers would easily become overloaded unless there are preventive measures, which may also lead to serious consequences like cascading failures. So in order to eliminate the overload, effective overload preventive control strategies must be developed through regulating the power output of the generator units at power sending end. On the other hand, from the aspect of economy, renewable (wind/PV) generator units could start/stop much faster and cost less than conventional thermal units. However, the individual wind/PV power plant has a relatively smaller capacity to regulate. If the overload problem is too large that we need to shed large amounts of power, and the control cost could be even higher due to the regulation of a plurality of wind/PV power stations. So, this control dilemma has to be resolved.

The conventional overload preventive control measures are based on whether the power flow results exceed the safe operational constraints or not. If yes, the power output of relevant generation units should be regulated or the reactive power compensation device should be adjusted. If it cannot reach a safe operational limit, the next step is to carry out load reduction or other measures of control. The objective is to maintain the system's stability margin under normal operational conditions and under contingencies as well as through the minimum of control costs. According to the different control methods, the cost and the priority are different too. Among them, the load-shedding is the last choice when there is no alternative, the cost of which is also the highest. Preventive control priority should be given to the active power redistribution of relevant generators, and then there are the other means of control. Hence, in this paper we only discuss the control strategy which involves the regulation of the relevant power outputs.

Basically, two categories of approach could be used considering overload preventive control: the optimal programming approach and the sensitivity approach. Normally the optimal programming approach has a complete mathematical model including the optimization objective and all kinds of constraints, which ensures the safety of the power system but may lead to convergence problem as so many devices need to be regulated. In contrast, the sensitivity approach has no iteration and focuses on objectives like the minimum amount of regulation or the minimum number of devices to be regulated, which enhance its practicability in actual practice. Deng et al. (1999) proposed earlier an active power control strategy, based on the sensitivity approach, and this was proposed to ensure the minimum regulation of the amount of output power. Zhang & Wu adopted in 2012 a control assessment index based on a sensitivity analysis of the active power in a transmission line and the active power in a generator bus. It determined the regulation order according to the sensitivity ranking results. The objective was to minimize the amount of regulation, but the intermittence and the control costs for different renewable energy sources were not taken into consideration. Then Song et al. (2013) added the control cost of different generator types on the basis of Zhang & Wu. In addition, the concept of overall sensitivity was developed by Cheng et al. (2011), so that the power regulations could be determined according to the overall judgment of the overloaded lines or lines close to overload. Moreover, Liu et al. (2008) established an overall performance index algorithm to calculate the optimal emergency control strategy.

Therefore, this paper proposes a preventive control method for overload elimination in a multi-source grid. Preventive control is relatively undemanding in terms of speed. So, this paper focuses on an economic-oriented method, which also meets the safety requirements. Based on control sensitivity, the effective control capacity is calculated, and the control performance index is formulated. After considering the control cost of each generator type, the overload elimination is tackled while keeping the system within the minimum control cost and post-control risks.

The paper is divided as follows. Section 2 presents the application of sensitivity in preventive control. Section 3 develops a complete model and procedure of overload preventive control. Lastly, Section 4 presents verification and comparison results using the IEEE 39-bus test system.

2 AN OVERLOAD PREVENTIVE CONTROL MODEL

2.1 *Application of sensitivity in preventive control*

Putting forward the concept of sensitivity is to quantify the effect of control measures on the system stability. So the sensitivity analysis has a significant impact on safety and stability assessments and preventive control of power systems. It can determine the variation trend of a controllable variable with respect to another corresponding variable. Then giving priority ranking for all available control strategies, so as to realize the unity of safety and economy.

For an individual generator unit, we introduce the sensitivity of control effect S_{ij}, which can quantify the influence of the power control amount of unit i on the overload line j:

$$S_{ij} = \frac{\Delta P_j}{\Delta P_i} \quad (1)$$

where ΔP_i = power control amount of unit i; and ΔP_j = power reduction amount of line j after regulating the active power of source i.

The bigger the control sensitivity is, the greater the amount of overload that can be reduced. In other words, the unit with biggest control sensitivity needs smallest regulation when the overload reduction remains unchanged.

2.2 *Overload assessment*

Assuming that the power grid has $j = 1, 2, \ldots, L$ monitoring lines, and the contingency set has $k = 1, 2, \ldots, N_c$ faults which could cause overload emergency. The overload assessment indicator are given by

$$M_j = \frac{P_j}{P_{j, rate}} > 1 \quad (2)$$

where P_j = actual active power of line j; and $P_{j,rate}$ = rated power of line j.

2.3 *A control Performance Index (PI)*

Based on the control effect sensitivity of the units, the effective control capacity of the controllable power sources can be obtained:

$$P_{i, eff} = P_i \cdot \frac{1}{S_{ij}} \quad (3)$$

where P_i = controllable capacity of unit i; and $P_{i,eff}$ = effective control capacity respectively of unit i.

According to Liu et al. (2008), the weight was multiplied by the sensitivity in the evaluation index. Using it for reference, for each controllable unit in the grid, we provide the following control performance index, in order to rank the priority of available control units. Under contingency, we suppose, there are $j = 1, 2, \ldots, L_k$ lines in the state of emergency k. The expression of PI presents as follows:

$$PI_i = \frac{1}{C_i} \sum_{k=1}^{N_c} \sum_{j=1}^{L_k} (W_j \cdot P^k_{i,eff}) \quad (4)$$

where $P^k_{i,eff}$ = the effective control capacity of unit i under contingency k; W_j = the weight of $P^k_{i,eff}$; $W_j = (W_j)^n$, n is positive, which indicates that the bigger the M_j value is, the more urgent the overload condition is; C_i = economic cost coefficient implementing the strategy i; and N_c = the total number of contingencies.

In fact, this control performance index PI_i is a cost-performance ratio, which helps us select the controllable nodes with the best control performance and lowest control costs. After sorting all the controllable units according to the index, we can obtain regulation plans conforming the contingency set. On this basis, the plans can be continued to the subsequent steps, if they pass the static safety verification.

2.4 Search of regulation amount

To determine the regulation amount of controllable units, we adopt the bisection method to search, which has a relative high search efficiency. We set the initial maximum regulation amount (ΔP_{max}) to the sum of the maximum controllable capacity of all the selected generator units (ΔP_M), and set the initial minimum regulation amount (ΔP_{min}) to 0. The convergence criterion is as follows, in which ΔP_{th} can be set according to the requirements of a practical engineering project (recommended between 1 MW and 20 MW).

$$\Delta P_{max} - \Delta P_{min} < \Delta P_{th} \quad (5)$$

Once the criterion is met, the calculation terminates. At this moment, ΔP_{min} corresponds to the critical state of overload and ΔP_{max} corresponds to the critical state of non-overload. The search program will output the section flow and the regulation amount of all participating units. Specific steps are as follows:

Step 1: Check initially whether the maximum regulation amount could eliminate the overload or not. If the answer is negative, exit with failure; otherwise, go to Step 2.
Step 2: Verify the power flow. If convergent, exit successfully; otherwise, go to Step 3 to search the required regulation amount through the bisection method.
Step 3: Check the regulation amount of step k.

$$\Delta P_k = \frac{\Delta P_{max} + \Delta P_{min}}{2} \quad (6)$$

Then distribute the total amount to all controllable units according to the previous ranking result, followed by the flow check as to whether there is still an overload problem or not. If there is, go to Step 4; otherwise, go to Step 5.
Step 4: Assign the value of ΔP_k to ΔP_{min}. Then go to Step 2 to repeat the above process until it completely eliminates the overload or attains the maximum iteration times (recommended 3 to 5 times).
Step 5: Assign the value of ΔP_k to ΔP_{max}. Then go to Step 2 to repeat the above process until it completely eliminates the overload or attains the maximum iteration times.

If all the optional control strategies cannot eliminate the overload, the user would get alarmed. When the user is alarmed, he should add other types of control measures (load-shedding, etc.). The brief flow chart describes as follows in Figure 1.

2.5 Objective function

In order to have better screening control strategies, we minimize the total control cost C_p as the objective. The control cost under a certain contingency consists of two parts: a cost (C_{Grid}) of implementing the selected strategy into the grid, and another cost (C_{After}) to control the post-strategy risk of overload. Hence, we

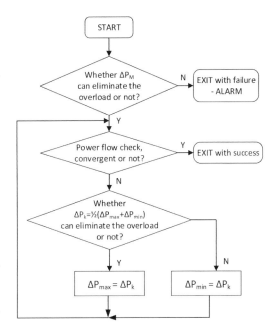

Figure 1. Brief flow chart of regulation amount search by bisection method.

establish the following objective function of overload preventive control model of a multi-source grid.

$$\min C_p = C_{Grid} + C_{After} \quad (7)$$

Supposing the system is under a contingency k, and i is the subscript to indicate a generation unit. The number of controllable wind power generator is A, the current power regulation amount of wind unit is $\Delta P_{i,W}$. For controllable photovoltaic generator units, the number and the current power regulation amounts are B and $\Delta P_{i,PV}$ respectively. For a hydroelectric power generation unit, the number and the current power regulation amounts are C and $\Delta P_{i,H}$ respectively. For thermal power generation units, the number and the current power regulation amounts are D and $\Delta P_{i,TH}$ respectively, and for gas turbine generators, E and $\Delta P_{i,G}$. Moreover, I_i is the binary state variable of each controllable unit. And C_W, C_{PV}, C_H, C_{TH} and C_G are the cost coefficient of wind, PV, hydro, thermal and gas units respectively. So C_{Grid} can be defined as:

$$C_{Grid} = \sum_{i=1}^{A}(C_W \cdot I_i \cdot \Delta P_{i,W}) + \sum_{i=A+1}^{B}(C_{PV} \cdot I_i \cdot \Delta P_{i,PV})$$
$$+ \sum_{i=B+1}^{C}(C_H \cdot I_i \cdot \Delta P_{i,H}) + \sum_{i=C+1}^{D}(C_{TH} \cdot I_i \cdot \Delta P_{i,TH}) \quad (8)$$
$$+ \sum_{i=D+1}^{E}(C_G \cdot I_i \cdot \Delta P_{i,G})$$

As for post-control cost C_{After}, it is essentially a value of risk. In other words, it is the product of the

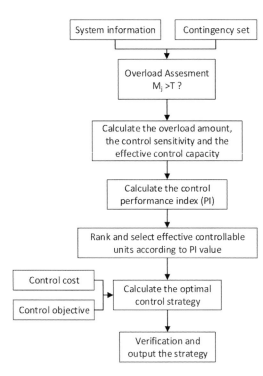

Figure 2. Preventive control process of overload.

cost and the probability of other contingencies occurring after a control strategy is implemented. So it can be defined as:

$$C_{After} = \sum_{k=1}^{N_c} (\alpha_k \cdot C_k) \quad (9)$$

where N_c = total number of contingency; α_k = probability that contingency k occurs; and C_k = its cost.

Considering the economic control of different types of unit, and based on the controllable variables selected through the control performance index, we can obtain the final optimal control strategy.

2.6 Preventive control procedure of overload

To sum up, the procedure of optimal preventive control method of overload in a multi-source grid is shown as Figure 2.

3 CASE STUDY

3.1 IEEE 39-bus test system

Simulations have been conducted with IEEE 39-bus test system in MATLAB. This system presents similar features to actual power systems. There are 10 generational units in total, buses 30 to 39 respectively. To be specific, bus 33, 34 and 35 are wind power units, bus 37 and 38 are PV units, and bus 32 and 36 hydroelectric units, bus 30 and 39 are thermal units, and bus 31

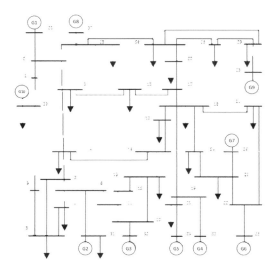

Figure 3. IEEE 39-bus test system.

Table 1. Unit type, number and current active power.

Unit type	Unit No.	Bus No.	Active Power MW
Wind	G4	33	632
	G5	34	508
	G6	35	650
PV	G8	37	540
	G9	38	830
Hydro	G3	32	650
	G7	36	560
Thermal	G1	30	250
	G10	39	1000
Gas	G2	31	250

Table 2. Contingency set & overload assessment.

Contingency No.	1	2	3	4	5
Line No.	10	46	33	37	8
Overload assessment (value of M_j)	1.295	1.051	1.195	1.233	1.082

is a gas unit. The schematic diagram of the system and the list of units are shown in Figure 3 and Table 1.

We set up a contingency set which includes 5 overload faults. The overload lines and the overload assessments are shown in Table 2.

According to the previous method, we can obtain the control performance index of each controlled unit, as shown in Table 3.

After the regulation amount search, we obtain 4 potential control strategies as shown in Table 4.

As for the control costs of regulating different types of generator units, we use relative cost coefficient to differentiate the type. For conventional thermal and

Table 3. Ranking result of control performance index.

Ranking	1	2	3	4	5	6	7	8	9	10
PI value	1.476	1.473	1.472	1.412	1.414	0.873	0.748	0.724	0.724	0.165
Unit No.	G9	G8	G1	G7	G6	G10	G4	G5	G3	G2

Table 4. Potential control strategies.

Strategy No.	Unit type	Unit to be regulated	Regulation amount MW
1	Wind	G4	+20
	PV	G9	−20
	Hydro	G3	+100
		G7	−20
	Thermal	G1	−40
		G10	−100
	Gas	G2	+60
2	PV	G8	−10
	Gas	G2	+10
3	Hydro	G3	+10
		G7	−10
4	PV	G8	−5
		G9	−35
	Hydro	G3	+40

Table 5. Cost coefficient (per unit).

Wind	PV	Hydro	Thermal	Gas
100	100	1	1.5	1.2

Table 6. Control cost of each strategy.

Strategy No.	1	2	3	4
C_{Grid}	4402	1012	20	4040
C_{After}	64.6	73.8	80.9	38.0
C_p	4466.6	1085.8	100.9	4078.0

hydro power, we use the grid purchase price for reference. For the renewable energy sources (wind/PV), however, we give them bigger enough coefficients according to Chinese renewable energy policy so that we could try to avoid regulating wind or PV units. So the cost coefficient (per unit) of those five types of generators are shown in Table 5.

In addition, the probability coefficient of post-control contingency occurrence (α_k) in the objective function is set as 0.007.

After substituting the relative cost coefficient and the control strategy into the objective function, we obtain the control cost of each strategy in Table 6.

From the table we can see that the difference between strategies is amplified after taking the unit type into consideration. As can be seen from the comparison between strategy No. 1 and No. 4, it is not advisable to regulate wind or PV units due to the

Table 7. Contingency No.5.

Contingency No.	Overload Line No.	Overload Assessment (value of M_j)
5	8	0.9795

extremely heavy cost. However, we should choose strategy No. 3, which gives us the minimum cost.

The proposed method can not only find out the optimal control strategy with small cost, but also adapt to other power systems with different penetration degrees of renewable energy by means of adjusting the cost coefficients.

3.2 Comparison with previous method

Zhang & Wu (2012) adopted the A control assessment index simply based on the sensitivity analysis of the active power in a transmission line compared to the active power in a generator bus. It determined the regulation order according to the sensitivity ranking result. The objective was to minimize the regulation amount.

We take a contingency from our contingency set as an example.

After integrating our case data in the method of Zhang & Wu (2012), we obtain in Table 8 the ranking result of all controllable units on the basis of sensitivity and in Table 9 the control strategy under two methods.

From Table 9, the number of units to be regulated is 4 and 3 respectively. And the total regulation amount is 120 MW and 80 MW respectively, which makes this method in this paper the best. Moreover, Zhang & Wu (2012) did not take the control cost of different unit type into the objective. If so, the total cost of the two methods is compared in Table 10.

It is clear to see that the method in this paper has advantages both in the regulation amount and the total control cost.

4 CONCLUSION

With the increasing integration of renewable energy sources into power grids all over the world, this paper proposes an optimal preventive control method for overload elimination in a multi-source grid. Based on control sensitivity, a control performance index is derived. Both optimal control cost and post-control system risks are taken into consideration as the objective. As evidenced by the result of the IEEE 39-bus

Table 8. Ranking result of sensitivity.

Ranking	1	2	3	4	5	6	7	8	9	10
Sensitivity	2.306	2.306	2.306	1.306	1.067	0.950	0.918	0.872	0.466	0
Unit No.	G6	G4	G5	G7	G9	G8	G1	G3	G10	G2

Table 9. Control strategy under the two methods.

Method	Unit type	Unit to be regulated	Regulation amount MW	Absolute value sum
Of Zhang & Wu	wind	G4	−20	120
		G6	−37	
	Hydro	G7	−3	
	Gas	G2	+60	
In this paper	PV	G8	−5	80
		G9	−35	
	Hydro	G3	+40	

Table 10. Total control cost under the two methods.

Method	Of Z. & W.	In the paper
Total C_p	4078	5775

system, the proposed method can provide several suggestions for multi-source coordinative control in an actual engineering application.

The difficulty of preventive control is that a certain generator unit may require regulations of different amounts or different directions when two or more contingencies coexist. To some extent, the proposed control performance index resolves the concurrence of a few numbers of contingencies. When a large number of contingencies occur at the same time, however, further research should be made in terms of making the index more practical and making the regulation amount search more rapid, using an intelligent algorithm.

ACKNOWLEDGEMENT

This work is supported by the National Key Technology Research and Development Program of China (No. 2013BAA02B01).

REFERENCES

[1] Abrantes, H. D. & Castro, C. A. 2000. New Branch Overload Elimination Method Using Nonlinear Programming. *Power Engineering Society Summer Meeting, IEEE* 1: 231–236.

[2] Cai, G., Zhang, Y., Yu, T. & Ren, Z. 2008. Static Voltage Stability Preventive Control by Unified Sensitivity Method. *High Voltage Engineering* 34 (4): 748–752.

[3] Chen, W. & Luo, L. 2008. Risk Assessment and Preventive Control for Power System Overload. *East China Electric Power* 36 (7): 42–45.

[4] Cheng, L., Hao, Z., Zhang, B., Li, J. et al. 2011. Fast Elimination of Overload in Transmission Line Section Based on Simplified Primal-dual Interior Point Method. *Automation of Electric Power Systems* 35 (17): 51–55.

[5] Cui, X., Li, W., Ren, X., Li, B., Xue, F. & Fang, Y. 2012. AGC and Artificial Dispatching Based Real-time Control Framework of Power System Thermal Stability. *Sustainable Power Generation and Supply (SUPERGEN 2012), International Conference*: 1, 5, 8–9.

[6] Deng, Y., Li, H., Zhang, B. et al. 1999. Adjustment of Equal and Opposite Quantities in Pairs for Strategy of Active Power Security Correction of Power Systems. *Automation of Electric Power Systems* 23 (18): 5–8.

[7] Hoji, E.S., Padilha-Feltrin, A. & Contreras, J. 2012. Reactive Control for Transmission Overload Relief Based on Sensitivity Analysis and Cooperative Game Theory. *IEEE Transactions on Power Systems* 27 (3): 1192–1203.

[8] Liu, H., Yang, W., Li, H., Meng, Z. et al. 2008. Centralized real-time decision-making and emergency control method for overload condition in large-scale power grids. *China patent, CN101299537*, issued November 5, 2008.

[9] Song, X., Fang, Y., Chang, K., Xue, F. et al. 2013. Optimal Emergency Control Device Layout Method under Overload Condition of Single Device. *China patent, CN103050977A*, issued April 17, 2013.

[10] Xue, F., Chang, K. & Wang, N. 2011. Coordinated Control Frame of Large-scale Intermittent Power Plant Cluster. *Automation of Electric Power System* 35 (22): 45–53.

[11] Yao, F., Zhang, B., Zhou, D. et al. 2006. Active Power Security Protection of Transmission Section and its Fast Algorithm. *Proceedings of the CSEE* 26 (13): 31–36.

[12] Zhang, Z. & Wu, X. 2012. Research on Line Thermal Stability Control Method Based on Sensitivity Analysis. *Jilin Electric Power* 40 (1): 8–10.

[13] Zhao, J., Chiang, H. D. & Zhang, B. 2005. A Successive Linear Programming Based on Line Static Security Corrective Control Approach. *Power System Technology* 29 (5): 25–30.

[14] Zhu, W., Hao, Y., Liu, G., Yu, E., Zhang, J. & Wang, H. 1994. Security Constrained Dispatch using Linear Programming. *Proceedings of the CSEE* 14 (4): 57–64.

Effects of thermal annealing on the tungsten/lanthanum oxide interface

Hei Wong & Jieqoing Zhang
Department of Electronic Engineering, City University of Hong Kong, Kowloon, Hong Kong SAR, China

Kuniyuki Kakushima & Hiroshi Iwai
Frontier Research Center, Tokyo Institute of Technology, Nagatsuta-cho, Yokohama, Japan

Jian Zhang & Hao Jin
Department of Information Sciences and Electronic Engineering, Zhejiang University, Hangzhou, China

ABSTRACT: Tungsten (W)/lanthanum oxide (La_2O_3) stack has been considered as the promising metal-gate/high-k gate material for future CMOS technology nodes. The requirement of subnanometer Equivalent Oxide Thickness (EOT) has called for some more stringent constraints on the high-k dielectric film properties as well as the high-k/silicon and high-k/metal interfaces. This work reports the reactions taken place at the W/La_2O_3 interface during thermal treatment at temperature above 500°C. Angle-Resolved X-ray Photoelectron Spectroscopy (ARXPS) measurements on the bonding structure and composition distribution of the W/La_2O_3 stack indicate that the sputtered tungsten film has high-oxygen content of over 40% which can be further enhanced with thermal annealing. The thermal treatment also enhances the out-diffusion of La atoms and forming La-O-W phases at the W/La_2O_3 interface. These effects would result in the increase of oxygen vacancies in the La_2O_3 film and need to be suppressed.

1 INTRODUCTION

In the decananometer era of CMOS technology, the gate dielectric thickness, in the sense of equivalent oxide thickness (EOT), will be scaled down to the subnanometer range (Wong 2012, Wong & Iwai 2006a). Higher dielectric constant (*k*) materials will be indispensable in order to have better controls on the process and the gate current leakage. Lanthanum oxide has been considered as the promising candidate for this application (Sen *et al.* 2007, Lichetnwalner *et al.* 2005, Wong 2012, Yamada *et al.* 2003). However, server challenges arise not only due to the material properties and deposition technology of the ultrathin high-*k* film themselves, but also because of the constraints of the interfaces between the high-*k*/Si and the high-*k*/metal (Lucovsky *et al.* 2000, Wong 2014, Yamada *et al.* 2002). The interfacial low-*k* layers at both interfaces may limit the thinnest achievable EOT of the high-*k* film in the metal/high-*k*/Si structure (Wong 2014). The transition low-*k* layer between the high-*k*/Si interface has caught lots of attentions over the decade (Wong & Iwai 2006b, Wong *et al.* 2004, Wong *et al.* 2010, Wong *et al.* 2007). Most high-k/Si, including La_2O_3/Si, interfaces are not stable (Lucovsky *et al.* 2000, Wong 2012, Yamada *et al.* 2002), a thermal treatment of the stack can lead to the formation of a low-*k* transition layer. The thickness of this transition layer depends on the processing temperature and the duration as well. It was also found that the interface layer at the high-*k*/silicon interface caused most of the performance degradations of the MOS transistors (Wong 2014, Wong & Iwai 2006b). The gate metal/high-*k* interface was not considered as an important factor that could affect the device performance significantly. This scenario has now been changed. The EOT and as well as the physical thickness are so small that any minor effects would become the limiting factors (Wong 2014). This work aims to study the interface bonding structure between the La_2O_3 and tungsten electrode by using angle-resolved XPS measurements. Special emphasis was placed on the interface reactions under thermal treatments at temperatures between 500 to 600°C.

2 EXPERIMENTAL

The tungsten/La_2O_3 metal-gate/high-*k* gate stacks used in this investigation were deposited on *n*-type Si substrate with (100) orientation. The La_2O_3 film of about 5 nm thick were first sputtered using an electron-beam evaporation chamber and the pressure of the chamber before sputtering was about 10^{-7} Pa. Tungsten gate electrode of about 3 nm thick was then deposited *in situ* using magnetron sputtering to avoid any moisture absorption and contamination. Some samples were further thermally annealed at 500°C or 600°C for 30 min in a rapid thermal annealing furnace. The chemical compositions as well as the

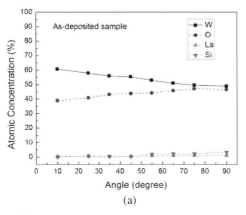

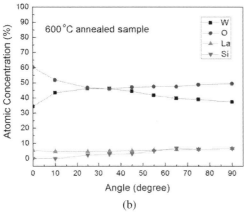

Figure 1. Depth concentration profiles of W, O, La, and Si in the W layer revealed by using angle-resolved XPS measurements: (a) as-deposited sample; and (b) sample with thermal annealing in nitrogen at 600°C for 30 min.

bonding structures of the as-prepared W/La$_2$O$_3$/Si stack at different depths were investigated in detail by using angle-resolved x-ray photoelectron spectroscopy (ARXPS) measurements. The apparatus used was a Physical Electronics PHI 5802 spectrometer. The monochromatic x-ray was generated from an Al Kα (1486.6 eV) radiation source.

3 RESULTS AND DISCUSSION

Figure 1 depicts the depth profile of the W/La$_2$O$_3$/Si stack at different depths by using ARXPS at take-off angle ranging from 10 to 90°. High amount of oxygen content (>40%) was always found in the tungsten film. For as-deposited W layer, the O content changed from 40 at.% on the surface (at take-off angle of 10°) to 45 at.% at a region closer to W/La$_2$O$_3$ interface (i.e. for the case with take-off angle of 90°).

The high oxygen content should be mostly come from the metal source used for the sputtering or the ambient. Higher O contents at larger take-off angles indicate that some of the oxygen should migrate from the La$_2$O$_3$. The existence of oxygen in the W film should not have any serious impact on the current conduction as the WO$_x$ phases are conductive. The conductive interface layer will not affect the EOT also. However, the oxidized W may be corroded easier and lead to some reliability issues in long-term operation. The second effect of high O-contenting W film is that it would help to suppress the oxygen vacancies, which are the major trapping sites, in the La$_2$O$_3$ film. It was found that the W/La$_2$O$_3$/Si structures had better electrical characteristics than other kinds of electrode and that observation was attributed the effect of oxygen in the W film (Kawanago 2012). For sample with thermal annealing at 600°C for 30 min, the O content on the surface increases to 60 at.%, indicating a significant oxidation of the tungsten occurred. As the annealing was taken place in the nitrogen ambient, the oxygen seems to mainly come from the La$_2$O$_3$ layer. That is, the thermal annealing may not improve the electrical characteristics of the La$_2$O$_3$, it might deteriorate the film by creating more oxygen vacancies in the film, instead. In addition, notable increases of both La and Si contents were found in the 600°C annealed sample (See Fig. 1b). It indicates that out-diffusion or some sorts of chemical reaction should have been taken place to make the La and Si being detectable in the W layer.

Figure 2 shows the W 4f spectra at different take-off angles for the three samples, namely, as-deposited (a), with 500°C annealing (b), and with 600°C annealing (c). The as-deposited W layer shows double peak elemental W 4f spectra with binding energies at 31.2 and 33.2 eV. The peak locations remain fairly unchanged at different take-off angles. For sample with 500°C, the binding energies in the surface region still remain unchanged but a slight high-energy shift was found at larger take-off angles. For 600°C annealing sample, both surface and bulk W 4f peaks shift to higher energy side. The energy shift should be due to the tungsten oxidation or the forming of other kinds of bonding. It is further noted that the bulk (at take-off angle of 45 to 55°) has the strongest elemental W 4f peaks. On the surface and at the interface, the peaks were broadened, indicating the peaks should be constituted by some other oxidation states. We further conducted a detailed study on the bonding structures of these samples. By decomposing the W 4f peaks using Gaussian de-convolution technique, several different kinds of bonding states can be found. The majority of the film is still the elemental W which has a W 4f doublet at 31.2 and 33.3 eV. Bonding states of W^{4+}, W^{5+} and W^{6+} can be found in the surface region and in the region near the W/La$_2$O$_3$ interface, indicating there exist of WO$_x$ phases. Figure 3 shows a typical W 4f XPS spectrum with peak decomposition. The sample was annealed at 500°C and the take-off angle was 45°. The markers represent the original data; broken lines indicate several decomposed peaks due to W^{4+}, W^{5+}, and W^{6+} oxidized states; and the solid line is the synthesized spectrum based on the given decomposed peaks.

Figure 4 shows the La 4d spectrum of the same sample as given in Figure 3. Gaussian decomposition of

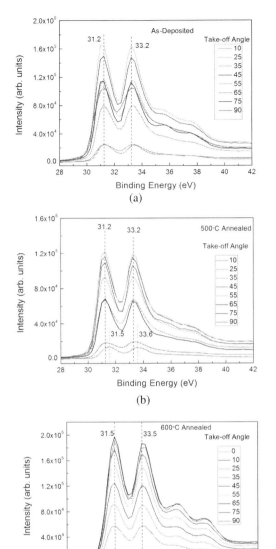

Figure 2. Angle-resolved W 4f XPS spectra of: (a) as-deposited; (b) 500°C annealed; and (c) 600°C annealed sample.

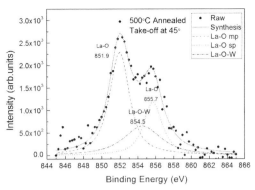

Figure 3. W 4f XPS spectrum taken at a take-off angle of 45° for sample with 500°C annealing. Elemental W together with W^{4+}, W^{5+} and W^{6+} oxidation states were found in the W film.

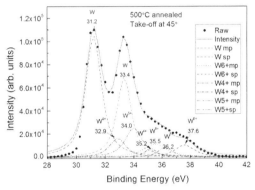

Figure 4. La 4d XPS spectrum taken at a take-off angle of 45° for sample with 500°C annealing. A board peak with peak energy of 854.5 is assigned to the La-O-W bonding.

the peak indicates that in addition to the La-O bonding with peak energy at 851.9 and 855.7 eV. A board peak with energy of 854.5 eV is needed in order to fit the measured XPS spectrum. This peak is attributed to La-O-W bonding. Together with the concentration profile as given in Figure 1, it confirms that the thermal annealing has resulted in the intermixing at the W/La$_2$O$_3$ interface and that explains the observation of the enhancement of La content in W and La-O-W bonding near the W/La$_2$O$_3$ interface. The second important implication, yet needs some further in deep investigation, is that the enhanced W oxidation during annealing and the reduction of oxygen content in La$_2$O$_3$ film as a result of oxygen out-diffusion. These reactions should have some adverse effects such as large gate leakage current, higher trap density, etc, in the MOS devices with these treatments.

4 CONCLUSION

Although W/La$_2$O$_3$ metal-gate/high-k stack has been identified as the promising candidate for future CMOS technology nodes, the stability of W/La$_2$O$_3$ interface has yet to be explored. In this work, by using angle-resolved XPS measurements, we found that some reactions took place when the W/La$_2$O$_3$ stack subjected to thermal treatment at processing temperature at 500°C or above. The thermal annealing may accomplish with the out-diffusion of oxygen from the La$_2$O$_3$ and results in the W oxidation which can lead to some performance degradations of the MOS devices. High temperature treatment should be avoided after the deposition of high-k and metal gate.

ACKNOWLEDGEMENT

This work is supported by the GRF Project #121212 of Research Grants Council of Hong Kong, Hong Kong.

REFERENCES

[1] Kawanago, T., Lee, Y., Kakushima, K., Ahmet, P., Tsutsui, K., Nishiyama, A., Sugii, N., Natori, K., Hattori, T. & Iwai, H. 2012. EOT of 0.62 nm and high electron mobility in La-silicate/Si structure based nMOSFETs achieved by utilizing metal-inserted poly-Si stacks and annealing at high temperature. *IEEE Trans. Electron Devices* 269–276.

[2] Lichtenwalner, D.J., Jur, J.S., Kingon, A.I. *et al.* 2005. Lanthanum silicate gate dielectric stacks with sub-nanometer equivalent oxide thickness utilizing an interfacial silica consumption reaction. *J. Appl. Phys.* 98: 024314, 2005.

[3] Lucovsky, G., Yang, H., Niimi, H., Keister, J.W., Rowe, J.E., Thorpe, M.F. & Phillips, J.C. 2000. Intrinsic limitations on device performance and reliability from bond-constraint induced transition regions at interfaces of stacked dielectrics. *J. Vac. Sci. Technol. B*, 18: 1742–1748.

[4] Sen, B., Wong, H., Molina, J., Iwai, H., Ng, J. A., Kakushima, K. & Sarkar C. K. 2007. Trapping characteristics of lanthanum oxide gate dielectric film explored from temperature dependent current-voltage and capacitance-voltage measurements. *Solid State Electron.* 51: 475–480.

[5] Wong, H. 2012. *Nano-CMOS Gate Dielectric Engineering*, Boca Raton: CRC Press.

[6] Wong, H. 2014. Lanthana and its interface with silicon. *Proceedings of 29th International Conference on Microelectronics*, Belgrade. pp. 35–41.

[7] Wong, H. & Iwai H. 2006a. On the scaling issues and high-k replacement of ultrathin gate dielectrics for nanoscale MOS transistors. *Microelectron. Engineer.* 83: 1867–1904.

[8] Wong, H. & Iwai, H. 2006b. Modeling and characterization of direct tunneling current in dual-layer ultrathin gate dielectric films. *J. Vac. Sci. Technol. B*, 24: 1785–1793.

[9] Wong, H., Ng, K. L., Zhan, N., Poon, M. C. & Kok, C. W. 2004. Interface bonding structure of hafnium oxide prepared by direct sputtering of hafnium in oxygen. *J. Vac. Sci. Technol. B*, 22: 1094–1100.

[10] Wong, H., Iwai, H., Kakushima, K., Yang, B.L. & Chu P.K. 2010. XPS study of the bonding properties of lanthanum oxide/silicon interface with a trace amount of nitrogen incorporation. *J. Electrochem Soc.* 157: G49–G52.

[11] Wong, H. Sen, B., Yang, B.L., Huang, A.P. & Chu, P.K. 2007. Effects and mechanisms of nitrogen incorporation in hafnium oxide by plasma immersion implantation. *J. Vac. Sci. Technol. B*, 25: 1853–1858.

[12] Yamada, H., Shimizu, T. & Suzuki, E. 2002. Interface reaction of a silicon substrate and lanthanum oxide films deposited by metalorganic chemical vapor deposition. *Jpn. J. App. Phys.* 41: L368–370.

[13] Yamada, H., Shimizu, T,, Kurokawa, A., Ishii, K. & Suzuki, E. 2003. MOCVD of high-dielectric-constant lanthanum oxide thin films, *J. Electrochem. Soc.* 150: G429–G435.

A study on the capacity optimization and allocation of wind/solar/diesel and energy storage hybrid micro-grid systems

Jinggang Zhang, Pengyue Liu & Hui Zhang
University of Technology, Xi'an, China

ABSTRACT: Micro-grids have become an important issue in the field of distributed energy systems This paper introduces the component of wind/solar/diesel hybrid micro-grid system and the capacity of primary resources. Based on the emissions of CO_2 and other pollutants as constraint conditions, and basing the renewable energy mix optimization model (HOMER) as the platform, a planning scheme of electricity generation from renewable energy resources in some areas was proposed. Using annual wind speed, light intensity, and diesel prices as the input and the yearly wind and diesel prices as sensitive variables, the expected investment costs and economic benefits of the hybrid power generation system were analysed synthetically by HOMER, and the optimal capacity configurations and final conclusions were obtained.

Keywords: Distributed energy system; Hybrid micro-grid system; HOMER; Sensitivity variables; The optimal capacity

1 INTRODUCTION

With the development of the economy and the constant progress of science and technology, the energy shortage has attracted more and more attention. Owing to exhaustion of fossil fuels, and the deterioration of the living environment, the development and utilization of renewable and sustainable energy sources has become the consensus of all countries and an inevitable development tendency.

In recent years, the concept of a micro grid has been proposed and attracted extensive attention worldwide especially a multiple distributed grid which includes solar and wind energy generation systems of mixed source integration and control has become an important issue.

Studies abroad into hybrid energy generation systems have mainly concentrated on large scale grids connected to wind farms. There has been less research on the wind/solar/diesel and energy storage hybrid micro-grid systems. However, initial results have been achieved. Our application in hybrid energy generation is not mature enough.

Research into hybrid power systems has focused on wind, photovoltaic hybrid power plant architecture optimization and system simulation. Literature[1] proposed a variety of complex energy-based models of distributed power generation system cost optimization Literature[2] established a distributed power as a backup power supply capacity optimization model, taking into account economic, environmental and reliability issues. However, the influence of a micro grid's optimization configuration caused by the randomness and volatility of renewable wind energy, solar energy, as well as energy storage capacity configuration problems was not considered.

In this paper, the capacity optimization and the allocation scheme of a micro-grid for year-round use including turbine, photovoltaic array, diesel engine, battery energy storage and power converter were studied. Based on the power supply reliability as constraint conditions, the optimization capacity allocation problem was solved by the HOMER software simulating the system's configuration.

2 HOMER SOFTWARE

The HOMER Micro-power Optimization Model is a computer model developed by the U.S. National Renewable Energy Laboratory (NREL). The HOMER simulate different system configurations to achieve simulation, optimization and sensitivity analysis and other functions, based on the net present that the cost renewable hybrid power system installation and operation in the life cycle of maintenance in the total cost basis. The optimization and sensitivity analysis of HOMER software can be used as an economic and technical feasibility assessment system, and for the availability of resources.[3–5]

3 RENEWABLE RESOURCES

3.1 *Solar resource*

Solar energy is a common renewable energy, and has no geographical restrictions, whether on land or sea,

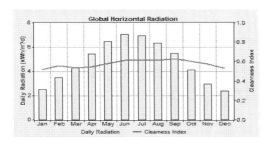

Figure 1. Monthly average light and clearness index.

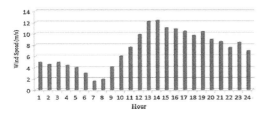

Figure 2. Surface wind speed at 40 m within 24 hours.

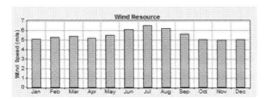

Figure 3. A month average wind speed.

mountains or islands, and everywhere therein. It is one of the cleanest sources of energy that can be directly developed and utilized without transportation. Figure 1 indicates monthly average solar radiation. It can be seen that in the area, there is adequate light, and in June the average illumination reaches 7.07 kwh/m^2/day. The appropriate power of PV modules configured to take advantage of local natural resources could be considered.

3.2 Wind resource

The local belongs to the windy area, and the wind which has the very high periodic law can be efficient utilization. Figure 2 indicates local wind speed variations at 40 meters within 24 hours. As can be seen, the wind speed increases after noon, to reach the largest at 14:00 and decreases slowly thereafter. From 11:00 to 24:00 the turbine can have effective output power. Figure 3 shows a local monthly average wind speed. The annual average wind speed is 5.08 m/s, the wind power density is more than 200 W/m^2, and the time when wind speeds are more than 3 m/s reach 6000 to 7500 hours. The development of this condition is relatively mature, and very suitable for wind power.[6]

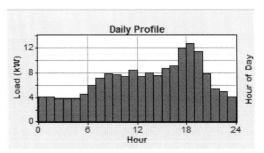

Figure 4. Daily load curve of system simulation.

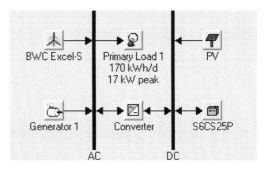

Figure 5. The structure of hybrid power system.

4 THE COMPOSITION OF A RENEWABLE ENERGY HYBRID POWER SYSTEM STRUCTURE

The load of a simulation system is shown in Figure 4. It can be seen from the load curve that the minimum requirement of the load is about the power required of 4 kw in the night time. The load steadily increased during the day, and the maximum load is about 13 kw between 17:00 to 19:00. Due to user demand fluctuations, HOMER gives a random variation to estimate the maximum and minimum daily load curve. In this paper the random change is set at 3%.

The structure of a hybrid power system is shown in Figure 5. The system consists of solar power, a wind power generation module, a diesel engine module, a battery module and a converter. Because solar energy is a direct current, therefore it must be changed by an inverter before delivering it to the load. Wind power can provide AC to the load, but because the wind speed is unstable, as radio frequency emitted is not stable, it can be converted through the rectifier/inverter and then transported to the load[7], in order to improve the reliability of the power supply. As a backup a diesel power generation group only works to meet the operational load when the photovoltaic and wind turbine power generation is insufficient. When the power system has an excess charge, surplus electricity charges the battery after rectification. When solar power, wind power, and diesel power generation do not meet the load demand, the energy stored in the battery inverts for the load to use.

	PV (kW)	XLS	Label (kW)	S6CS25P	Conv. (kW)	Disp. Strgy	Initial Capital	Operating Cost ($/yr)	Total NPC	COE ($/kWh)	Ren. Frac.	Diesel (L)	Label (hrs)
⚠🌬👤🔋📊✏	10	2	15	20	10	CC	$ 166,000	20,580	$ 352,802	0.626	0.46	12,243	3,187
🌬 🔋📊✏	20		15	30	10	CC	$ 177,000	21,856	$ 375,388	0.666	0.33	14,226	3,173

Figure 6. Simulation optimization results.

Production	kWh/yr	%
PV array	15,857	21
Wind turbines	24,946	33
Generator 1	33,674	45
Total	74,477	100

Pollutant	Emissions (kg/yr)
Carbon dioxide	32,240
Carbon monoxide	79.6
Unburned hydrocarbons	8.81
Particulate matter	6
Sulfur dioxide	64.7
Nitrogen oxides	710

Figure 7. Renewable energy proportion and CO_2 emission of scheme 1.

The following will introduce the module parameters of each part and the capacity of primary.

Diesel engine module: The main work of diesel engines is at night, in cloudy or windless weather, when solar modules with the turbine module are not working properly, which can greatly reduce the battery capacity requirements, and to some extent, cost savings. Diesel power is preset at 15 kW, with the initial cost of capital being assumed to be 6000 yuan/kW, the replacement and operating costs are assumed to be 5400 yuan/kW or 0.12 yuan/hour. Diesel engine service life is 15,000 hours. Diesel price is 7.24 yuan/litre.

Solar modules: a solar power module is one of the main power system modules. The total capacity of a PV module is determined according to its photovoltaic power energy and the PV equivalent full load hours. However, due to the high cost of solar panels, solar panels are not suitable for large installations, and their installed capacity is less than the peak load capacity generally. This will be the three photovoltaic array of different size (10, 15 and 20 kW) which were analysed. Each 1 kW solar panel installed, had a capital cost and replacement cost of 36,000 yuan and 30,000 yuan respectively.

Wind power modules: Wind power generation used excel-s fixed pitch wind turbines with a rated capacity of 10 kW and an AC output of 220 V. According to their geographical location and the abundance of wind energy, the height of wind turbine chooses 10 meters. The simulation will be three cases (0 units, 1, 2) for analysis. The capital cost is 180,000 yuan for each turbine, and replacement and operational and maintenance costs were 156,000 yuan and 900 yuan/year, and the turbine's life is 15 years.

The battery module: The battery's main task is to store excess electricity when excess energy was produced in the power system, and release battery storage of electricity to supply power to the load when other modules cannot meet the load demand. This paper used Surrette 6cs25p model according to the system's requirements (1156Ah, 6 V, 6.945 kW). Capital costs per unit of battery replacement costs and operation and maintenance costs were 6600 yuan, 6000 yuan and 60 yuan/year. To find an optimal configuration, the numbers in the battery pack is assumed to be 10, 20 and 50 units.

5 CONCLUSIONS

HOMER software simulates all qualifying system configuration and lists. The optimal configuration simulated by HOMER software is shown in Figure 6, and there are two optimal configurations. Table 1 lists the detailed parameters of the two programs.

In a comparative analysis of options 1 and 2, option 1 is clearly the most economical, and the reasons are as follows:

1) The cost of the turbine, the photovoltaic array and the battery is significantly higher than the other power supplies, affecting the economy of the entire program. At the current market price, the large capacity of micro-grid solar and wind power is not economical. But as technology advances, the cost of wind power and solar power are expected to decline significantly, the price of natural gas and diesel fuel will continue to rise, fuel costs will continue to rise, so the economics of wind turbines and solar energy will be reflected.

2) Scheme 2 countermanded the turbines, and the capacity of batteries and photovoltaic increased slightly.

Although the initial investment is reduced, this increases the output of diesel generators. Figure 7 and Figure 8 shows that in scheme 2 the proportion of renewable energy decrease by 12%, compared with scheme 1, which means that the program modules 2 with diesel will increase by 12% the proportion of electricity generated. Diesel power output increases, which is going to consume more fuel and more operational and maintenance costs. In scheme 2 the NPC total value was more than option 1, at 135,516 yuan, and the levelled cost of energy (levelled cost of energy, COE) has increased from 3.756 yuan/kWh to 3.996 yuan/kwh. Of course, emissions of CO_2 and other pollutants will also increase, and there are more than 5220 kg of pollutants discharged into the atmosphere compared with scheme 1, which is inconsistent with sustainable development and protection of the ecological environment, consistent with science policy. The scheme 2 failed to make full use of local natural resources, and is not the best choice for a hybrid renewable energy power generation system.

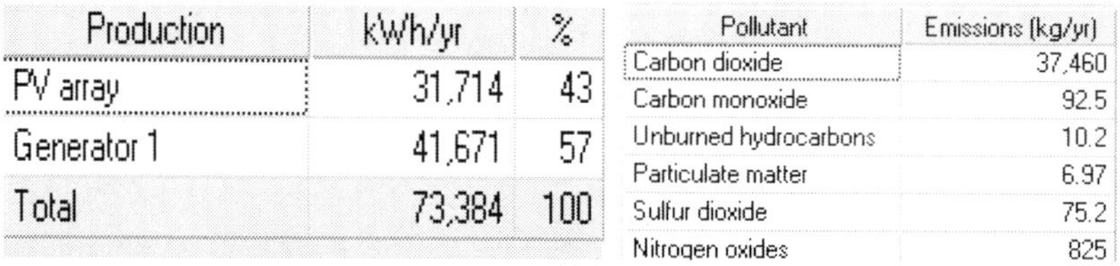

Production	kWh/yr	%
PV array	31,714	43
Generator 1	41,671	57
Total	73,384	100

Pollutant	Emissions (kg/yr)
Carbon dioxide	37,460
Carbon monoxide	92.5
Unburned hydrocarbons	10.2
Particulate matter	6.97
Sulfur dioxide	75.2
Nitrogen oxides	825

Figure 8. Renewable energy proportion and CO_2 emission of scheme 2.

Table 1. A hybrid renewable energy system cost and configuration.

DG	Unit capacity/kW	Installation cost/ (RMB/kW)	Replacement cost/ (RMB/kW)	Operation and maintenance/ (RMB/kW·y)	Scheme 1/ one	Scheme 2/ one
Turbine	10	180000	156000	900	2	0
PV	0.01	360	300	0	1000	2000
Battery	6.94	6600	6000	60	20	30
Diesel engine	15	6000	5400	43.8	1	1

Table 2. Two-dimensional sensitive variable simulation results.

Wind speed m/s	Diesel price yuan/L	PV power/ kW	Turbine power/kW	Diesel engine power/kW	Number of battery/a	Initial capital/yuan	COE yuan/ kW.h
4	7.24	20	10	15	30	2295198	4.074
4	7.92	20	10	15	30	2373846	4.212
4	8.7	20	10	15	30	2459046	4.368
5	7.24	10	20	15	20	2116812	3.756
5	7.92	15	20	15	30	2179464	3.87
5	8.7	15	20	15	30	2245368	3.984
6	7.24	10	20	15	30	1882176	3.342
6	7.92	10	20	15	30	1937382	3.438
6	8.7	10	20	15	30	1998060	3.546

In a real system, the price of diesel prices changes regularly as the market changes, and wind is a variable quantity. It can also set the wind and diesel price as sensitive variables. Table 2 shows the simulation results of different wind speeds and under different systems of diesel prices. As can be seen in the table, the status of different wind speeds (4–6 m/s) and diesel prices are configured under different system's needs and economic situations.

6 RESULTS AND DISCUSSIONS

From the simulation results with the HOMER software, we can draw the following conclusions:

1) Under the same premise supply of resources (such as light intensity, the average wind speed diesel prices, etc.), photovoltaic module configuration capacity is bigger, the average 1 kWh power cost is higher, because a photovoltaic cell at the present stage of its technical level of production its cost is high. According to the economic principle of priority, there is a need to choose a reasonable selection of photovoltaic arrayed in size, to guarantee the effective use of solar energy resources.

2) Reasonable control of the diesel power generation. It was found in the simulation process that, if the power of diesel generator module is too large, the system will need a large amount of the electric power from the diesel engine to send, but contrary to the original intention of the system design, it does not meet the requirements of the system's design. In the system, in order to ensure the PV module and wind power module power generation accounted for the proportion, a value can be defined by a renewable energy system by setting the proportion of renewable energy proportion.

3) The generation from photovoltaic modules and wind power modules is linked closely with the condition of natural resources PV module power output relates to the solar radiation of the locality[8] At longitude 58°48′, north latitude 36°12′, in May to September, illumination is enough, and in August is the largest amount of radiation Therefore, in these several months, the proportion of photovoltaic power generations is higher. Similarly, the average wind speed in June, July and August is larger, so during these months wind power generation accounts for a relatively high proportion of power generation.

REFERENCES

[1] Liu mengxuan, Guo li, Wang chengshan. The design of wind/diesel/storage isolated micro-grid system coordination control strategy [J]. Automation of electric power systems. 2012 (15).

[2] Li xiaogang. Wang shuai. Chen simin. The realization of micro grid back-up power system based on distributed photovoltaic power station [J]. Journal of Shunde Polytechnic, 2014 (01).

[3] HOMER energy Software [EB/OL]. 2012-02-01. http://homerenergy.com/ software.

[4] HOMER energy. HOMER Version history [EB/OL] 2012-02-01 http://homerenergy.com/version_history.html.

[5] HOMER energy. Energy modeling software for hybrid renewable energy systems [EB/OL]. 2012-02-01. http://homer energy.com/index.html.

[6] Ji pin. Zhou xiaoxin. Wu shouyuan. A regional wind resource assessment employed average wind speed participation factor method [J]. Chinese Journal of Electrical Engineering. 2012, 32 (19): 10–15.

[7] Wang haiying. Research on power system reliability problems with large-scale renewable energy [J]. Huazhong University of Science and Technology 2012 (06).

[8] Ma ning. Overview and development prospects on solar photovoltaic [J]. Intelligent Electrical Building Technology. 2011 (02).

A neural network Proportional-Integral-Differential (PID) control based on a genetic algorithm for a coupled-tank system

Yu-Shan Li
Institute of Information, Beijing Union University, Beijing, P.R. China

Hong-Xing Li
Institute of Automation, Beijing Union University, Beijing, P.R. China

ABSTRACT: For the complexity of the Coupled-Tank system, the traditional control method is difficult to obtain a high quality control. In this paper, an effective control method of the neural network Proportion Integration Differentiation (PID) based on a Genetic algorithm (GA-NN PID) is proposed for the coupled-tank system. The neural network is used to adjust and optimize the parameters of a PID controller. The genetic algorithm is employed to optimize the weights of the neural network. The experimental results show that the method is efficient and practical for the Coupled-Tank system.

1 INTRODUCTION

Water level control has always been one of the most important areas for living, industrial control, and chemical process. However, this kind of system mostly has large time-delays, and nonlinear, time-varying characteristics. Considering that with the traditional control methods was difficult to obtain good control, intelligent control such as neural networks, fuzzy control, genetic algorithms, etc, can be good ways to solve such problems.

A neural network is a nonlinear dynamic system, widely connected by a large number of neurons with a nonlinear mapping ability. It has been applied in various fields so far and achieved encouraging progress. A multilevel feed-forward neural network is a powerful learning system that has the advantages of a simple structure and easy programming. Though the BP network model has been applied in many areas in recent years as an important model of the artificial neural network, its speed of convergence is slow, which is related to the weights and tends to be trapped in a minimum point. Thus, the BP network is not a highly practical method.

But with a genetic algorithm, a random search and optimization algorithm developed by simulating genetic biological and long-term evolution, such limitations are overcome to a certain extent. It imitates the mechanism of the biological world in which only the fittest survive and uses the successive iteration method to get the optimal solution.

In this paper, the BP neural network and a genetic algorithm were combined on the basis of a conventional PID controller, using the Coupled-Tank system as the research object. Through learning the performance of the system, the neural network is able to identify the optimal PID controlling parameters with which the system performs at its full capacity.

2 THE STRUCTURE AND MATHEMATICAL MODEL OF A COUPLED-TANK SYSTEM

The Coupled-Tank plant is a 'Two-Tank' module, consisting of a pump with a water basin and two tanks. A schematic drawing of the Coupled-Tank plant is represented in Figure 1.

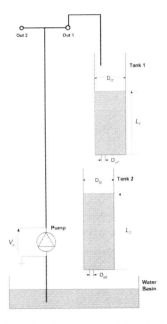

Figure 1. The schematic of the Coupled-Tank plant.

Table 1. Coupled-Tank system model parameters.

Symbol	Description	Value	Unit
K_p	Pump flow constant	3.3	cm³/s/v
D_{o1}	Tank 1 outlet diameter	0.47625	cm
D_{o2}	Tank 2 outlet diameter	0.47625	cm
A_{o1}	Tank 1 outlet area	$\pi D_{o1}^2/4$	cm²
A_{o2}	Tank 2 outlet area	$\pi D_{o2}^2/4$	cm²
g	Gravity constant	981	cm²/s
D_{t1}	Tank 1 inside diameter	4.445	cm
D_{t2}	Tank 2 inside diameter	4.445	cm
A_{t1}	Tank 1 inside cross-section area	$\pi D_{t1}^2/4$	cm²
A_{t2}	Tank 2 inside cross-section area	$\pi D_{t2}^2/4$	cm²
V_{P0}	Steady-state pump voltage	9.26	V
L_{10}	Steady-state water level in Tank 1	15	cm
L_{20}	Steady-state water level in Tank 2	15	cm

It can be described as a state-coupled SISO system. The pump feeds into Tank 1, which in turn feeds into Tank 2. Different inlet and outlet diameters in Tank 1 and Tank 2 can be set up and tried. In this paper, the medium outflow orifices (i.e. outlet) were chosen. The water flows into Tank 1 from 'Out 1', then through the medium outflow orifice it flows into Tank 2. From Tank 2, the water flows from the same size of outlet to the water basin.

2.1 A Coupled-Tank model

2.1.1 Tank 1 model

Table 1, below, lists and characterizes some of the system model parameters, which can be used for mathematical modelling of the system as well as to obtain the water level's Equation Of Motion (EOM).

Applying Bernoulli's equation for small orifices, the outflow velocity from Tank 1, v_{o1} is expressed by the following relationship:

$$v_{o1} = \sqrt{2} * \sqrt{gL_1} \tag{1}$$

So, the outflow rate from Tank 1, F_{o1} can be expressed by:

$$F_{o1} = A_{o1}v_{o1} = A_{o1} * \sqrt{2} * \sqrt{gL_1} \tag{2}$$

Moreover using the mass balance principle for Tank 1, the following first-order differential equation in L_1 can be obtained:

$$A_{t1}(\frac{\partial}{\partial t}L_1) = F_{i1} - F_{o1} \tag{3}$$

where $F_{i1} = K_p V_p$.

In the case of the water level in Tank 1, the operating range corresponds to small departure heights, L_{11}, and small departure voltages, V_{p1}, from the desired equilibrium point (L_{10}, V_{p0}). Therefore, L_1 and V_p can be expressed as the sum of two quantities, as shown below:

$$L_1 = L_{10} + L_{11} \text{ and } V_p = V_{p0} + V_{p1} \tag{4}$$

Applying the Taylor's series approximation about (L_{10}, V_{p0}), equation can be linearized as represented below:

$$\frac{\partial}{\partial t}L_1 = \frac{K_p V_{p0} - A_{o1}\sqrt{2}\sqrt{gL_{10}}}{A_{t1}} - \frac{1}{2}\frac{A_{o1}\sqrt{2}L_{11}}{\sqrt{gL_{10}}A_{t1}} + \frac{K_p V_{p1}}{A_{t1}} \tag{5}$$

Applying the Laplace transform to Equation 5, the desired open-loop transfer function for the Coupled-Tank's Tank 1 system is shown as:

$$G_1(s) = \frac{K_{dc\_1}}{\tau_1 s + 1} \tag{6}$$

with:

$$K_{dc\_1} = \frac{K_p\sqrt{gL_{10}}\sqrt{2}}{A_{o1}g} \text{ and } \tau_1 = \frac{\sqrt{gL_{10}}A_{t1}\sqrt{2}}{A_{o1}g} \tag{7}$$

According to the system's parameters and desired requirements, gives:

$$K_{dc\_1} = 3.2 \text{ and } \tau_1 = 15.2 \tag{8}$$

2.1.2 Tank 2 model

According to the above method, Tank 2 water level's EOM and transfer function can be derived. Thus, the Tank 2 system's EOM can be expressed under the following format:

$$\frac{\partial}{\partial t}L_2 = f(L_2, L_1) \tag{9}$$

The Coupled-Tank's Tank 2 system is shown as:

$$G_2(s) = \frac{K_{dc\_2}}{\tau_2 s + 1} \tag{10}$$

with:

$$K_{dc\_2} = 1 \text{ and } \tau_2 = 15.2 \tag{11}$$

Therefore, the transfer functions of the system are:

$$G(s) = \frac{3.2}{15.25 * 15.25 s^2 + 30.5 s + 1} \tag{12}$$

3 SYSTEM DESIGN OF CONTROLLER

Since the PID control effects depend on three control actions: proportional, integral, and differential control actions, it is of great importance as to how to adjust the relationship between the three. As the neural network has an ability of approach to arbitrary nonlinear mapping, it can thus achieve control of a PID controller with the best combinations through a study on

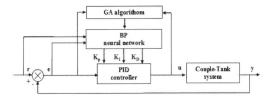

Figure 2. The controller structure.

the function of the system. In order to overcome the limitations of the BP algorithm, a genetic algorithm is combined with a neural network in PID control in this paper. The PID controller is based on a BP neural network, which is composed of a classical incremental digital PID controller and a BP neural network, to realize the online real-time adjustment of PID parameters. By using the global searching ability of GA, it optimizes the initial weights so that the learning efficiency of weights is improved and the possibility of falling in the local minimum is reduced. The controller structure of a PID controller, a combined genetic algorithm with a BP neural network, is shown in Figure 2.

The controller structure in Figure 2 comprises three parts:

(1) A classical incremental PID controller that can be directly used to control the object in a close-loop control.
(2) The optimization module of the genetic algorithm that would optimize the weights of a BP neural network.
(3) A BP neural network that can change its weight coefficients and realize the real-time adjustment of three parameters (K_P, K_I, K_D) in the system.

The incremental digital PID control algorithm is:

$$u(k) = u(k-1) + K_P(error(k) - error(k-1)) + K_I(error(k)) + K_D(error(k) - 2error(k-1) + error(k-2)) \quad (13)$$

where K_P, K_I, K_D are respectively the coefficient of proportional, integral and differential.

3.1 Control algorithm

The parameters of the PID controller are adjusted in real-time by a BP neural network while the system is running, in order to realize the self-learning of the PID parameters. But when the network is learning under a BP algorithm, once the weight values are not appropriate, it's possible to cause system oscillation and further divergence. Even if it is able to converge, the training time may be too long to get perfect weights, because of the slow convergence rate. To solve these problems, the method utilizing a genetic algorithm to optimize the connection weights in the BP system, can improve the system performance effectively.

In this paper a three-layer BP neural network is established, the structure being 4-5-3. The BP-PID controller module is built by using the S-function module in Simulink. The number of neurons in input layer is 4 (value r, error e, output y, threshold $\theta = 1$ were given); Then the neuron number in the hidden layer is 5. Three output values correspond to three adjustable parameters of the PID controller. As the actual values of K_P, K_I, and K_D cannot be negative, the non-negative Sigmoid function is selected as the activation function of the network output layer:

$$g(x) = \frac{1}{2}(1 + \tanh(x)) = \frac{e^x}{e^x + e^{-x}} \quad (14)$$

While the hidden layer's activation function is a symmetrical Sigmoid function:

$$f(x) = \tanh(x) = \frac{e^x - e^{-x}}{e^x + e^{-x}} \quad (15)$$

3.1.1 Forward propagation algorithm

The input and output of the BP neural network's hidden layer can be expressed as follows:

$$\begin{cases} O_j^{(1)} = x(j) \\ net_i^{(2)}(k) = \sum_{j=1}^{m} w_{ij}^{(2)} o_j^{(1)} \quad i=1,2,3...n, j=1,2,3...m \\ O_i^{(2)} = f(net_i^{(2)}(k)) \end{cases} \quad (16)$$

where the $w_{ij}^{(2)}$ is weights from input layer to hidden layer. The superscripts (1), (2), (3) here represent the input layer, hidden layer, and output layer respectively.

The input and output of BP neural network's output layer are:

$$\begin{cases} net_l^{(3)}(k) = \sum_{i=1}^{n} w_{li}^{(3)} o_i^{(2)} \\ O_l^{(3)} = g(net_l^{(3)}(k)) \end{cases} \quad l=1, 2, 3 \quad (17)$$

3.1.2 Back propagation algorithm

The performance index function of the BP neural network is taken as:

$$E(k) = \frac{1}{2}(r(k) - y(k))^2 \quad (18)$$

The network weights are modified by the gradient descent method and have added an inertia term to speed up the convergence rate of the network and let the search converge to the global minimum quickly.

$$\begin{cases} \Delta w_{li}^{(3)}(k) = -\eta \frac{\partial E(k)}{\partial w_{li}^{(3)}} + \alpha \Delta w_{li}^{(3)}(k-1) \\ \frac{\partial E(k)}{\partial w_{li}^{(3)}} = \frac{\partial E(k)}{\partial y(k)} \cdot \frac{\partial y(k)}{\partial u(k)} \cdot \frac{\partial u(k)}{\partial o_l^{(3)}(k)} \cdot \\ \frac{\partial o_l^{(3)}(k)}{\partial net_l^{(3)}(k)} \cdot \frac{\partial net_l^{(3)}(k)}{\partial w_{li}^{(3)}} \end{cases} \quad (19)$$

where η is the learning rates, while α is the inertial coefficient.

According to Equation 13, we can obtain that:

$$u(k) = u(k-1) + o_1^{(3)}(error(k) - error(k-1)) + o_2^{(3)}(error(k)) + o_3^{(3)}(error(k) - 2error(k-1) + error(k-2)) \quad (20)$$

After arrangement, the adjustment formula of the BP neural network's output layer weights can be shown as:

$$\Delta w_{li}^{(3)}(k) = \eta e(k) \frac{\partial(k)}{\partial u(k)} \frac{\partial u(k)}{\partial o_l^{(3)}(k)} \cdot \quad (21)$$
$$g'(net_l^{(3)}(k)) o_i^{(2)}(k) + \alpha \Delta w_{li}^{(3)}(k-1)$$

where $\frac{\partial u(k)}{\partial o_l^{(3)}(k)}$ is determined by taking the partial derivatives of Equation 20, $\frac{\partial y(k)}{\partial u(k)}$ can be replaced by the sgn function.

By the same token, it can be drawn that the adjustment formula of the hidden layer weights are:

$$\begin{cases} \Delta w_{li}^{(2)}(k) = \eta \delta_i^{(2)} o_j^{(1)}(k) + \alpha \Delta w_{ij}^{(2)}(k-1) \\ \delta_i^{(2)} = \sum_{l=1}^{3} \delta_l^{(3)} w_{li}^{(3)} f'(net_i^{(2)}(k)) \\ \delta_l^{(3)} = e(k) \text{sgn}\left(\frac{y(k)-y(k-1)}{u(k)-u(k-1)}\right) \frac{\partial u(k)}{\partial o_l^{(3)}(k)} \cdot \\ g'(net_l^{(3)}(k)) \end{cases} \quad (22)$$

3.2 Optimizing the network weights by GA

3.2.1 Coding

Before solving practical problems with a genetic algorithm, the problem should be analyzed to identify the range of solutions and feasible coding methods so that a genetic algorithm can be applied. The coding methods that are commonly used nowadays are binary coding, real number coding, and hybrid coding. In consideration of its higher efficiency, real number coding is adopted in this paper.

3.2.2 Fitness function

In the genetic algorithm, fitness function is used to describe the problem-solving ability of chromosomes, the same as the individual ability to adapt to the environment in the natural world. The greater the fitness function value is, the more excellent the chromosome is, and the more should it be retained and passed on to the next generation. The individual fitness function is:

$$Fit(f(x)) = \frac{1}{f(x)+c} \quad c \geq 0, \; c+f(x) \geq 0 \quad (23)$$

3.2.3 The selection operator

The selection operation is a kind of genetic algorithm that is used to determine which individuals from the parent group should be chosen to pass on their chromosomes to the next generation through certain methods. Commonly used selection strategies are roulette wheel selection, an elitism preserving strategy, an expected value model, and a rank-based model.

3.2.4 A crossover operator

Crossover strategy can be divided into single point crossover, two-point crossover, and uniform crossover. In this paper, a uniform arithmetic crossover is adopted, in which each point is regarded as a potential intersection and where it perform the crossover operation on two individuals is determined by randomly generated probability.

3.2.5 The mutation operation

Mutation is an important way to keep a population's diversity in genetic algorithms. It would choose a locus with a certain probability and achieve the aim of mutation via changing the value of the locus.

The mutation probability P_m usually takes between 0.001–0.1. If the mutation rate is too large, it may destroy many good varieties and it may not be possible to obtain the optimal solution. Supposing individuals of variation are $P = (p_1, p_2, \ldots, p_n)$, and the gene to mutate is $p_i \in [X_{max}, X_{min}]$, the mutated genes can be generated by:

$$p_i' = (X_{max} + X_{min})/2 + (X_{max} - X_{min}) \bullet \quad (24)$$

The process of optimizing weights by GA, is as follows:

(1) Supposing $W_{ij}(k)$ is the actual weight in the solution space, $w_{ij}(k) = (w_{ij})_{min} \sim (w_{ij})_{max}$ signifies its variation range; the real number coding is adopted;
(2) Determine GA's parameters: $(N, P_c, P_m) = (50, 0.9, 0.1)$; the genetic manipulation: roulette wheel selection and uniform crossover;
(3) The space in GA: initial populations ($N = 50$) are generated randomly. As the structure of the network is $N_{4,5,3}$, the weight number of the network is 35;
(4) Extract all the individuals to the solution space in GA. After the neural network determines error functions and identify fitness values, the genetic individual that has the maximum fitness values would pass them on to the next generation directly; the current generation group would be operated by crossover and mutation operators until the next generation group is generated;
(5) If the termination condition is satisfied, proceed with step 6, otherwise turn to execute step 4;
(6) Chose the best individual that has the maximum fitness value as the optimal solution.

The whole procedure can be described as Figure 3.

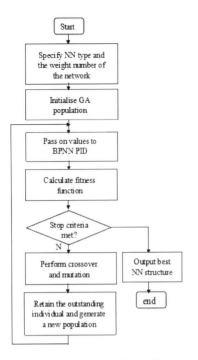

Figure 3. The flowchart of a genetic algorithm.

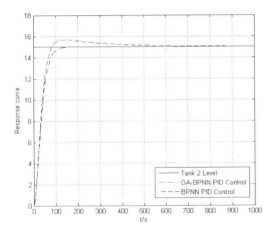

Figure 4. The simulation curves.

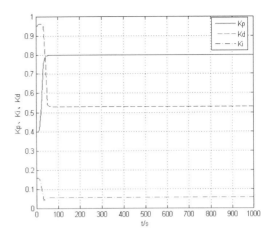

Figure 5. The change tendency of parameters.

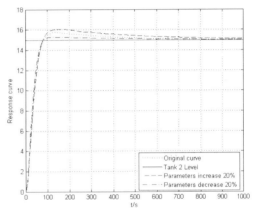

Figure 6. The response of BPNN PID control, when the parameters increase/decrease by 20%.

4 SYSTEM SIMULATION

Based on a Couple-Tank System, the simulation of liquid level control system is taken respectively in the PID controller, based on the BP neural network and the BP neural network PID controller, which are based on a genetic algorithm to verify the effectiveness of the controller. As shown in Equation 12, the control object is determined. The neural network is chosen as 4-5-3. After repeated debugging, the parameters are set to $\eta = 0.00028$, $\alpha = 0.1$. Furthermore, the BPNN PID controller model is established by S-function. In the calculation of the weights' updating, the last moment and the moment before the last moment weights, and the last controlled quantity output, are applied. Thus there are (14*hideNums + 1) state variables in total.

In Figure 4, there is shown a small overshoot and the whole transition process is smooth, but the stable time is long. Whereas, after the optimization by a genetic algorithm, the overshoot is zero, the stable time is shorter than before, and especially there is no steady-state error basically. In the method of PID control combined with the BP neural network and GA, the simulation result also shows the change tendency of the parameters according to time, as shown in Figure 5. It is clearly seen that a GA-BPNN PID control could satisfy the system's requirements: zero-shoot and fast stable.

Due to the presence of a certain error in system modelling, the parameters of the model are respectively increased by 20% and decreased by 20% to analyse the robustness of the controller. The comparative analysis of system response curves are shown in Figure 6 and Figure 7. It can be seen that the control method of GA-BPNN PID, which optimize BP neural network's connection weights by using a genetic algorithm, can

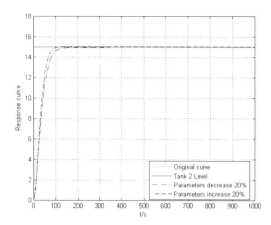

Figure 7. The response of GA-BPNN PID control, when the increase/decrease by 20%.

still maintain a good working condition and control effect, and the shape of the curve changes are small when the disturbance of the internal parameters of system's model occurs. So this kind of control method for nonlinear, time-varying system has a good robust performance.

5 CONCLUSION

In this paper, the BP network's strong ability of self-learning and nonlinear approximation property are applied. By combining a digital incremental PID controller, the PID controller can adjust parameters online. But in the study by using a BP algorithm, if the weight values are not suitable, it may cause oscillations of the network and cannot converge. Even if it can converge, the training time may be too long because of the slow convergence speed. In order to obtain the best weights' distribution, the optimization of the BP neural network's connection weights by using a genetic algorithm, is considered. The simulation results show that the method proposed for controlling the Coupled-Tank system is robust and effective. And its controlling quality is better than the BP-PID controller.

REFERENCES

[1] Fei He, Dong-feng He, An-jun Xu, etc, Hybrid Model of Molten Steel Temperature Prediction Based on Ladle Heat Status and Artificial Neural Network, Journal of Iron and Steel Research, International. 21 (2014) 181–190.
[2] Jia Chun-yu, Shan Xu, Cui Yan-cao, etc, Modeling and Simulation of Hydraulic Roll Bending System Based on CMAC Neural Network and PID Coupling Control Strategy, Journal of Iron and Steel Research, International. 20 (2013) 17–22.
[3] Karl Henrik Johansson, The Quadruple-Tank Process: A Multivariable Laboratory Process with an Adjustable Zero, IEEE Transactions on Control Systems Technology. 8 (2000) 456–465.
[4] Wang Lei, Tuo Xianguo, Yan Yucheng, etc, A Genetic-Algorithm-based Neural Network Approach for Radioactive Activity Prediction, Journal of Nuclear Science and Techniques. 24 (2013) 060201.
[5] Zeng-Hui Zhu, Hui-Ying Sun, Application of PID Controller Based on BP Neural Network In Export Steam's Temperature Control System, Journal of Measurement Science and Instrumentation. 5 (2011) 81–85.

Research into the reactive power compensation of a new dual buck non-isolated grid inverter

Peng Sun, Yong Xie, Yu Fang, Lijuan Huang & Yong Yao
College of Information Engineering, Yangzhou University, China

ABSTRACT: A new type of modulation strategy is presented to suppress the current distortion when the grid voltage crossing zero in this paper. The concrete method is that the Uni-Polar Pulse-Width Modulation (UP-PWM) is employed when the inverter in the positive power generation, while the Edge-Uni-Polar Pulse-Width Modulation (EUP-PWM) is adopted when the inverter in the negative power generation. The proposed modulation strategy can improve the current waveforms and meet the specification requested by VDE-AR-N 4105 standard when the grid-connected inverter serves as reactive power compensation device. Finally, an 1800W dual-Buck grid inverter is developed and the experimental results verify the permanent.

1 INTRODUCTION

In order to achieve a low leakage current and high efficiency for a non-isolated photovoltaic grid inverter, researchers have done a lot of work. It is well welcome to use a traditional uni-polar pulse-width modulation (UP-PWM), because that UP-PWM is helpful to improve the efficiency of the converter. However, the current distortion occurs at the point of voltage zero-crossing when the grid-connected inverter provides reactive power support. In recent years, Germany has issued the standard of VDE-AR-N 4105, and, according to this standard, a grid-connected inverter should have the function of a reactive power compensator when the total apparent power of the power station is equal to or greater than 3.68 kVA ($\sum S_{Emax} \geq 3.68$ kVA). Reactive power generation is usually implemented by regulating the phase angle between the output current and the grid voltage. And there are two requirements, i.e., one is that the inverter can generate both the lagging and leading reactive power; the other is that the inverter only supplies lagging reactive power. However, the grid current distortion can be produced when the reactive power is generated by the UP-PWM. This increases Total Harmonic Distortion (THD) of the grid current, and meanwhile the results in failure in 0.01 error tolerance is according the standard (requirement of power factor ($\cos\varphi$) is given). This paper will discuss the implementation of the modulation EUP-PWM that can cancel off the above mentioned current distortion. Of course, we will analyse EUP-PWM and reference the contents based on a new Dual Buck Inverter (DBI), as shown in Figure 1. This DBI has low common leakage current and high efficiency.

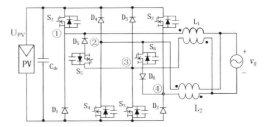

Figure 1. Topology of DBI non-isolated grid inverter.

2 ANALYSIS OF CURRENT ZERO-CROSSING DISTORTION OF UP-PWM IN REACTIVE POWER COMPENSATION

As shown in Figure 1, the topology of the DBI is composed of two Buck converters which are in parallel, and one of them operates in the positive half cycle of the grid voltage, while another is in the negative half cycle. When the grid voltage in the positive half cycle, Buck (I) consisting of S1, S3, S5, D5 and L1 works, and S1 and S3 work with high frequency, while S5 works with a power frequency. When S1 and S3 are on, and D5 is off, photovoltaic cells transfer energy to the grid; when S1 and S3 are off, S5 and D5 start to freewheel. Besides, Buck (II) consisting of S2, S3, S6, D6 and L2 starts to work when the grid voltage is in the negative half cycle. S2 and S4 work with high frequency while S6 works with power frequency, and S6 and D6 will freewheel. This new topology of the DBI has the advantages of a high-usage of the input DC voltage, and there does not exist common conducting in all bridge legs.

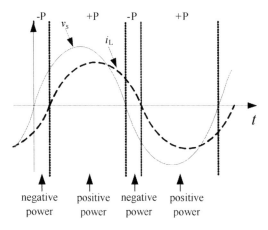

Figure 2. Relationship between grid voltage and current with the lagging PF.

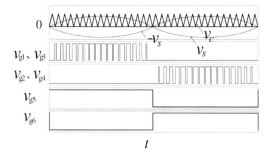

Figure 3. Drive signal of the new DBI.

As shown in Figure 2, when the inverter injects reactive power to the grid, it needs to regulate phase difference between the reference current and the grid voltage by control, and this will make the phase of the grid-connected current shift. In Figure 2, positive power here is defined when the polarities of the grid-connected current and the grid voltage are the same, or the negative power is defined. And then the phase angle between the grid-connected current and the grid voltage can be expressed as follows:

$$\theta = cos^{-1}PF \qquad (1)$$

where, θ is the phase angle between grid-connected current and grid voltage, and PF is the power factor.

Figure 3 shows the drive waveforms of the new DBI with UP-PWM. Obviously, distortion of grid-connected current cannot occur when the grid voltage crossing zero point if reactive power is not required. In this case, only positive power is being generated. Therefore, this modulation is usually called as active modulation.

When reactive power compensationis needed, high frequency switches in the part of negative power cannot get drive signals if the UP-PWM is still used. Taking the grid voltage in the positive half cycle as an example, referring to Figure 1 and Figure 2, it can

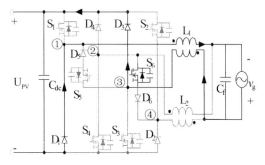

Figure 4. Operating state diagram of the inverter circuit of a negative power part when the power switches are turned off.

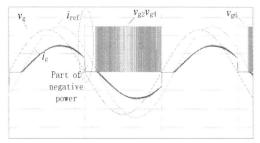

Figure 5. Simulation waveforms in active modulation.

be divided into three situations. ① In the part of negative power, grid voltage is negative while the current is still positive, if S1 and S3 are still on with high frequency at this time, and both photovoltaic cells and the grid will output energy so that the current of Buck (I) may be high enough to damage the power switches; ② if S1 and S3 are turned off at this time, while S2 and S4 are conductedwith high frequency, Buck (I) may return the grid energy to the photovoltaic side through D1 and D3. In contrast, Buck (II) will transmit the photovoltaic cells energy to the grid, which cannot effectively implement reactive power compensation; ③ if all of the high frequency power switches are not driven with a high frequency in the part of negative power, the path of the current flowing is shown in Figure 4. Also, the grid-connected current is not distorted at the zero-crossing point of the grid voltage as is shown in Figure 5.

It is known from Figure 4 that the grid-connected current flows through L1, grid vg, diodes D3 and D1, the DC side capacitor C_{dc}. Due to that U_{pv} is higher than the grid voltage, the inductor current dropsto zero quickly, and the grid-connected current distortion appears at the zero-crossing point of the grid voltage.

3 A EUP-PWM METHOD FOR REACTIVE POWER COMPENSATION

According to the analysis of the above, the grid-connected current in lagging reactive power compensation can achieve the ideal and stable control in the

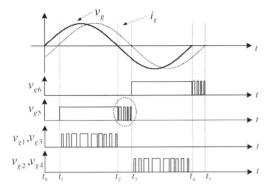

Figure 6. Driving waveforms under the proposed modulation.

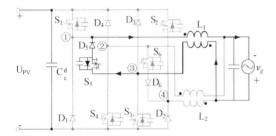

Figure 7. Inverter circuit operating diagram of the part of reactive power when S_5 conducts.

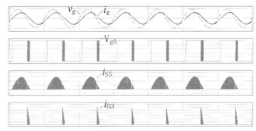

Figure 8. Simulation waveforms by using the EUP-PWM.

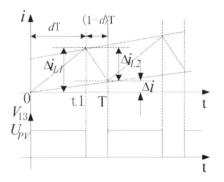

Figure 9. Waveforms of the output voltage V_{13} and of the inductor current.

part of positive power, if regulated by UP-PWM. However, the current distortion occurs when it works in the negative power. Therefore, we consider that the drive way of power switches S_5 (S_6) should be changed in the part of negative power. This is just an Edge-Uni-Polar Pulse-Width Modulation (EUP-PWM), which is proposed in this paper.

The EUP-PWM is shown in Figure 6. The active modulation is adopted when the referenced current has the same polarity as the grid voltage; while a high frequency modulation strategy is employed when working in the negative power stage. It is noted that a low-frequency modulation is given originally, and that only the part of high-frequency modulation is adopted so as to raise the power efficiency. It is seen from Figure 6 that the driving signal of S_5 (S_6) has already been changed partly into a high-frequency modulation, and is still of the uni-polar modulation, so we call this modulation EUP-PWM. It can also be seen that the current waveforms are perfect and the current distortion at the zero-crossing point of grid voltage has already been cancelled.

Figure 7 describes the operating mode in the part of negative power when S_5 conducting. When S_5 turns on, the grid-connected current flows through S_5, the diode D_5, the coupling inductance L_1 and the grid, while the grid is still charging the coupling inductance L_1 at this moment. When S_5 turns off, the operating mode is the same as in Figure 4, the grid-connected current flows through L_1, grid v_g, the diodes D_3, D_1 and the DC side U_{pv}, at this time, and the current flows from the inductor L_1 to the DC side. It is obvious from the above analysis that the grid along with the DC side charges the inductor in the part of the negative power. This changes the previous phenomenon that the grid only charges the inductor in the part of negative power, and supplies enough energy to shave the sinusoidal current at the zero-crossing point of grid voltage. As a result, the grid-connected current can well follow the reference current in the part of negative power.

Figure 8 represents the simulation waveforms by using EUP-PWM, and the grid-connected current almost has no distortion in the grid voltage zero-crossing.

4 THE REALIZATION OF A EUP-PWM

To realize the modulation scheme proposed in the paper, it needs to work out the duty cycle of S_1 (S_2), and S_3 (S_4) in a positive power modulation and that of S_5 (S_6) in a negative power modulation. And then it can be programmed with DSP to achieve the output of the control signals.

Figure 9 illustrates the waveforms of the inductor current and inverter output voltage within the switching cycles in a positive power modulation.

In mode $[0\ t_1]$, the switches S_1 and S_3 are on while D_5 is off, the voltage of inductor L_1 is $U_{PV} - |v_g|$, the inductor current is a linear ascent, and its variation is Δi_{L1}, $\Delta t = dT$, filter inductor $L_s = L_1$, $V_{13} = U_{PV}$,

T is the switching cycle and v_g is the grid voltage. Therefore:

$$dT = \frac{L_s \Delta i_{L1}}{U_{PV} - |v_g|} \quad (2)$$

In mode [t_1 T], the switches S_1 and S_3 are off while D_5 and S_5 are on, the voltage of inductor L_1 is $-|v_g|$, inductor current is a linear decline, its variation is Δi_{L2}, $\Delta t = (1-d)T$, $V_{13} = 0$. Therefore:

$$T - dT = \frac{L_s \Delta i_{L2}}{|v_g|} \quad (3)$$

Using equation (2) and (3), Δi can be gotten as equation (4):

$$\Delta i = \Delta i_{L1} - \Delta i_{L2} = \frac{U_{PV} - |v_g|}{L_s} dT - \frac{|v_g|}{L_s}(1-d)T \quad (4)$$

As is shown in the following equation (5), we can get the duty cycle of S_1 and S_3 in positive power modulation as the equation (4).

$$d = \frac{\Delta i \times L_s}{U_{PV} T} + \frac{|v_g|}{U_{PV}} = \frac{(i_{ref} - i_g) \times L_s}{U_{PV} T} + \frac{|v_g|}{U_{PV}} \quad (5)$$

where, Δi is the difference between the reference current i_{ref} of the next switching cycle and the present sampled feedback current i_g, that is $\Delta i = i_{ref} - i_g$.

In the part of negative power, according to Figure 4 and Figure 7, the change of inductor current in a switching cycle can be deduced.

$$\Delta i_L = \Delta i_{L1} - \Delta i_{L2} = \frac{|v_g|}{L_s} dT - \frac{U_{PV} - |v_g|}{L_s}(1-d)T \quad (6)$$

From the equation (6), we can obtain the duty cycle of switch S_5 in negative power modulation as equation (7):

$$d = \frac{\Delta i \cdot L_s}{U_{PV} T} + \frac{U_{PV} - |v_g|}{U_{PV}} \quad (7)$$

In fact, the duty cycle is generated by DSP and Figure 10 presents the flowchart scheme. Here, n_{PF} is the required compensation points of power frequency switches S_5 and S_6 in EUP-PWM. In this paper, the phase angle between the grid-connected current and the grid voltage depends on the required power factor and 400 is the switching periods in the grid cycle (Given 20 kHz/50 Hz = 400, 20 kHz is the switching frequency, 50 Hz is the grid frequency). Hence, n_{PF} can be expressed as equation (8):

$$n_{PF} = (cos^{-1} PF) \frac{400}{360°} \quad (8)$$

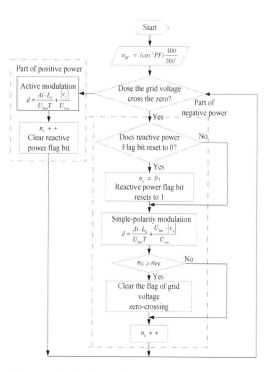

Figure 10. A flowchart of the reactive power modulation.

5 EXPERIMENTAL RESULTS

On the basis of this analysis and simulation, some experiments have been done to verify this paper. The main parameters are as follows: maximum input dc voltage is 550V, switching frequency $f_s = 20$ kHz, MOSFETs choose IPW60R041C6, diodes are APT60DQ60B, C_f is 6.8 μF and the filter inductor is 1.2 mH.

Figure 11(a) shows the waveforms of the grid voltage and grid-connected current in positive power modulation when $cos\varphi = -0.97$ and the output power is 1800W. It can be seen from Figure 11 that the new DBI, in positive power modulation, will generate grid-connected current distortion in the grid voltage zero-crossing.

Figure 11(b) shows the experimental waveforms with the proposed modulation when $cos\varphi = -0.97$ and output power is 1800W. It can be seen that EUP-PWM can effectively solve the problem of grid-connected current distortion when generating reactive power. And the total current harmonic is less than 5% which can meet the power factor ($cos\varphi$) error range of $+/-0.01$. Although the proposed method in this paper aims at lagging reactive compensation, it is also effective in leading reactive compensation.

6 CONCLUSIONS

Taking the new DBI as an example, A EUP-PWM strategy has been proposed to effectively suppress the

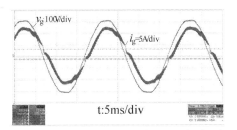

(a) Grid-connected current waveforms in the lagging reactive power compensation of the positive modulation

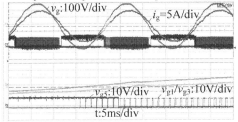

(b) Grid-connected current waveforms in the lagging reactive power compensation of the new modulation

Figure 11. Experimental waveforms.

current distortion at the zero-crossing point of grid voltage when the inverter generates negative power. This modulation strategy is effective in implementing reactive power compensation. Finally, the experimental results verify the feasibility of the proposed method in this paper. The proposed modulation method can meet the relevant requirements of the standard of VDE-AR-N 4105 and has a good application value.

ACKNOWLEDGEMENTS

This work is supported by the Prospective Research Project of Jiangsu Province (BY2013063-01), the college students' innovation and entrepreneurship training plan of Jiangsu Province (201411117030Z), and the Science and Technology Cooperation Fund of Yangzhou City Hall (project 2012038-2).

REFERENCES

[1] Baifeng Chen, Pengwei Sun, Chuang Liu. High Efficiency Transformerless Photovoltaic Inverter with Wide-Range Power Factor Capability [J]. IEEE Trans. Power Electron, 2008, 23(3): 1320–1333.
[2] Bin Gu, Dominic J, Jih-Sheng Lai. High Reliability and Efficiency Single-Phase Transformerless Inverter for Grid-Connected Photovoltaic Systems [J]. IEEE Trans. on Power Electronics, 2013, 28(5): 2235–2245.
[3] B. Yang, W. Li, Y. Gu, W. Cui and X. He. Improved transformerless inverter with common-mode leakage current elimination for a photovoltaic grid-connected power system [J]. IEEE Trans. Power Electron, 2012, 27(2): 752–762.
[4] H. Xiao, S. Xie. Transformerless split-inductor neutral point clamped three-level PV grid-connected inverter [J]. IEEE Trans. Power Electron, 2012, 27(4): 1799–1808.
[5] T. Kerekes, R. Teodorescu, P. Rodriguez and G. Vazquez. A new high-efficiency single-phase transformerless PV inverter topology [J]. IEEE Trans. Ind. Electron, 2011, 58(1): 184–191.
[6] Y. Gu, W. Li, Y. Zhao and B. Yang. Transformerless inverter with virtual DC bus concept for cost-effective grid-connected PV power systems [J]. IEEE Trans. Power Electron, 2013, 28(2): 793–805.

Modelling condition monitoring inspection intervals

A. Raza
Department of the President's Affairs, Overseas Projects and Maintenance Group, Abu Dhabi, UAE

V. Ulansky
National Aviation University, Kiev, Ukraine

ABSTRACT: This paper reports on the modelling of condition monitoring intervals. The proposed model is based on a regenerative stochastic process involving change in states of a single-unit system with imperfect inspections and perfect repair. An inspection consists of checking the system state parameters against the critical threshold levels. The decision rule used for inspecting the system condition is considered, and the probabilities of the correct and incorrect decisions are derived. The system operation process is analyzed in a finite time interval as a sequence of varying states. Expressions are derived for determining the mean times the system stays in various states during operation. The proposed model can be used for both sequential and periodic inspection policies, which allows maximization of system availability or minimization of the expected average cost per unit time.

Keywords: Condition-based maintenance; imperfect inspection; decision rule; false failure; undetected failure; unrevealed failure; corrective repair

1 INTRODUCTION

Condition-based maintenance is widely used for stochastically deteriorating technical systems. A basic maintenance operation of this kind is condition monitoring. Obviously, condition monitoring is preferred among other maintenance techniques in cases where system parameter deterioration can be measured and where the system enters the failed state when at least one state parameter deteriorates beyond the level of functional failure. Condition-based maintenance allows for system state parameter assessment via continuous monitoring or periodic inspections. Condition-monitoring-based maintenance will be effective if it decreases maintenance costs and increases availability and safety (in the case of critical systems). The growing interest about condition monitoring is evident from the large number of reports of development related to various relevant mathematical models and optimization techniques.

1.1 Review of models based on perfect inspection

The following references do not relate to the models with imperfect inspections, which is the subject of this paper. However, these references are important for understanding the proposed approach to determining condition monitoring inspection intervals.

Barzilovich et al. (1980) analyzed an adaptive model for determining condition monitoring intervals. Their proposed policy minimizes the expected maintenance cost per unit time. Ito & Nakagawa (1997) considered a sequential inspection policy with a finite number of checks for minimizing the total expected cost. Abdel-Hameed (1987 & 1995) analyzed an optimal periodic inspection policy model based on the class of increasing pure jump Markov processes. In his model, failure can only be detected by inspection. Park (1988) considered the same model as Abdel-Hameed, but failure is assumed to be discovered immediately without the need for inspection, which is equivalent to perfect continuous condition monitoring. In addition, Park considered a model with a fixed failure level. Kong & Park (1997) later generalized this model by including a random failure level. An adaptive Bayesian version of Park's model was presented by van Noortwijk et al. (1995). In this model, the average rate of deterioration is uncertain. Jia & Christer (2002) proposed almost the same model as Park (1988), but their model included one more decision variable, i.e., optimal time of the first inspection. The optimal time interval for periodic inspections then starts from the time of the first inspection. Grall et al. (2002) and Dieulle et al. (2003) proposed a more extensive inspection model based on that of Park (1988). This model includes an optimization for aperiodic inspection policy, which is scheduled on the basis of a function of the deterioration state. Failure is detected only by inspection, and the cost of unavailability of the system per unit time is included to account for repair cost. Ferreira et al. (2009) proposed a decision model that can simultaneously determine inspection intervals for condition monitoring regarding the failure behaviour of equipment to be inspected. The model features maintainability and decision-maker preferences pertaining to cost and downtime. Golmakani & Fattahipour (2011) proposed an approach in which preventive and failure replacement costs as well as inspection costs are considered for determining the optimal replacement policy in addition to an age-based inspection scheme in which the total average costs of replacement and inspection are minimized.

1.2 Review of models based on imperfect inspection

Maintenance models with imperfect inspections usually consider two types of errors: 'false positives' (false alarms) with probability α and 'false negatives' (i.e. non-detecting of failure) with probability β; for example, Berrade et al. (2012). Such models are not condition-based maintenance models because the error probabilities are not constant coefficients but depend on time and the parameters of the deterioration process. Moreover, such models depend on the results of multiple previous inspections as shown by Ulansky (1987 & 1992). Therefore we analyze only those studies in which the inspection error probabilities depend on the deterioration process parameters. Ulansky (1987) analyzed condition-based maintenance policies with imperfect operability checks. The proposed probabilities of correct and incorrect decisions depend on the deterioration process parameters and the results of previous inspections. Probabilistic and cost maintenance measures were analyzed. Newby & Dagg (2002) considered the result of a measurement, which includes the original deterioration process along with a normally distributed measurement error. Based on this model, a decision rule was analyzed and optimal monitoring policies were found. The same approach was used by Whitmore (1995) to include measurement error in a Wiener diffusion process-based degradation model. Kallen & van Noortwijk (2005) also used a similar approach to find the likelihood for more than one inspection. They proposed a simple extension to the Bayesian updating model, such that the model can incorporate the results of inaccurate measurements.

In this paper, a new condition-based maintenance model is developed for determining inspection intervals. The model assumes imperfect inspections based on a decision rule dependent on both the functional failure level and the critical threshold level. Using the developed mathematical model, equations are proposed for such maintenance policy measures as average availability and average cost per unit time. A linear deterioration stochastic model is used for illustrating the efficiency of the proposed maintenance policy.

2 IMPERFECT INSPECTIONS

2.1 Decision rule

Assume that the state of a system is fully identified by the value of one parameter $X(t)$, which is a non-stationary stochastic process with continuous time. A system should operate for a finite time interval T and be inspected at successive times t_k ($k = 1, 2, \ldots, N$), where $t_0 = 0$. When the system state parameter exceeds its functional failure level FF, the system passes into the failed state. The measured value of $X(t)$ at time t_k is expressed as follows:

$$Z(t_k) = X(t_k) + Y(t_k), \quad (1)$$

where $Y(t_k)$ is the measurement error of the system state parameter at time t_k.

We introduce the following decision rule for inspecting the system condition at time t_k. If $z(t_k) < PF$, the system is said to be suitable over the interval $(t_k, t_{k+1}]$, where PF ($PF < FF$) is the critical threshold level equivalent to the potential failure level of the system state parameter $X(t)$. If $z(t_k) \geq PF$, the system is said to be unsuitable and it should not be used in the interval $(t_k, t_{k+1}]$. Thus, this decision rule is aimed towards the rejection of systems that are unsuitable for use in the next operation interval.

2.2 Space of events

Suppose that a random variable Ξ ($\Xi \geq 0$) denotes the failure time of a system with failure density function $\omega(\xi)$. We introduce two new random variables associated with the critical threshold level PF. Let Ξ_0 denote a random time of a system operation until it exceeds the critical threshold level PF by the parameter $X(t)$, and let Ξ_k denote a random assessment of Ξ_0 based on the results of inspection at time t_k. From the definition of the random variable Ξ_k, it follows that

$$\Xi_k = \begin{cases} t_k, \text{if } z(t_k) \geq PF \ (k=1,2,\ldots,N), \\ T, \text{if } \bigcap_{k=1}^{N} z(t_k) < PF. \end{cases} \quad (2)$$

The random variables Ξ, Ξ_0 and Ξ_k are determined as the smallest roots of the following stochastic equations:

$$X(t) - FF = 0, \quad (3)$$

$$X(t) - PF = 0, \quad (4)$$

$$Z(t_k) - PF = 0. \quad (5)$$

From (2) and (5), it follows that the previously introduced decision rule can be converted to the following form: the system is judged to be suitable at time point t_k if $\xi_k > t_k$; otherwise (i.e. if $\xi_k \leq t_k$), the system is judged to be unsuitable, where ξ_k is the realisation of Ξ_k for the considered system. Mismatch between the solutions of (3) and (5) results in the appearance of one of the following mutually exclusive events when inspecting system suitability at time t_k:

$$h_1(t_k) = \left\{ \Xi > t_{k+1} \cap \left(\bigcap_{i=1}^{k} \Xi_i > t_i \right) \right\}, \quad (6)$$

$$h_2(t_k) = \left\{ \Xi > t_{k+1} \bigcap \Xi_k \le t_k \bigcap \left(\bigcap_{i=1}^{k-1} \Xi_i > t_i \right) \right\}, \quad (7)$$

$$h_3(t_k) = \left\{ t_k < \Xi \le t_{k+1} \bigcap \left(\bigcap_{i=1}^{k} \Xi_i > t_i \right) \right\}, \quad (8)$$

$$h_4(t_k) = \left\{ t_k < \Xi \le t_{k+1} \bigcap \Xi_k \le t_k \bigcap \left(\bigcap_{i=1}^{k-1} \Xi_i > t_i \right) \right\}, \quad (9)$$

$$h_5(t_k) = \left\{ \Xi \le t_k \bigcap \left(\bigcap_{i=1}^{k} \Xi_i > t_i \right) \right\}, \quad (10)$$

$$h_6(t_k) = \left\{ \Xi \le t_k \bigcap \Xi_k \le t_k \bigcap \left(\bigcap_{i=1}^{k-1} \Xi_i > t_i \right) \right\}. \quad (11)$$

From (8) and (9), we see that in terms of system suitability for use over the interval $(t_k, t_{k+1}]$, the event $h_3(t_k)$ corresponds to the incorrect decision, and the event $h_4(t_k)$ corresponds to the correct decision. When the event $h_3(t_k)$ occurs, the unsuitable system is mistakenly allowed to be used over the time interval $(t_k, t_{k+1}]$. From the system operability checking standpoint, the event $h_3(t_k)$ corresponds to the correct decision, and the event $h_4(t_k)$ corresponds to the incorrect decision.

Event $h_2(t_k)$ is the joint occurrence of two events: the system is suitable for use over the interval $(t_k, t_{k+1}]$ and the system is judged as unsuitable. Furthermore, event $h_2(t_k)$ is called a 'false failure', and events $h_3(t_k)$ and $h_5(t_k)$ are called 'undetected failure 1' and 'undetected failure 2', respectively. Events $h_1(t_k)$, $h_4(t_k)$ and $h_6(t_k)$ correspond to the correct decisions pertaining to system suitability and unsuitability.

Note that even when $Y(t) = 0$, erroneous decisions are possible when checking system suitability. Indeed, if $Y(t_i) = 0$ $(i = 1, 2, \ldots, k)$, expressions (6)–(11) are converted to the following form:

$$h_1(t_k) = \{\Xi > t_{k+1} \bigcap \Xi_0 > t_k\}, \quad (12)$$

$$h_2(t_k) = \{\Xi > t_{k+1} \bigcap \Xi_0 \le t_k\}, \quad (13)$$

$$h_3(t_k) = \{t_k < \Xi \le t_{k+1} \bigcap \Xi_0 > t_k\}, \quad (14)$$

$$h_4(t_k) = \{t_k < \Xi \le t_{k+1} \bigcap \Xi_0 \le t_k\}, \quad (15)$$

$$h_5(t_k) = \emptyset, \quad (16)$$

$$h_6(t_k) = \{\Xi \le t_k \bigcap \Xi_0 \le t_k\}, \quad (17)$$

where Ø denotes the impossible event.

The errors arising at $Y(t_k) = 0$ are methodological in nature and non-removable with the decision rule used herein.

2.3 Inspection policy

Suppose that the operation of a new system begins at time $t_0 = 0$ and sequential inspections are planned at times $t_1 < t_2 < \ldots < t_N < T$. When inspecting system suitability at time t_k $(k = 1, 2, \ldots, N)$, the following decisions are possible:

If the system is judged suitable, it is allowed to be used over the interval $(t_k, t_{k+1}]$.

If the system is judged unsuitable, it is repaired and allowed to be used over the interval $(t_0, t_1]$.

In the case of periodic inspections, interval T is divided equally into $N + 1$ subintervals and the system is inspected periodically at time $k\tau (k = 1, 2, \ldots N)$, where $(N + 1)\tau = T$. At time point T, the system is not inspected.

3 MAINTENANCE ASSUMPTIONS

It is assumed that the system state can only be inspected at discrete moments t_k $(k = 1, 2, \ldots, N)$. The inspections are not perfect but are nondestructive. The inspections are not instantaneous, but the time required for inspection is considerably less than any interval between inspections. Therefore, it is assumed that the system failure rate does not change during any inspection. If the system fails between inspections, the failure is unrevealed. Therefore, inspections provide the opportunity for failure detection only.

Two types of repairs are possible: false corrective repair owing to event $h_2(t_k)$ and true corrective repair owing to event $h_4(t_k)$ or $h_6(t_k)$. After any type of repair, the system becomes as good as new. The repairs are not identical in terms of down time and cost.

Both repairs and inspections increase the total down time, thereby decreasing system availability and increasing maintenance cost. The objective of the proposed maintenance policy is to find a set of sequential or periodic times for inspecting or repairing the system that maximizes system availability or minimizes total maintenance cost.

4 MAINTENANCE MODEL

4.1 Space of system states

For determining the maintenance measures, we use a well-known property of regenerative stochastic processes, Barlow & Proschan (1975), which is based on the fact that the fraction of time for which the system is in the state S_i $(i = 1, \ldots, n)$ is equal to the ratio of the average time spent in state S_i per regeneration cycle to the average cycle duration. The system operation process can be considered as a sequence of various changing states in a finite time interval. Therefore, in the range of maintenance planning $(0, T)$, the system behaviour can be described using a stochastic process $L(t)$ with a finite space of states:

$$\bigcup_{i=1}^{n} S_i = \vec{S}. \quad (18)$$

The process $L(t)$ changes only stepwise, with each jump due to the system transitioning to one of the possible states.

Let us define random process $L(t)$. At any given time t, the system can be in one of the following states: S_1, if at time point t, the system is used as intended and is in the operable state; S_2, if at time point t, the system is used as intended and is in an inoperable state (unrevealed failure); S_3, if at time point t, the system is not used for its intended purpose because of suitability inspection; S_4, if at time point t, the system is not used for its intended purpose because 'false corrective repair' is performed; S_5, if at time point t, the system is not used for its intended purpose because 'true corrective repair' is performed.

Let TS_i be the time in state S_i ($i = 1, 2, \ldots, 5$). Obviously, TS_i is a random variable with the expected mean time of $E[TS_i]$. The average duration of system regeneration cycle is determined by the following formula:

$$E[TS_0] = \sum_{i=1}^{5} E[TS_i] \quad (19)$$

4.2 Probabilities of correct and incorrect decisions

To determine the maintenance indicators, we use the joint probability density function (PDF) of random variables $\Xi, \Xi_1, \ldots, \Xi_k$, which we denote as $\omega_0(\xi, \xi_1, \ldots, \xi_k)$. From (5), it follows that Ξ_k is a function of random variables Ξ and $Y(t_k)$. The presence of $Y(t_k)$ in (5) leads to a random measurement error with respect to time to failure at time point t_i, which is defined as follows:

$$\Lambda_i = \Xi_i - \Xi, \; i = \overline{1, k}. \quad (20)$$

The additive relationship between random variables Ξ ($0 < \Xi < \infty$) and Λ_i ($-\infty < \Lambda_i < \infty$) leads to $-\infty < \Xi_i < \infty$. We denote the conditional PDF of random variables $\Lambda_1, \ldots, \Lambda_k$ as $f_0(\lambda_1, \ldots, \lambda_k | \xi)$ under the condition that $\Xi = \xi$. The following statement allows us to express PDF $\omega_0(\xi, \xi_1, \ldots, \xi_k)$ using PDFs $\omega(\xi)$ and $f_0(\lambda_1, \ldots, \lambda_k | \xi)$.

Theorem 1. If $Y(t_1), \ldots, Y(t_k)$ are independent random variables, then

$$\omega_0(\xi, \overline{\xi_1, \xi_k}) = \omega(\xi) f_0[(\xi_1 - \xi), (\xi_k - \xi) | \xi]. \quad (21)$$

Proof. Using the multiplication theorem of the PDFs, we can write

$$\omega_0(\xi, \overline{\xi_1, \xi_k}) = \omega(\xi) \omega_1(\overline{\xi_1, \xi_k} | \xi), \quad (22)$$

where $\omega_1(\xi_1, \ldots, \xi_k | \xi)$ is the conditional PDF of random variables $\Xi_1, \ldots, \Xi_k$ under the condition that $\Xi = \xi$. When $\Xi = \xi$, random variables $\Xi_1, \ldots, \Xi_k$ can be represented as $\Xi_1 = \xi + \Lambda_1, \ldots, \Xi_k = \xi + \Lambda_k$.

By virtue of the additive relationship between random variables Ξ and Λ_i ($i = 1, \ldots, k$), the following equality holds:

$$\omega_1(\overline{\xi_1, \xi_k} | \xi) = f_0[(\xi_1 - \xi), (\xi_k - \xi) | \xi]. \quad (23)$$

Substituting (23) in (22), we obtain (21).

In practice, the condition of independence of random variables $Y(t_1), \ldots, Y(t_k)$ is usually adopted because the correlation intervals of the measurement errors are considerably smaller than the intervals between inspections.

We introduce below the conditional probabilities of correct and incorrect decisions when checking system suitability; these are needed to determine the expected mean times $E[TS_i], i = 1, \ldots, 5$. Assume that a system failure occurs at time ξ, where $t_k < \xi \leq t_{k+1}$. Then, the conditional probability of the event 'false failure' when checking system suitability at time t_ν ($\nu = 1, \ldots, k-1$) is formulated as follows:

$$P_{NS|S}(\overline{t_1, t_{\nu-1}}; t_\nu | \xi) = P\left\{ \bigcap_{i=1}^{\nu-1} \Xi_i > t_i \cap \Xi_\nu \leq t_\nu | \Xi = \xi \right\}. \quad (24)$$

The conditional probability of the event 'operable system is correctly judged unsuitable' when checking system suitability at time t_k ($k = 1, \ldots, N$) is formulated as follows:

$$P_{NS|O}(\overline{t_1, t_{k-1}}; t_k | \xi) = P\left\{ \bigcap_{\nu=1}^{k-1} \Xi_\nu > t_\nu \cap \Xi_k \leq t_k | \Xi = \xi \right\}. \quad (25)$$

The conditional probability of the event 'undetected failure 1' when checking system suitability at time t_k is formulated as follows:

$$P_{S|O}(\overline{t_1, t_{k-1}}; t_k | \xi) = P\left\{ \bigcap_{\nu=1}^{k} \Xi_\nu > t_\nu | \Xi = \xi \right\}. \quad (26)$$

The conditional probability of the event 'suitable system is correctly judged suitable' when checking system suitability at time t_ν is formulated as follows:

$$P_{S|S}(\overline{t_1, t_{\nu-1}}; t_\nu | \xi) = P\left\{ \bigcap_{i=1}^{\nu} \Xi_i > t_i | \Xi = \xi \right\}. \quad (27)$$

The conditional probability of the event 'inoperable system is correctly judged unsuitable' when checking system suitability at time t_j ($j = k+1, \ldots, N$) is formulated as follows:

$$P_{NS|NO}(\overline{t_1, t_{j-1}}; t_j | \xi) = P\left\{ \bigcap_{i=1}^{j-1} \Xi_i > t_i \cap \Xi_j \leq t_j | \Xi = \xi \right\}. \quad (28)$$

The conditional probability of the event 'undetected failure 2' when checking system suitability at time t_N ($N = 1, 2, \ldots$) is formulated as follows:

$$P_{S|NO}(\overline{t_1, t_{N-1}}; t_N | \xi) = P\left\{ \bigcap_{i=1}^{N} \Xi_i > t_i | \Xi = \xi \right\}. \quad (29)$$

The probabilities (24)-(29) are determined by integrating PDF $f_0(\lambda_1,\ldots,\lambda_k|\xi)$ over the corresponding limits:

$$P_{NS|S}(\overline{t_1,t_{v-1}};t_v|\xi) = \int_{t_1-\xi}^{\infty}\ldots\int_{t_{v-1}-\xi}^{\infty}\int_{-\infty}^{t_v-\xi} f_0(\overline{u_1,u_v}|\xi)du_1du_v, \quad (30)$$

$$P_{NS|O}(\overline{t_1,t_{k-1}};t_k|\xi) = \int_{t_1-\xi}^{\infty}\ldots\int_{t_{k-1}-\xi}^{\infty}\int_{-\infty}^{t_k-\xi} f_0(\overline{u_1,u_k}|\xi)du_1du_k, \quad (31)$$

$$P_{S|O}(\overline{t_1,t_{k-1}};t_k|\xi) = \int_{t_1-\xi}^{\infty}\ldots\int_{t_k-\xi}^{\infty} f_0(\overline{u_1,u_k}|\xi)du_1du_k, \quad (32)$$

$$P_{S|S}(\overline{t_1,t_{v-1}};t_v|\xi) = \int_{t_1-\xi}^{\infty}\ldots\int_{t_v-\xi}^{\infty} f_0(\overline{u_1,u_v}|\xi)du_1du_v, \quad (33)$$

$$P_{NS|NO}(\overline{t_1,t_{j-1}};t_j|\xi) = \int_{t_1-\xi}^{\infty}\ldots\int_{t_{j-1}-\xi}^{\infty}\int_{-\infty}^{t_j-\xi} f_0(\overline{u_1,u_j}|\xi)du_1du_j, \quad (34)$$

$$P_{S|NO}(\overline{t_1,t_{N-1}};t_N|\xi) = \int_{t_1-\xi}^{\infty}\ldots\int_{t_N-\xi}^{\infty} f_0(\overline{u_1,u_N}|\xi)du_1du_N. \quad (35)$$

4.3 Expected mean times of the system staying in various states

The expected mean times $E[TS_i]$ ($i=1,\ldots,5$) are determined as follows:

$$E[TS_1] = \sum_{k=0}^{N}\int_{t_k}^{t_{k+1}}\left[\sum_{v=1}^{k-1} t_v P_{NS|S}(\overline{t_1,t_{v-1}};t_v|\vartheta) + t_k P_{NS|O}(\overline{t_1,t_{k-1}};t_k|\vartheta) + \vartheta P_{S|O}(\overline{t_1,t_{k-1}};t_k|\vartheta)\right]\omega_0(\vartheta)d\vartheta + \int_{T}^{\infty}\left[\sum_{k=1}^{N} t_k P_{NS|S}(\overline{t_1,t_{k-1}};t_k|\vartheta) + T P_{S|S}(\overline{t_1,t_{N-1}};t_N|\vartheta)\right]\omega_0(\vartheta)d\vartheta, \quad (36)$$

$$E\{TS_2\} = \sum_{k=0}^{N-1}\int_{t_k}^{t_{k+1}}\left[\sum_{j=k+1}^{N}(t_j-\vartheta)P_{NS|NO}(\overline{t_1,t_{j-1}};t_j|\vartheta) + (T-\vartheta)\times P_{S|NO}(\overline{t_1,t_{N-1}};t_N|\vartheta)\right]\omega_0(\vartheta)d\vartheta + \int_{t_N}^{T}(T-\vartheta)P_{S|O}(\overline{t_1,t_{N-1}};t_N|\vartheta)\times \omega_0(\vartheta)d\vartheta, \quad (37)$$

$$E\{TS_3\} = \tau_{si}\sum_{k=0}^{N-1}\int_{t_k}^{t_{k+1}}\left[\sum_{v=1}^{k-1}vP_{NS|S}(\overline{t_1,t_{v-1}};t_v|\vartheta) + kP_{NS|O}(\overline{t_1,t_{k-1}};t_k|\vartheta) + \sum_{j=k+1}^{N}jP_{NS|NO}(\overline{t_1,t_{j-1}};t_j|\vartheta) + NP_{S|NO}(\overline{t_1,t_{N-1}};t_N|\vartheta)\right]\omega_0(\vartheta)d\vartheta + \tau_{si}\int_{t_N}^{\infty}\left[\sum_{k=1}^{N-1}kP_{NS|S}(\overline{t_1,t_{k-1}};t_k|\vartheta) + NP_{S|S}(\overline{t_1,t_{N-2}};t_{N-1}|\vartheta)\right]\times \omega_0(\vartheta)d\vartheta, \quad (38)$$

$$E\{TS_4\} = \tau_{fr}\left[\sum_{k=1}^{N}\int_{t_k}^{t_{k+1}}\sum_{v=1}^{k-1}P_{NS|S}(\overline{t_1,t_{v-1}};t_v|\vartheta)\omega_0(\vartheta)d\vartheta + \int_{T}^{\infty}\sum_{k=1}^{N}P_{NS|S}(\overline{t_1,t_{k-1}};t_k|\vartheta)\omega_0(\vartheta)d\vartheta\right], \quad (39)$$

$$E\{TS_5\} = \tau_{tr}\left[\sum_{k=1}^{N}\int_{t_k}^{t_{k+1}}P_{S|S}(\overline{t_1,t_{k-2}};t_{k-1}|\vartheta)\omega_0(\vartheta)d\vartheta + \int_{0}^{t_1}\omega_0(\vartheta)d\vartheta\right], \quad (40)$$

where τ_{si} is the mean time for checking system suitability, τ_{fr} is the mean time for 'false corrective repair', and τ_{tr} is the mean time for 'true corrective repair'.

4.4 Maintenance policy measures

Knowing expected mean times $E[TS_i]$ ($i=1,\ldots,5$) allows us to determine the maintenance policy measures. The average availability is given by

$$A(\overline{t_1,t_k}) = E\{TS_1\}/E\{TS_0\}. \quad (41)$$

The average cost per unit time is expressed as

$$E[C(\overline{t_1,t_k})] = \frac{1}{E\{TS_0\}}\sum_{i=2}^{5}C_i E\{TS_i\}, \quad (42)$$

where C_i is the cost per unit time of maintaining the system in state S_i.

5 EXPERIMENT

5.1 Modelling of deterioration process

Assume that the deterioration process of a one-parameter system is described by the following monotonic stochastic function:

$$X(t) = a_0 + A_1 t, \quad (43)$$

where a_0 is the initial parameter value and A_1 is the random rate of parameter deterioration defined in the interval from 0 to ∞. It should be pointed out that a linear model of a stochastic deterioration process was used in many previous studies for describing real physical deterioration processes. For example, the linear regressive model studied by Ma & et al. (2013) describes a change in radar supply voltage with time, and Kallen & van Noortwijk (2005) used a linear model for representing a corrosion state function.

The following theorem allows us to find conditional PDF $f_0(\lambda_1,\ldots,\lambda_k|\xi)$ for the stochastic process given by (43).

Theorem 2. If $Y(t_1),\ldots,Y(t_k)$ are independent random variables and the system deterioration process is described by (43), then

$$f_0(\overline{\lambda_1,\lambda_k}|\xi) = \left(\frac{FF-a_0}{\xi}\right)^k \prod_{i=1}^{k} \varphi\left[\frac{(a_0-FF)\lambda_i}{\xi} + PF - FF\right]. \quad (44)$$

where $\varphi(y_i)$ is the PDF of the random variable $Y(t_i)$ at time point t_i.

Proof. Since the random variables $Y(t_1),\ldots,Y(t_k)$ are assumed to be independent, according to the well-known theorem of probability theory, the random variables $\Lambda_1 = \Psi[Y(t_1)],\ldots,\Lambda_k = \Psi[Y(t_k)]$ are independent as well. Hence, based on the multiplication theorem of PDFs, we have

$$f_0(\overline{\lambda_1,\lambda_k}|\xi,x(t)) = \prod_{i=1}^{k} f(\lambda_i|\xi,x(t)). \quad (45)$$

In general, for obtaining PDF $f_0(\lambda_1,\ldots,\lambda_k|\xi)$, it is necessary to integrate PDF $f_0[\lambda_1,\ldots,\lambda_k|\xi,x(t)]$ over all possible realisations with the time to failure ξ by considering the occurrence probability of these realisations. However, in the case of model (43), there is only one realisation for any ξ because a_0 is a non-random coefficient. Therefore, for the deterioration process (43), we have the following:

$$f_0(\overline{\lambda_1,\lambda_k}|\xi) = \prod_{i=1}^{k} f(\lambda_i|\xi). \quad (46)$$

Let us denote $Y_i = Y(t_i)$ ($i=1,\ldots,k$). Solving the stochastic equations

$$a_0 + A_1\Xi = FF, \quad (47)$$

$$a_0 + A_1\Xi_i + Y_i = PF \quad (48)$$

gives

$$\Xi = (FF-a_0)/A_1, \quad (49)$$

$$\Xi_i = (PF-Y_i-a_0)/A_1. \quad (50)$$

Substituting (49) and (50) in (20) results in

$$\Lambda_i = (PF-FF-Y_i)/A_1. \quad (51)$$

By combining (49) and (51), we find that

$$\Lambda_i = \Xi(PF-FF-Y_i)/(FF-a_0). \quad (52)$$

For any value $Y_i = y_i$ and $\Xi = \xi$, the random variable Λ_i with probability 1 has only one value; then, the conditional PDF of Λ_i with respect to Y_i and Ξ is the Dirac delta function:

$$f(\lambda_i|y_i,\xi) = \delta[\lambda_i - \xi(PF-FF-y_i)/(FF-a_0)]. \quad (53)$$

Using the multiplication theorem of PDFs, we find the joint PDF of the random variables Λ_i, Y_i and Ξ

$$f(\lambda_i,y_i,\xi) = f(y_i,\xi)f(\lambda_i|y_i,\xi) =$$
$$f(y_i,\xi)\delta[\lambda_i - \xi(PF-FF-y_i)/(FF-a_0)]. \quad (54)$$

Integrating the PDF (54) with variable y_i gives

$$f(\lambda_i,\xi) = \int_{-\infty}^{\infty} f(u_i,\xi)\delta\left[\lambda_i - \frac{\xi(PF-FF-u_i)}{(FF-a_0)}\right]du_i. \quad (55)$$

Since random variables Y_i and Ξ are independent,

$$f(y_i,\xi) = \varphi(y_i)\omega(\xi). \quad (56)$$

Considering (56), PDF (55) is transformed into

$$f(\lambda_i,\xi) = \omega(\xi)\int_{-\infty}^{\infty} \varphi(u_i)\delta\left[\lambda_i - \frac{\xi(PF-FF-u_i)}{(FF-a_0)}\right]du_i. \quad (57)$$

Using the shifting property of the Dirac delta function, PDF (57) is represented as

$$f(\lambda_i,\xi) = \omega(\xi)\left(\frac{FF-a_0}{\xi}\right)\varphi\left[\frac{(a_0-FF)\lambda_i}{\xi} + PF - FF\right]. \quad (58)$$

Finally, by applying the multiplication theorem of PDFs to (58), we get

$$f(\lambda_i|\xi) = \frac{f(\lambda_i,\xi)}{\omega(\xi)} =$$
$$\left(\frac{FF-a_0}{\xi}\right)\varphi\left[\frac{(a_0-FF)\lambda_i}{\xi} + PF - FF\right]. \quad (59)$$

Substituting (59) in (46), we obtain (44).

5.2 Numerical example

Assume that system state parameter $X(t)$ is a stochastic deterioration process described by (43). Let A_1 be a normal random variable. In this case, the PDF of the random variable Ξ is given by

$$\omega(t) = \frac{m_1\sigma_1^2 t^2 + \sigma_1^2 t(FF-a_0-m_1 t)}{\sqrt{2\pi}\sigma_1^3 t^3} \times$$
$$\exp\left\{-\frac{(FF-a_0-m_1 t)^2}{2\sigma_1^2 t^2}\right\}, \quad (60)$$

where $m_1 = E[A_1] = 0.002$; $\sigma_1 = \sqrt{Var[A_1]} = 0.00085$; $a_0 = 16$; $FF = 20$.

Let $Y(t)$ be a normal variable with $E[Y(t)] = 0$ and $\sigma_y(t) = \sqrt{Var[Y(t)]} = \sigma_y = 0.25$. Assume $T = 3000\,h$, $\tau_{si} = \tau_{fr} = 3\,h$, $\tau_{tr} = 10\,h$, $PF = 18.5$, and a periodic inspection policy with periodicity τ. Let us now find the optimum number of inspections that maximizes the system's average availability. By using a special program, it was determined that when $PF = 18.5$, the optimal solution was $N^* = 5$, $\tau = 500\,h$, and $A(N^*) = 0.990$. If $PF = FF$, then $N^* = 25$, $\tau = 115.4\,h$, and $A(N^*) = 0.928$. Thus, the use of a critical threshold level $PF < FF$ increases system availability and significantly reduces the number of inspections.

6 CONCLUSIONS

In this paper, we have described a mathematical model that allows us to find the optimal condition monitoring intervals for a discretely monitored one – unit system with imperfect inspections. A new formulation of the decision rule for inspecting system suitability has been proposed. Based on this decision rule, new equations have been derived for calculating the probabilities of correct and incorrect decisions, which are sensitive to both the functional failure level and the critical threshold level. Expressions have been derived for determining the mean times the system stays in various states during operation. Based on the developed mathematical model, equations have been proposed for such maintenance policy measures as average availability and average cost per unit time. A linear deterioration stochastic model was used for determining the joint PDF of the random measurement errors of time to failure. The numerical example considered shows the efficiency of the proposed maintenance policy with a critical threshold level, which differs from the functional failure level.

REFERENCES

[1] Abdel-Hameed, M. 1987. Inspection and maintenance policies of devices subject to deterioration. Advances in Applied Probability 19: 917–931.

[2] Abdel-Hameed, M. 1995. Correction to: "Inspection and maintenance policies of devices subject to deterioration". Advances in Applied Probability 27: 584.

[3] Barlow, R. & Proshan, F. 1975. Statistical theory of reliability and life testing: probability models. Holt, Rinehart and Winston. New York: 290.

[4] Barzilovich, E.Y., et al. 1980. Optimal control in condition-based maintenance of complex systems. In N.G. Bruevich (ed.), Basic theory and practice of reliability: 190–202. Moscow: Sov. Radio (in Russian).

[5] Berrade, M., Cavalcante, A. & Scarf, P. 2012. Maintenance scheduling of a protection system subject to imperfect inspection and replacement. European Journal of Operational Research 218:716–725.

[6] Dieulle, L., B'erenguer, C., Grall, A. & Roussignol, M. 2003. Sequential condition-based maintenance scheduling for a deteriorating system. European Journal of Operational Research 150 (2): 451–461.

[7] Ferreira, R., de Almeida, A. & Cavalcante, C. 2009. A multi-criteria decision model to determine inspection intervals of condition monitoring based on delay time analysis. Reliability Engineering and System Safety 94(5): 905–912.

[8] Golmakani, H. & Fattahipour, F. 2011. Age-based inspection scheme for condition-based maintenance. Journal of Quality in Maintenance Engineering 17(1): 93–110.

[9] Grall, A., Dieulle, L., B'erenguer, C. & Roussignol, M. 2002. Continuous-time predictive-maintenance scheduling for a deteriorating system. IEEE Transactions on Reliability 51 (2): 141–150.

[10] Dieulle, L., B'erenguer, C., Grall, A & Roussignol, M. 2003. Sequential condition-based maintenance scheduling for a deteriorating system. European Journal of Operational Research 150 (2): 451–461.

[11] Ito, K. & Nakagawa, T. 1997. An optimal inspection policy for a storage system with finite number of inspections. Journal of Reliability Engineering Association. Japan. 19: 390–396.

[12] Jia, X. & Christer, A.H. 2002. A prototype cost model of functional check decisions in reliability-centered maintenance. Journal of the Operational Research Society 53 (12): 1380–1384.

[13] Kallen, M. & Noortwijk, J. 2005. Optimal maintenance decisions under imperfect inspection. Reliability Engineering and System Safety 90(2–3): 177–185.

[14] Kong, M.B. & Park, K.S. 1997. Optimal replacement of an item subject to cumulative damage under periodic inspections. Microelectronics Reliability 37(3): 467–472.

[15] Ma, C., Shao, Y. & Ma, R. 2013. Analysis of equipment fault prediction based on metabolism combined model. Journal of Machinery Manufacturing and Automation 2(3): 58–62.

[16] Newby, M. & Dagg, R. 2002. Optimal inspection policies in the presence of covariates. In: Proc. of the European Safety and Reliability Conf. ESREL'02, Lyon, 19-21 March 2002: 131–138.

[17] Park, K.S. 1988. Optimal continuous-wear limit replacement under periodic inspections. IEEE Trans. on Reliability 37 (1): 97–102.

[18] Ulansky, V. 1987. Optimal maintenance policies for electronic systems on the basis of diagnosing. Collection of Proceedings: Issues of Technical Diagnostics: 137–143. Rostov na Donu: RISI Press (in Russian).

[19] Ulansky, V. 1992. Trustworthiness of multiple-monitoring of operability of non-repairable electronic systems. Collection of Proceedings: Saving Technologies and Avionics Maintenance of Civil Aviation Aircraft: 14–25. Kiev: KIIGA Press (in Russian).

[20] Van Noortwijk, J.M. et al. 1995. A Bayesian failure model based on isotropic deterioration. European Journal of Operational Research 82 (2): 270–282.

[21] Whitmore, G. 1995. Estimating degradation by a Wiener diffusion process subject to measurement error. Lifetime Data Analysis 1: 307–319.

A coordinated voltage control strategy for a Doubly-Fed Induction Generator (DFIG) wind farm system

Jingjing Zhao, Xiaoguang Hu & Xue Lv
Shanghai University of Electric Power College of Electrical Engineering, Shanghai, China

Xuhang Zhang
State Grid Shanghai Electric Power Company State Power Economic Research Institution, Shanghai, China

ABSTRACT: In order to enhance the stability of the voltage of grid-connected wind farm, a coordinated control strategy of a Doubly-Fed Induction Generator (DFIG) and wind farm with FACTS device-STATCOM is proposed. In this strategy, DFIG grid-side converter and the stator side reactive power generation ability are utilized. With this method, the transient voltage quality of the Point of Common Coupling (PCC) is improved, and the voltage recovery process is speeded up. The proposed coordinated control strategy is simulated by DIgSILENT/Power Factory, and the simulation results verified that the method is correct and feasible.

1 INTRODUCTION

With the increasing installed capacity of wind farms, the stability and the security operation of regional grids, which wind farms connect to, are giving a growing concern. Due to the stochastic variability of the wind energy output, the high level of wind energy penetration will lead to grid voltage fluctuations, and the voltage and reactive power problems caused by grid-connected wind farms has become one of the most prominent problems for network operations.

In the literature, there are a few studies related to the reactive power management of wind farms. Cao Jun et al. have analysed the Doubly-Fed Induction Generator (DFIG) wind farms and proposed a voltage control strategy following the principle of controlling the grid voltage from a high side [1]. Song Wang et al. proposed a wind farm voltage control strategy which only considers the wind farm in voltage regulation tasks by the DFIG, but did not consider the coordination with reactive power compensation devices [2]. Miad Mohaghegh Montazeri et al. analysed the difference of the impact among reactive power compensation equipment when participating in voltage regulation, but did not propose a corresponding reactive power control strategy [3].

Based on the structural characteristics of doubly-fed wind turbines, this paper presents an advanced coordinated voltage control with the DFIG wind farm unit and a Static Compensator device-STATCOM. The simulations system is established in DIgSILENT / Power Factory, and the simulation results verify the effectiveness of the proposed control strategy.

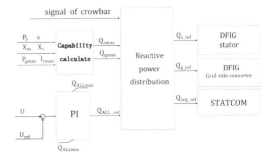

Figure 1. A schematic diagram of the coordinate control.

2 A COORDINATED CONTROL STRATEGY

A DFIG can transfuse reactive power to the grid through the stator and the grid side converter. But now, the grid side converter of the double-fed wind turbine usually works in unity power factor mode, thus, the advantage of the DFIG to regulate reactive power is underutilized.

To maximize the potential of the DFIG's reactive product ability, both the DFIG stator side and the grid side converter are used as sources of reactive power in this paper. The reactive power compensation equipment—STATCOM is also taking into account to coordinate with the DFIG in the task of regulating wind farm voltage. As the DFIG stator's side response is slower than the grid-side converter and STATCOM in reactive power regulation, thus, the DFIG grid side converter acts as the primary source of reactive power in the transient process.

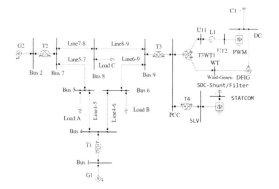

Figure 2. A schematic diagram of the studied grid-connected wind farm.

The block diagram of the proposed voltage control scheme is shown in Figure 1. The control system measures real-time wind farm voltage and monitors the network protection facilities switching states. The wind farm maintains its voltage by the control of reactive power production, according to the required grid voltage value from the system scheduler. When a system fault occurs and the crowbar protection is activated, the reactive power sources' priority order is a grid side converter, STATCOM. If the crowbar protection is not activated, the priority order of the wind farm's reactive power sources to support the grid is as follows: a grid-side converter, a DFIG stator, and STATCOM.

3 SIMULATION RESULTS

3.1 Studied system

As shown in Figure 2, the model of IEEE three machines nine nodes simulation systems with one wind farm including DFIG is established with the software DIgSILENT / Power Factory.

This wind farm consists of 18 double-fed wind turbine units, rated at 5 MW, which contain crowbar protection, replacing the original synchronous generator connected to a PCC bus. The rated capacity of each DFIG unit grid side converter is 2 MVA, with the grid side converter connected to the three-winding transformers through the reactor. The transformer uses a set of 30 kV/3.3 kV/0.69 kV windings.

The compensation device, STATCOM, is connected to the PCC bus via a step-up transformer. The wind speed is kept at a fixed value of about 13.8 m/s during the simulation.

The capacity of STATCOM is set at 25 MVA, which is 30% of the wind farm capacity, considering the common practice.

3.2 Simulation and analysis

The coordinate control model of the DFIG grid-side converter, the stator side, and STATCOM (referred

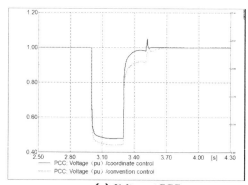

(a) Voltage at PCC

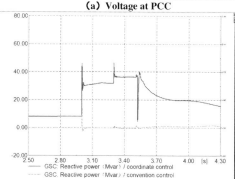

(b) Reactive power output of GSC

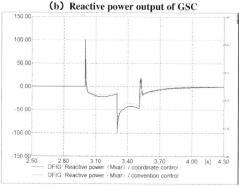

(c) Reactive power output of DFIG stator

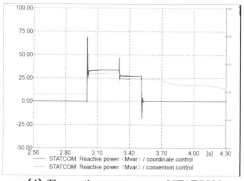

(d) The reactive power output of STATCOM

Figure 3. Simulated transient responses of the studied system in case 1.

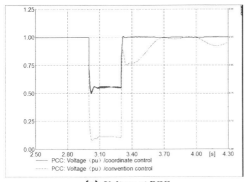

(a) Voltage at PCC

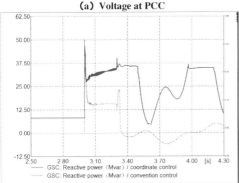

(b) Reactive power output of GSC

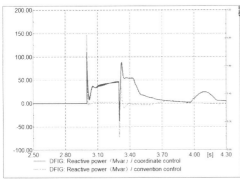

(c) Reactive power output of DFIG stator

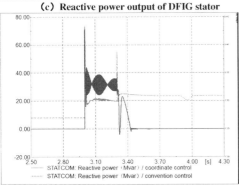

(d) The reactive power output of STATCOM

Figure 4. The simulated transient responses of the studied system in case 2.

to as the coordination control) are established in this simulation.

3.2.1 Case 1: A system fault applied with crowbar protection activated

A three-phase fault is applied at the PCC bus. The fault is initiated at $3.0\,s$ and cleared after $300\,ms$.

The simulation results displayed above show that the network is hardly affected in such a fault condition, which results in the activation of crowbar protection. Comparing the voltage variation at PCC, in conventional control, STATCOM is the sole reactive power source to provide voltage support during a fault, when the minimum voltage drops to $0.43\,pu$. When utilizing coordinated control, a DFIG grid-side converter, together with a STATCOM, injects lots of reactive power to the grid, and the lowest voltage reaches a value above $0.50\,pu$. Thus the coordinated control plays an important role in improving the transient voltage level and the voltage recovery speed.

3.2.2 Case 2: A system fault applied with crowbar protection not being activated

A three-phase fault is applied at the point $0.2\,km$ of the transmission line from Bus 6 to Bus 9. The fault is initiated at $3.0\,s$ and cleared after $300\,ms$.

In this simulation, the crowbar protection is not active with this fault. The minimum voltage at the PCC drops to $0.2\,pu$ in a conventional control system, when the STATCOM is a controllable reactive power source in the wind farm.

The lowest magnitude of voltage is raised to $0.5\,pu$ by a coordinated control with the reactive power support from a DFIG (including a grid-side converter and a stator side) and, thus, wind farms can stay connected to the network and keep operating. It can be concluded that, the coordinate control can effectively improve the transient voltage level and accelerate the recovery process after grid faults.

The simulation experiments prove that the coordinated control strategy takes full advantage of the reactive power capability of a DFIG grid-side converter and stator side. Due to the additional reactive power support compared with conventional controls, the voltage sag during the fault has been significantly improved.

4 CONCLUSION

This paper analyses the voltage and reactive power control of wind farms. This coordinated control strategy rationally allocates reactive power between the DFIG unit and a STATCOM on the basis of the state of crowbar protection. Simulation results show that the proposed coordinated control strategy can make full use of the reactive generation capability of the DIFG wind turbines and the compensation equipment, STATCOM, thus providing more dynamic reactive power to the grid for voltage support during the system's fault.

ACKNOWLEDGMENTS

This work is supported by National Natural Science Foundation of China (51207087), Innovation Program of Shanghai Municipal Education Commission (12YZ143) and Shanghai Green Energy Grid Connected Technology Engineering Research Center (13DZ2251900).

REFERENCES

[1] Cao Jun, Rong Lin Zhang, Guoqing Lin, Hongfu Wang, Jiaju Qiu. VSCF doubly fed wind farm voltage control strategy [J]. Automation of Electric Power Systems, 2009. 87–91.
[2] Wang Song, Li Gengyin, Zhou Ming, The Reactive Power Adjusting Mechanism & Control Strategy of Doubly Fed Induction Generator [J] Proceedings of the CSEE, 2014. 2724–2720.
[3] Miad Mohaghegh Montazeri, David Xu. Coordination of DFIG and STATCOM in a Power System Industrial Electronics (ISIE), IEEE International Symposium 28–31 May 2012, Hangzhou: 993–998.
[4] B. Pokharel; Gao Wenzhong, "Mitigation of disturbances in DFIG-based wind farm connected to weak distribution system using STATCOM," North American Power Symposium (NAPS) 26–28 Sept. 2010, 1–7.
[5] Liu Yu, Huang A.Q., Song Wenchao, et al. Small-signal model-based control strategy for balancing individual do capacitor voltages in cascade multilevel inverter-based STATCOM. IEEE Trans. on Industrial Electronics, 2009, 56(6): 2259–2269.
[6] Mohaghegh Montazeri, M.; Xu, D.; Bo Yuwen, "Improved Low Voltage Ride Thorough capability of wind farm using STATCOM," Electric Machines & Drives Conference (IEMDC), IEEE International 5–18 May 2011, 813–818.
[7] Qin Tao, Gang Lvyue, Daping Xu. Reactive Power Control of Wind Farm Adopting Doubly-Fed Induction Generators [J]. Power System Technology, 2009, 33(2): 105–110.

SOM-based intrusion detection for SCADA systems

Huang Wei, Hao Chen, Yajuan Guo, Guo Jing & Jianghai Tao
Jiangsu Electric Power Research Institute Co. Ltd., Nanjing, China

ABSTRACT: Due to standardization and connectivity to the Internet, Supervisory Control and Data Acquisition (SCADA) systems now face the threat of cyber attacks. Thus, there is an urgent need to study viable and suitable Intrusion Detection System (IDS) techniques to enhance the security of SCADA systems. This paper proposes an intrusion detection approach for SCADA systems, which applies Self-Organizing Maps (SOM) to capture system behaviors from collected status information and takes a neighborhood area density of a winning neuron as an anomaly quantification metric to detect abnormal system states. Experimental results show that the algorithm has a high sensitivity as well as a high specificity.

Keywords: intrusion detection, SCADA, security, self-organizing map

1 INTRODUCTION

The role of Supervisory Control and Data Acquisition (SCADA) systems has increased in the past decades in many fields especially in critical infrastructure sectors. SCADA systems monitor and control physical processes such as electrical power grids, oil and natural gas pipelines, chemical processing plants, water distribution and wastewater collection systems, nuclear power plants, traffic lights, etc. First generation SCADA networks operate in isolated environments, with no connectivity to any system outside the network. Nowadays, the extensive use of Information and Communication Technologies (Internet, wireless networks, cell phones) in critical infrastructures has made SCADA networks more and more interconnected with the outside world, and therefore SCADA systems are exposed to electronic attacks nowadays more than ever.

For the purpose of implementing an efficient defense of SCADA systems, it is necessary to research on novel security approaches, implement them and measure their suitability in terms of efficiency and effectiveness. In this paper, we focus on an intrusion detection approach for SCADA systems using the self-organizing map (SOM). Our work belongs to the model-based intrusion detection approach [1, 2]. Its basic idea is to construct models that characterize the expected/acceptable behavior of the entities. A behavior is typically an anticipated sensor signal (pressure, temperature, flow, level etc.). This approach detects attacks that cause the system to behave outside of the models, and is applicable for attacks on the protocol, on the OS platforms, and on the networking infrastructure. The whole detection approach is characterized by two phases: in the training phase, the behavior of the system is observed in the absence of attacks, and machine learning techniques [3, 4] used to create a profile of such normal behavior. In the detection phase, this profile is compared against the current behavior of the system, and any deviations are flagged as potential attacks. Unfortunately, systems often exhibit legitimate but previously unseen behavior, which leads behavior detection techniques to produce a high degree of false alarms. Thus, this paper proposes a SOM-based intrusion detection approach with a higher accuracy.

Section 2 presents an overview of SCADA systems and attacks. Section 3 proposes a SOM-based intrusion detection approach and experiment results are analyzed in Section 4. Section 5 provides conclusion and future works.

2 SCADA SYSTEMS AND ATTACKS

In this section we give a brief overview of the characteristics and attacks of a typical SCADA architecture. As shown in Figure 1, SCADA can be a mix of networked devices (switches, firewalls, servers, databases, etc.), embedded controllers, actuators, sensors, physical networks and bus technologies.

The HMI is used by the operator to perform any number of tasks including requests for historical data, real-time system status information, and to command control changes to the system. The SCADA Server, also known as a Master Terminal Unit (MTU), is used to interface with the SCADA network elements. Remote Terminal Units (RTUs) and Intelligent Electronic Devices (IEDs) are generally small dedicated devices designed for rough field or industrial environment. An RTU typically has a network interface

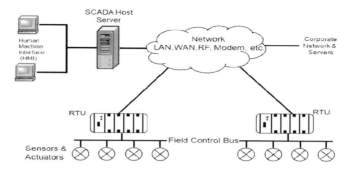

Figure 1. A typical SCADA architecture.

and a control bus interface. The network interface can be anything from a dial-up modem, an RF link, or an IP network connection. The field control bus is the primary interface between the RTU and field data elements including controlled elements (actuators, switches, motors, valves, etc.) and sensors (temperature, pressure, level, etc.).

In general SCADA systems are based on Master-Slave architectures. A single master MTU or multiple masters are allowed and the RTUs, PLCs or other controller devices are the slaves. Slaves may perform local control of a set of actuators (pumps, valves, motors, etc.) based on the state of one or more sensors (temperature, pressure, flow, velocity, etc.). The transmission of data and control commands between an MTU and an RTU, designated as SCADA communications, are carried over a variety of media, including Ethernet, corporate frame relay, fiber channel, microwave signals, direct satellite broadcast and many licensed or unlicensed radio systems. Common open communication protocols include IEC 60870 standards, Distributed Network Protocol Version 3 (DNP3), and Modbus, in addition to several other private protocols.

The main attacks on SCADA systems include command injections, response injections, electronic man-in-the-middle (MITM) attacks and Denial-of-Service (DoS) attacks. Command injection attacks inject false control commands into a control system. For example, remote terminals and intelligent electronic devices are generally programmed to automatically monitor and control the physical process directly at a remote site. However, response injection attacks inject false responses into a control system. Since control systems rely on feedback control loops which monitor physical process data before making control decisions protecting the integrity of the sensor measurements from the physical process is critical. Many SCADA network standards do not include mechanisms to support response integrity. The next attacks identified are MITM attacks. The attacker intercepts and modifies messages to cause system erroneous operation. MITM and command injection attacks can cause damage and serious safety issues when used to attack SCADA systems. DoS attacks provide multiple master (operator) commands to a RTU to cause it to fail. This type of attack could cause an input buffer overflow or cause resource demands that exceed the RTUs processing capabilities, which results in the breakdown of the communication link between RTUs and MTU or HMI.

3 ONLINE INTRUSION DETECTION APPROACH

Self-organizing map (SOM) has been widely used to detect anomalies in several applications [5]. We also chose to use the SOM learning technique in this work to achieve efficient system behavior learning by taking the following features it owns into consideration. First, the SOM can handle multi-variant system behavior learning well without missing any representative behaviors because that the SOM can map a high dimensional input space into a low dimensional map space while preserving the topological properties of the original input space. In our work, we continuously collect a vector of a SCADA system runtime measurements $D(t) = [r_1, r_2, \ldots, r_n]$, where r_i denotes a system-level metric, and use the measurement vectors as inputs to train SOMs. Second, system metric measurements are often fluctuating due to dynamic workloads and measurement noises. However, using the neighborhood area size or density to describe system states can make the trained SOM adapt to these fluctuations. Third, the neighborhood of a neuron share similar characteristics which can tolerate a certain degree of variance between different patterns through smooth transition. This property is useful for describing multimodality of normal patterns or abnormal patterns.

A SOM is composed of a set of neurons arranged in a lattice. In our work, a two-dimensional lattice with $K \times K$ neurons is used, in which each neuron is associated with a coordinate (u, v), where $0 \leq u, v \leq K - 1$, and a weight vector W_{uv} in the map. Weight vectors should be the same length as the measurement vectors (i.e., $D(t)$), which are dynamically updated based on the values of the measurement vectors in the training samples. We use the SOM to model the system behaviors in two different phases: learning and detection.

For the learning phase, a competitive learning process is used to adjust the weight vectors of different neurons, which works by comparing the Euclidean distance of the input measurement vector to each neuron's weight vector in the map and selecting the neuron with the smallest Euclidian distance as the winning neuron. The winning neuron's weight values along with its neighbor neurons are then updated. The intuition behind this approach is to make the currently trained neuron and the neurons in its neighborhood converge to the input space.

Definition 1 (r-neighbor) For an arbitrary neuron with the coordinate (u,v) in the map, named $N_{u,v}$, its neighbor neurons in a radius of r is a set $N = \{N_{u-a,v-b}, N_{u-a,v+b}, N_{u+a,v-b}, N_{u+a,v+b}\}$, where $a+b = r$ and $0 \leq u-a, u+a, v-b, v+b \leq k-1$.

The neighbor radius r is a critical parameter in the learning phase. The general formula for updating the weight vector of a given neuron i at time t is given in Equation (1).

$$W(t+1) = W(t) + h(j,t) \times l(t) \times (D(t) - W(t)) \quad (1)$$

where $W(t)$ and $D(t)$ denote the weight vector and the input vector at time instance t respectively, neuron j is a neighbor of the given neuron i, $h(j,t)$ denotes a neighborhood function, a Gaussian function in the basic SOM is used as the neighborhood function, $l(t)$ denotes a learning coefficient that can be applied to modify how much each weight vector is changed as learning proceeds.

After learning, each neuron in the SOM represents a certain system state. For a long running system, its system state is in a normal state for most of the time since its startup. Thus, most of system measurement vector is mapped to neurons representing a normal state. These winning neurons are frequently trained, which results in the weight vector values of the winning neurons and their neighboring neurons have been frequently modified with the same input measurement vectors. As a result, the weight vectors of the neurons will be similar to the weight vectors of their neighboring neurons. The similarity of the weight vectors of two neighboring neurons will be helpful to determine system state. We propose a neighborhood area density-based anomaly detection and quantification algorithm for the detection phase, as shown in Figure 2. The idea behind this algorithm is that the neighborhood area density of a neuron representing the abnormal state is lower than the density of the neuron representing the normal state. We call the neighborhood area density as an anomaly quantification metric (AQM), and its definition is as follows.

Definition 2 (Anomaly quantification metric, AQM). Given a winning neuron i, its neighborhood area density is defined as $ND_r(i) = 1/(\sum_{j \in N} W(j) * M(i,j)/|N|)$, where N is the neighbor set of neuron i, $W(j)$ that depends on the step length to neuron i is the weight value of neuron j and $M(i,j) = \sum_{p=1}^{k} |W_p^i - W_p^j|$, where W_p^i and W_p^j denotes the pth weight vector element of neuron i and neuron j, respectively.

For an arbitrary input sample, the algorithm maps it to a neuron using the same distance measurement method as the learning phase. And then the neighborhood area density of the winning neuron is calculated. If the neighborhood area density exceeds a neighborhood area density threshold for the map, that means the input sample has mapped to a neuron which is close to many other neurons. We consider this sample to be a normal sample and do not raise an anomaly alarm. However, if the sample maps to a neuron with a density value lower than or equal to our threshold value, this sample represents something we rarely see during learning. We consider this type of sample to be anomalous. Determining a neighborhood area density threshold to differentiate normal and abnormal neurons is integral to the accuracy of our anomaly detection algorithm. So, we sort all calculated neighborhood area density in a descending order and set the threshold value to be the value at a selected percentile T_p. The threshold value should be a higher percentile because that system stays in a normal state for most of the time. We will further examine the effect of the threshold on accuracy through experiments.

4 EXPERIMENTS AND RESULTS

In the following, the dataset, training procedure and evaluation of our approach are described. To provide quantitative results for the proposed approach, we use a real data set comes from the daily measures of sensors in an urban waste water treatment plant (referred to as DUWWTP), and it consists of 38 data points (attributes) [6]. This data set consists of approximately 527 instances, while 14 instances are labeled as abnormal.

The goal of our intrusion detection is to separate SCADA system states into two classes: the normal state and the abnormal state. Performance of our intrusion detection approach is evaluated in terms of *sensitivity* and *specificity* widely used metrics to measure a binary classification test. *Sensitivity* is the proportion of correct abnormal samples classifications to the number of actual abnormal samples, while *Specificity* is the proportion of correct normal samples classifications to the number of actual normal samples. A good mechanism should provide a high value (close to 1.0) for both metrics. A sensitivity of 1.0 means that the mechanism recognizes all the samples in the abnormal state, while a specificity of 1.0 means that the mechanism identifies all the samples in the normal state. High sensitivity is expected to avoid the cases, where humans have to reprocess all the data to find out those missed anomalies. In essence, it determines whether the proposed automated mechanism is useful in practice. Low specificity should be avoided as it leads to nontrivial human effort to remove false alarms via manual processing.

We analyze the impacts of several key parameters on detection accuracy, respectively. Specifically, we analyze *sensitivity* and *specificity* by adjusting the neighborhood radius r from 1 to 4, the number of

Algorithm 1. Anomaly detection based on SOM
Input: new observation samples: $T=\{t_1,t_2,....t_{ws}\}$; neuron neighbor radius: r;
    the base line value of AQM at a selected percentile : T_p ; trained SOM with K*K neurons:
Output: a specific status: st
1.   Initialize neighor area density, $Td = 0$; st = false; count = 0; a K*K AQM matrix to zero
2   Normalize all metric values in new observation samples to the range [0,100];
3.   For (q=0; q < ws; q++)   // detect every new sample
4.   For (i = 0; i < n; i++)
5.   For (j = 0; j < n; j++)
6.   compute the Euclidean distance d_{qij} between sample t_q and Neuron $N_{i,j}$;
7.   compute the AQM value of neuron $N_{i,j}$ according to the definition 2
8   and fill it to the AQM matrix;
9   END_For
10.   END_For
11.   Select the neuron with the smallest Euclidean distance to the sample tq as the
12.   mapped neuron, named N_o, and record its coordinate (u,v) and get the $AQM_{u,v}$;
13.   Get the set N of neighbors of the mapped neuron N_o according to the definition 1;
14.   Sort these neurons' AQM values in the descending order;
15.   Get the AQM value in the position of $\lfloor K\times K\times T_p \rfloor$ from the sorted AQM values
16.   as the threshold, T_{aqm}
17.   If $AQM_{u,v} < T_{aqm}$
18. st = true;
19.   Else
20.   break;
21.   END_If
22. END_for

Figure 2.   The process of the anomaly detection algorithm.

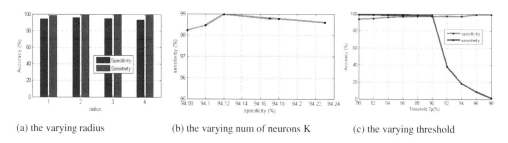

(a) the varying radius   (b) the varying num of neurons K   (c) the varying threshold

Figure 3.   Detection accuracy with the varying parameters.

neurons K from 20 to 40 with step by 5 and the neighborhood area density threshold T_p from 80% to 89%. The default parameter settings are as followings: $r = 2$, $k = 25$ and $T_p = 86\%$. We only change one parameter value and keep the rest two parameters fixed each time. The results are shown in Figure 2.

We begin with the results of varying radius. With the increasing radius, we were able to achieve a higher sensitivity as well as a decreasing and relatively acceptable specificity. Similarly, it happens to the varying numbers of neurons. These two parameters have no obvious impact on the detection accuracy. However, the threshold parameter has a great impact on the sensitivity. Specifically, when the threshold is greater than 90%, the sensitivity rapidly decreases. The main cause is that in our dataset used in experiments, the normal samples are the majority of all samples. Moreover, with the increasing threshold, the specificity decreases monotonously. Thus, we should select a higher threshold close to the point that makes the sensitivity decreasing rapidly. Generally speaking, our detection approach has a high sensitivity as well as a high specificity.

5   CONCLUSIONS

In this paper, we have presented an intrusion detection approach for SCADA systems. Our approach leverages the SOM learning technique to capture dynamic system behavior from the collected data and can recognize the system's abnormal state with a high sensitivity as well as a high specificity, which is helpful to reduce the cost of management.

ACKNOWLEDGMENT

This work is funded by a grant from State Grid Corporation of China under the project "Research on New generation Smart Substation Network Communication and In-Depth-Protection Technology".

REFERENCES

[1] S. Cheung, B. Dutertre, M. Fong, U. Lindqvist, K. Skinner, and A. Valdes, "Using model-based intrusion detection for SCADA networks", In Proc. the SCADA Security Scientific Symposium, 2007, pp. 127–134.

[2] Goldenberg N, Wool A. Accurate modeling of Modbus/TCP for intrusion detection in SCADA systems [J]. International Journal of Critical Infrastructure Protection, 2013, 6(2), 63–75.

[3] Linda, T. Vollmer, and M. Manic, "Neural network based intrusion detection system for critical infrastructures," In International Joint Conference on Neural Networks, 2009, pp. 1827–1834, June 2009.

[4] Jianmin Jiang, Lasith Yasakethu: Anomaly Detection via One Class SVM for Protection of SCADA Systems. CyberC 2013, 82–88.

[5] V. Chandola, A. Banerjee, V. Kumar, "Anomaly Detection: A Survey," ACM Computing Surveys, 2009, 41(3), Article 15.

[6] A. Frank and A. Asuncion, "UCI machine learning repository [http://archive.ics.uci.edu/ml]. University of California, Irvine, School of Information and Computer Sciences," 2014.

Nonlinear and adaptive backstepping speed tracking control of a permanent magnet synchronous motor despite all parameter uncertainties and load torque disturbance variation

H.I. Eskikurt & M. Karabacak
Department of Electrical and Electronics Engineering, Sakarya University Technology Faculty, Sakarya, Turkey

ABSTRACT: In this paper, a new nonlinear and adaptive backstepping speed tracking control design was proposed for a Permanent Magnet Synchronous Motor (PMSM). All the parameters in both the PMSM and load dynamics were considered unknown. It was assumed that all state variables are measurable and available for feedback in the controller design. The final control and parameter estimation laws were derived from the design of virtual control inputs and a Lyapunov function candidate. Simulation results clearly showed that the controller guarantees the tracking of a time varying sinusoidal and a ramp reference speed owing to the fact that the speed and current tracking errors converge to zero asymptotically. All parameter and load torque disturbance estimation errors converged to zero asymptotically except that the stator inductance estimation error converged to a constant for a sinusoidal reference trajectory. The proposed controller ensured strong robustness against all parameter uncertainties/perturbations and load torque disturbance variation.

Keywords: Permanent magnet synchronous motor; Speed tracking; Nonlinear control; Adaptive control; Lyapunov methods; Robustness; Uncertain nonlinear system; Parameter uncertainty; Backstepping

1 INTRODUCTION

The PMSM is drawing increased attention for electric drive applications because of its higher efficiency, its power factor, power density, reliability, and its larger torque to an inertia ratio and having a longer life over other kinds of motors, such as DC motors and induction motors (Leonhard, 1985). However, the main disadvantage of the PMSM is the need for a more complex controller for electric drive applications due to its highly nonlinear characteristic. Conventional fixed gain PI, PID controllers are widely used for reasons of simplicity, applicability, and robustness in most industrial drive applications (Ang & Chong, 2005). However, such kind of controllers may fail to meet the high performance requirements of industrial servos and speed tracking drive applications since it is high vulnerable to parameter perturbations and unknown external disturbances if the controlled system is uncertain and highly nonlinear (Parasiliti et al. 1996, Tursini et al. 2002, Jian-Xin Xu et al. 2005, Gadoue et al. 2009). The control strategies, based on recent modern control theories, are put forward to meet the high performance application requirements of industrial drive applications. Fuzzy logic based controllers are robust to parameter variations and external disturbances, because their design is independent of the controlled system (Kovacic & Bogdan 1994, Elmas & Ustun 2008, Yu et al. 2010). However, expert knowledge and heavy computational effort is necessary for real time application of these controllers.

Backstepping control is a newly developed technique for the control of uncertain nonlinear systems, particularly those systems that do not satisfy matching conditions (Kokotovic 1992, Krstic et al. 1995). The most appealing point of it is to use the virtual control variable to make the original high order system simple, and, thus, the final control outputs can be derived systematically through suitable Lyapunov functions. An adaptive robust nonlinear controller, straightforwardly derived, using this control method, is proposed in (Zhou & Wang 2002, Zhou & Wang, 2005). The controller is robust against stator resistance, viscous friction uncertainties, and unknown load torque disturbance. However, this approach uses feedback linearization, which its utilization may cause to cancel out some useful nonlinearities (Rahman et al. 2003). Another adaptive nonlinear backstepping design method is proposed for speed control of a PMSM without utilization of feedback linearization. An exact model and adaptive controllers, despite viscous friction uncertainty and unknown load torque disturbance, are compared to each other in a PMSM drive system for step reference speeds (Ouassaid et al. 2004). It is seen that an adaptive controller exhibits higher performance results than an exact model controller. An improved backstepping control technique is proposed by inserting the integral of velocity error into a Lyapunov function. It is proved that the tracking error converges to zero for a step reference speed under variations of the parameter and load torque disturbance (Ouassaid et al. 2005).

Among adaptive nonlinear control methods in literature, backstepping design on control of uncertain highly nonlinear systems has excellent performance in terms of adaptation ability to parameter uncertainties, transient and steady state behaviours, disturbance rejection capacity, and suitability for real time implementation (Kokotovic 1992, Krstic et al. 1995). All nonlinearities have to be taken into account for the stable performance control of highly nonlinear systems, consisting of parameter uncertainties, unknown external disturbances, and a variation of system parameters, with temperature and saturation effects. Therefore, nonlinear and adaptive backstepping control that does not employ linearization theorems and it becomes an important design tool for the control of highly nonlinear systems. Linearization theorems, based on exact cancellation of nonlinearities, limit the operating range of a controlled system (Rahman et al. 2003). Moreover, exact cancellation of nonlinearities is not completed and results in a cancellation error if parameter uncertainties and unknown external disturbances exist in the controlling system. Consequently, this case causes an error in controlled variables (Zhou & Wang 2002). In this study, a new nonlinear and adaptive backstepping speed tracking control design for PMSMs is proposed under all parameter uncertainties/perturbations and unknown bounded load torque disturbance variation without the utilization of linearization theorems. Stability results, adaptation, and control laws are derived using the appropriate Lyapunov function. The asymptotic stability of the resulting closed loop system is ensured in the sense of the Lyapunov stability theorem. The controller guarantees convergence of the speed and current tracking errors to zero asymptotically and it does not have the drawbacks of linearization methods. The work is organized as follows: in the next section, the mathematical model of a PMSM is introduced and it is pointed out that the PMSM is highly nonlinear. In the third section, the overall control design is given in connection with Lyapunov's stability theory. Section 4 comprises the simulations results and a discussion about the results. The achievements obtained with the proposed controller are interpreted in the last section.

2 MATHEMATICAL MODEL OF A PMSM

The mathematical model of a typical surface mounted PMSM can be described in the d-q frame as follows (Song et al. 2006):

$$\frac{d\omega}{dt} = \frac{3P\lambda_m}{2J}i_q - \frac{B}{J}\omega - \frac{T_L}{J} \quad (1)$$

$$\frac{di_q}{dt} = -\frac{R}{L}i_q - P\omega i_d - \frac{P\lambda_m}{L}\omega + \frac{1}{L}V_q \quad (2)$$

$$\frac{di_d}{dt} = -\frac{R}{L}i_d + P\omega i_q + \frac{1}{L}V_d \quad (3)$$

where i_d and i_q are the d-q axis currents, V_d and V_q are the d-q axis voltages, R and L are the stator resistance and the inductance per phase respectively, P is the number of pole pairs, ω is the rotor speed, λ_m is the permanent magnet flux, J is the rotor moment of inertia, B is the viscous friction factor, T_L is also the applied load torque disturbance. From the equations above, it is understood that a PMSM is a highly nonlinear system owing to the cross coupling between electrical currents and speed state equations. It is should be noted that all parameters vary with operating conditions, primarily applied load torque disturbance, temperature, and saturation. Thus, if high performance speed control of a PMSM is necessary, all nonlinearities, parameter uncertainties and unknown external disturbances, have to be taken into any account of the controller design (Rahman et al. 2003).

3 THE PROPOSED CONTROL DESIGN STRATEGY

3.1 *Control objective*

The main control objective is to design an asymptotically stable speed tracking controller for PMSM so as to make the rotor speed track the reference trajectory correctly under all parameter uncertainties/perturbations and unknown bounded load torque disturbance variation. This objective can be reached to only make the design of the speed tracking controller become independent of all parameters of the PMSM and load torque disturbance. Thus, all parameters and external disturbances have to be estimated adaptively. The parameters B, J, R, L and bounded by load torque disturbance, while T_L is assumed to be unknown in the proposed control design.

$$\tau = \frac{T_L}{J}; \ \tilde{\tau} = \hat{\tau} - \tau; \ \frac{d\tilde{\tau}}{dt} = \frac{d\hat{\tau}}{dt}, \ b = \frac{B}{J}; \ \tilde{b} = \hat{b} - b; \ \frac{d\tilde{b}}{dt} = \frac{d\hat{b}}{dt},$$
$$\tilde{R} = \hat{R} - R; \ \frac{d\tilde{R}}{dt} = \frac{d\hat{R}}{dt}, \ \tilde{L} = \hat{L} - L; \ \frac{d\tilde{L}}{dt} = \frac{d\hat{L}}{dt}, \ \tilde{J} = \hat{J} - J; \ \frac{d\tilde{J}}{dt} = \frac{d\hat{J}}{dt} \quad (4)$$

The definitions with cap depict the estimations of the parameters and with tilde depict estimations of errors individually.

3.2 *Nonlinear and adaptive backstepping controller design*

The aim of adaptive backstepping is to identify a virtual control state and force it to become a stabilizing function. This procedure generates a corresponding error variable (Kokotovic 1992). As a result, the error variable can be stabilized by a proper selection of control inputs by the Lyapunov stability theory (Rahman et al. 2003). The overall control design was arranged in the following order.

The speed tracking error can be given as:

$$e = \omega - \omega_d \quad (5)$$

ω is the actual rotor speed and ω_d is the desired reference speed trajectory. For stabilizing speed component, the speed tracking error dynamics derived from (1) and (4) can be obtained as follows:

$$\dot{e} = \dot{\omega} - \dot{\omega}_d = \frac{3P\lambda_m}{2J} i_q - \frac{B}{J}\omega - \frac{T_L}{J} - \dot{\omega}_d \tag{6}$$

The current tracking errors have to be defined to develop their dynamics:

$$e_q = i_q - i_{q\_d} \tag{7}$$

$$e_d = i_d - i_{d\_d} \tag{8}$$

i_{d_d} and i_{q_d} describe the desired trajectory of the three phase currents on d-q axis respectively. These virtual control inputs have to be designed to guarantee that rotor speed ω tracks desired reference speed trajectory ω_d. Using backstepping procedure, the speed tracking error dynamics can be restated from that (7) is substituted into (6):

$$\dot{e} = \frac{3P\lambda_m}{2J} e_q + \frac{3P\lambda_m}{2J} i_{q\_d} - b\omega - \tau - \dot{\omega}_d \tag{9}$$

The speed tracking error dynamics can be stabilized if the virtual control inputs i_{q_d} and i_{d_d}, as the stabilizing functions, are designed as follows:

$$i_{q\_d} = \frac{2\hat{J}}{3P\lambda_m}\left(\hat{b}\omega + \hat{\tau} + \dot{\omega}_d - k_1 e\right) \tag{10}$$

$$i_{d\_d} = 0 \tag{11}$$

Then, the resultant speed tracking error dynamics is obtained by that (10) being substituted into (9):

$$\dot{e} = \frac{3P\lambda_m}{2J} e_q + \tilde{b}\omega + \tilde{\tau} - k_1 e + \frac{\tilde{J}}{J}\left(\hat{b}\omega + \hat{\tau} + \dot{\omega}_d - k_1 e\right) \tag{12}$$

$$\dot{e}_q = -\frac{R}{L} i_q - P\omega i_d - \frac{P\lambda_m}{L}\omega + \frac{1}{L}V_q - \frac{2\dot{\hat{J}}}{3P\lambda_m}\left(\hat{b}\omega + \hat{\tau} + \dot{\omega}_d - k_1 e\right) - \frac{2\hat{J}}{3P\lambda_m}\left(\dot{\hat{b}}\omega + \hat{b}\dot{\omega} + \dot{\hat{\tau}} + \ddot{\omega}_d - k_1(\dot{\omega} - \dot{\omega}_d)\right) \tag{13}$$

Using (4), the final q axis current tracking error dynamics can be given as the following:

$$\dot{e}_q = -\frac{\hat{R}}{L} i_q + \frac{\tilde{R}}{L} i_q - \frac{P\omega\hat{L}i_d}{L} + \frac{P\omega\tilde{L}i_d}{L} - \frac{P\lambda_m}{L}\omega + \frac{1}{L}V_q$$
$$-\frac{2\dot{\hat{J}}}{3P\lambda_m}\left(\hat{b}\omega + \hat{\tau} + \dot{\omega}_d - k_1 e\right) - \frac{2\hat{J}}{3P\lambda_m}\left(\dot{\hat{b}}\omega + \dot{\hat{\tau}} + \ddot{\omega}_d + k_1\dot{\omega}_d\right)$$
$$-\left(\hat{b} - k_1\right)i_q - \frac{\tilde{J}}{J}\left(\hat{b} - k_1\right)i_q + \frac{2\hat{J}}{3P\lambda_m}\left(\hat{b} - k_1\right)\left(\hat{b}\omega + \hat{\tau}\right)$$
$$-\frac{2\hat{J}}{3P\lambda_m}\left(\hat{b} - k_1\right)\left(\tilde{b}\omega + \tilde{\tau}\right) \tag{14}$$

The final d axis current tracking error dynamics, using (3), (4), (8), and (11), can be developed as:

$$\dot{e}_d = \frac{\tilde{R}}{L} i_d - \frac{P\omega\tilde{L}}{L} i_q + \frac{P\omega\hat{L}}{L} i_q - \frac{\hat{R}}{L} i_d + \frac{1}{L}V_d \tag{15}$$

To realize the speed tracking objective, the current tracking errors dynamics are stabilized by the control inputs, designed in the following:

$$V_q = \hat{R} i_q + P\omega\hat{L} i_d + P\lambda_m\omega - k_2 \hat{L} e_q$$
$$+ \hat{L}\left[\begin{array}{l}\frac{2\dot{\hat{J}}}{3P\lambda_m}\left(\hat{b}\omega + \hat{\tau} + \dot{\omega}_d - k_1 e\right)\\ + \frac{2\hat{J}}{3P\lambda_m}\left(\dot{\hat{b}}\omega + \dot{\hat{\tau}} + \ddot{\omega}_d + k_1\dot{\omega}_d\right)\\ + \left(\hat{b} - k_1\right)i_q - \frac{2\hat{J}}{3P\lambda_m}\left(\hat{b} - k_1\right)\left(\hat{b}\omega - \hat{\tau}\right)\end{array}\right] \tag{16}$$

$$V_d = \hat{R} i_d - P\omega\hat{L} i_q - k_3 \hat{L} e_d \tag{17}$$

After (16) and (17) are substituted into (13) and (15) respectively, the resultant current errors dynamics develop into following form:

$$\dot{e}_q = -k_2 e_q - \frac{k_2 \tilde{L} e_q}{L} + \frac{\tilde{R}}{L} i_q + \frac{P\omega\tilde{L}i_d}{L} - \frac{\tilde{J}}{J}\left(\hat{b} - k_1\right)i_q$$
$$- \frac{2\hat{J}}{3P\lambda_m}\left(\hat{b} - k_1\right)\left(\tilde{b}\omega + \tilde{\tau}\right)$$
$$+ \frac{\tilde{L}}{L}\left[\begin{array}{l}\frac{2\dot{\hat{J}}}{3P\lambda_m}\left(\hat{b}\omega + \hat{\tau} + \dot{\omega}_d - k_1 e\right)\\ + \frac{2\hat{J}}{3P\lambda_m}\left(\dot{\hat{b}}\omega + \dot{\hat{\tau}} + \ddot{\omega}_d + k_1\dot{\omega}_d\right) + \left(\hat{b} - k_1\right)i_q\\ - \frac{2\hat{J}}{3P\lambda_m}\left(\hat{b} - k_1\right)\left(\hat{b}\omega - \hat{\tau}\right)\end{array}\right] \tag{18}$$

$$\dot{e}_d = -k_3 e_d - \frac{k_3 \tilde{L} e_d}{L} + \frac{\tilde{R}}{L} i_d - \frac{P\omega\tilde{L}}{L} i_q \tag{19}$$

Then, the positive definite Lyapunov function candidate can be defined for the overall control system in order to determine parameter adaptation laws and stability:

$$V = \frac{1}{2}e^2 + \frac{1}{2}e_d^2 + \frac{1}{2}e_q^2 + \frac{1}{2J\theta_1}\tilde{J}^2 + \frac{1}{2L\theta_2}\tilde{R}^2$$
$$+ \frac{1}{2L\theta_3}\tilde{L}^2 + \frac{1}{2\theta_4}\tilde{b}^2 + \frac{1}{2\theta_5}\tilde{\tau}^2 \tag{20}$$

$\theta_1, \theta_2, \theta_3, \theta_4$ and θ_5 are positive adaptation gains. The time derivative of the Lyapunov function candidate is expressed as using (4):

$$\dot{V} = e\dot{e} + e_d \dot{e}_d + e_q \dot{e}_q + \frac{1}{J\theta_1}\tilde{J}\dot{\tilde{J}}$$
$$+ \frac{1}{L\theta_2}\tilde{R}\dot{\tilde{R}} + \frac{1}{L\theta_3}\tilde{L}\dot{\tilde{L}} + \frac{1}{\theta_4}\tilde{b}\dot{\tilde{b}} + \frac{1}{\theta_5}\tilde{\tau}\dot{\tilde{\tau}} \tag{21}$$

After some mathematical manipulation, the time derivative of the Lyapunov function candidate is clearly expressed again by (12), (18) and (19) being substituted into (21):

$$\dot{V} = \frac{3P\lambda_m}{2J}e_q e - k_1 e^2 - k_3 e_d^2 - k_2 e_q^2$$

$$+ \frac{\tilde{J}}{J}\left[\frac{1}{\theta_1}\dot{\tilde{J}} - (\hat{b} - k_1)i_q e_q + \left(\hat{b}\omega + \dot{\hat{\tau}} + \dot{\omega}_d - k_1 e\right)e\right]$$

$$+ \frac{\tilde{R}}{L}\left[\frac{1}{\theta_2}\dot{\tilde{R}} + (i_d e_d + i_q e_q)\right]$$

$$+ \frac{\tilde{L}}{L}\begin{bmatrix} \frac{1}{\theta_3}\dot{\tilde{L}} - e_d(k_3 e_d + P\omega i_q) \\ + e_q \begin{pmatrix} P\omega i_d - k_2 e_q + \frac{2\hat{J}}{3P\lambda_m}\left(\hat{b}\omega + \dot{\hat{\tau}} + \dot{\omega}_d - k_1 e\right) \\ + \frac{2\hat{J}}{3P\lambda_m}\left(\dot{\hat{b}}\omega + \dot{\hat{\tau}} + \ddot{\omega}_d + k_1\dot{\omega}_d\right) + (\hat{b} - k_1)i_q \\ - \frac{2\hat{J}}{3P\lambda_m}(\hat{b} - k_1)(\hat{b}\omega - \hat{\tau}) \end{pmatrix} \end{bmatrix} \quad (22)$$

$$+ \tilde{b}\left[\frac{1}{\theta_4}\dot{\hat{b}} + \omega e - \frac{2\hat{J}}{3P\lambda_m}(\hat{b} - k_1)\omega e_q\right]$$

$$+ \tilde{\tau}\left[\frac{1}{\theta_5}\dot{\hat{\tau}} + e - \frac{2\hat{J}}{3P\lambda_m}(\hat{b} - k_1)e_q\right]$$

Then, the adaptation laws for the estimated parameters can be determined by the following expressions:

$$\dot{\hat{J}} = \theta_1\left((\hat{b} - k_1)i_q e_q - \left(\hat{b}\omega + \dot{\hat{\tau}} + \dot{\omega}_d - k_1 e\right)e\right) \quad (23)$$

$$\dot{\hat{R}} = -\theta_2(i_d e_d + i_q e_q) \quad (24)$$

$$\dot{\hat{L}} = -\theta_3 e_q \begin{pmatrix} (\hat{b} - k_1)i_q - P\omega i_d e_d - P\omega i_q - k_2 e_q - k_3 e_d^2 \\ + \frac{2\hat{J}}{3P\lambda m}\begin{pmatrix} \hat{b}\omega + \dot{\hat{\tau}} + \dot{\omega}_d - k_1 e \\ + \left(k_1\dot{\omega}_d + \dot{\hat{b}}\omega + \dot{\hat{\tau}} + \ddot{\omega}_d\right) \\ - (\hat{b} - k_1)(\hat{b}\omega - \hat{\tau}) \end{pmatrix} \end{pmatrix} \quad (25)$$

$$\dot{\hat{b}} = \theta_4\left(\frac{2\hat{J}}{3P\lambda m}(\hat{b} - k_1)\omega e_q - \omega e\right) \quad (26)$$

$$\dot{\hat{\tau}} = \theta_5\left(\frac{2\hat{J}}{3P\lambda m}(\hat{b} - k_1)e_q - e\right) \quad (27)$$

Therefore, the time derivative of the Lyapunov function candidate is obtained through (23), (24), (25), (26), and (27) is substituted into (18):

$$\dot{V} = \frac{3P\lambda_m}{2J}e_q e - k_1 e^2 - k_3 e_d^2 - k_2 e_q^2 \leq 0 \quad (28)$$

k_1, k_2 and k_3 are the positive finite feedback gains. There is no restriction on choices of the feedback gains other than that which has to force the time derivative of the Lyapunov function candidate to be a negative semi definite. Then, the equality in (29) is ensured. Otherwise, the asymptotic stability of the overall control system is not guaranteed.

$$\frac{3P\lambda_m}{2J}e_q e \leq k_1 e^2 + k_3 e_d^2 + k_2 e_q^2 \quad (29)$$

Table 1. PMSM specifications.

Parameter	Value
Rated power	1.1 kW
Rated speed	3000 rpm
DC voltage	220 V
Phase inductance	0.0085 H
Phase resistance	2.875 Ω
Rotor moment of inertia	0.0008 kgm^2
Friction factor	0.001 Nms
Magnetic flux	0.175 Wb
Number of pole pairs	4

Using the LaSalle Yoshizawa's theorem, it is proved that the overall control system is asymptotically stable (Kokotovic 1992). In other words, this means that the speed and current tracking errors, e, e_d, and e_q, converge to zero asymptotically as time goes to infinity. It is should be noted the magnitudes of the feedback and the adaptive gains also directly specify the convergence rates of the speed and current tracking errors as well.

4 SIMULATION RESULTS AND DISCUSSION

The ratings and nominal parameters of the PMSM used in the simulations are given in Table 1. The block diagram of the overall control system is given in Figure 1 and the detailed block diagram that depicts nonlinear and adaptive backstepping control schemes is shown in Figure 2.

The simulation of the proposed controller based on Matlab/Simuling is carried out on three different cases in order to examine the behaviours of the controller. The feedback and adaptation gains chosen are $k_1 = 2500$, $k_2 = 10000$, $k_3 = 50000$ and $\theta_1 = 0.0001, \theta_2 = 25000, \theta_3 = 0.0001, \theta_4 = 1200, \theta_5 = 100000$ respectively.

The reference speed profile is chosen as a time varying ramp profile. Disturbance rejection capacity and speed tracking performance at zero and near zero speed are examined in detail under load torque disturbance variation for such kind of speed reference. Figure 3 plots the reference and actual velocities under load torque disturbance variation to investigate speed tracking performance of the controller. The ramp reference profile and actual motor speed are given together in Figure 3a. The load torque suddenly applied to PMSM is 0.5 Nm is between 0–3 sec, 2 Nm is between 3–6 sec, 3 Nm between 6–9 sec, and 1.5 Nm between 9–12 sec and no load between 12–15 sec. The corresponding dynamic speed tracking responses of the controller are zoomed in Figure 3b–e. It is observed that the actual motor speed takes 20 msec in Figure 3b–c, 5 msec in Figure 3d and 20 msec in Figure 3e to again track the reference profile very closely.

The variation of the load torque disturbance applied to a PMSM and the corresponding tracking errors are

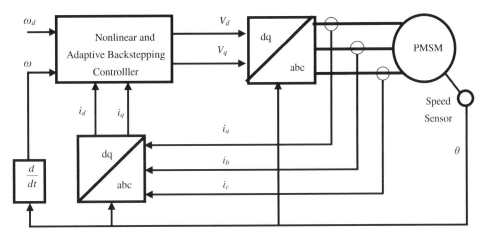

Figure 1. The overall diagram of the proposed control scheme.

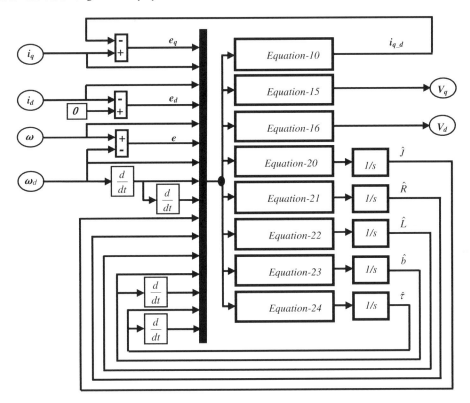

Figure 2. A detailed diagram of the nonlinear and adaptive backstepping controller.

shown in Figure 4 to examine the disturbance rejection capacity of the controller. The applied load torque disturbance variation is shown in Figure 4a, the corresponding speed tracking error e and current tracking errors e_q and e_d are plotted in Figure 4b–d individually. As can be seen from Figure 4, the tracking errors stably remained at zero despite the presence of load torque disturbance. As shown in Figure 3 and Figure 4, the controller guarantees strong robustness in regard to speed tracking performance and disturbance rejection capacity at zero and near zero speed under load torque disturbance variation.

Parameter and load torque disturbance estimations and estimation errors are showed in Figure 5 for the purpose of showing that the overall control system is asymptotically stable. Figure 5a–k plots the load torque estimation in Figure 5a and the estimation error in Figure 5b, stator resistance estimation in Figure 5c, and estimation error in Figure 5d, stator inductance estimation in Figure 5e, and estimation

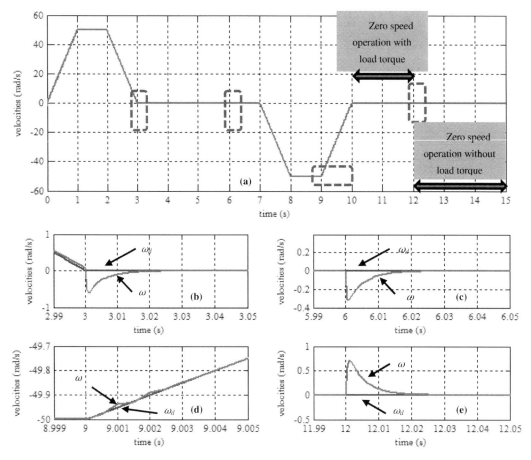

Figure 3. A speed tracking response for ramp reference profile under load torque disturbance variation: (a) reference and actual velocities, (b) zoom around 3 sec, (c) zoom around 6 sec, (d) zoom around 9 sec, (e) zoom around 12 sec.

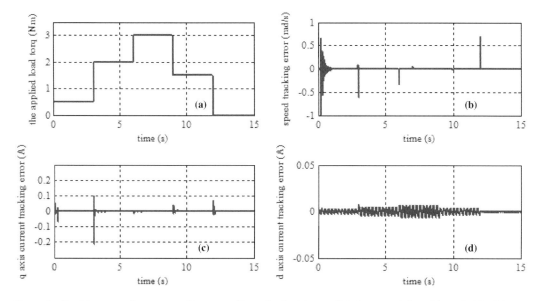

Figure 4. Tracking errors for a ramp reference profile under load torque disturbance variation: (a) applied load torque disturbance variations, (b) speed tracking error, (c) d axis current tracking error, (d) q axis current tracking error.

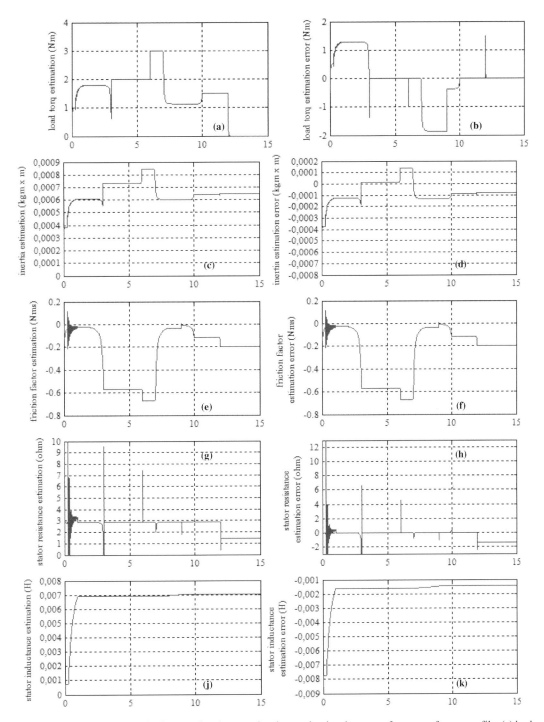

Figure 5. The parameter and load torque disturbance estimations and estimation errors for ramp reference profile: (a) load torque disturbance estimation, (b) load torque disturbance estimation error, (c) rotor moment of inertia estimation, (d) rotor moment of inertia estimation error, (e) friction factor estimation, (f) friction factor estimation error, (g) stator resistance estimation, (h) stator resistance estimation error, (j) stator inductance estimation, (k) stator inductance estimation error.

error in Figure 5f, friction factor estimation in Figure 5g, and estimation error in Figure 5h, moment of inertia estimation in Figure 5j and estimation error in Figure 5k. It is seen from Figure 5 that the parameter and load torque disturbance estimation errors converge to a constant but not to its own actual values.

5 CONCLUSIONS

The main drawback of most nonlinear control method is that some parameter and/or external disturbances in controlled system dynamics are assumed to be known in the design of them. Consequently, suchlike control systems are sensitive to the parameter uncertainties/perturbations and/or unknown external disturbances used in the design of them inherently. Another important drawback is also utilization of the linearization methods, based on an exact cancellation of all inherent system nonlinearities in nonlinear control design since it limits the operational range of a controlled system. Furthermore, exact cancellation of nonlinearities is not achieved if parameter uncertainties and unknown external disturbances exist in a controlled system. Nonlinear and adaptive backstepping speed tracking control of uncertain PMSMs, in the sense of Lyapunov stability theory, is addressed in this work as a complete solution to these drawbacks.

In this work, the proposed speed tracking control design method ensures that the speed and current tracking errors converge to zero asymptotically under all parameter uncertainties/perturbations and load torque disturbance variation for sinusoidal and ramp reference speed. As can be observed from the simulation results, it is obvious that the proposed speed tracking controller guarantees strong robustness with regard to all parameter uncertainties/perturbations and unknown bounded load torque disturbances in both the PMSM and load dynamics respectively. Simulations based on Matlab/Simulink prove the validity of the proposed nonlinear and adaptive backstepping speed tracking control design for uncertain PMSMs.

REFERENCES

Ang, K.H., Li, G. & Chong, Y. "PID control system analysis, design, and technology", IEEE Transactions on control Systems Technology, 2005, 13 (July): 559–76.

Elmas, C. & Ustun, O. "A hybrid controller for the speed control of a permanent magnet synchronous motor drive", Control Engineering Practice 2008, 16 (March): 260–70.

Gadoue, S.M., Giaouris, D. & Finch, J.W. "Artificial intelligence-based speed control of DTC induction motor drives–A comparative study", Electric Power Systems Research, 2009, 79 (January): 210–9.

Hasirci, U. & Balikci, A. "Nonlinear and adaptive state feedback control of variable speed PMSM drives", 7th Asian Control Conference (ASCC 2009), 2009, p. 1605–10.

Jianhui, H., Yongxiang, X. & Z. Jibin, "Design and Implementation of Adaptive Backstepping Speed Control for Permanent Magnet Synchronous Motor", The Sixth World Congress on Intelligent Control and Automation (WCICA 2006), vol. 1, 2006, p. 2011–5.

Jian-Xin, X., Panda, S.K., Ya-Jun, P., Tong Heng, L. & Lam, B.H. "A modular control scheme for PMSM speed control with pulsating torque minimization", IEEE Transactions on Industrial Electronics, 2005, 51 (June): 526–36.

Kim, S.K. "Speed and current regulation for uncertain PMSM using adaptive state feedback and backstepping control", IEEE International Symposium on Industrial Electronics (ISIE 2009), 2009, p. 1275–80.

Kokotovic, P.V. "The joy of feedback: nonlinear and adaptive", IEEE Control Syst., 1992, 12:7–17.

Kovacic, Z. & Bogdan, S. "Model reference adaptive fuzzy control of high-order systems", Engineering Applications of Artificial Intelligence, 1994, 7 (October): 501–11.

Krstic, M., Kanellakopoulos, I. & Kokotovic, P.V. "Nonlinear and adaptive control design", Wiley, New York, 1995.

Leonhard, W. "Control of Electrical Drives", Springer-Verlag, Berlin, 1985.

Liu, D., Yan, W. & He, Y. "Speed Tracking Control of PMSM with Adaptive Backstepping", The Sixth World Congress on Intelligent Control and Automation (WCICA 2006), 2006, vol. 1, p. 1986–9.

Ouassaid, M., Cherkaoui, M. & Zidani, Y. "A nonlinear speed control for a PM synchronous motor using an adaptive backstepping control approach", International Conference on Industrial Technology (IEEE ICIT '04), vol. 3, 2004, p. 1287–92.

Ouassaid, M., Cherkaoui, M. & Maaroufi, M. "Improved nonlinear velocity tracking control for synchronous motor drive using backstepping design strategy", IEEE Russia Power Tech, 2005, p. 1–6.

Parasiliti, F., Tursini, M. & Zhang, D.Q. "Adaptive fuzzy logic control for high performance PM synchronous drives", in: 8th Mediterranean Electrotechnical Conference (MELECON '96), vol. 1; 1996, pp. 323–7.

Rahman, M.A., Vilathgamuwa, D.M., Uddin, M.N. & King-Jet, T. "Nonlinear control of interior permanent-magnet synchronous motor", IEEE Transactions on Industry Applications, 2003, 39 (Mar/Apr): 408–16.

Song, Z., Hou, Z., Jiang, C. & Wei, X. "Sensorless control of surface permanent magnet synchronous motor using a new method", Energy Conversion and Management, 2006, 47 (September): 2451–60.

Tursini, M., Parasiliti, F. & Zhang, D.Q. "Real-time gain tuning of PI controllers for high-performance PMSM drives", IEEE Transactions on Industry Applications 2002, 38 (Jul/Aug): 1018–26.

Yu, J., Chen, B., Yu, H. & Gao, J. "Adaptive fuzzy tracking control for the chaotic permanent magnet synchronous motor drive system via backstepping", Nonlinear Analysis: Real World Applications, In Press, Corrected Proof, Available online 16 July 2010.

Zhou, J. & Wang, Y. "Adaptive backstepping speed controller design for a permanent magnet synchronous motor", IEE Proceedings Electric Power Applications, vol. 149, 2002, p. 165–72.

Zhou, J. & Wang, Y. "Real-time nonlinear adaptive backstepping speed control for a PM synchronous motor", Control Engineering Practice 2005, 13 (October): 1259–69.

The convolution theorem associated with fractional wavelet transform

Y.Y. Lu, B.Z. Li & Y.H. Chen
School of Mathematics and Statistics, Beijing Institute of Technology, Beijing, China

ABSTRACT: Novel Fractional Wavelet Transform (NFRWT) is a comparatively new and powerful mathematical tool for signal processing. Many results from the Wavelet Transform (WT) domain have currently been extended to NFRWT. However, there are no results from the convolution theorem of the NFRWT. In this paper, we first study the convolution theorem for continuous wavelet transform, and then we derive the convolution theorem of fractional wavelet transform.

Keywords: Continuous wavelet transform; fractional wavelet transform; convolution

1 INTRODUCTION

Wavelet transform can provide a way to classical linear time-frequency representations, so it is useful in the analysis of non-stationary signals [1]. On the other hand, wavelet transformation is also useful in removing the non-stationary properties of the involved signals. As a result, the conventional estimation algorithms for stationary signal processing can be employed in each scale of the wavelet domain [2].

As we know, wavelet transform is very important in signal processing systems. Shi, et.al introduces a new transform- a fractional wavelet transform in [7], It is more useful in some cases than wavelet transform, but until now there is no results published about the convolution theorem for it. It is therefore interesting and worthwhile to investigate the associated results about fractional wavelet transform.

2 BACKGROUND

In this section, we will review some known topics about continuous wavelet transform and the fractional wavelet transform [3, 4].

2.1 Convolution theorems

Continuous Wavelet Transform (CWT) stands for a scale-space representation and is an effective way to analyse non-stationary signals and images. In [5], A.R. Lindsey has given a function $g(x)$ as the convolution of another function $f(x)$ and a filter $h(x)$, and according to the classical convolution formula, we can receive that:

$$g(t) = f(t) * h(t) = \int_{-\infty}^{+\infty} f(\tau)h(t-\tau)d\tau \quad (1)$$

where $*$ in the subscript denotes the classical convolution operator, let $\hat{g}(a,b)$ and $\hat{f}(a,b)$ denote the CWT with the same wavelet of $g(x)$ and $f(x)$, respectively. Then, it is shown that

$$\hat{g}(a,b) = \int_{-\infty}^{+\infty} dt \hat{f}(a,t)h(b-t) \quad (2)$$

From this formula we can see that the CWT of the convolution between a function and a filter is the convolution in time or space variables, at every fixed scale, of the CWT of the function and the filter.

If f and h are complex-valued functions defined on $\mathbb{R}$, their convolution $f \otimes h$ is defined as a new function:

$$g(x) = (f \otimes h)(x) = \int_{-\infty}^{+\infty} dt \hat{f}(t)h(x-t) \quad (3)$$

From this formula we know that the integral exists. If $f \in L^2(\mathbb{R})$ and $h \in L^1(\mathbb{R})$, then such an integral exists almost everywhere and $f \otimes h \in L^2(\mathbb{R})$ with $\|f \otimes h\|_2 = \|f\|_2 \cdot \|g\|_1$.

If f and h are complex-valued functions defined on $\mathbb{R}^2$, It has two formalizations to convolve them in just one variable as follows:

$$g_1(x,y) = (f \otimes_1 h)(x,y) = \int_{-\infty}^{+\infty} dt f(t,y)h(x-t,y),$$

$$g_2(x,y) = (f \otimes_2 h)(x,y) = \int_{-\infty}^{+\infty} dt f(x,t)h(x,y-t), \quad (4)$$

From these formulas we also know that both integrals exist. If $f \in L^2(\mathbb{R}^2)$ and $h \in L^1(\mathbb{R}^2)$, then such integrals exist almost everywhere and $g_1, g_2 \in L^2(\mathbb{R}^2)$ [6].

Now we define a fixed function $\psi \in L^2(\mathbb{R})$, called a wavelet, and it is just as follows:

$$\psi_{ab}(x) = \frac{1}{|a|^{1/2}} \psi\left(\frac{x-b}{a}\right) \quad (5)$$

with $a, b \in \mathbb{R}$ and $a \neq 0$. This function is scaled so that it is independent of a. The CWT of a function $f \in L^2(\mathbb{R})$ with wavelet ψ is now defined as

$$\hat{f}^{\psi}(a,b) = \int_{-\infty}^{+\infty} dt f(x) \psi_{ab}^*(x)$$

$$= \frac{1}{|a|^{1/2}} \int_{-\infty}^{+\infty} dt f(x) \psi^*\left(\frac{x-b}{a}\right) \quad (6)$$

We say that a wavelet function $\psi(x)$ is admissible wavelet if [6]

$$0 < C_\psi = \int_{-\infty}^{+\infty} dw \frac{|\tilde{\psi}(w)|^2}{|w|} < \infty \quad (7)$$

While from [6], Antonio F. Perez-Rendon and Rafael Robles introduced the convolution theorem for the continuous wavelet transform:

Let $\psi_f \in L^2(\mathbb{R})$, $\psi_h \in L^1(\mathbb{R}) \cap L^2(\mathbb{R})$ two admissible wavelets, and let $\hat{f}^{\psi_f}$ and $\hat{h}^{\psi_h}$ denote the CWT of two functions $f \in L^2(\mathbb{R})$ and $h \in L^1(\mathbb{R}) \cap L^2(\mathbb{R})$ with wavelets ψ_f and ψ_h, respectively. If $g = (f \otimes h)$ and $\psi_g = (\psi_f \otimes \psi_h)$, then [6]

$$\hat{g}^{\psi_g}(a,b) = \frac{1}{|a|^{1/2}} \left(\hat{f}^{\psi_f} \otimes_2 \hat{h}^{\psi_h}\right)(a,b) \quad (8)$$

2.2 The novel fractional wavelet transform (NFRWT)

As we know, wavelet transform and the fractional Fourier transform are powerful tools for many applications in field of signal processing, but they have their disadvantages: wavelet transform is limited in the time-frequency plane, although the Fourier transform overcomes such limitations, and in the fractional domain it can provide signal representations, but it cannot obtain the local structures of the signal. While fractional wavelet transform not only rectifies the limitation of wavelet transform and the fractional Fourier transform, but it also inherits the advantages of multiresolution analysis of wavelet transform and the capability of signal representations in the fractional domain which is similar to the fractional Fourier transform. So the study of fractional wavelet transform is important in any signal processing system.

The subject of fractional wavelet transform was first raised in 1997 by Mendlovic and Zalevsky, and, from then on, there were three types in the definition of fractional wavelet transform [8]. Now let me introduce one which is the most important and widely used.

A novel fractional wavelet transform with an order α of a square integrable signals $x(t)$ as [7]

$$W_x^\alpha(a,b) = x(t) \Theta_\alpha \left(a^{-\frac{1}{2}} \psi^*(-t/a)\right)$$

$$= e^{-(j/2)b^2 \cot\theta} \cdot \left\langle x(t) e^{(j/2)(.)^2 \cot\theta} \cdot \psi_{a,b}(.)\right\rangle \quad (9)$$

$$= \int_{-\infty}^{+\infty} x(t) \psi_{\alpha,a,b}^*(t) dt$$

while the kernel $\varphi_{\alpha,a,b}(t)$ satisfies

$$\psi_{\alpha,a,b}(t) = e^{-(j/2)(t^2-b^2)\cot\theta} \psi_{a,b}(t) \quad (10)$$

$$\psi_{a,b}(t) = a^{-\frac{1}{2}} \psi\left(\frac{t-b}{a}\right) \quad (11)$$

Furthermore, the format of (9) can be rewritten as

$$W_x^\alpha(a,b) = e^{-(j/2)b^2 \cot\theta} \int_{-\infty}^{+\infty} \left(x(t) e^{(j/2)t^2 \cot\theta}\right) \psi_{a,b}^*(t) dt \quad (12)$$

From this formula we can obtain that the computation of the NFRWT corresponds to the following steps:

1) for any continuous function $x(t)$, multiply a chirp signal, i.e., $x(t) \rightarrow \hat{x}(t) = x(t) e^{(j/2)t^2 \cot\theta}$
2) a traditional wavelet transform,

$$\hat{x}(t) \rightarrow W_{\hat{x}}(a,b)$$

3) then multiply the chirp signal again,

$$W_{\hat{x}}(a,b) \rightarrow W_x^\alpha(a,b) = W_{\hat{x}}(a,b) e^{-(j/2)b^2 \cot\theta}$$

Since the input signal (e.g. a digital signal) is processed by a digital computing machine, it is prudent to define the discrete version of each step.

3 THE CONVOLUTION THEOREM OF NFRWT

Based on these conditions above, we can get the following conclusion:

Theorem 1. Suppose that $\psi_f \in L^2(\mathbb{R})$ and $\psi_h \in L^1(\mathbb{R}) \cap L^2(\mathbb{R})$ are two admissible wavelets, and let $\hat{f}^{\psi_f}$ and $\hat{h}^{\psi_h}$ denote the NFRWT of two functions $f \in L^2(\mathbb{R})$ and $h \in L^1(\mathbb{R}) \cap L^2(\mathbb{R})$ with wavelets ψ_f and ψ_h, respectively. If $g = (f \otimes h)$ and $\psi_g = (\psi_f \otimes \psi_h)$, then

$$\hat{g}_\alpha^{\psi_g}(a,b) = \frac{1}{|a|^{1/2}} \left(\hat{f}^{\psi_f} \otimes_2 \hat{h}^{\psi_h}\right)(a,b) \quad (13)$$

Proof. The NFRWT of $g(x)$, given by (6), may be rewritten as

$$\hat{g}_\alpha^{\psi_g}(a,b) = a^{-\frac{1}{2}} \iiint f(t) h(x-t) e^{-(j/2)(t^2-b^2)\cot\theta}$$
$$\times e^{-(j/2)((x-t)^2-b^2)\cot\theta} \psi_f^*(y) \psi_h^*\left(\frac{x-b}{a}-y\right) dxdydt. \quad (14)$$

Performing the change of variables $\alpha = x - t$, $\beta = b + ay - t$ and $\beta^2 - b\beta + \alpha t = 0$, (14) may be rewritten as

$$\hat{g}_\alpha^{\psi_g}(a,b) = a^{-\frac{3}{2}} \iiint f(t) h(\alpha)$$
$$\times e^{-(j/2)(t^2-(b-\beta)^2)\cot\theta} e^{-(j/2)(\alpha^2-\beta^2)\cot\theta}$$
$$\times \psi_f^*\left(\frac{(t-(b-\beta))}{a}\right) \psi_h^*\left(\frac{\alpha-\beta}{a}\right) d\alpha d\beta dt \quad (15)$$

from which it is easy to find

$$\hat{g}_\alpha^{\psi_g}(a,b) = a^{-\frac{1}{2}} \left(\int_{-\infty}^{+\infty} d\beta\right)$$
$$\times \left[a^{-\frac{1}{2}} \int_{-\infty}^{+\infty} f(t) e^{-(j/2)(t^2-(b-\beta)^2)\cot\theta} \psi_f\left(\frac{t-(b-\beta)}{a}\right) dt\right]$$
$$\times \left[a^{-\frac{1}{2}} \int_{-\infty}^{+\infty} h(\alpha) e^{-(j/2)(\alpha^2-\beta^2)\cot\theta} \psi_h\left(\frac{\alpha-\beta}{a}\right) d\alpha\right]$$
$$= a^{-\frac{1}{2}} \int_{-\infty}^{+\infty} d\beta \hat{f}_\alpha^{\psi_f}(a,b-\beta) \hat{h}_\alpha^{\psi_h}(a,\beta)$$
$$= \frac{1}{|a|^{1/2}} \left(\hat{f}_\alpha^{\psi_f} \otimes_2 \hat{h}_\alpha^{\psi_h}\right)(a,b) \quad (16)$$

4 CONCLUSIONS

In this paper we have studied the application of the NFRWT to perform signal filtering processes. We have first show the convolution of two wavelet functions and then we showed the convolution theorem for CWT, and in the same way, we derived the convolution for the NFRWT.

ACKNOWLEDGMENTS

This work was supported by the National Natural Science Foundation of China (no. 61171195) and was also supported by Program for New Century Excellent Talents in University (no. NCET-12-0042).

REFERENCES

[1] O. Rioul, M. Vetterli, Wavelets and signal processing, IEEE Signal Process. Mag. 8 (October 1991) 14–38.
[2] B.S. Chen, C.W. Lin, Multiscale Wiener filter for the restoration of fractal images: wavelet filter bank approach, IEEE Trans. Signal Process. 42 (November 1994) 2972–2982.
[3] C.E. Heil, D.F. Walnut, Continuous and discrete wavelet transform, SIAM Rev. 31 (December 1989) 628–666.
[4] G.Kaiser, A Friendly Guide to Wavelet, Birkhauser, Boston, 1994.
[5] A.R. Lindsey, The non-existence of a wavelet function admitting a wavelet transform convolution theorem of the Fourier type, Technical report, Ohio University, Athens, Ohio, August 1994.
[6] Antonio F. Perez-Rendon, Rafael Robles, The convolution theorem for the continuous wavelet transform, Signal Processing 84 (2004) 55–67.
[7] Shi Jun, Zhang .NaiTong, Liu XiaoPing, A novel fractional wavelet transform and its application, Science China, June 2012 Vol. 55 No. 6: 1270–1279.
[8] Shang Yu, Liu Suqin, Denoising of ECG signals processing research based on fractional wavelet transform, Journal of xi'an University of Technology, Vol. 34 No.9, September 2014.

Robustness testing method for intelligent electronic devices

H.T. Jiang, Y. Yang, W. Huang & Y.J. Guo
Jiangsu Electric Power Company Research Institute, Nanjing, China

ABSTRACT: This paper presents a Device Robustness Testing Method based on Abnormal Messages (DRTMAM), in order to detect the cyber vulnerabilities for Intelligent Electronic Devices (IEDs) in industry control systems (ICS). This approach presents an Abnormal Messages Generating Method based on Genetic Algorithm (AMGMGA) to create test cases, which can optimize test results. The proposed method has been verified and validated by robustness testing experiments. The experimental results show that the proposed method can effectively detect cyber vulnerabilities of IEDs.

Keywords: Robustness testing; genetic algorithm; cyber vulnerability

1 INTRODUCTION

Nowadays industrial control systems (ICS) have become cornerstones in facilities such as electric power, water power, natural gas, and traffic transportation. The cyber security of ICS is directly related to the national security and social stability. Intelligent electronic devices (IEDs) based on transmission control protocol/internet protocol (TCP/IP) are widely used in the ICS, which not only enhances interoperability and accessibility between devices, but also brings cyber vulnerabilities into networks and increases the possibility of malicious incidents [1]. In terms of the robustness testing for IEDs, this paper adopts abnormal messages to detect the fault-tolerant and recovery ability of the IED's communication module, which can be utilized to improve the safety and stability of ICS.

Many robustness testing methods have been proposed in the past. B. Lei et al [2] proposes a testing framework based on a formal semantics of a component system, and implements a prototype tool named RoTesCo. By adding states, events, and transitions to resist various impacts, the robustness finite state machines (RFSM) is constructed in [3]. By modeling software products using finite state machines, Z.H. Zhou et al constructs increased complete finite state machines by through of extending states and associated responses to the various unusual events for the robustness testing [4]. In commercial applications, a lot of robustness testing tools have appeared. For examples, BackTrack is an operating system penetration test tool, and Appscan is a web application vulnerability scanning tool. However, there is little robustness testing research for IEDs in the ICS.

The ICS is a relatively closed environment; therefore there are few threats from the outside. However, due to the popularity of mobile storage media and wireless devices, the closure of ICS has been broken. For example, the Stuxnet virus which destroys centrifuges in Iran's nuclear power station is brought through the USB flash drives [5]. Thus, there is already some cyber security related research about the ICS. By analyzing the security objectives and threats of industrial wireless networks, M.Wei realizes security architecture based on key distribute center (KDC) for wireless industry control networks [6]. D. Dzung et al analyzes the security of automation communication systems in substations and makes some effective ways to improve cyber security [7]. Y. Yang et al proposes an intrusion detection approach of the industrial IEC 60870-6-104 protocol by analyzing protocol models and traffic patterns [8].

The aforementioned research is based on cyber security architectures or detection methods in ICS, which has not focused on the robustness of the IEDs. In this paper, an abnormal messages generating method based on genetic algorithm (AMGMGA) is used to create test cases, and the devices robustness testing method based on abnormal messages (DRTMAM) is utilized to realize the robustness testing of IEDs. A test-bed is built for robustness testing experiments including several IEDs. The experimental results verify the feasibility of test methods and rationality of test cases, which can provide references for detecting the robustness of devices.

2 DEVICES ROBUSTNESS TESTING METHOD BASED ON ABNORMAL MESSAGES (DRTMAM)

The core idea of the IEDs robustness testing is to send the abnormal message test cases to an IED, and then check the communication status of the IED. The robustness of the IED depends on its response to the abnormal messages. In terms of the same test cases, the

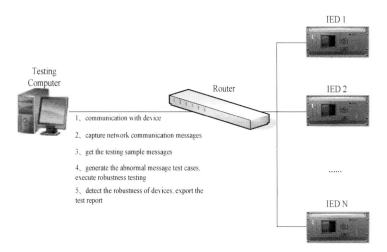

Figure 1. The test-bed.

more the number of communication status exceptions are, and the worse the robustness of the IED is. This is a black box test method, which only needs to know the standard communication protocol of IEDs, rather than the specific implementation of communication module inside the IEDs.

2.1 *Test-bed and testing process*

To implement the above core idea, a test-bed is built, as shown in Figure 1. There are three kinds of equipment in the figure, as follows,

(1) The IEDs are servers of the protocol.
(2) Testing computer runs the robustness testing method proposed in this paper.
(3) The router connects IEDs and the testing computer.

The general testing processes are as follows:

(1) As protocol client, the testing computer accesses the IED. In this case, the normal communication can be built between the client and sever.
(2) The packet capture software such as *Wireshark* is used to capture network communication messages between the testing computer and the IED in the test-bed.
(3) After the collected messages are pre-processed, useless and redundant data are removed, and the test sample messages are saved into files.
(4) According to the test sample messages, abnormal message test cases are generated, and the robustness testing is executed. A number of different abnormal message test cases can be generated from a test sample message. During the testing process, the abnormal messages are sent to the IED one by one, and abnormal messages which result in device communication status exception will be recorded.
(5) After the testing, the robustness report of detected IEDs is attained according to the ability of IEDs to addressing abnormal messages.

In the above processes, the first three steps are the preparation phase. The main purpose is to get test sample messages. The same sample messages can be used to do multiple robustness tests with different devices by repeating the last two steps. In these steps, the robustness testing is the most import part of the implementation process. The next section will give the detailed design and implementation of the robustness testing system.

2.2 *Architecture of the robustness testing system*

The core function of the robustness testing system is to read the file including test sample messages, generate abnormal message test cases, send test cases to the device, check the communication status of the device, and record the test cases which lead to exception of devices. Figure 2 shows the architecture of the robustness testing system, which includes five functional modules and two files. The main function of each module is as follows:

(1) Test sample messages loading module: This module is used to load test sample messages from files, and remove the IP addresses, media access control (MAC) addresses and other header information in the messages, and then extract the message contents.
(2) Test case generation module: It is utilized to generate abnormal message test cases by the AMGMGA.
(3) Test cases sending module: The module is used to establish the TCP connection between the testing computer and the IED, and send the test cases to the device. If the communication connection between the testing computer and the IED

is interrupted and cannot be recovered after the IED receives the test case, in this case the device is abnormal and the test case can be recorded as a cyber-vulnerable point of the IED. A test case which lead to the device exception is called a validation message in this paper.

(4) Graphical user interfaces (GUI) module: The module can be used to configure the testing controller, such as IP addresses of IEDs, as well as testing start/stop/pause command. In addition, real-time testing information is demonstrated in the GUI, such as the number of sent test cases and the remainder testing time.

(5) Test report generation module: This module is to generate test reports which mainly include device information, test execution information and validation message information.

2.3 Execution flow of test system

The test execution flow is shown in Algorithm 1. In this algorithm, the preparation stage is to read the test sample messages from files and set the IP address of the device, as shown in lines 1–2. The double loops in lines 3–20 is the main process of the execution. The outer loop iterates the test sample messages, and generates abnormal message test cases from each sample message. The inner loop is to iterate the test cases, send test cases to the device, check the device status, and record the test cases which bring the device exception. When an exception occurs, the testing process is suspended till the device resumes normal (lines 12–15). Lines 21–28 show the verification process about abnormal messages recorded. The validation messages are created by sending the recorded test cases to the device once again and checking the device status. The purpose of validating the abnormal messages is to exclude mistakes caused by objective factors such as communication interruption. Finally, the line 30 generates the test report according to the information of validation messages.

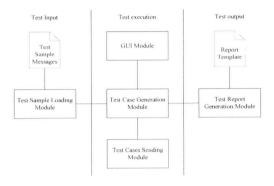

Figure 2. Architecture of robustness testing system.

Algorithm 1 test execution flow	
01. get the test sample messages;	//Test sample loading module
02. enter the IP address of the IED;	//GUI module
03. for each(test sample : test sample messages)	
04. {	
05. generate abnormal message test cases;	//Test case generation module
06. for each(test case : test cases)	
07. {	
08. send test case to the device;	//Test cases sending module
09. check device status;	//Test cases sending module
10. if(device status exception)	
11. {	
12. while(device status exception)	
13. {	
14. check device status;	//Test cases sending module
15. }	
16. record abnormal message into a collection;	
17. }	
18. update interface;	//GUI module
19. }	
20. }	
21. for each(abnormal message: abnormal message collection)	
22. {	
23. sent abnormal message to the device;	//Test cases sending module
24. check device status;	//Test cases sending module
25. if(device status exception)	
26. {	
27. record validation message in a collection;	
28. }	
29. }	
30. generate and show test report;	//Test report generation module

3 ABNORMAL MESSAGES GENERATING METHOD BASED ON GENETIC ALGORITHM (AMGMGA)

In five modules of the robustness testing system, the test case generation module determines the result and effectiveness of testing. Any test sample message (referred to TM) is made up of some fields. Each field (f) has a corresponding length (l) and value (v), thus any TM can be expressed as follows: $TM = f_1 l_1 v_1 f_2 l_2 v_2 \cdots f_k l_k v_k$, where k is the number of fields. An abnormal message (referred to AM) is generated using changing any field in a TM.

Assuming the of a TM is 20 and the average length of fields is 3, the number of AM generated are approximately 3×10^8. For this order of magnitude data, a limited number of test cases are needed to select from the whole test cases to constitute an optimal or near-optimal Test Cases Set (referred to TCS). Currently, heuristic algorithms have the potential to address the problem.

Among current heuristic algorithms, the genetic algorithm has obvious advantages to solve the complex issue that lack dedicated research [9]. Therefore, this paper presents an abnormal messages generating method based on genetic algorithm (AMGMGA). According to a single TM, AMGMGA generates an optimal or near-optimal TCS which contains n AMs. Any TCS can be expressed as follows, $TCS_i = \{AM_i^1 AM_i^2 \cdots AM_i^n\}$.

3.1 Basic operations and fitness function

Genetic algorithm is a global optimization heuristic search algorithm based on natural selection. It includes four basic operations (initialization, crossover, mutation and choice) and a fitness function.

3.1.1 Initialization

In the initialization phase, m individuals ($TCSs$) are randomly generated as the initial population, where m is the number of individuals at each iteration round. AM is generated from TM according to equation (1).

The function $replace(M, flv_{iold}, flv_{jnew})$ means to use flv_{jnew} field to replace flv_{iold} field in message M, and the two fields have the same length.

Function $rndInt(num1, num2)$ means to randomly get an integer value between $num1$ and $num2$ ($num1 < num2$). l_i is the length of the field i.

$$AM = replace(TM, flv_i, flu_i) \quad (1)$$
$$[(i = rndInt(1,k)][u = rndInt(0, 256^{l_i} - 1)]$$

The initial population is generated by calling equation (1) $n \times m$ times, which contains m individuals ($TCSs$) randomly generated, and each individual ($TCSs$) contains n AMs.

3.1.2 Crossover

Equation (2) is used to execute the crossover operation, which will create new individuals TCS_c and TCS_d from individuals TCS_i and TCS_j.

$$\begin{aligned}
TCS_c &= \{AM_i^a AM_i^b \cdots AM_i^{n/2}\} \cup \{AM_j^a AM_j^b \cdots AM_j^{n/2}\} \\
TCS_d &= TCS_i \cup TCS_j - TCS_c \\
&\text{where } \{AM_i^a AM_i^b \cdots AM_i^{n/2}\} \in TCS_i \\
&\{AM_j^a AM_j^b \cdots AM_j^{n/2}\} \in TCS_j \\
&a,b \cdots n/2 = rndInt(1,n)
\end{aligned} \quad (2)$$

At each iteration of the crossover operation, equation (2) is called $m/2$ times, which will generate m new individuals, and retain the original individuals, thus the number of individuals are $2m$.

3.1.3 Mutation

For any AM, equation (3) is called to carry out the mutation operation, which will create a new AM.

$$AM_{new} = replace(AM_{old}, flv_i, flu_i) \quad (3)$$
$$[(i = rndInt(1,k)][u = rndInt(0, 256^l - 1)]$$

At each round of iteration, each individual (TCS_i) executes mutation operation independently. For each AM in TCS_i, pv is the probability of getting a new AM by executing equation (3).

3.1.4 Choice

This operation requires to keep m individuals ($TCSs$) form $2m$ individuals ($TCSs$) for the next round of iteration. This paper uses the classic roulette wheel selection algorithm to execute the choice operation. Assuming that the fitness of TCS_i is F_i, the probability that TCS_i being reserved is p_i, which is calculated by equation (4).

$$p_i = F_i \Big/ \sum_{i=1}^{2m} F_i \quad (4)$$

3.1.5 The fitness function

This function is utilized to evaluate the fitness of an individual, which determines the probability of an individual being kept. In this paper, the evaluation of an individual's fitness is from two indexes: history information and the uniformity. Uniformity means that, the number of abnormal messages in each field is proportional to the length of this field in a TM, thus the comprehensiveness can be ensured when running the test process. Equation (5) is used to calculate the uniformity of a TCS, where L is the length of AM, and T_i is number of AM which changes the field i in the TCS.

$$Eve_{TCS} = L \Big/ \sum_{i=1}^{k}(l_i \cdot e^{|T_i - nl_i/L|}) \quad (5)$$

The history validation messages are important references for generating abnormal messages because devices from the same manufacturer may have the same problems. Equation (6) is used to calculate the history information reference value, where t_i is the number of validation messages of the field i about TM, which is obtained from the history statistical information.

$$His_{TSC} = 1/\sum_{i=1}^{k}(f_p^i \cdot e^{|T_i-t_i|}) \text{ where}(f_p^i = t_i/\sum_{i=1}^{k}t_i) \quad (6)$$

Combining above two evaluation indexes, the ultimate individual's fitness is calculated by equation (7), where the parameters ω_1 and ω_2 are decided according to the actual conditions. If the device is the first time to be tested, no history information is available, thus the ω_2 is zero. If there is history information about the device from the same manufacturer, ω_2 can be increased to an appropriate value.

$$F_i = \omega_1 Eve_{TCS}^i + \omega_2 His_{TCS}^i \quad (\omega_1 + \omega_2 = 1) \quad (7)$$

The algorithm is terminated when the fitness of any TCS reaches the threshold Th. The TCS with highest F_i is being selected to run the robustness testing. As the theoretical optimal value of individual's fitness is one, the value of Th is less than one in actual scenarios.

3.2 Process of test case set generation algorithm

Combining the basic elements of the algorithm described in Section 3.1, Algorithm 2 shows the TCS generation process.

In the algorithm, the lines 2–11 are the population initialization phase. In this phase, m individuals are created and stored in a *group*. The lines 12–40 are the iteration process of Genetic Algorithm. The lines 15–21 execute crossover operation between random two individuals in the *group*, and create $2m$ new individuals, then store new individuals into a *newGroup* and clear *group*. The lines 15–21 execute mutation operation for each individual in the *newGroup*, and then calculate fitness for each individual. The lines 31–35 select m individuals from *newGroup* by the choice operation, and save the selected individuals into a *group*. The lines 36-39 are the algorithm termination condition. If the condition is not satisfied, the algorithm will use the individuals in the *group* to perform a new round of iteration. The algorithm finally returns the individual with maximum fitness at line 41.

Algorithm 2 process of test case set generation		
01.	Input TM and algorithm parameters: Th, ω_1, ω_2, m, n, pv;	
02.	Set<TCS> *group* = new Set<TCS>(*m*);	//initialization population
03.	for(int i = 0 ; i < m ; i++)	
04.	{	
05.	TCS *tcs* = new TCS(*n*);	//initialization individuals
06.	for(int j = 0 ; j < n ; j++)	
07.	{	
08.	*tcs*.add(createAM(*TM*));	//initialization abnormal messages,
09.	}	equation(1)
10.	*group*.add(*tcs*);	
11.	}	
12.	while(true)	//begin iteration
13.	{	
14.	Set<TCS> *newGroup* = new Set<TCS>(*2m*);	
15.	for(int i = 0 ; i < m ; i++)	
16.	{	
17.	TCS *tcs1* = *group*.remove(random(0,*group*.size()));	//random select two individuals
18.	TCS *tcs2* = *group*.remove(random(0,*group*.size()));	
19.	TCS[] *tcsNew* = cross(*tcs1* , *tcs2*);	//crossover, equation(2)
20.	*newGroup*.add(*tcs1*, *tcs2* , *tcsNew*);	//add individual into new group
21.	}	
22.	for(int i = 0 ; i < $2m$; i++)	
23.	{	
24.	TCS *tcs* = *newGroup*.get(*i*);	
25.	for(int j = 0 ; j < n ; j++)	
26.	{	
27.	*tcs*.replace(mutate(*tcs*.get(j) , *pv*);	//mutation, equation(3)
28.	}	
29.	calFit(*tcs*);	//calculate fitness, equation(7)
30.	}	
31.	for(int i = 0 ; i < m ; i++)	
32.	{	
33.	TCS *tcs* = rws(*newGroup*);	//choice, roulette wheel selection
34.	*group*.add(tcs);	//algorithm and equation(4)
35.	}	
36.	if(maxFit(*group*) > *Th*)	//algorithm termination condition
37.	{	
38.	break;	
39.	}	
40.	}	
41.	return *group*.getMaxFitTCS();	//return TCS with maximum fitness

Table 1. Experiment parameters.

Parameter Name	Value
Number of test cases/n	10000, 50000
Initialization population size/m	20
Individual mutation probability/pv	0.05
Uniformity proportion/ω_1	0.5
History information proportion/ω_2	0.5
Algorithm termination threshold /Th	0.95

4 ROBUSTNESS TESTING EXPERIMENTS AND RESULTS ANALYSIS

In order to verify the validity of AMGMGA and the feasibility of DRTMAM, use five devices based on IEC61850 protocol [10] from different manufacturers for experiments. Two approaches of generating test cases are compared in the experiments. During the testing process, each test case generation method adopts the same sample messages, and creates the same number of abnormal messages.

(1) Random Abnormal Messages Generating Method (referred to RAMGM): This method randomly changes a field value in sample messages to generate abnormal messages. It is the simplest method to generate multiple abnormal messages.
(2) Field Random Abnormal Messages Generating Method (referred to FRAMGM): first of all, the number of abnormal messages for each field in a sample message is allocated averagely, and then the field value is randomly revised to generate abnormal messages. The method can ensure that each field contains abnormal messages.

Since all the test case generation methods have certain randomness, the average result of five independent experiments is taken as the final result. The parameter settings of AMGMGA are as follows:

In the experiments, the more the number of validation messages that cause the device exception are, the better the method of generating the test cases is. Figure 3 and Figure 4 separately show the experimental results of different test case generation method, when n are 10000 and 50000, respectively.

The following conclusions can be obtained from the experimental results:

(1) Whatever test case generation method is used, validation messages can be found in the testing process, which illustrates the effectiveness of DRMAM proposed in this paper.
(2) Increasing test cases will help to find more validation messages, however the execution time of the testing process is also increased at the same time.

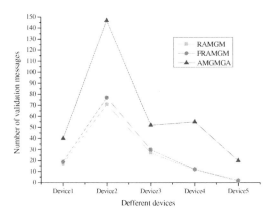

Figure 3. Results of different test case generation methods (n = 10000).

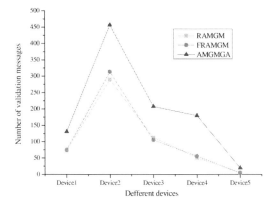

Figure 4. Results of different test case generation methods (n = 50000).

(3) The differences of the robustness between different IEDs are distinct, which are related to the message parsing programs in IEDs.

In Comparison with three test cases generation methods, the conclusions are as follows,

(1) FRAMGM can find more validation messages than RAMGM. The reason is that FRAMGM can ensure all fields have the same number of abnormal messages when generating test case set, therefore the coverage of test set is more comprehensive.
(2) AMGMGA can find most validation messages due to AMGMGA considering the history information when generating the test case set. AMGMGA can generate more validation messages in subsequent testing processes for the same device.

5 CONCLUSIONS

In order to test the robustness of communication modules in IEDs, this paper designs an abnormal messages generating method based on genetic algorithm, and proposes the robustness testing method based on abnormal messages. A test-bed is built to execute the robustness experiments and verify the proposed approaches. The experimental results show that the designed test cases generation method can effectively explore the abnormal messages which may result in device communication faults, and the proposed method can effectively detect the robustness of IEDs. The next work plans are follows: (1) This paper focuses on the robustness testing method based on the TCP/IP protocol. In the near future, the robustness testing methods for MAC layer protocols will be presented; (2) Robustness testing methods can only detect the cyber vulnerabilities of IEDs from point of view of communication protocols, therefore, the robustness of ICS for multiple cyber-attacks is still needed to be evaluated in the future, in order to improve the ability of cyber security for ICS.

ACKNOWLEDGMENTS

This work was supported by Jiangsu Provincial Science Project of State Grid under Grant No. J2014033, and Natural Science Foundation of Jiangsu Province, China under Grant No. BK20140114.

REFERENCES

[1] Peng Jie, Liu Li. Analysis of Information Security for Industrial Control System[J]. Process Automation Instrumentation, 33 (12) 2012: 36–39.

[2] Lei Bin, Wang Linzhang, Bu Lei, etc. Robustness Testing for Components Based on State Machine Model[J]. Journal of Software, 21 (5), 2010: 930–941.

[3] Wang Lechun, Zhu Peidong, Gong Zhenghu. BGP robustness testing based on RFSM[J]. Journal on Communications, 26(9), 2005: 21–29.

[4] Zhou Zhanghui, Wang Tongyang, Wu Junju etc. Research of Robustness Testing Based on FSM[J]. Computer Engineering & Science, 31(5), 2009: 93–97.

[5] Guo Qiang. Information Security Cases of Industrial Control System[J]. Information Security and Communications Privacy, 33(12), 2012:68–70.

[6] Wei Min, Wang Ping, Wang Quan. Research and implementation of security strategy for wireless industry control network[J]. Chinese Journal of Scientific Instrumen, 30(4), 2009: 679–684.

[7] Dzung D, Naedele M, Vobhoff T. Security for industrial communication systems[J]. Proceedings of the IEEE, 2005, 96(6): 1152–1177.

[8] Yi Yang, K. McLaughlin, T. Littler, S. Sezer, B. Pranggono, H. F. Wang. Intrusion Detection System for IEC 60870-5-104 Based SCADA Networks. Proceeding of the IEEE Power & Energy Sociey General Meeting[C] 2013.6.

[9] Deng Liang, Zhao Jin, Wang Xin. Genetic Algorithm Solution of Network Coding Optimization[J]. Journal of Software, 20(8),2009: 2699–2279.

[10] Gao Lei. Research and Implementation of Comparison Tool for IEC61850 SCL configuration File[J]. Automation of Electric Power Systems, 37(20), 2013: 88–91.

Design and analysis of quasi-optics for a millimeter wave imaging system

Nannan Wang, Jinghui Qiu, Yuwei Zhang, Yang Zhang, Pengyu Zhang & Hua Zong
School of Electronics and Information Engineering, Harbin Institute of Technology, Harbin, Heilongjiang, China

Liyi Xiao
Microelectronics Center, Harbin Institute of Technology, Harbin, Heilongjiang, China

ABSTRACT: High spatial resolution is one of the most important design targets in the millimeter-wave imaging system. In this paper, quasi-optics design has been used for high resolution. A novel dielectric rod antenna was employed as the feed antenna. The proposed antenna has good performance with low return loss, a low side-lobe level, symmetrical radiation patterns, and it is easy to form feed antenna arrays with low mutual coupling. Owing to the large diameter and small F/D ratio of regular lens antennae, loss and cost could be quite high. In order to overcome these disadvantages, a new bias-placed ellipsoid antenna was used. The focusing characteristics of the bias-placed ellipsoid antenna, with bias-feed, was analysed by the Multilevel Fast Multipole Method (MLFMM) method.

1 INTRODUCTION

In recent years, research into Millimeter-Wave Imaging Systems (MMWIS) has been developed rapidly. MMWIS are widely used in security-checking systems, which require MMWISs of the highest performances. Spatial resolution is one of the most important indexes of MMWISs, which is in direct proportion to the diameter of the antenna aperture and in inverse proportion to a working wavelength. Therefore, when the working frequency has been determined, enlarging the diameter of the antenna will be the most direct and effective method of raising a spatial resolution.

MMWIS requires the regular lens antenna to have a very large aperture and a small F/D ratio to meet the high resolution request, which leads lens antennae to be too thick and heavy, with low transmission rate, poor temperature sensitivity, and they are costly to manufacture.

To solve these problems, a bias-placed ellipsoid reflector antenna is proposed to be used as the focusing antenna of MMWIS. The electromagnetic wave that comes from one focus point can be focused onto the other focus point of the antenna. Due to the bias-placed structure, there is no block in the optical path caused by the source array, which raises the efficiency of the system.

In order to improve the imaging speed, antennae in the source array are arranged in one dimension and the other dimension is covered by mechanical scanning. A novel dielectric rod antenna is used as the feed antenna unit. Under the condition that the feed is also bias-placed, the focusing characteristic of the ellipsoid reflector antenna is analysed.

2 QUASI-OPTICS DESIGN OF THE MMWIS

As shown in Figure 1, the Ellipsoid Antenna (EA) with diameter of D, radiates to the minus z-axis direction. Radiometer and source antennae are placed along the x-direction at the first focal point (F_1 point) of the EA. The human body under checking is at the second focal point (F_2 point). The imaging area around the body is $0.8\,\text{m} \times 2.0\,\text{m}$.

$$\delta = 1.22 \frac{\lambda \cdot f}{D} \qquad (1)$$

The system works at Ka-band. According to the spatial resolution (δ) formula 1, when the wavelength is λ, the antenna diameter is D, the distance between the antenna aperture and F_2 point f is equal to 8.57 mm, 1300 mm and 2800 mm respectively, δ equals to 23 mm. The biased angle of the feed corresponds to the body edge and is equal to 5.5°, and the biased angle of the feed corresponds to the background edge and equal to 8°. The feeding antenna array is arranged in the range of 349 mm.

According to the Nyquist sampling theorem, the azimuth sampling overlap rate is chosen as 0.5, so the sampling interval is 11.5 mm. 70 samples are demanded in the 800 mm horizontal field of view, i.e. 70 feeds should be placed within the range of 349 mm where the interval between feeds is approximately 5 mm.

The construction of the ellipsoid is shown in Figure 2. A folded optical path is adopted in the system as shown in Figure 3.

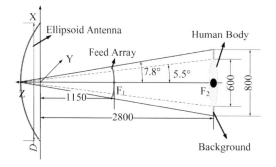

Figure 1. Quasi-optics design of ellipsoid antenna.

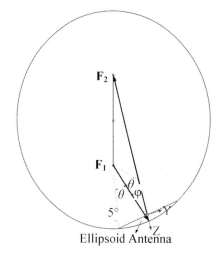

Figure 2. Construction of the ellipsoid antenna.

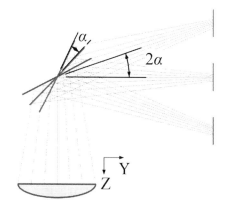

Figure 3. A folded optical path.

3 FEEDING ARRAY DESIGN

According to the quasi-optical design, the interval between feeds should be 5.06 mm. However, since the dimension of the Ka-band waveguide is 9.112 mm, the feeds should be arranged in 2 or 3 rows.

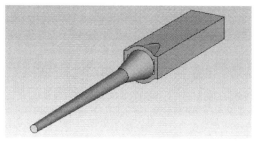

Figure 4. Model of dielectric rod antenna.

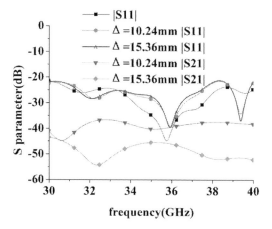

Figure 5. S-parameter of feed antenna array.

Various horn antennae, printed dipoles or gradient slot antennas, are proposed in the MMWIS by a lot of authors. However, disadvantages such as bad illumination for focusing an antenna, impossibility of tight arrangement or bad cross-polarization negate the applications of these antennae in such systems. The novel dielectric rod antenna is a kind of efficient feed. By reasonable design, high radiation gain can be obtained by extending the length of the rod instead of enlarging the cross section.

The structure of the novel dielectric rod antenna is shown in Figure 4. The dielectric rod is made of PTFE. A cylindrical platform structure is adopted as the transition between a square waveguide and a circular waveguide. The radiation part is divided into several sections in order to lower the side-lobe lever.

Return loss and mutual coupling between feeds are shown in Figure 5. The return loss is higher than 18 dB and the mutual coupling is lower than −30 dB in the frequency range of 26.5 GHz–40 GHz.

The radiation patterns of dielectric rod antennae are shown in Figure 6. Radiation patterns are distorted due to serious mutual coupling when the feed interval Δ is 10.24 mm. The distortion is eliminated when the interval changes to 15.36 mm. The edge illumination level of the ellipsoid from the feed is approximately −15 dB, and the symmetrical E-plane and the H-plane radiation patterns are guaranteed.

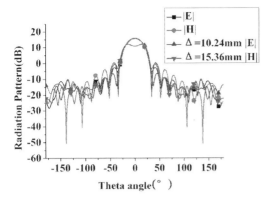

Figure 6. Radiation pattern of feed antenna unit.

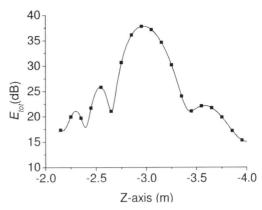

Figure 8. Field strength distribution of the ellipsoid antenna on the main radiation direction.

Figure 7. Model of ellipsoid antenna in simulation.

4 FOCUSING CHARACTERISTIC OF ELLIPSOID ANTENNA ANALYSIS

The radiation pattern of the dielectric rod antenna is used as the primary feed, while the edge of its illumination level equals to −10 dB in the numerical simulation in FEKO simulation software by the method of MLFMM. The structure is shown in Figure 7.

4.1 Field strength distribution of the ellipsoid antenna on the main radiation direction

The field strength distribution along the main radiation z-axis direction is calculated and shown in Figure 8 where the biased angle of the feed equals to 0°. The maximal field strength equals to 37.9 dB at the field point 2.95 m away from the antenna aperture.

4.2 Field strength distribution on the cross section when the partial angle equals to 0°

When the biased angle of the feed θ equals to 0°, the spatial resolution along the main radiation direction, i.e. the beam radius where the field strength is 3 dB lower than the maximum in the cross section, in the range of 2.8 m–3.2 m away from the antenna aperture, i.e. z ranges from −2.8 m to −3.2 m which meets the focusing depth requirement of the focusing antenna in the millimeter-wave near-field imaging system design, is shown in Figure 9. The spatial resolution is approximately 24 mm at the body part where z ranges from

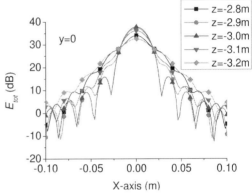

Figure 9. Field strength distribution on the cross section when the partial angle equals to 0°.

−2.85 m to −3 m; the resolution gets worse around the head, shanks and feet, with the approximate value of 26 mm.

4.3 Field strength distribution on the cross section when the partial angle equals to 5.5°

Since the horizontal field of view is covered by radiometers and feeding antenna array and the vertical field of view is covered by mechanical scanning, the antenna unit on the edge of the array focuses partially. The partial angle of the feed equals to 5.5° regarding the edge of body with the width of 600 mm.

Figure 10 shows the field distribution on the cross section perpendicular to the main radiation direction with z ranging from −2.8 m to −3.2 m where the biased angle equals to 5.5°. The electromagnetic wave is focused at −0.3 m in the horizontal direction; the spatial resolution is approximately 24 mm on both edges of the body with z ranging from −2.85 m to −3 m; the resolution gets worse around the head, shanks and feet, equalling to around 38 mm.

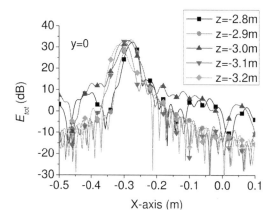

Figure 10. Field strength distribution on the cross section when the partial angle equals to 5.5°.

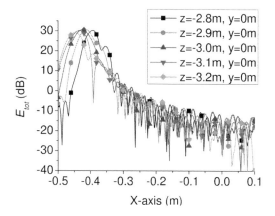

Figure 11. Field strength distribution on the cross section when the partial angle equals to 8°.

4.4 *Field strength distribution on the cross section when the partial angle equals to 8°*

The biased angle corresponding to the edge of 0.8-meter-wide background is 8°. The simulation results are shown in Figure 11. The electromagnetic wave is focused at −0.4 m in the horizontal direction; the spatial resolution is approximately 26 mm–38 mm on the edges of the image corresponding to the body with z ranging from −2.85 m to −3 m; the resolution gets worse around the head, shanks and feet on the edges of the image, approximately 38 mm–40 mm.

5 CONCLUSION

A quasi-optical design of near-field MMWIS is discussed in this paper. The proposal of a biased-placed ellipsoid reflector antenna as a focusing antenna overcomes the disadvantages of high loss, high weight, and the high cost of lens antenna, applied in high-resolution near-filed imaging systems. Novel dielectric rod antenna are used as feeds, and the focusing characteristic of the ellipsoid reflector antenna with different biased angles is analysed by simulation. The results show that when the ellipsoid reflector antenna with an aperture diameter of 1.3 m focuses at 2.95 m, the spatial resolution is around 24 mm–26 mm; when the biased angle equals 5.5°, the spatial resolution is approximately 24 mm–38 mm; and when the biased angle equals to 8°, the spatial resolution is approximately 26 mm–40 mm. Therefore, the application of an ellipsoid reflector antenna in near-field MMWIS, guarantees better focusing characteristics.

ACKNOWLEDGEMENT

This project is funded by National Major Scientific Equipment Development Special Project-Research on millimeter wave imaging detector and industrialization demonstration projects with item number 2012YQ140037.

REFERENCES

C. Chen, C. A. Schuetz, R. D. Martin, "Analytical Model and Optical Design of Distributed Aperture System for Millimeter-Wave Imaging," Proc. of SPIE Vol. 7117, Millimetre Wave and Terahertz Sensors and Technology, pp. 711706-1–711706-11, 2008.

S.V. Boriskin and A.I. Nosich, "Exact near-fields of the dielectric lenses for sub-mm wave receivers," 33rd European Microwave Conference, Munich, Germany, pp. 65–68, Oct. 2003.

S. S. Qian, X. G. Li, "Research on the Tapered Slot Antenna Fed by Microstrip-Slotline Transition and the Array Application," Modern Defence Technology, Vol. 35, No. 6, 2007, pp. 2397–2402.

A simply fabricated hybrid type metal based electrode for application in supercapacitors

Su Chan Lee, U.M. Patil & Seong Chan Jun
Department of Mechanical Engineering, School of Engineering, Yonsei University, Republic of South Korea

ABSTRACT: Higher performance portable energy storage devices are required for the increased use of portable mobile electronics such as smart phones, tablets, and wearable devices. Recent portable energy storage should have not only a large charging capacity but also a fast charging-discharging cycle. There are two key factors, which represent the performance of energy storage devices. One is energy density, which means how much energy can be stored. The other is power density, which is related to charging-discharging time. Recently, a lithium ion battery has been the most widely used energy storage device because it has a large energy density. Lithium ion battery, however, has a small power density. This defect causes many inconveniences for customers such as long battery charging time. Supercapacitors are one of the best candidates for replacing current lithium ion battery technology. A supercapacitor has a very large power density and this is the most important advantage of the supercapacitor. Supercapacitors also have a long life and less risk of exploding. Until now, the supercapacitor has less energy density than a lithium ion battery, so many researchers have studied new methods for installing a high energy density into supercapacitors. In this study, we suggest a new simple method for constructing a metal based supercapacitor electrode coated with a metal hydroxide. The performance of our device was tested with various techniques.

1 INTRODUCTION

1.1 Energy storage device

Energy storage devices have become a most important part of portable electronic devices. Most other parts of electronic devices have been developed sufficiently. For more performance development of other parts such as Central Processing Unit energy storage devices should be researched in advance of that because current energy storage devices do not provide enough power for CPUs. At present, higher performance portable energy storage devices have been studied by many researchers. Currently, there are many energy storage device types, but still those have some defects. The most widely used type is a lithium ion battery. It has a large charged storage capacity and a small packing size. That is the reason why the lithium ion battery is the most popular. However, the lithium ion battery has significant disadvantages such as a low charging-discharging speed and the risk of exploding. So many researchers have studied how to solve those problems. On the other hand, alternatives to the lithium ion battery have also been researched. The Supercapacitor is one of the best candidates for replacing current lithium ion battery technology.

1.2 Supercapacitor

The Supercapacitor, also known as electrochemical capacitor, has a larger power density than the lithium ion battery and this is most important advantage of the supercapacitor. The supercapacitor also has a long life and less risk of exploding. There are two types of conventional supercapacitors. One is the Electric Double Layer Capacitor (EDLC). The other is a pseudo capacitor, with an EDLC store charge on its surface with an electric double layer. A pseudo capacitor can store charge by surface redox reaction. The EDLC has a larger power density than the pseudo capacitor, but the pseudo capacitor has a larger energy density than the EDLC. Many hybrid types of supercapacitors have been studied recently in order to increase power and energy density.

In this study, we report a new simple method for metal base electrode fabrication for supercapacitors. Photolithography, a widely used technique in the semiconductor industry, was used for making rod shaped electrodes. Then, metal hydroxide was deposited on the metal surfaces to increase their capacity. Finally, the performance of our device was tested with electrochemical methods.

2 EXPERIMENTS & RESULTS

2.1 The fabrication of a supercapacitor electrode

A metal electrode was made by a photolithography process. Briefly, photoresist (AZ 5214) was coated on a PET substrate by spin coating and heating. Then, the substrate was exposed to UV and developed. After the

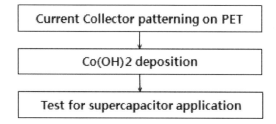

Figure 1. A schematic flow of our work.

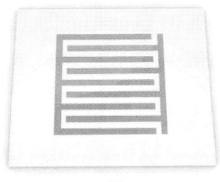

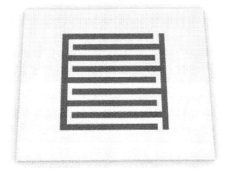

Figure 2. The electrode fabrication method. First, metal electrode was made by photolithography. Then, Co(OH)2 was deposited on the metal surface by electrophoretic force.

photolithography process, gold was deposited on the substrate by an E-beam evaporator with a Ti adhesion layer.

The Co(OH)2 was coated on the gold surface by an electrophoretic force. Cobalt nitrate was used as source material for the Co(OH)2. We used a potentially static method.

2.2 Characterization

The shape and quality of the Co(OH)2 and the metal electrode was checked by Scanning Electron Microscopy (SEM). Figure 3 shows well the shaped metal electrode and the Co(OH)2 on its surface.

Figure 3. A SEM image of our electrode.

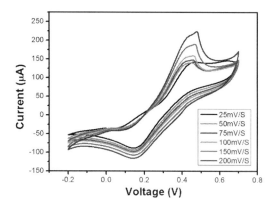

Figure 4. CV performance of our electrode.

2.3 Performance test

The performance of our electrode was tested by an electrochemical method. Figure 4 shows the CV test data. It shows well a discrete oxidation and reduction peaks because it is a pseudo capacitor. The potential window was [−0.2–0.7].

3 CONCLUSION

A Co(OH)2 nanosheet was successfully coated on the electrode by electrophoretic force. Electrophoretic deposition is one of the best method for making nano materials due to its low cost.

REFERENCES

[1] Li, X.; Wei, B., Supercapacitors based on nanostructured carbon. Nano Energy Vol. 2, No. 2, 2013, pp. 159–173.
[2] Davies, A.; Yu, A., Material advancements in supercapacitors: From activated carbon to carbon nanotube and graphene. The Canadian Journal of Chemical Engineering, Vol. 89, No. 6, 2011, pp. 1342–1357.

Joint scheduling based on users' correlation in MU-CoMP

Yafeng Wang & Danlin Wang
Beijing University of Posts and Telecommunications, Beijing, China

ABSTRACT: Scheduling is the key to resource allocation in a CoMP system. In this paper, we research joint scheduling by different pairing algorithms of users, in order to find a multi-user scheduling algorithm which can keep the balance between the average spectrum efficiency and the cell-edge spectrum efficiency, at the same time, with low complexity. The two scheduling algorithms based on the angle and the correlation coefficient are compared. Simulation results show that the joint scheduling based on the correlation coefficient can bring better experience for users in MU-CoMP.

1 INTRODUCTION

Inter-cell interference is an important reason which restricts the system throughput, especially in the Multi-Input Multi-Output (MIMO) system. Therefore, Coordinated Multi-Point (CoMP) transmission was proposed in the study of 4G standard. CoMP connects the multiple base stations by backbone, and they can share the channel information and the data information of users to some degree.

Scheduling algorithm is a key technology for achieving higher system performance. In the CoMP system, the base stations can complete independent scheduling or joint scheduling. For independent scheduling, the base stations in a collaborative cluster do not impact on each other in the scheduling process. For joint scheduling, the base stations in a collaborative cluster complete the scheduling by cooperating with each other. It can assign the system resource more effectively. At the same time, it can avoid conflict between users on time, space and frequency domain.

In order to improve the spectrum efficiency and the throughput of system, multiple users are sending and receiving data in the same Resource Block (RB) for joint scheduling. We called them the pairing users. In multi-MIMO (MU-MIMO) system, pairing users are in the same cell, but in the multi-CoMP (MU-CoMP) system, the selection of the pairing users expands to the collaborative cluster consisting of neighbouring cells. The interference between pairing users impacts on the system performance seriously.

In this paper, we consider the problem of joint scheduling in CoMP system, trying to find a multi-user scheduling algorithm which can keep the balance between the average spectrum efficiency and the cell-edge spectrum efficiency, at the same time, with low complexity. This paper studies the correlation-based joint scheduling, including the scheduling algorithm

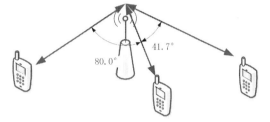

Figure 1. The constraint of the angle.

based on the users' angle and the optimized scheduling algorithm based on the correlation coefficient.

The rest of this paper is organized as follows: in Section 2 we present a scheduling algorithm based on the angle of users and the optimized algorithm based on the correlation coefficient. In section 3 we present the simulation results and finally we conclude in section 4.

2 THE CORRELATION-BASED JOINT SCHEDULING

2.1 *The scheduling based on the angle of users*

If pairing users are selected at random, there may be a certain correlation between their channels. Thus it will result in severe interference. Therefore, if making the angle of users as a constraint condition, it will ensure the correlation of users isn't too strong. In Figure 1, the angle between pairing users should be greater than a certain threshold.

The steps of algorithm implementation are as follows:

1) U represents the set of candidate users. First, users in U should be sort in accordance with the priority

order of Proportional Fairness (PF). Selecting the user with the highest priority as the first user u_1:

$$u_1 = \arg\max_{i \in U}\left(\text{Priority}(i)\right) \quad (1)$$

Recording the angle of arrival from the user u_1 to the base station:

$$a = \{a_1\} \quad (2)$$

C represents the set of scheduled users. Adding the user u_1 to C, and removing the user u_1 from the set U:

$$U = U - \{u_1\}, \quad C = C + \{u_1\} \quad (3)$$

2) Selecting the user with highest priority in the current set U as the next user u_i, the angle between the user u_i and users in set C should be greater than the specified threshold θ:

$$\begin{cases} u_i = \arg\max_{i \in U}\left(\text{Priority}(i)\right) \\ |a_k - a_i| > \theta, a_k \in a \end{cases} \quad (4)$$

If the user u_i meet the constraint in (4), adding the user u_i to the set C, and removing it from the set U:

$$U = U - \{u_i\}, \quad C = C + \{u_i\} \quad (5)$$

Recording the angle of arrival a_i for the user u_i:

$$a = \{a_1, a_2, \ldots, a_i\} \quad (6)$$

If the user u_i doesn't meet the constraint in (4), only removing it from the set U:

$$U = U - \{u_i\} \quad (7)$$

3) Repeating the steps 2) until the number of selected users in set C reaching the maximum number M of pairing users, or meeting other constraints.

2.2 The scheduling based on the correlation coefficient

As described before, the purpose of the angle-based scheduling algorithm is to reduce the correlation between users. But it isn't the best algorithm. So we introduce an optimized scheduling algorithm based on the correlation coefficient. By further constraining the correlation of users, it will be effective to improve the performance of the multi-user system.

The steps of algorithm implementation are as follows:

1) U is the set of candidate users. First, users in U should be sorted in accordance with the priority order of Proportional Fairness (PF). Selecting the user with the highest priority as the first user u_1:

$$u_1 = \arg\max_{i \in U}\left(\text{Priority}(i)\right) \quad (8)$$

Recoding the channel matrix's eigenvector W_1 of the user u_1, the W_1 can get by (9–11):

$$R_1 = \left(\sum_{t=1}^{T} h_{1,t}^H h_{1,t}\right)/T \quad (9)$$

Where $h_{1,t}$ is the channel matrix of user u_1 in the t-th subcarrier, T is the number of subcarriers in one RB.

Decomposing R_1 based on SVD:

$$R_1 = U_1 S_1 V_1^H \quad (10)$$

Taking the right singular vector, which corresponds to the largest eigenvalue as W_1:

$$W_1 = V_1(:,1) \quad (11)$$

C represents the set of scheduled users. Adding the user u_1 to the set C, and deleting the user u_1 from the set U:

$$U = U - \{u_1\}, \quad C = C + \{u_1\} \quad (12)$$

2) Selecting the user with highest priority in the current set U as the next user u_i, the correlation coefficient between user u_i and users in set C should be less than the specified threshold ρ:

$$\begin{cases} u_i = \arg\max_{i \in U}\left(\text{Priority}(i)\right) \\ W_k^H W_i < \rho, \forall k \in C \end{cases} \quad (13)$$

W_i is got as the same method as W_1. If the user u_i meet the constraint in (13), adding the user u_i into the set C, and removing it from the set U:

$$U = U - \{u_i\}, \quad C = C + \{u_i\} \quad (14)$$

Recoding the feature vectors W_i of user u_i. If the user u_i doesn't meet the constraint in (13), only removing it from the set U:

$$U = U - \{u_i\} \quad (15)$$

3) Repeating the steps 2) until the number of selected users in set C reaching the maximum number M of pairing users, or meeting other constraints.

3 THE CORRELATION-BASED JOINT SCHEDULING

3.1 The comparison on performance of the two algorithms

We compared the scheduling based on the angle and based on the correlation coefficient by the simulation. The 3GPP Case 1-3D and ITU UMi-3D are two different simulation scenarios. MU-BF as a contrast was compared with MU-CoMP. The protocol of LTE-Advanced provides the number of independent data streams for MU-MIMO and it is at most 4 for every cell. For MU-CoMP, if the number of collaborative cells is 3, it can support up to 12 pairing users in theory. The angle threshold θ is 40°, the specified threshold ρ in the scheduling based on a correlation coefficient of 0.4. For the current LTE-A system, the

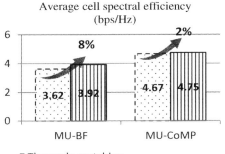

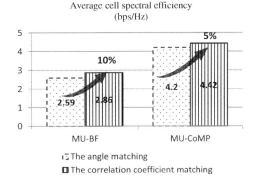

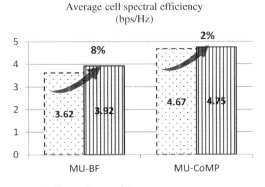

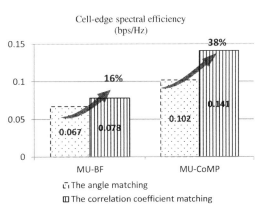

Figure 3. The comparison between the performances of scheduling algorithms based on correlation (ITU UMi-3D 4Tx).

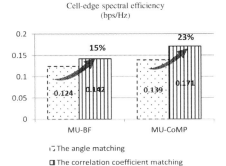

Figure 2. The comparison between the performances of the two scheduling algorithms based on correlation (3GPP Case1-3D 4Tx).

configuration with 8 antennas gradually become into the mainstream. So we simulated two cases with 4/8 transmitting antennas respectively.

1) There are 4 transmitting antennas in the base station and the 2 receiving antennas, with double polarization for users. MU-BF schedules 2 users, the collaborative cluster of MU-CoMP schedules 2 to 6 users adaptively. The 3GPP Case1-3D and ITU UMi-3D are two different simulation scenarios.

2) Setting the transmitting antennas of the base station are 8 and the receiving antennas of the user are 2, with double polarization. MU-BF schedules 2 to 4 users, the collaborative cluster of MU-CoMP schedules 2 to 12 users adaptively.

From the simulation results, we can see that MU-CoMP outperform MU-BF apparently. For MU-BF, the scheduling based on correlation coefficient has improved both the average cell spectral efficiency and cell-edge spectrum efficiency compared with the scheduling based on the angle. And for MU-CoMP, the improvement in the average spectral efficiency and edge spectrum efficiency is more obvious. If the number of transmitting antennas is 4, the improvement of the performance in UMi-3D compared with Case1-3D is more. If the number of transmitting antennas is 8, compared with UMi-3D, the improvement in performance in the scene of Case1-3D is more. But the difference is not so obvious. We see that the gain in the system with 8 antennas is 40 % more than the system with 4 antennas for the joint scheduling based on correlation coefficient.

3.2 *The performance analysis of the scheduling algorithm based on the correlation coefficient*

In order to study the performance of the scheduling algorithm based on the correlation coefficient in greater depth, we also researched the Probability

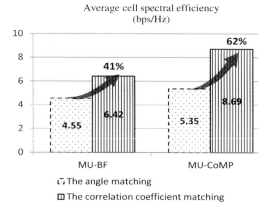

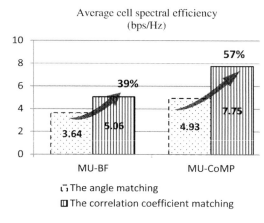

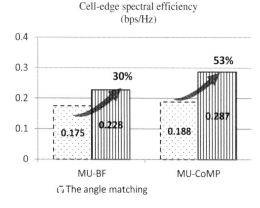

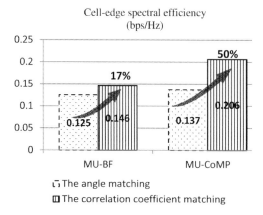

Figure 4. The comparison between the performances of scheduling algorithms based on correlation (3GPP Case1-3D 8Tx).

Figure 5. The comparison between the performances of scheduling algorithms based on correlation (ITU UMi-3D 8Tx).

Dense Function (PDF) and the Cumulative Distribution Function (CDF) for the number of scheduled users when the number of antennas is 4 or 8 in a CoMP.

For the system with 8 antennas, it can provide greater degrees of freedom than the system with 4 antennas, with the number of scheduled users concentrating on 8 to 12. The probability of full scheduling (scheduling 12 users) can reach about 40 percent. But for the system with 4 antennas, it is more inclined to schedule about 8 users. Through the simulation results, we find that the scheduling algorithm based on the correlation coefficient can implement the adaptive pairing algorithm better than the scheduling based on the angle. If the system configures 4 antennas, the probability reaches the highest when the number of pairing users is 8, and the number of scheduling users approximates Gaussian distribution. If the system configures 8 antennas, it can support more pairing users.

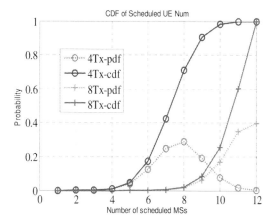

Figure 6. The CDF and PDF of the number of scheduled users.

4 CONCLUSIONS

In this paper, we have researched a joint scheduling and pairing algorithm, especially put forward an optimized algorithm based on the correlation coefficient. We have simulated this in different scenarios. Both the theoretical analysis and the simulation results show that the joint scheduling based on the correlation coefficient can bring better experience for the users, especially for the edge-cell users in MU-CoMP.

ACKNOWLEDGMENT

This paper is supported by National Key Technology R&D Program of China under grant No. 2013ZX03003002.

REFERENCES

[1] 3GPP TR 36.814 v9.0.0, "Further Advancements for E-UTRA; Physical Layer Aspects," Mar. 2010.
[2] 3GPP TR 36.814 v9.0.0, "Further Advancements for E-UTRA; Physical Layer Aspects," Mar. 2010.
[3] Alcatel Shanghai Bell, Alcatel-Lucent R1-082812: Collaborative MIMO for LTE-Advanced Downlink, Meeting #54, Jeju, Korea, August 18–22, 2008.
[4] Hardjawana, W.; Vucetic, B.; Yonghui Li; "Multi-User Cooperative Base Station Systems with Joint Precoding and Beamforming," IEEE Journal of Selected Topics in Signal Processing, vol. 3, no. 6, 2009, pp. 1079–1093.
[5] Jia Liu, Krzymien, W.A., "A Novel Nonlinear Joint Transmitter-Receiver Processing Algorithm for the Downlink of Multiuser MIMO Systems," IEEE Transactions on Vehicular Technology, vol. 57, no. 4, 2008, pp. 2189–2204.
[6] Sigdel, S.; Krzymie, W. A.; "Simplified Fair Scheduling and Antenna Selection Algorithms for Multiuser MIMO Orthogonal Space-Division Multiplexing Downlink" IEEE Transactions on Vehicular Technology, vol. 58, no.3, March 2009, pp. 1329–1344.
[7] Wang Zhijie; Wang Yafeng; Lin Chongsheng; "Enhanced Downlink MU-CoMP Schemes for TD-LTE-Advanced," IEEE Wireless Communications and Networking Conference (WCNC), 2010, pp.1–6.
[8] Yao Ma; "Rate Maximization for Downlink OFDMA with Proportional Fairness," IEEE Transactions on Vehicular Technology, vol. 57, no. 5, 2008, pp. 3267–3274.

A monitoring system of harmonic additional loss from distribution transformers

Z. Liu & Y. Liu
Jiangjin Power Supply Branch, State Grid Chongqing Electric Power Company, Chongqing City, China

Q.Z. Cao & Z.L. Zhang
Shapingba, Chongqing, China

ABSTRACT: The Harmonic Additional Loss (HAL) in distribution transformers changes with load fluctuations, so on-line monitoring is needed to truly reflect the HAL from distribution transformers, caused by load fluctuation. Therefore, this paper develops a monitoring system of harmonic additional loss in distribution transformers, combining computer techniques with wireless communication techniques. This system can exchange information with monitoring personnel through the Global System of Mobile (GSM) networks, and reflect the harmonic additional loss from distribution transformers in operation, which can help to reduce operating costs and waste in conveying electrical energy, and increase the economic benefits of power-supply enterprises.

Keywords: Distribution transformers; harmonic; additional loss; monitoring; device; system

1 INTRODUCTION

The harmonic loss, caused by a non-linear load on a power grid, will aggravate the loss in distribution transformers, which results in great waste of electric energy. There are very well-informed methods for calculating additional loss caused by harmonic loss. However, the HAL from distribution transformers does not have a fixed value, which changes dynamically with loss fluctuation. Even if the distribution transformer models are exactly same, the HAL will vary with different types of loss and capacities. So only on-line monitoring could truly reflect the HAL from distribution transformers, caused by loss fluctuation. Therefore, this paper tries to develop a monitoring system of harmonic additional loss from distribution transformer, integrating computer techniques and wireless communication techniques. This system can communicate with monitoring personnel through wireless networks and can show the harmonic additional loss from distribution transformers in operation, by nice interactive interface, which can provide a reference for monitoring personnel to know the harmonic status of distribution transformers so that an improvement can be adopted.

2 OVERALL SCHEME DESIGN OF SYSTEM

To achieve those purposes mentioned above, this paper designs the monitoring system of harmonic additional loss in distribution transformers, of which the overall structural block diagram is shown in Figure 1.

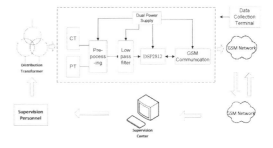

Figure 1. Structure of a distribution transformer's harmonic loss with an online monitoring system.

This monitoring system consists of two parts: a data collection terminal and background data analysis software. The data collection terminal is installed in the secondary side of the distribution monitoring transformer, which is mainly used to collect voltage and currents for the distribution transformer. This terminal is composed of a Current Transformer (CT), a Potential Transformer (PT), a Digital Signal Processor (DSP2812), a Global System for Mobile communications (GSM), an () module, etc. A signal of voltage and current is scaled-down by the transformer, and is digitized and collected by the A/D convertor, integrated in the DSP2812 after simple preprocessing. It is then transmitted wirelessly to a remote supervision centre by a GSM module. After receiving valid data, the background data analysis software analyses it and intuitively displays the HAL of the distribution transformer by the nice interactive display system,

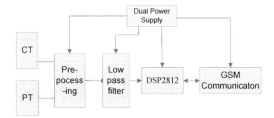

Figure 2. Structure diagram of data collection terminal.

so that the monitoring personnel can have a good knowledge of the distribution transformer's status and make further improvement [1–3]. In the following section, this paper will respectively illustrate clearly the data collection terminal and background data analysis software.

3 DESIGN OF DATA COLLECTION TERMINAL

The structural block diagram of the data collection terminal is shown in Figure 2. Its main program first needs to initialize the DSP, including the PIE register, the PIE interrupt vector table, flash, the ADC, and the serial port etc. The GSM module is then initialized and runs into the main loop program to wait for short messages after setting a phone number for the supervision centre. If there is short message arrived and texts of short message accord with the protocol requirements, the data collection terminal executes the related instruction. otherwise, the terminal returns to GSM module and standby. The main program flow chart is shown in Figure 3.

In the following section, this paper will introduce a low-pass filter and dual-power supply.

3.1 A low-pass filter [4–5]

A low-pass filter is used to filter high-frequency components of signals, avoiding aliases and distracting signals. This design adopts the Butterworth second order low-pass filter, of which the characteristic is that its frequency's response curve can be flat to the utmost within the pass band, while gradually reducing to zero within the suppressed band. The amplitude-frequency characteristics of the filter is shown in Figure 4.

Figure 4 illustrates the corresponding relationship between the input signal frequency and the output's signal attenuated amplitude, from which we can find that a signal with a frequency above 1 kHz will gradually attenuate at a rated speed. Because harmonic current frequency collected cannot exceed 750 Hz, a low-pass filter with a cut-off frequency 1 kHz will not affect accurate measurements of harmonic currents. Every A/D convertor channel adopts the same filter so that any phase deviation caused by filters is too small to affect accurate measurements of the harmonic current.

3.2 A dual power supply

A dual power shift device is made up of a standby lithium battery, a relay set, a switching power supply

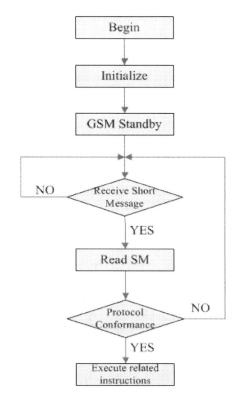

Figure 3. The main program flow chart of the data collection terminal.

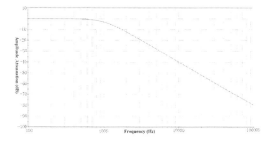

Figure 4. The amplitude-frequency characteristics of a filter circuit.

and a lithium battery charging circuit. Normally, a data collection terminal takes power from a certain phase of the distribution transformer. When the switching power supply breaks down, the relay set will automatically switch to a lithium battery to supply power to the data collection terminal. Its structure block diagram is shown in Figure 5.

4 THE DESIGN OF BACKGROUND DATA ANALYSIS SOFTWARE

Background data analysis software is mainly used to supply real-time data, data storage, and a query for monitoring personnel. To achieve validly reading,

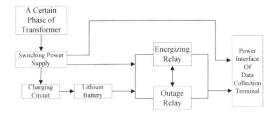

Figure 5. Data collection terminal dual power supply structure.

Figure 6. Data analysis software structure.

analysis and display of collected data, this software is divided into four parts: a data display module, a serial port operation module, a transformer's parameter setting module, and a GSM control module. Its structure block diagram is shown in Figure 6.

4.1 Serial port operation module

Before controlling the GSM module by serial port in data analysis software, the serial port should first be initialized. There are two methods of controlling a serial port with Visual C++ under Windows: one is with an API function for serial ports, supplied by Windows; the other is directly with MSCOMM.OCX, which is one ActiveX supplied by Microsoft company. The latter method has a simpler program design, while the former method is more flexible. This paper adopts the API function to control the serial port, including an open and close serial port, and send and receive data [6–7].

4.2 GSM control module

To achieve normal communication between the supervision centre and the terminal, the data analysis software must reasonably control the GSM module to send-receive short messages by the serial port. While the software is reading information from the data collection terminal, in the meantime, the supervising personnel would like to send messages to query it. This may result in a conflict with the serial port operation. In order for it to be more stable in running the software, it is necessary to control the reading and sending threads of the GSM module. The detailed flow chart is shown in Figure 7.

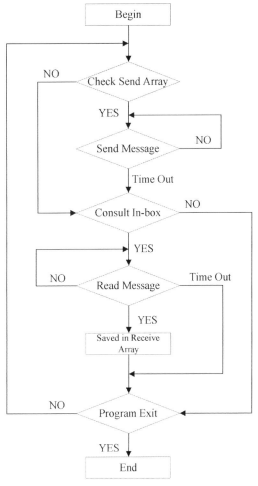

Figure 7. GSM control flow chart.

4.3 A transformer's parameter setting module

After setting the transformer's parameters, the data analysis software can backstage build the harmonic the additional loss model from this transformer. It also can modify the parameters to build a new corresponding harmonic additional loss model to adapt on-line monitoring of distribution transformers with different versions. Its flow chart is shown in Figure 8.

4.4 Data display module

Information from the data collection terminal contains the harmonic voltage and current of the transformer, which needs 58 characters to show the status every time, so it can be sent by using a short message. After the supervision centre receives this short message, it needs to correctly extract the real-time data of every electric parameter according to the related protocol in order to calculate the harmonic additional loss of the distribution transformer. In this paper, each order of harmonic current and harmonic additional loss is displayed with a table, while the total harmonic additional

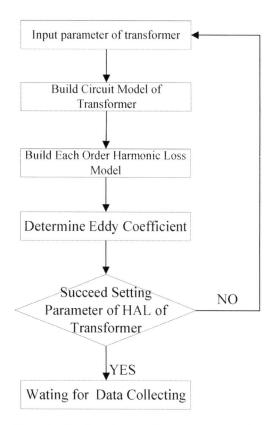

Figure 8. The flow chart transformer parameter setting module.

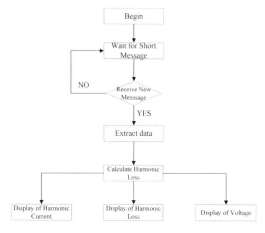

Figure 9. Data display module flow chart.

loss is displayed with a figure. Its flow chart is shown in Figure 9.

5 EXPERIMENT AND DATA ANALYSIS

To verify the feasibility and validity of this on-line monitoring system of harmonic additional loss from the distribution transformer, we built a simulated

Figure 10. Simulation experimental platform.

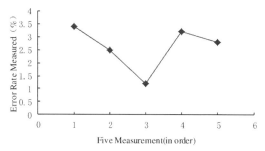

Figure 11. Error rate curve.

experimental platform. The rate capacity of the experimental transformer was 5 kVA, the short-circuit loss 160 W, the no-load loss was 75 W and the rate frequency 50 Hz. It could adjust the voltage secondary side of the transformer by three single-phase voltage regulators to imitate different states of the transformers in service. The wiring of the experimental platform is shown in Figure 10.

Meanwhile, this paper adopts a Fluck 434 power quality analyser to measure input power and output power of the distribution transformer, then calculates the total harmonic additional loss of the transformer with a classical algorithm. The final result is calculated by a power quality analyser, with measured data as the reference, and compared with the results of the monitoring system, mentioned above. The error rate curve is shown in Figure 11.

Comparing the data from this monitoring system with that of the power quality analyser shows that the difference between them is not obvious. In the figure above, the error rate of this monitoring system is within 5.5%, though this is a little different from standard values. If we consider that the input power of the experimental transformer contains some harmonic components, some of which cannot be inducted at the secondary side of the transformer, which results in interior loss of the transformer, we can say that it is receivable for harmonic additional loss from the experimental transformer as measured by this monitoring system.

6 CONCLUSION

The monitoring system developed in this paper has important significance for status monitoring of transformers, loss reduction, and energy conversation of the power grid, which can be expected to be widely used. Research of this paper is summarized as follows:

1) Developed a data collection terminal of the distribution transformer, which achieved signal conditioning of harmonic current, digital collecting, FFT spectral analysis and remote data transmitting. Collected data which can not only be used to analyse harmonic additional loss from the transformer, but also offers a data origin for on-line monitoring of the capacity, no-load loss, and with-load loss of the transformer.
2) Developed background data analysis software of harmonic additional loss of the distribution transformer, which mainly consisted of a serial port operation module, transformer's parameters, a setting module, a GSM control module, and a display module, which can provide the human-computer interaction interface with good visibility and operability for harmonic additional loss from the transformer.
3) Experimented with the contrast between the on-line monitoring of harmonic additional loss from transformers and this developed system. The results indicated that this monitoring system can be applied to the on-line monitoring of harmonic additional loss from transformers and that it has its own high accuracy with measurements.

REFERENCES

[1] Li G.C. et al. 2000. Information acquisition and communication protocols of distribution automation feeder terminal. *Power system technology*, 24(7): 55–58.
[2] Hammous & James T. 2004. Impact of new technology on Asian power generation transmission and distribution development in the 21st century. *Part B China and Japan, International Journal of Power and Energy Systems*, (24): 118–124.
[3] Wang H.P. et al. 2002. Feasibility analysis of remote data of distribution network by GSM, *High-voltage electrical apparatus*, 38 (5): 12–15.
[4] Wang P.H. & Pei S.C. 2000. Closed-form Design of Generalized Maximally Flat Low-Pass FIR Filters Using Generating Functions. *IEEE Transactions on Instrument and Measurement*, 32: 472–475.
[5] Williams A.B. & Taylor F.J. 2008. Electronic filter design. *Beijing: Science Press*.
[6] Zeng Z.Q & Wang Y.H. 2005. Three method of communication between microcontroller and serial port of PC by VC++. *Automation and Instrument*, (3): 60–63.
[7] Zhou R.Y. & Shang B. 2009. Introduction to Visual C++ serial port communication development and programming practice. Beijing: Electronic Industry Press.

A performance evaluation of higher order modulation considering Error Vector Magnitude (EVM) in a Long Term Evolution (LTE) advanced downlink

Xiaosong Liu & Zhigang Wen
School of Electronic Engineering, Beijing University of Posts and Telecommunications, Beijing, P.R. China

ABSTRACT: Recently, small cell deployment has received a lot of attention from 3GPP. Assuming that new indoor or high profile small cells are deployed, the geometry of certain User Equipment (UE) can be greatly improved and 256 QAM can be employed to explore the channel capacity for small cell enhancement. In real telecommunication systems, Error Vector Magnitude (EVM) is inevitable and it is significant to consider the impact of EVM when evaluating the performance of 256 QAM under small cell deployment. In this paper, we evaluate the performance of higher order modulations, while considering the impact of EVM under small cell deployment. The EVM can be defined as Tx EVM and Rx EVM. The model of Tx EVM and Rx EVM is proposed. Simulation results show that 256 QAM can effectively improve the system performance in a small cell indoor sparse scenario with a low traffic load.

Keywords: LTE-A, Small Cell, Higher Order Modulation, 256 QAM, EVM

1 INTRODUCTION

Long Term Evolution (LTE) is the famous project name of an excellent performance for cellular telephones. The LTE specifications have been finalized as Release 8 in 3GPP [2]. Then the specifications on LTE-Advanced [3] were initiated as Release 10 and beyond. Recently, small cell deployment has received a lot of attention from 3GPP. The continuing deployment of macro BS will lead to increased inter-cell interferences and reduced cell capacity. Therefore, 3GPP TR 36.932 [4] has proposed new Small Cell Enhancement (SCE) scenarios. Considering that the channel quality can be quite high and stable in these scenarios, especially the indoor scenario, higher order modulation, ie 256 QAM can be beneficial in the evaluated small cell scenarios with low mobility.

In practice, Error Vector Magnitude (EVM) always existed in the telecommunication system. The details of EVM have been identified in [8]. Therefore, non-ideal EVM is inevitable and it is significant to consider the impact of EVM when analyzing the performance of 256 QAM in SCE scenarios.

This paper investigates the impact of EVM, employing higher order modulation for small cell enhancement and evaluates the performance of 256 QAM under small cell scenario 3 in LTE-A downlink transmission. The reminder of this paper is organized as follows. Section 2 gives an overview of the LTE system with a focus on the high order modulation and the detailed concept of EVM. In Section 3, the algorithm models of EVM are given. Then, the simulation configuration is given in Section 4. Section 5 summarizes and analyses the performance of the impact of EVM employing higher order modulation in SCE scenario 3. Finally, Section 6 concludes this paper.

2 LTE SYSTEM OVERVIEW

2.1 Error Vector Magnitude (EVM)

Error Vector Magnitude (EVM) is the primary metric that represents the quality of the transmitted signal in many telecommunication systems. EVM defines the difference between the ideal symbol and the actual received symbol. Figure 1 illustrates the relationship between the actual received symbol and the ideal symbol. The error vector is the deviation in magnitude and phase between the received symbol and the ideal symbol. The greater the value of the EVM means the worse the quality of the transmitted symbols.

In LTE and LTE-Advanced systems, the calculation of EVM is based on a large number of symbols and the EVM is defined as the square root value of the ratio of the mean error vector power to the mean ideal vector power expressed in a percentage. The calculation of EVM can be given as:

$$EVM(\%) = \sqrt{\frac{P_{error}}{P_{ideal}}} * 100\% \qquad (1)$$

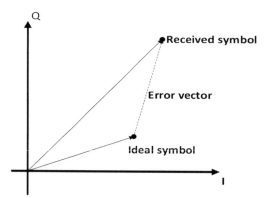

Figure 1. Definition of EVM.

$$EVM(\%) = \sqrt{\frac{\frac{1}{N}\sum_{n=1}^{N}|S_{received}(n) - S_{ideal}(n)|^2}{\frac{1}{N}\sum_{n=1}^{N}|S_{ideal}(n)|^2}} * 100\% \qquad (2)$$

In (2), N is the number of symbol in a period of time, $S_{received}(n)$ denotes the actual received symbol and $S_{ideal}(n)$ denotes the ideal symbol.

2.2 Channel Quality Indicator (CQI) definition

In LTE/LTE-A systems, there are various levels of Modulation and Goding Schemes (MCS) according to different channel conditions. To select an appropriate MCS level, a UE should make a measurement of the channel condition and feedback the Channel Quality Information (CQI) to the Base Station (BS) in the serving cell. In this way, the BS can select the MCS level to use. By adaptively selecting the appropriate modulation order and coding rate, the wireless communication system can provide a much better system performance than the fixed modulation and coding schemes. More details about CQI can be found in [5].

2.3 Higher order modulation

The introduction of small cell deployment provides the possibility for using higher order modulation schemes (i.e. 256 QAM) for the downlink transmission. The channel capacity is positively related with spectrum bandwidth and channel quality, as is shown in equation (3). In small cell scenarios, the SINR is generally between 20 dB and 50 dB.

$$C = B\log(1 + SINR) \qquad (3)$$

In LTE Rel. 8, QPSK, 16 QAM and 64 QAM have been specified for data transmission. eNodeBs can transmit data to UEs in good channel conditions (SINR $\approx$ 20 dB) with the most efficient MCSs. The channel quality in the small cell scenario can be even better and 256 QAM can be employed to further exploit the channel capacity of these UEs. These UEs can expect at most 33% throughput gain. The 256 QAM

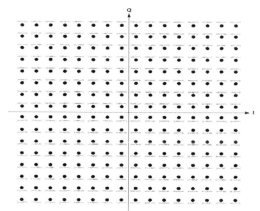

Figure 2. Constellation of 256 QAM.

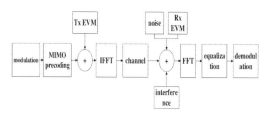

Figure 3. Illustration of the EVM system model.

constellation mapping is illustrated in Figure 2. Each modulated symbol represents 8 binary bits. Adjacent symbols should have only 1 different bit for optimal Bit Error Rate (BER).

We assume the symbol after modulation as:

$$\begin{cases} s = (x + yi)/\sqrt{170} \\ x, y \in \{\pm 1, \pm 3, \pm 5, \pm 7, \pm 9, \pm 11, \pm 13, \pm 15\} \end{cases} \qquad (4)$$

3 ALGORITHM MODEL

3.1 Definition and system model of EVM

In this section, the definitions and system models of Tx EVM and Rx EVM are demonstrated. As is referenced in [7] and [8], Tx EVM is well defined. Tx EVM means that Transmitter EVM and Tx EVM for 256 QAM can be modeled as an AWGN component. In [7], there is no clear definition of Rx EVM. However, the impairments at the receiver side have been mentioned and discussed in [9]. As mentioned in [9], Rx local oscillator phase noise, Rx dynamic range, I/Q imbalance, Carrier leakage, Carrier frequency offset are the main Rx impairments in RAN4 discussions. In [8], the applicable Rx impairments can be modeled as an equivalent AWGN component at the receiver side. In this way, Tx EVM and Rx EVM are all defined.

The achievable EVM values for higher order modulations (i.e. 256 QAM) have been concluded in 3GPP TSG RAN4 Meeting #68 (referred in [8]). As is defined in [8], for Tx EVM, low power BS such as

20 dBm and 24 dBm may achieve a better EVM such as 3–4% with power back-off and/or relaxed clipping at the cost of decreased coverage, and an increase in price and size. Rx EVM includes the UE receiver impairments and Rx EVM can become 1.5–4%. However, the guaranteed minimum performance of Tx EVM and Rx EVM is not evaluated by RAN4. The details about EVM values can be seen in [7] and [8].

3.2 Algorithm model of EVM

To evaluate the impact of Tx EVM and Rx EVM on 256 QAM, the EVM model should be considered. As is mentioned in [8], the noise variance of the modeled Tx EVM should be defined as follows:

$$\sigma_{TxEVM}^2 = I_0 * EVM_{Tx}^2 \tag{5}$$

σ_{TxEVM}^2 represents the AWGN noise variance, and I_0 represents the transmit power spectral density of the downlink signal, as measured at the eNodeB antenna connector. In this paper, I_0 is normalized in the simulation.

In the actual situation, Tx EVM is a random error. Therefore, the AWGN noise power which is added at the transmitter side $P_{TxEVM(transmitter)}$ should be as follows:

$$P_{TxEVM(transmitter)} = \left(\sigma_{TxEVM} * Random\right)^2 \tag{6}$$

Random represents the pseudorandom value drawn from the standard normal distribution.

Because Tx EVM is modeled as an AWGN component, Tx EVM should experience the large-scale fading and small-scale fading in the downlink channel as the downlink signal. At the receiver side,

$$P_{TxEVM(receiver)} = P_{TxEVM(transmitter)} * P_{large-scale\ fading} * P_{small-scale\ fading} \tag{7}$$

where $P_{large-scale\ fading}$ represents the large-scale fading power in the downlink channel and $P_{small-scale\ fading}$ represents the small-scale fading power.

In [8], applicable Rx impairments can be regarded as Rx EVM at the receiver side and Rx EVM can also be modeled as an AWGN component.

$$I_1 = \frac{1}{N_{t,f}} \sum_{l,k} |y(k,l)|^2 \tag{8}$$

I_1 represents the average power spectral density of the received downlink signal based on one resource block (RB) at the receiver side. $y(k,l)$ denotes the received signal in one resource element (RE) of the RB and k denotes the RE index of the time domain in the RB and l denotes the RE index of frequency domain in the RB.

$$\sigma_{RxEVM}^2 = I_1 * EVM_{Rx}^2 \tag{9}$$

σ_{RxEVM}^2 is the AWGN noise variance of the Rx EVM added to the receiver.

Table 1. CQI table for 256 QAM.

CQI index	modulation	Code rate *1024	efficiency
Original CQIs			
15	64 QAM	948	5.5547
Newly Defined CQIs			
16	256 QAM	776	6.0625
17	256 QAM	822	6.4219
18	256 QAM	881	6.8828
19	256 QAM	925	7.2266
20	256 QAM	948	7.4063

In the actual situation, Rx EVM is a random error. Therefore, P_{RxEVM} should be as follows:

$$P_{RxEVM} = \left(\sigma_{RxEVM} * Random\right)^2 \tag{10}$$

3.3 SINR calculation including EVM

Since Tx EVM and Rx EVM are considered in the LTE-Advanced system, the signal interference and noise ratio (SINR) calculation should be changed and contains the interference of Tx EVM and Rx EVM. The new SINR calculation including Tx EVM and Rx EVM should be defined as follows:

$$SINR = \frac{P_{signal}}{P_{noise} + P_{interference} + P_{TxEVM(receiver)} + P_{RxEVM}} \tag{11}$$

P_{signal} denotes the power of the downlink signal at the receiver. P_{noise} represents the noise power including the noise in the channel and other kinds of noise, such as thermal noise, at the receiver. $P_{interference}$ represents the power of different kinds of interference, such as inter-layer interference, inter-user interference, and the interference from other cells. $P_{TxEVM(receiver)}$ is the power of Tx EVM modeled as the AWGN component and is defined in (7). P_{RxEVM} is the power of Rx EVM and is defined in (10).

4 SIMULATION CONFIGURATION

In the simulation, we defined new Modulation and Coding Schemes (MCSs) and extended the current CQI table with additional entries for 256 QAM. Table 1 gives the details of the CQI tables defined in the simulation. Table 2 shows the simulation parameters. As shown in the table, we can check all the main parameters.

5 THE SIMULATION RESULTS AND EVALUATION

In this section, we present the system level performance for 256 QAM in a Small Cell Enhancement

Table 2. System level simulation assumptions.

Parameter	Values used for evaluation
Deployment scenarios	Small Cell Enhancement 3 sparse
Simulation case	• ITU InH between small cell and UE • UE speed: 3 km/h • Carrier Frequency: 2 GHz for macro layer (if exists), 3.5 GHz for small cell layer • UE noise figure: 9 dB
Small cell TX power (Ptotal)	24 dBm
UE distribution	For scenario 3, randomly and uniformly distributed over area;
System bandwidth	10 MHz
Possible transmission schemes in DL	SU-MIMO without CoMP
Antenna configuration	2 × 2, cross polarized
Antenna gain + connector loss	For small BS: 5 dBi
Feedback scheme	PUSCH mode 3-1, 5 ms period. New CQIs introduced to support 256 QAM. See details in table 3
Channel estimation	Non-ideal

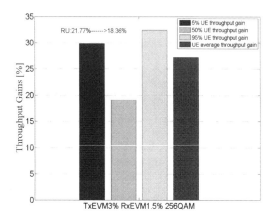

Figure 5. System level results for 3% Tx EVM 1.5% Rx EVM.

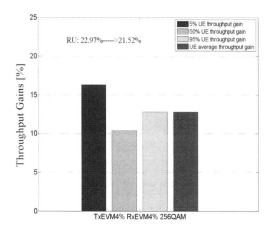

Figure 4. System level results for 4% Tx EVM 4% Rx EVM.

(SCE) scenario 3 (sparse) under the interference of Tx EVM and Rx EVM impairments.

As is discussed in Section IV, the Tx EVM and Rx EVM values can impact the actual SINR and result in the decrease of SINR. In order to verify the impact of Tx EVM and Rx EVM on the system throughput result, we used 4 different combinations of Tx EVM and Rx EVM in our simulations: {Tx EVM, Rx EVM} = {4%, 4%}, {3%, 1.5%}.

In the SCE scenario 3 (sparse), the interference to UEs in the indoor scenario comes only from the small cell in the same indoor scenario. In this scenario, some UEs can have good channel conditions. In this way, 256 QAM can be used by those UEs and the simulation results can be more obvious. In these simulation results, the respective baseline is the situation employing a 64 QAM with the same combination of Tx EVM and Rx EVM.

From Figure 4, it is observed that with Tx EVM 4%, and Rx EVM 4% assumptions, the 5% UE shows nearly 15% gain and the 95% UE shows more than 10% gain. For the cell center UEs, the gain is easily understood because the SINR is always good enough for 256 QAM based MCSs. And for the 5% 'cell edge' UEs, it is observed that in this low traffic load indoor sparse scenario, the cause of the low throughput is generally not the interference from the neighboring small cell, but the fact that several packets arrive at the same small cell continuously and the chance of each UE being scheduled in subframe is reduced. In this case, the new MCSs can be still beneficial to exploit the channel capacity when the inter-cell interference does not exist, which according to the RU, this is quite a normal case. Compared with the 5% UE and 95% UEs, the performance gain to the median UEs is around a 10% gain, while the UE average throughput shows around a 13% gain.

With the assumptions of Tx EVM at 3%, and Rx EVM at 1.5% in Figure 5, the 5% UE shows nearly 30% gain, and the 50% UE shows around a 20% gain and the 95% UE shows more than 30% gain. The UE average throughput in this case shows around a 25% gain.

Moreover, since the EVM is not considered when the UE generates the CQIs, this means that the CQIs reported may be always higher than the achievable SINR. When the interfering cell is not active, the initialization offset of OLLA must be carefully selected because longer response time may fully mitigate the gain brought by more efficient MCSs. This is especially important for the Tx EVM 4%, Rx EVM 4% case based operations.

6 CONCLUSION

In this paper, we have simulated the system under different Tx EVM values and different Rx EVM values in system levels. Based on the simulation results, it is observed that 256 QAM can effectively improve the system's performance in the small cell indoor sparse scenario with a low traffic load.

REFERENCES

[1] 3GPP36.201, Mar. 2009 "Evolved Universal Terrestrial Radio Access (E-UTRA); LTE physical layer; General description," 3GPP, Sphia Antipolis, Technical Specification 36.201 v8.3.0.
[2] 3GPP, TS36.201 (V8.1.0), Nov, 2007.
[3] 3GPP, TR36.913 (V8.0.0), June 2008.
[4] 3GPP TR 36.932 V12.0.0, Dec. 2012, "Scenarios and Requirements for Small Cell Enhancements".
[5] 3GPP, Sept, 2013, "Evolved Universal Terrestrial Radio Access (E-UTRA); Physical layer procedures", 3GPP, 6TS 36.213 v11.4.0.
[6] NTT DOCOMO, Jan. 28 to Feb. 1, 2013 "Text Proposal for TR36.923 on Small Cell Enhancement Scenarios," R1-130748.
[7] 3GPP TS36.104, Sept. 2013, "Base Station (BS) radio transmission and reception".
[8] R4-134571,"Reply LS on Higher Order Modulation Evaluation Assumptions", RAN4#68
[9] R4-132019, "LS on 256 QAM Support" 3GPP TSG-RAN4, RAN4#66bis.

Research on the control method of the driving system for the in-wheel driven range-extended electric vehicle

Songtao Wang & Xin Zhang
School of Mechanical, Electronic and Control Engineering, Beijing Jiaotong University, Beijing, China

ABSTRACT: A research on the energy management and electric differential control method for the in-wheel driven extended-range heavy-duty commercial vehicle was proposed. A vehicle energy management strategy was proposed for the energy management via constant power control method. Considering the tire characteristics, the centripetal force and the load shift during turns, the electric differential control strategy was presented. And the co-simulation was implemented in the AMESim-Simulink. The simulation results show that good performances are achieved based on the energy management strategy. And the electric differential control strategy can keep the tire slip ratio within an appropriate range, ensuring the stability of the vehicle.

1 INTRODUCTION

Range-extended electric vehicle (REV) can not only make up for the disadvantages of pure electric vehicle in mileage, but also can reduce the fuel consumption, so it is one of the effective schemes of vehicle energy saving and emission reduction at present. And the reasonable control of the stroke increasing device power and the motor torque is the key to improve vehicle dynamic performance and the energy recovery efficiency. Besides, different from the traditional centralized driving, electric vehicle that use the distributed driving is superior to concentrated driven electric vehicle in mobility, brake energy recovery and interior body space utilization. During steering, the motor torque needs to be reasonable controlled, so it is necessary to develop a proper electronic differential control strategy. This paper aiming at the in-wheel driven extended-range heavy-duty commercial vehicle carried out the energy management strategy and the electric differential control strategy for REV.

2 STRUCTURE OF THE DRIVING SYSTEM

2.1 Driving mode

The structure of researched in-wheel driven extended-range heavy-duty commercial vehicle (REV) is shown in figure 1. It uses the in-wheel driven series hybrid mode; the powertrain system mainly contains the in-wheel motor, the power battery, stroke increasing device and the in-wheel driving axle. Engine and generator form the stroke increasing device cascading in the DC bus, charge the battery and provide auxiliary power to the driving motor. Two large torque permanent magnet synchronous motors respectively connected to the rear wheels through in-wheel reducer, thus form the in-wheel driven driving mode.

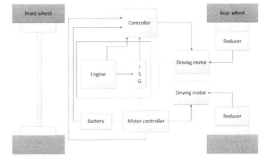

Figure 1. Structure of the REV.

Table 1. Basic parameters of the vehicle.

Parameters	Values	Unit
Maximum speed	80	km/h
Climbing ability	20	%
Acceleration time	≤20	s
Wheel base	6100	mm
Track	2400: 2400	mm
Wheel radius	478	mm

2.2 The basic parameters

Basic parameters and performance indexes of vehicle are shown in table 1. The engine was selected to the cummins ISF2.8S4129V Turbocharged diesel engine, and the emission meets the euro IV emission regulation. The generator was selected to the R90WA ISG, and the driving motor was the 80 kW permanent magnet synchronous motor. This vehicle uses electric differential controller. While steering, the controller receives information from the driver, calculating the torque that each motor needs and send instruction to the motor controller to regulate the wheel speed.

3 DYNAMIC MODEL OF THE IN-WHEEL REV

In order to study the static and dynamic performances of the in-wheel driven REV, a vehicle dynamics model of 15 degree of freedom was proposed. The kinetic equations as follows:

$$\begin{cases} m(\dot{u} + \omega_y w - \omega_z v) = \sum F_{xsij} mg\sin\theta \\ m(\dot{v} + \omega_z u - \omega_x w) = \sum F_{ysij} - mg\sin\varphi\cos\theta \\ m(\dot{w} + \omega_x v - \omega_y u) = \sum (F_{zsij} + F_{dzij}) - mg\cos\varphi\cos\theta \\ J_x\dot{\omega}_x = \sum M_{xij} + \frac{c}{2}(F_{zsfl} + F_{zsrl} - F_{zsfr} - F_{zsrr}) \\ J_y\dot{\omega}_y = \sum M_{yij} + (F_{zsrl} + F_{zsrr})b - (F_{zsfl} + F_{zsfr})a \\ J_z\dot{\omega}_z = \sum M_{zji} + (F_{ysfl} + F_{ysfr})a - (F_{ysrl} + F_{ysrr})b + \frac{c}{2}(-F_{xsfl} + F_{xsfr} - F_{xsrl} + F_{xsrr}) \end{cases} \quad (1)$$

Among them, θ is the pitch angle of car-body, φ is the roll angle, ψ is the yaw angle; X_G, Y_G, Z_G are respectively the absolute displacement of front car-body center of gravity on the x-axis y-axis and z-axis; θ_{rel11}, θ_{rel12}, θ_{rel21}, θ_{rel22} are respectively the relative rotation angle of the four wheels; Z_{rel11}, Z_{rel12}, Z_{rel21}, Z_{rel22} are respectively the vertical lift of four spindles; y_{carv} is the relative displacement steering rack on y-car-body-axis.

The vertical tire loads are calculated as follows:

$$\begin{cases} N_{fl} = \frac{l_r}{2(l_f+l_r)}(mg - \frac{2hF_c}{d_f}) \\ N_{fr} = \frac{l_r}{2(l_f+l_r)}(mg + \frac{2hF_c}{d_f}) \\ N_{rl} = \frac{l_f}{2(l_f+l_r)}(mg - \frac{2hF_c}{d_r}) \\ N_{rr} = \frac{l_f}{2(l_f+l_r)}(mg - \frac{2hF_c}{d_r}) \end{cases} \quad (2)$$

While steering, the centripetal acceleration will lead to load shift, thus affects the slip ratio of tires. The slip angles can be calculated as follows:

$$\begin{cases} \alpha_{fl} = \beta + \frac{l_f\gamma}{v} - \delta_{in} \\ \alpha_{fr} = \beta + \frac{l_f\gamma}{v} - \delta_{out} \\ \alpha_{rl} = \alpha_{rr} = \beta - \frac{l_r\gamma}{v} \end{cases} \quad (3)$$

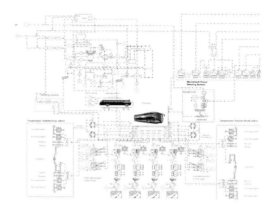

Figure 2. Vehicle model of the in-wheel driven REV.

Among them the car-body sideslip angle: $\beta = \arctan(v_y/v_x)$.

The longitudinal forces and lateral forces are calculated as follows:

$$\begin{cases} F_{xi} = \mu_{xi}N_i \\ F_{yi} = c_i\alpha_i \end{cases} \quad (4)$$

Among them i stands for fl, fr, rl, rr; μ_{xi} is a constant which is related to road surface. The slip ratio of tires can be calculated as:

$$S_i = \frac{\omega_{wi}r - v}{\omega_{wi}r} \quad (5)$$

Based on the software AMESim, the established vehicle model of 15 degree of freedom shown in figure 2 contains the following sections:

a) Powertrain system mainly contains the driving motors and stroke increasing device (including the energy management strategy).
b) Car-body, includes the car body and steering system.
c) Elastokinematics modular, mainly, includes the suspensions and tires.
d) Data collecting module, mainly includes kinds of sensors.

4 DRIVING CONTROL STRATEGIES OF THE IN-WHEEL DRIVEN REV

Shown in figure 3, The REV uses the traction hierarchical control, the upper control is the energy control that manage the power via constant power control method; the lower control is the coordinated control of driving motor torques, according to the vehicle running state, it implements the electric differential control, the regenerative braking and the anti-slip control to ensure the driving more safe, stable and energy-efficient.

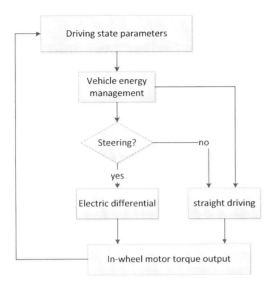

Figure 3. Driving control strategies of the in-wheel driven REV.

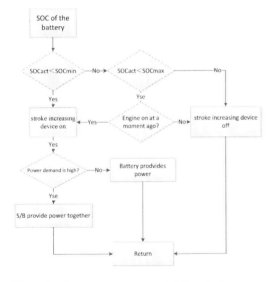

Figure 4. Energy management control flow chart.

4.1 Vehicle energy management

The vehicle energy management control strategy divides into the power consumption stage and the charging driving stage. In the power consumption stage, the battery provides power supply and the stroke increasing device doesn't work. In the charging driving stage, when the SOC of battery reduces to the set lower limit value, the stroke increasing device starts to work, providing power with battery together to the driving motors. Thus, the mileage can be extended and the engine can work in the best condition with high efficiency and low emission.

Shown in figure 4, when the SOC of battery reduces to the set lower limit value, the stroke increasing device

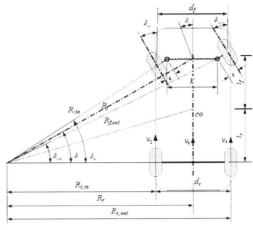

Figure 5. Ackerman model of vehicle steering.

starts to work, charging the battery, and if the power demand is high, it will provide energy with the battery together. When SOC is greater than the set value, the stroke increasing device stop working, it switches to the pure electric mode. To avoid frequent shut down of the engine, a certain range of state holding area was set. If the engine was working in a moment ago it will continue to work, and vice versa.

4.2 Design of electronic differential control strategy

4.2.1 Vehicle steering model

The steering state follows the Ackerman model, shown as figure 5.

According to the geometric relations:

$$\begin{cases} R_r = l/tan\left(\frac{\delta_1+\delta_2}{2}\right) \\ R_{rin} = R_r - d_r/2 \\ R_{rout} = R_r + d_r/2 \end{cases} \quad (6)$$

R_{rin} and R_{rout} are respectively radius of the two rear wheels, δ_1, δ_2 are front wheel steering angles.

According to the input of steering wheel angle, the steering angles of the front wheels can be calculated as follows.

$$\begin{cases} \delta_{in} = arctan\left(\frac{(l_f+l_r)tan\delta}{l_f+l_r-\frac{K}{2}tan\delta}\right) \\ \delta_{out} = arctan\left(\frac{(l_f+l_r)tan\delta}{l_f+l_r+\frac{K}{2}tan\delta}\right) \end{cases} \quad (7)$$

In the formula, δ is the steering angle of knuckle, K is the distance between the pins center extend line and the ground.

Relative steering radius of vehicle center relative to the rotation center is:

$$R_{cg} = \sqrt{\left(\frac{l_f + l_r}{2}\right)^2 + R_r^2} \quad (8)$$

The rotating angular velocity is:

$$\omega_0 = \frac{v}{R_{cg}} \quad (9)$$

According to the fact that each point of the ideal vehicle model has the same angular velocity, it can be get that $v_l = \omega_0 R_{r,in} = \omega_{wl} r$, $v_r = \omega_0 R_{r,out} = \omega_{wr} r$. Thus angular velocity of two driving wheels ω_{wl}, ω_{wr} can be calculated. So when steering, considering the vehicle centroid speed as reference, calculate the wheels line velocity around the center, thus get demand angular velocity of each driving motor to distribute the torque output of each driving motor reasonably.

4.2.2 Electronic differential control strategy

Wheel speed is related to road surface, tire slip ratio and wheel steering angle, so the Electronic differential control must consider the influence of the above factors comprehensively. Normally, the initial slip of tire is caused by the elastic deformation of the tire. So for the start the force and torque of tire increase linear with the slip ratio. When the wheel torque increasing further, the driving force and slip rate turn to a non-linear relationship. The experimental data show that when driving on the good roads, the driving force usually reaches a maximum value in the slip rate 15%–20%. So, the electric differential control strategy was designed in the logic threshold control theory to keep the tire slip ratio within an appropriate range. The control strategy flow chart is shown in figure 6.

After get the target motor speed, the demand torque and power of driving motor at any speed can be calculated by using the one dimensional interpolation. Then input the demand torque instruction to the motor controller to control the torque of driving motor. The motor torque control can use PID closed loop control.

Shown in figure 7, for the proposed electric differential control strategy, the control modular was written in MATLAB/Simulink, and then used AMESim-Simulink to implement the co-simulation and analysis.

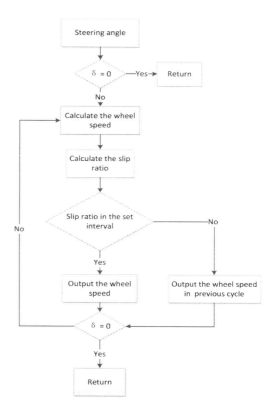

Figure 6. Electric differential control flow chart.

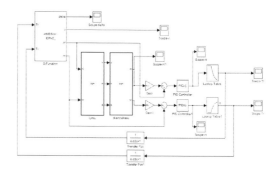

Figure 7. Electric differential control modular.

5 THE SIMULATION RESULTS AND ANALYSIS

5.1 Straight driving simulation

In order to test the energy management control strategy and performance of the vehicle model, firstly selected the typical UDC driving cycle (European bus cycle) to implement the simulation, and calculated the pure electric mileage and range-extended mileage.

It was shown in figure 8 that the variations of velocity, engine power, and engine speed, SOC and driving motor torque with time in the UDC driving cycle. The initial value of SOC is 80%, at this time the stroke increasing device didn't work; the in-wheel driven motors provided torque to the vehicle. At time 59 s, SOC reduced to 49.9% that below the SOC_{min} value, the stroke increasing device started to work, the engine worked at a constant speed 1700 r/min, and the engine power was 29.38 kW. The engine worked in the high efficiency range and charged the battery to extend the mileage. In the case that oil tank volume is 200 L, the pure electric mileage and range-extended mileage were calculated and shown in table 2.

5.2 Electric differential simulation

The co-simulation was implemented in AMESim-Simulink, and the simulation parameter settings referred to 《GB/T 6323.2-1994》. Through adjusting the parameters, the initial vehicle speed was 20 km/h, the vehicle turned left, and the step input of steering wheel angle, and the whole simulation time was 10 s.

Figure 9 showed the simulation results, known from 9(a) to 9(c) that after a short adjustment, Vehicle side slip angle and yaw rate reached a stable value thus making the vehicle has a good stability. Figure 9(d) and 9(e) were respectively the slip ratio of rear left wheel and rear right wheel. It was shown that, through the regulation of electric differential system, the slip ratios of two wheels both were kept in the 15%–20% interval, avoiding the vehicle slip. Figure 9(f) and 9(g) were respectively the speed and torque of two driving wheels. Due to the vehicle driving a left shift, the speed and torque of right rear wheel were greater than the left rear wheel The control system achieved a good function of electric differential, ensuring the stability of vehicle.

6 CONCLUSION

The in-wheel driven range-extended electric vehicle has good prospects for development. This paper

Figure 8. Parametric variations in UDC driving cycle.

Table 2. Mileage in the UDC driving cycle.

Parameters	Values	Unit
Distance	1.014	km
Oil consumption	40.88	L/100 km
SOC initial value	49.9	%
SOC final value	50.71	%
ΔSOC	0.81	%
Pure electric mileage	122.6	km
Range-extended mileage	724.9	km

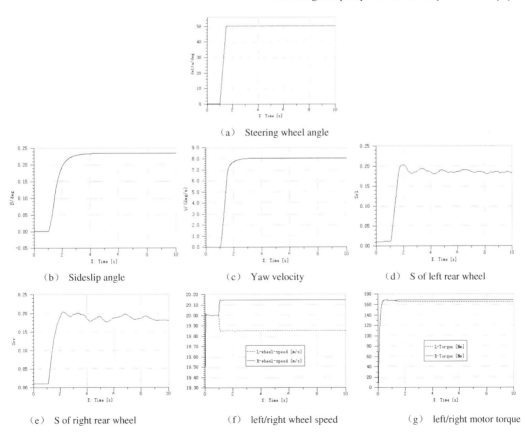

(a) Steering wheel angle
(b) Sideslip angle
(c) Yaw velocity
(d) S of left rear wheel
(e) S of right rear wheel
(f) left/right wheel speed
(g) left/right motor torque

Figure 9. The simulation results

researched the energy management strategy of range-extended heavy-duty commercial vehicle. Besides, aiming at the in-wheel driven form, this paper researched the electric differential control strategy, considering the load shift during turns, established an electric differential control system to reasonably control the speed and torque of driving motor. The simulation results show that good performances are achieved based on the energy management strategy. And the electric differential control strategy can keep the tire slip ratio within an appropriate range, avoiding excessive tire wear and ensuring the stability of the vehicle.

ACKNOWLEDGEMENTS

This research is supported by The National High Technology Research and Development Program of China (863 Program) (Grant No. 2012AA111106), and the Fundamental Research Funds for the Central Universities (Grant No. 2013JBM073).

REFERENCES

M.Z. Ahmad, E. Sulaiman, Z.A. Haron, T. Kosaka, Impact of Rotor Pole Number on the Characteristics of Outer-rotor Hybrid Excitation Flux Switching Motor for In-wheel Drive EV, Procedia Technology, Volume 11, 2013, Pages 593–601, ISSN 2212-0173.

Lin hui, Research on Composite ABS Control Strategy of Fuzzy Self-adjusting PID for Electric-wheel Vehicle [D]. Jilin University, 2013.

Hu M.Y., Yang F.Y., Ou M.G., Xu L.F., Yang X.Q., Fang C. A Research on the Distributed Control System for Extended-range Electric Vehicle [J]. Automotive Engineering 2012, 03:197–202.

Ge Y.H., Ni G.Z. A Novel Electronic Differential Algorithm for In-wheel Motor Driven EV [J]. Automotive Engineering 2005,03:340–343.

Zhao Y.E., Zhang J.W. Study on Electronic Differential Control System of Independent In-wheel Motor Drive Electric Vehicle [J]. Journal of System Simulation 2008,18:4767–4771+4775.

Jin Li-qiang, Wang Qing-nian, Yue Wei-qiang, Song Chuan-xue. Dynamic Model for Simulation of EV with 4 Independently-driving Wheels [J]. Journal of System Simulation, 2005, 17(12): 3053–3055.

Zhi J.N., Xiang C.L., Zhu L.J., Ma Y., Xu B., Li M.X. A Study on the Performance Simulation and Control Scheme for In-wheel Motor Driven Vehicles [J]. Automotive Engineering 2012,05:389–393.

Wu Z.H., Guo Y., Zhu Y., Haung T., Design and Implement of Central Control Unit in Electric Vehicle Application [J]. Machinery & Electronics, 2008, 07:28–31.

Zhou S, Niu J.G, Chen F.X, Pei F.L, A Study on Powertrain Design and Simulation for Range-extended Electric Vehicle [J]. Journal of System Simulation, 2011,11:924–929.

GB/T 6323.2-1994, Controllability and stability test procedure for automobiles-Steering transient response test (Steering wheel angle step input) [S].

Research on TSP based two phases path planning method for sweep coverage of a mobile wireless sensor network

Zhenya Zhang, Weili Wang & Qiansheng Fang
Anhui Provincial Key Laboratory of Intelligent Building, Anhui Jianzhu University, Hefei, China

Hongmei Cheng
School of Management, Anhui Jianzhu University, Hefei, China

ABSTRACT: A sweep coverage from a Mobile Wireless Sensor Network (MWSN) is one kind of dynamic coverage technology for a sparing use of mobile nodes in MWSN. In order to use as few mobile nodes in MWSN as possible, moving paths for each mobile node for sweep coverage should be planned rationally. To design moving paths for each mobile node in a sweep coverage with MWSN, a fast moving path planning method based on two phases Travel Salesman Problem (TSP) is presented in this paper. To measure the balanceable features of moving paths for all mobile nodes in MWSN, a balance factor is also discussed in this paper. Experimental results show that moving path planning methods, based on two phases TSP, is faster than moving path planning method based on Multiple Travel Salesman Problem (MTSP) and the balance of a moving path, planned by a two phases TSP based path planning method, is better than an MTSP based path planning method.

1 INTRODUCTION

In applications for building energy management, building energy efficiency evaluation, building energy efficiency audit or building energy saving [1–6], some important environment information, such as temperature, humidity, wind speed, and air pressure, etc [5–7] in a building, can be monitored validly by the sweep coverage of a mobile wireless sensor network [1,2,9]. In the sweep coverage of a MWSN, each mobile node should move along one interesting point sequence in its moving path [1,2]. To sample building environment data with the least number of mobile nodes, the moving path for each mobile node should be planed rationally. Some research results on sweep coverage of MWSNs indicate that the path plan problem for mobile nodes in low cost sweep coverage is one Multiple Travel Salesman Problem (MTSP) [8] and path plan problem for mobile nodes in low cost effective sweep coverage is one MTSP problem with the maximum visiting deadline time T as a constraint [2].

Because the price of mobile nodes for MWSN is high and the maintenance of mobile nodes is complicated, it is necessary to use as few mobile nodes as possible in sweep coverage of MWSN. With less mobile nodes in MWSNs, the cost for a MWSN based system can be reduced but the complexity of that system can become greatly degraded. A MTSP is used to model the path plan problem for low cost effective sweep coverage in MWSN and a binary search based algorithm is given to solve the problem of having the minimum number of mobile nodes in a sweep coverage, in our research [1,2,7].

Although a binary search based algorithm can plan a moving path for each node in MWSN quickly, the differences in the cost of mobile nodes may be very evident. To measure the differences of mobile node in a cost balancing factor is defined in this paper. To plan a better moving path for mobile nodes in MWSN, the process for the solution of a path plan problem is divided into two phase in this paper and the Travelling Salesman Problem (TSP) is used to model the sub-problem in each phase. In the following paper, the balance factor for path planning in sweep coverage is discussed in Section 2. The TSP is based on a two phase path planning algorithm for sweep coverage of a WSN and is presented in Section 3. Experimental results are shown in Section 4. The conclusions and future work are given in Section 5.

2 THE LOADS IN SWEEP COVERAGE

Let $\mathbf{P} = \{P_i \in \mathbf{\Omega}, i = 1, 2, \ldots, n\}$ be set for Point Of Interest (POI) and $\mathbf{S} = \{s_i, i = 1, 2, \ldots, m\}$ be the set of all mobile nodes in MWSN. And let $PP_j = <P_{j1}, P_{j2}, \ldots, P_{js}>$ be the moving path of mobile node $s_j \in \mathbf{S}$ in one low cost effective sweep coverage. Because the sweep coverage is effective, $P = \bigcup PP_j, PP_i \cap PP_j = \emptyset$ and $P_{j1} = P_{js}$ where $1 \le j \le m$, and because the sweep coverage is also low cost, the moving path for each mobile node in a sweep coverage of an MWSN is one Hamilton loop.

In MWSN, there are three kinds of behavioural actions for mobile nodes. The first is a moving action which is the moving behaviour of one mobile node from one POI to another adjacent POI. The second is an information sampling action which is the information sampling behaviour of one mobile node near one POI. The third is other auxiliary actions which are some other behaviour in one mobile node. Because there are some loads happening in each kind of action, some cost needs to be paid for the maintenance of sweep coverage by the MWSN. Because each kind of action of one mobile node is happening during the sweep process when the mobile node is moving along its moving path in sweep coverage, the load of one mobile node is related to its moving path. Let $PP_j = <P_{j1}, P_{j2}, \ldots, P_{js}>$ be the moving path of a mobile node s_j in one low cost effective sweep coverage and let $mLoad(i,j) \geq 0$ be the moving action load of one mobile node when the mobile node moves from POI P_i to POI P_j, $sLoad(i) \geq 0$ be the information sampling load when the mobile node samples data near one POI, and $oLoad(i) \geq 0$ be the loads for other auxiliary actions of another mobile node. Let Lm, Ls and Lo be the summation of loads for moving action, information sampling action, and other auxiliary actions of one mobile node, when the mobile node moves along its moving path. If $Load$ is the summation of all load of mobile node s_i along moving path PP_j, some equations can be defined as follows:

$$Lm(PP_j) = \sum_{i=j_1}^{j_s-1} mLoad(i, i+1) \qquad (1)$$

$$Ls(PP_j) = \sum_{i=j_1}^{j_s-1} sLoad(i) \qquad (2)$$

$$Lo(PP_j) = \sum_{i=j_1}^{j_s-1} oLost(i) \qquad (3)$$

$$Load(s_i, PP_j) = \begin{pmatrix} c_1(s_i) & c_2(s_i) & c_3(s_i) \end{pmatrix} \begin{pmatrix} Lm(PP_j) \\ Ls(PP_j) \\ Lo(PP_j) \end{pmatrix} \qquad (4)$$

$$Load(s_i, PP_j) = \begin{pmatrix} c_1 & c_2 & c_3 \end{pmatrix} \begin{pmatrix} Lm(PP_j) \\ Ls(PP_j) \\ Lo(PP_j) \end{pmatrix} \qquad (5)$$

$$B^*(L) = \frac{\max(L) - \min(L)}{\max(L)} \qquad (6)$$

In equation (4), $c_1(s_i)$, $c_2(s_i)$ and $c_3(s_i)$ are the respective coefficients for a moving action load, information sampling action load, and other auxiliary actions in mobile node s_i. In equation (4), $c_1(s_i) \geq 0$, $c_2(s_i) \geq 0$ and $c_3(s_i) \geq 0$.

Proposition 1: In a sweep coverage of MWSN, if more POIs are in a moving path for one mobile node, the load of the mobile node is greater.

Proposition 2: In a sweep coverage of MWSN, if the length of a moving path for one mobile node is longer, the load of the mobile node is greater.

If structure, function and life-span of each mobile node in a MWSN are homothetic, each load coefficients is independent of a mobile node. With this assumption, $c_1(s_i)$, $c_2(s_i)$ and $c_3(s_i)$ in equation (4) can be denoted as c_1, c_2 and c_3 and equation (4) can be simplified as equation (5). In equation (5), $c_1 \geq 0$, $c_2 \geq 0$, $c_3 \geq 0$.

Because Lm is caused by mechanical motion meanwhile Ls and Lo are mainly caused by electronic operations, Lm is the main parts of load of one mobile node in a sweep coverage. If any load, caused by an electronic operation, is ignored, only Lm needs to be considered when the load of one mobile node is calculated, according to equation (4) or equation (5).

In sweep coverage of MWSN, effective sweep coverage means that each POI is visited by at least one mobile node along its moving path before the end of the coverage time. Let PP_i be the moving path of mobile node s_i and L_i be the abbreviation of $Load(s_i, PP_i)$. If $L = \{L_i \mid i = 1, 2, \ldots, m\}$ is the load set for mobile node set $S = \{s_i \mid i = 1, 2, \ldots, m\}$ in a sweep coverage, $E(L) = \frac{1}{m}\sum_{j=1}^{m} L_j$ is the mean load of a mobile node in a sweep coverage $D(L) = \frac{1}{m}\sum_{i=1}^{m}(L_i - E(L))^2$ is the variance for mobile node's load. $D(L)$ can be used to measure whether the load of each mobile node is balanceable.

Definition 1: Let $S = \{s_i, i = 1, 2, \ldots, m\}$ be the mobile node set for a sweep coverage of MWSN and $L = \{L_i \mid i = 1, 2, \ldots, m\}$ be the load set for each mobile node. If $d_{ij} = |L_i - L_j|$ where i, j = 1, 2, \ldots, m, $B(L) = \frac{1}{m^2}\sum_{i=1}^{m}\sum_{j=1}^{m} d_{ij}$ is the balance factor for a sweep coverage of MWSN.

If the balance factor is defined as at definition 1, three conclusions can be proved. Firstly, $B(L) \geq 0$. Secondly, $B(L) = 0$ if the load of each mobile node is same. Thirdly, if the difference for load of each mobile node is greater/less, B(L) is greater/less. The same conclusions can be proved for $B^*(L)$, as defined in equation (6). It is clear that load of a mobile node in a sweep coverage of MWSN is more balanceable if the value of $B(L)$ or $B^*(L)$ is more or less.

If there are m mobile nodes in a sweep coverage of MWSN, the time complexity of B(L) computing should be $O(m^2)$ according to equation (4). On the other hand, because $B^*(L)$, as defined by equation (6), has the same characteristics as B(L), on the measuring of balance for sweep coverage, $B^*(L)$ can be treated as one kind of balance factor for sweep coverage. Also, because the time complexity of $B^*(L)$ is $O(m)$ when there are m mobile nodes in a sweep coverage, $B^*(L)$ is the fast balance factor for sweep coverage of MWSN. It is obvious that $B^*(L) \in [0\ 1]$. Also, the more or the less the value of $B^*(L)$ is, the more balanceable is the load on a mobile node in a sweep coverage.

3 PATH PLANNING FOR SWEEP COVERAGE

If there are n POIs which should be visited in each period by m mobile nodes in a sweep coverage of MWSN, the number of all possible moving paths of each mobile node is $m! n!$ (If there is no difference between any two mobile nodes, the number is $n!$). It is obvious that best balanceable path planning problem in a sweep coverage of MWSN is one NP hard problem. It is necessary to design an approximate algorithm to solve the best balanceable path planning problem.

Although path planning problems for low cost effective sweep coverage are one of MTSP's problems, a MTSP based path planning algorithm can design the right moving path for each mobile node, but the load of each mobile node may be in a serious imbalance if each mobile node moves along its moving path which has been planned by a MTSP based path planning algorithm. Experimental results have discovered that this kind of imbalance phenomenon is one unavoidable problem for MTSP based path plan algorithms.

The balance status of mobile node's load in sweep coverage of MWSN can be measured by $D(L)$ which is the variance of load sequence $L = \{L_i, i = 1, 2, \ldots, m\}$ for all m mobile nodes in sweep coverage. Because $D(L)$ is equivalent to the balance factor $B(L)$ which is defined at definition 1, $B(L)$ is the quantitative indicator for balance status of the mobile node's load in a sweep coverage of MWSN. Because the fast balance factor $B^*(L)$ defined by equation (6) has same characteristic with $B(L)$ on the measuring of the balance status of sweep coverage, $B^*(L)$ can be used to measure the balance status of a sweep coverage. Also, because the time complexity of $B^*(L)$ is $O(m)$ which is less than either the time complexity of $B(L)$ or than the time complexity of $D(L)$, the balance status of the sweep coverage, measured by $B^*(L)$, can be computed quickly.

For a sweep coverage of MWSN with m mobile nodes and n POIs, the path planning process can be executed with two steps intuitively. Firstly, one Hamilton loop with the least load among all n POIs is constructed. Secondly, m path frames are partitioned off from the least loaded Hamilton loop and one Hamilton loop with the least load among POIs in each path frame is constructed. It is obvious that a moving path designed as above for each mobile node is one feasible solution for a best balanceable path planning problem of sweep coverage in MWSN. To find the approximate best solution, constraints on balance can be considered, when m path frames are partitioned off from the least costly Hamilton loop. Because the key technology for the implementation of those two phases is the construction of the least costly Hamilton loop, a TSP approximate algorithm can be used by a two phase based solution for path planning problems where the balance conditions are treated as constraints. A TSP based two phases path planning method for sweep coverage of MWSN is give in algorithm 1.

Algorithm 1: TSP based two phases Path planning Method for sweep coverage of MWSN
Input: POI set $\mathbf{P} = \{P_i \in \Omega, i = 1, 2, \ldots, n\}$, distance matrix $\mathbf{D} = (d_{ij})_{n \times n}$ where d_{ij} is the distance from P_i to P_j, m: the number, DS: partition tactics
Output: **path** which is the moving path set for each mobile node in a sweep coverage of MWSN.

1) Using an approximate algorithm for TSP, construct T as the shortest traversal sequence for all POIs in P where D is the cost matrix.
2) According to DS, partition T into m frame as PS_1, $PS_2 \ldots PS_m$
3) For each PS_i
4) Let SD be the sub matrix of D which is corresponding to POIs in PS_i. Using a TSP approximate algorithm, construct $path_i$ as the least cost traversal sequence for all POIs in PS_i where SD is the cost matrix.
5) path $= \bigcup_{i=1}^{m} path_i$

In algorithm 1, DS is function for partition tactics. Different partition tactics DS is corresponding to different requirements on path planning for sweep coverage. If the balance of a mobile's load is a requirement for path planning in a sweep coverage, the DS can be implemented according to a fast balance factor defined by equation (6). When the DS is constructed according to equation (6), mean (***L***) should be the expectation of the mobile node's movement along each path frame. Because max(***L***)-min(***L***) is not less than 0, max(***L***)-min(***L***) should be as little as possible if the mobile node's load is more balanceable. Thus if the load of a mobile node is merely that of the node caused by its mechanical motion, the partition tactics for balance requirement is divided in the first Hamilton loop into m frames averagely.

4 EXPERIMENTAL RESULTS

To test the performance of algorithm 1, the moving path among some POIs in a 10mx10m area for each mobile nodes in sweep coverage of MWSN is planned by algorithm 1 and MTSP based path planning algorithm for sweep coverage of MWSN (algorithm 2) respectively. Because an approximate algorithm of PSO is used in the implementation of algorithm 2 [2, 7], the same approximate algorithm of PSO is used in the implementation of algorithm1 in our experiment.

In our experiment, 30/50/100/1000 POIs are generated randomly in a 10×10 area respectively. The speed of each mobile node is 0.5/s and the maximum visiting deadline is 60. In the implementation of the PSO algorithm, the number of particles are 100 and $p_i = p_g = 0.85$. The fast balance factor, defined at equation (6), is used to evaluate the balance status of sweep coverage.

Firstly, a task which is a moving path for 5 mobile nodes among 100 POIs and is planned, by algorithm 1 and algorithm 2 respectively, is executed 100 times. Figure 1 shows the distribution of the fast balance factor of path planning by algorithm 1 and algorithm 2. By comparison the modalities of part (a) and part (b) in Figure 1, algorithm 1 is better than algorithm 2 if a fast balance factor is used as the criterion.

According to the data in Figure 1, as to path planning by algorithm 2, the minimum, maximum, mean and variance of the moving path's length are 180.8903 m,

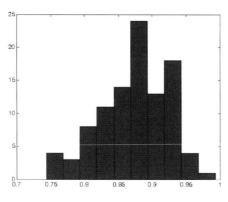

(a) algorithm 2

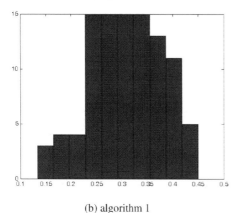

(b) algorithm 1

Figure 1. Distribution of the fast balance factor.

371.9354 m, 309.5582 m and 30.2053 m; the minimum, maximum, mean and variance of fast balance factors are 0.7429, 0.9929, 0.8734 and 0.0506; the minimum, maximum, mean and variance of time consuming are 9.079 s, 20.516 s, 19.4131 s, and 0.2833 s. Meanwhile, as to path planning by algorithm 1, the minimum, maximum, mean and variance of the moving path's length are 85.7516 m, 106.7741 m, 96.2148 m and 4.2348 m; the minimum, maximum, mean and variance of fast balance factor are 0.1331, 0.4501, 0.3105, 0.07086; the minimum, maximum, mean and variance of time consuming are 7.407 s, 12.421 s, 7.9192 s and 0.7892 s. All these data as before imply that the performance of algorithm 1 is superior to the performance of algorithm 2.

5 CONCLUSION AND FUTURE WORK

The loads of mobile nodes in sweep coverage in MWSN can cause a heavy effect on the life of mobile nodes, the cost of the MWSN, and the complexity of sweep coverages of MWSN based systems. In this paper, a TSP based two phases path planning method is presented for sweep coverage of MWSN and the method is implemented based on a PSO algorithm in our experiment. Experimental results show that the balance of path planning for a sweep coverage of MWSN is better and the summation of each moving path's length is short, which means that our proposed method is beneficial for the practical use of mobile wireless sensor networks.

With technologies for sweep coverage of MWSN, POIs, which are sparsely placed over a wide area, can be monitored with few mobile nodes. The application of our proposed method for resource monitoring in large scale spaces is one of our ongoing work.

ACKNOWLEDGEMENTS

This work is partially supported by the National Natural Science Foundation of China (No.61340030, 61300060), Special Foundation for Young Scientists of Anhui Province (2013SQRL043ZD) and Science and technology project of Ministry Housing and Urban-Rural Development Department in (2014-K8-061).

REFERENCES

[1] Zhenya Zhang, Jia Liu, Hongmei Cheng, Qiansheng Fang, Research on the Minimum Number of Mobile Node for Sweep Coverage in Wireless Sensor Network based on Genetic Algorithm [J], Journal of Chinese Computer Systems, 2013.10, v34n10: 2388–2392.

[2] Zhenya Zhang, Yan Chen, MTSP based solution for minimum mobile node number problem in sweep converge of wireless sensor network, International Conference on Computer Science and Network Technology, 2011, v1: 1827–1830.

[3] Jin Woo Moon, Seung-Hoon Han, Thermostat strategies impact on energy consumption in residential buildings [J], Energy and Buildings, 2011, 43 (2–3): 338–346.

[4] Zhenya Zhang, Hongmei Cheng, Shuguang Zhang, Research on the performance of feed forward neural network based temperature field identification model in intelligent building [C], proceedings of 3rd International Conference on Computer Research and Development, 2011, v4: 219–223.

[5] Zhenya Zhang, Hongmei Cheng, Shuguang Zhang, An Optimization Model for the Identification of Temperature in Intelligent Building [J], Journal of Information Technology Research, 2011.4, 4(2): 61–69.

[6] Huang Li, Zhu Ying-xin, Ou-yang Qin, Cao Bin. Field survey of indoor thermal comfort in rural housing of northern China in heating season [J]. Journal of Southeast University (English Edition). 2010, 26(2): 169–17.

[7] Yan Cheng, Research on methods for getting the minimum number of mobile sensors in WSN sweep coverage, Anhui University of Architecture, 2012.

[8] Alok Singh, Anurag Singh Baghel, A new grouping genetic algorithm approach to the multiple traveling salesperson problem [J], Soft Computing, 2009, v13n1: 95–101.

[9] Mo Li, Cheng, Weifang Cheng, Kebin Liu, Yunhao Liu, Sweep coverage with mobile sensors [J], IEEE Transactions on Mobile Computing, v2011.11, v10n11: 1534–1545.

A monitor method of a distribution transformer's harmonic wave compatible to its loss

D. Yu, Y. Zhao, Y. Zhang & Y. Tan
Suining City, Sichuan Province, China

J.M. Zhang & Z.L. Zhang
Shapingba District, Chongqing City, China

ABSTRACT: Nowadays, the phenomenon of harmonic wave pollution is causing various power quality problems and the most significant one is power loss and heat on the distribution transformer. This will lead to plenty of power loss and harm the safe operation of the distribution transformer. As a result, it is necessary to calculate and monitor the distribution transformer's additional loss. Aiming at this, this paper proposes a monitor method of a distribution transformer's harmonic wave compatibility to a calculation of harmonic power loss. The method is able to do real-time calculation and monitor harmonic order, harmonic components, and the additional loss in a distribution transformer, so that we can ensure the safety of a distribution transformer when there is heavy harmonic wave pollution. After researching the method, a laboratory experiment was conducted and the results of the analysis prove the feasibility and correctness of the method.

Keywords: harmonic wave; real-time monitor; distribution transformer; additional loss; experimental analysis

1 INTRODUCTION

Recently, it has been common sense to realize the intelligent management of power grids with modern information and control technology. However, since the difference among countries all over the world on energy distribution, user distribution, and the situation of grids, domestic researches mainly focus on satisfaction of increasing power demand and seem relatively lagging in the development and research on distribution network areas.

Along with the increase of the grid, there is a significant rising trend in the number of transformers in operation. At the same time, the harmonic wave problem has been one of the biggest dangers to distribution transformers and consumers' safe operation and will cause a large power loss to distribution transformers[1]. According to relative materials, power loss caused by harmonic waves to distribution transformer can be as high as 8% of total electrical generation. In additional, some reports provided by Tokyo Electric Power Company shows that the power loss in a transformer will be raised by 10% when the harmonic ratio of a fifth harmonic wave is 10%[2]. Therefore, it is an important issue to monitor and analyse the power loss of distribution transformers. On the other hand, as a characteristic parameter, harmonic wave additional loss is of great reference value to the online monitoring of distribution transformers[3].

2 MECHANISM ANALYSIS OF DISTRIBUTION TRANSFORMER'S HARMONIC WAVE LOSS

2.1 *Components of distribution transformer's power loss*

According to the actual results of research on the component of harmonic wave loss, the power loss from distribution transformers mainly contains iron loss, copper loss, electrolyte loss, and stray losses. Since electrolyte loss and stray losses are usually small and negligible, the distribution transformer's harmonic wave loss can be considered as a combined effect of winding loss (copper loss) and eddy current loss (iron loss).

Iron loss is defined as a power loss caused by an exciting current resulting in periodical changes of flux in the core at the rated voltage, including basic iron loss and additional iron loss. Iron loss can be calculated as the formula below[4]:

$$P_{Fe} = P_b + P_h + P_s = K_h f B_m^n V + K_e f^2 B_m^2 V + P_s \quad (1)$$

In the formula, P_{Fe} stands for the iron loss, K_h and K_e are constants, f stands for fundamental frequency, B_m stands for the maximum magnet flux density of the core, n stands for hysteresis factor, V stands for the volume of the core.

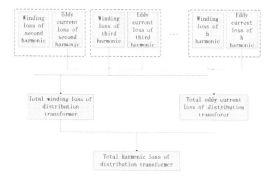

Figure 1. Flow diagram of calculating method of harmonic loss.

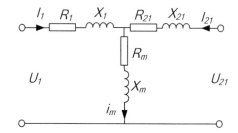

Figure 2. The equivalent T circuit of transformer.

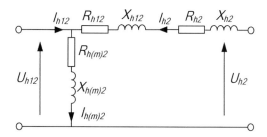

Figure 3. Simplified h sequence harmonic equivalent circuit model of transformer.

Copper loss is defined as power loss on the winding of the distribution transformer, including basic copper loss P_r, and additional copper loss P_s. Copper loss can be calculated as below:

$$P_{cu} = P_r + P_s = I_1^2 r_1 + I_2^2 r_2 + P_s \qquad (2)$$

In the formula, I_1 stands for the current of the distribution transformer's primary side, I_2 stands for the current of distribution transformer's secondary side, r_1 stands for the resistance of distribution transformer's primary side, and r_2 stands for the resistance of the distribution transformer's secondary side.

Harmonic waves caused by nonlinear load will increase the winding loss (copper loss) and eddy current loss (iron loss) in the distribution transformer[5].

3 CALCULATING METHOD OF HARMONIC WAVE POWER LOSS

On the basis of content and formation mechanism, with superposition theorem we can treat the total harmonic wave loss as a summation of winding loss and eddy current losses, caused by each harmonic waves, which are considered to be independent excitation sources[6]. Thus we can find the way to calculate harmonic wave power loss as shown in Figure 1.

From the flow chart we can find that the calculating method of harmonic loss firstly decomposes the harmonic wave loss into losses of every harmonic wave, and then further decomposes them into winding loss and eddy current loss, and finally a parameter calculation solves the size of each component and forms a total harmonic loss with a superposition principle.

3.1 Model building and parameter solution of distribution transformer

It is necessary to complete model building and parameter solutions if the harmonic wave loss of a distribution transformer is to be calculated, so we can build the equivalent T circuit of a distribution transformer as shown in Figure 2.

In the model above, it is needed to collect a voltage and current signal in the primary side to calculate the harmonic additional loss of the distribution transformer. It is of great difficulty to collect a voltage as high as 10 kV and a current at this situation from an actual distribution transformer.

Taken this into consideration, this paper builds a simplified harmonic loss model with Γ equivalent circuit, does reduction on resistance, and reactance from primary side to secondary side, as shown in Figure 3. Measuring amplitude and phase of current I_{h2} and I_{h12}, solves $I_{hm(2)}$ with vector relations between them. Finally we find the additional loss caused by harmonic waves on a distribution transformer according to the relationship I_{h2}, $I_{hm(2)}$, and their respective resistance. However, it is a must to measure harmonic current in the primary side I_{h1} and do a reduction to the secondary side, and the process needs to measure not only the current signal from the high voltage side but also the phases of the high voltage side's current. If calculating harmonic loss by this method, we will meet a problem of huge calculation and complex hardware.

In Figure 3, I_{h2} refers to current of h harmonic wave on the secondary side of the distribution transformer, I_{h12} refers to a current of h harmonic wave on the secondary side of the transformer after reduction from the primary side, R_{h12} refers to winding resistance of secondary side of the transformer after reduction from the primary side, X_{h2} refers to winding reactance of transformer's secondary side under h harmonic wave, R_{h2} refers to winding resistance of transformer's secondary side under h harmonic wave, X_{h12} refers to winding reactance of secondary side of the transformer after reduction from the primary side, $R_{h(m)2}$ refers to excitation resistance of transformer's secondary side of distribution transformer under h harmonic wave, $X_{h(m)2}$ refers to excitation reactance of transformer's secondary side under h harmonic wave, U_{h12} refers to voltage of h harmonic wave of secondary side of the transformer after reduction from the primary side, I_{h12} refers to current of h harmonic wave of secondary side of the transformer after reduction from the primary side, and U_{h2} refers to a voltage of an h harmonic wave from the transformer's secondary side.

Take a basic parameter of the distribution transformer as a benchmark; thus we can get various parameters in Figure 3 as below:

$$R_{h12} = \sqrt{h}R_1/k^2 \quad R_{h2} = \sqrt{h}R_2 \quad R_{h(m)12} = \frac{\sqrt{h}R_m}{k^2} \quad (3)$$
$$X_{h12} = \frac{hX_1}{k^2} \quad X_{h2} = hX_2 \quad X_{h(m)12} = \frac{hX_m}{k^2}$$

Among them, k refers to the transformation ratio of the distribution transformer.

3.2 Calculation of distribution transformer's winding loss

In the simplified h sequence harmonic equivalent circuit model of the transformer, we can conveniently measure a secondary harmonic current I_{h2}, and then calculate winding loss of the transformer under h sequence harmonic wave as below:

$$P_{w(h)} = I_{h2}^2(r_{h2} + r_{h12}) \quad (4)$$

In the formula, I_{h2} refers to h harmonic wave current of transformer's secondary side.

3.3 Calculation of distribution transformer's eddy current loss

Compensation calculation is necessary to solve eddy current loss.

The formula to solve per unit power loss of the transformer with load is shown as[7]:

$$P_{LL(pu)} = \sum_{h=1}^{h=\max} I_h(pu)^2 + P_{EC-R(pu)} \sum_{h=1}^{h=\max} I_h(pu)^2 \times h^2 \quad (5)$$

In the formula above, $I_{h(pu)}$ refers to per unit h harmonic wave current of the transformer's secondary side, $P_{EC-R(pu)}$ refers to an eddy current loss factor in the transformer, h refers to harmonic order.

As we can find from formulas in this paper, $P_{LL(pu)}$ is made up with two components: $\sum_{h=1}^{h=\max} I_h(pu)^2$ refers to per unit winding loss of the transformer under harmonic wave, while $\sum_{h=1}^{h=\max} I_h(pu)^2$ refers to per unit eddy current loss of transformer under harmonic wave. It is obviously that there are positive relationships between per unit winding loss and per unit harmonic current, per unit eddy current loss and $P_{EC-R(pu)}$, per unit harmonic current's square, harmonic order's square.

$P_{LL(pu)}$ is a superposition value of harmonic waves of different orders, so we can approximate the ratio winding loss to eddy current loss under h harmonic wave to be:

$$\frac{I_h(pu)^2}{P_{EC-R(pu)} \times I_h(pu)^2 \times h^2} = \frac{1}{P_{EC-R(pu)} \times h^2} \quad (6)$$

Thus we can solve an approximate eddy current loss under h harmonic wave $P_{EC(h)}$ as below:

$$P_{EC(h)} = I_{h2}^2 \times (r_{h2} + r_{h12}) \times P_{EC-R(pu)} \times h^2 \quad (7)$$

3.4 Calculation of distribution transformer's total harmonic additional loss

Treat harmonic waves on distribution transformer's secondary side as an independent power sources, and we can find the calculation formula of the distribution transformer's total harmonic additional loss:

$$P_{总} = \sum_{h=2}^{\infty} (I_{ah2}^2 + I_{bh2}^2 + I_{ch2}^2) \times (r_{h2} + r_{h12}) \times (P_{EC-R(pu)} \times h^2 + 1) \quad (8)$$

In the formula, I_{ah2} refers to effective value of h harmonic wave current on phase a of transformer's secondary side, I_{bh2} refers to effective value of h harmonic wave current on phase b of transformer's secondary side, I_{ch2} refers to effective value of h harmonic wave current on phase c of transformer's secondary side.

4 LABORATORY EXPERIMENT AND ANALYSIS

4.1 Running and commissioning of system

In order to verify the feasibility and availability of online monitor methods of the distribution transformer's harmonic wave's additional loss, we set up a simulated platform. The experimental transformer's rated capacity is 5 kVA, short-circuit loss is 160 W, no-load loss is 75 W, and rated frequency is 50 Hz. By means of three single-phase voltage regulators adjusting voltage of the transformer's secondary side, so as to simulate the different situations of transformer's online operation, connection of experiment platform as shown in Figure 4.

Integrated debugging is conducted as a monitoring system based on the simulation experiment platform, then, the changes of the transformer's harmonic wave additional loss under different load conditions are observed. Achieve a variation of harmonic additional loss of the transformer by changing the load type and adjusting the voltage regulator. This shows that the variation in the total harmonic wave additional loss of the transformer with linear load trends is quite gentle, around 20 W, as shown in Figure 5(a). While the nonlinear load is connected, the total harmonic wave additional loss increases significantly, of which the maximum value can reach 47.2 W, as shown in Figure 5(b).

A power flow diagram of the distribution network is shown in Figure 6, the total loss P_3, the input power P_1 and the output power P_2 of the transformer meet the requirements of formula (9), and P_3 comprises the fundamental wave loss P_{21} and each harmonic wave additional loss of the distribution transformer. P_{21} can be calculated by formula (10).

$$P_1 - P_2 = P_3 = P_{21} + Harmonic\ wave\ total\ additional\ loss \quad (9)$$

$$P_{21} = P_{Fe} + \frac{I_{21}^2}{I_{2N}^2} P_{Cu} \quad (10)$$

Among them, P_{Fe} is iron loss of the distribution transformer; P_{Cu} is copper loss of distribution transformer; I_{21} is fundamental wave current of secondary side of the distribution transformer; and I_{2N} is rated current for distribution transformer's secondary side;

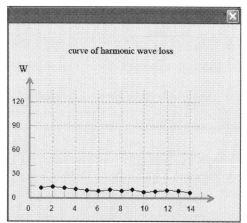

Figure 4. Simulation experimental platform.

When the transformer is under the same condition, on the one hand, by means of the instrument measuring P_1, P_2 and I_{21}, we can calculate the total harmonic wave additional loss of the transformer, measured by instruments with formula (9) and (10), as a standard to compare the results of measuring and monitoring the system.

4.2 Analysis of experimental measurement data

Five sets of data are measured under different load conditions, and measurement results of Fluck434 are respectively compared with those of this monitoring system.

Comparing measuring data of a monitoring system with those of Fluck434 instrument for analysing power quality can find that the differences between the two is not significant. As shown in Figure 7, the measuring error rate of the monitoring system is within 5.5%. It shows that experimental transformer harmonic additional loss measured by this system is credible. With the measured data, an analysis of the proportional relations between the experimental transformer's harmonic additional loss and the fundamental wave loss, is shown in Figure 8, and it can be show that the transformer's additional loss, caused by a nonlinear load, can be up to 63.2% of the transformer loss, which

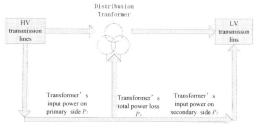

Figure 6. A power flow diagram of distribution transformers.

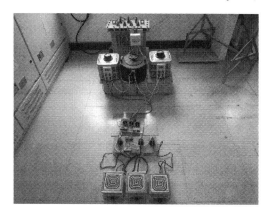

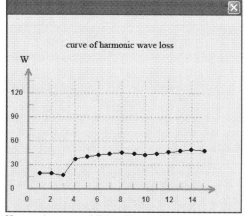

(a) Harmonic wave additional loss with linear load
(b) Harmonic wave additional loss with non-linear load

Figure 5. Total additional harmonic loss of the transformers under different load curves.

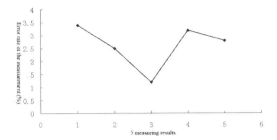

Figure 7. Error rate curve.

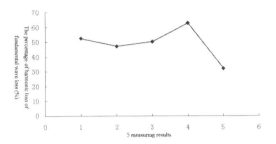

Figure 8. The ratio between additional harmonic additional loss and fundamental loss.

means transformer harmonic additional loss should not be ignored.

5 CONCLUSION

This paper is based on a method of calculating a transformer's harmonic additional loss, and takes the skin effect theories and circuit superposition principles into consideration, and builds a simplified model of the distribution transformer's harmonic additional loss, and finds the calculation formula of distribution transformer's harmonic additional loss. After that, it sets up an experimental platform in the laboratory environment, making experiments on the calculation and monitoring of the distribution transformer's harmonic additional loss and analysing the results of the experiments. After those researches, we get conclusions as below:

1) Harmonic wave pollution will affect a distribution transformer and cause additional loss on it;
2) It is available to build simplified equivalent circuit model based on a distribution transformer and it solves each characteristic parameter of the model under harmonic waves of different orders;
3) This paper proposes a calculation method of distributing a transformer's harmonic additional loss, treats harmonic waves as a composition of a series of independent excitation sources, and calculates their additional loss separately, thus finally solving the total additional loss caused by harmonic waves.

REFERENCES

[1] Hu J.S. & Zhao Y.J. 2007. Implementation guidelines of distribution transformer's efficiency standard. Beijing: China Standards Press.
[2] Zhang Z.P. & Li F.C. 2001. Manual of Urban Grid Harmonic. Beijing: China Electric Power Press.
[3] Tan J.Y. 2008. The influence of harmonic of transformer and its suppression measures. Electric Age. Vol. 9, p. 100–102.
[4] Seo J.H. et al. 2009. Harmonic Iron Loss Analysis of Electrical Machines for High-Speed Operation Considering Driving Condition, IEEE Transactions on Magnetics, Vol. 45: 4656–4658.
[5] Pierce L.W. 1996. Transformer Design and Application Considerations for Nonsinusoidal Load Currents. IEEE Trans on Industry Application, Vol. 32: 35–47.
[6] Liu C.J. & Yang R.G. 2008. Calculation and analysis of transformer's harmonic loss, Power System Protection and Control. Vol. 36: 33–36, 42.
[7] Wang K & Shang Y. 2009. Transformer overheating analysis and derating calculation under harmonic. Power System Protection and Control. Vol. 37: 50–53.

Identification of a gas-solid two-phase flow regime based on an electrostatic sensor and Hilbert–Huang Transform (HHT)

Jinxiu Hu, Xianghua Yao & Tao Yan
School of Electronics and Information Engineering, Xian, Shanxi, China

ABSTRACT: In the gas-solid two-phase system, particles can accumulate a large number of electrostatic charges because of the collision, friction, and separation between particles or between particles and wall. When particles are moving through an electrostatic sensor, the sensor induces a fluctuant signal because of the constant fluctuations in the quasi-electrostatic filed charges around the sensor. Through detecting and processing the induced fluctuant charged signals, a measuring system can obtain two-phase flow parameters, such as concentration, flow regimes, etc. An electrostatic sensor is used for detecting the moving electric charge carried by solid particles in pneumatically conveying pipelines. And a microcomputer and NI PCI 6024-based measuring system has been built for data acquisition. Then the Hilbert–Huang Transform (HHT) is used to process and analyse the electrostatic fluctuant signals with different flow regimes. The result shows that Hilbert marginal spectrums of the electrostatic signal can be used for identifying the gas-solid two-phase flow regimes.

1 INTRODUCTION

The transportation of a wide variety of particulate materials by pneumatic means is becoming increasingly widespread in order to achieve greater productivity, improve product quality, and increase process efficiency [1]–[2]. The development of online instrumentation for particle flow measurement has been of great interest to many academic institutions and industrial organizations [3]–[5]. In order to achieve the pneumatic conveying systems' full potential, it is necessary to measure the parameters of the gas-solid two-phase flow without affecting the flow in any way. Varieties of non-intrusive measurement method based on the phenomenon of electrostatic induction have been specially developed to measure the electric charge carried on solid particles in pipes of pneumatic transport, as well as to indirectly evaluate, determine, or estimate the following mechanical parameters of the two-phase flows: mass flow rate, concentration, volume loading, and velocity [7]–[12]. The correct identification of two-phase flow regimes is the basis for the accurate measurement of other flow parameters in a two-phase flow measurement. However, there is no valid method to identify flow regimes according to the nonlinearity and the nonstationarity of the electrostatic fluctuated signal.

The Hilbert–Huang Transform (HHT) is a new time–frequency analysis method [13]. The main difference between the HHT and all other methods is that the elementary wavelet function is derived from the signal itself and is adaptive. Recently, there are some researches using HHT to analyse the signals in two-phase flow. Ding et al. have used HHT to analyse the differential pressure fluctuation signals of a gas–liquid two-phase flow [14]. They have found that the extracted energy characteristics give a good indication of the dynamic state of the gas–liquid two-phase flow and thus can be used for flow regime recognition. Chuanlong Xu et al. have used HHT to analyse the electrostatic fluctuation signals in a dense-phase pneumatic conveyance of pulverized coal at high pressure [15]. They have investigated the relations between the energy distribution transmissions in Intrinsic Mode Functions (IMFs) with different orders and the flow characterizations of a dense phase gas–solid flow.

Furthermore, some researchers have also used HHT to study flow regimes of two-phase flow. Lu Peng et al. have used HHT to study flow regimes of high-pressure and dense-phase pneumatic conveyance [16]. They have found that the EMD characteristics of electrostatic fluctuation signal are correlated strongly with the flow regimes. SUN Bin et al. have used HHT to identify a flow regime of gas-liquid two-phase flow [17].

However, they all have not used the Hilbert marginal spectrums of electrostatic fluctuation signal to identify flow regimes of a gas-solid two-phase flow. In this paper, an electrostatic sensor is used in the measurements of moving electric charge carried by solid particles in pipelines of pneumatic transport. Then Hilbert marginal spectrum is adopted by processing the sensor signals with the HHT, and applied to identify flow regimes. This effort aims to provide a novel way to identify the gas-solid two-phase flow regimes.

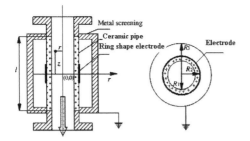

Figure 1. The construction of electrostatic sensor.

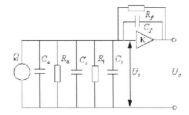

Figure 2. Equivalent circuit of electrostatic sensor.

2 MEASUREMENT SYSTEM

The measurement system consists of an electrostatic sensor, a preamplifier, and a PC and a NI PCI6024-based data acquisition system. The output signal of the electrostatic sensor is amplified by the preamplifier, and then is sampled by NI PCI6024, saved, and analysed by HHT. The sampling rate is 10000 Hz and the number of sampling points is 10000.

2.1 Electrostatic sensor

During the pneumatic transport process of powder particles, the particles carry charges due to frictional contact charging between particles and the pipe wall and between particles and the airflow. An electrostatic sensor can be used to detect the charges carried by the moving particles. The electrostatic sensor is enclosed in the measuring head which is illustrated in Fig. 1 [18], [19].

In this paper, the inner diameter of the test ceramic pipe R1 is 100 mm, and the outer diameter R2 is 105 mm. The ring shaped sensor is made of copper sheeting, and the axial length W_e and the thickness of the sensor are 10 mm and 0.2 mm, respectively. The inner screen is made of copper sheeting. Its inner diameter R_3 is 120 mm, the axial length l is 50 mm, and the thickness is 2 mm. A supporting steel tube outside the inner screen acts as the outer screen. A follower is connected between the inner and outer screens, which forms an equal potential between the screens. This specific design is used to prevent leakage of a weak signal and to eliminate the effect of electromagnetic interference [20].

2.2 Measurement circuit

The electrostatic sensor can be simplified as a charge source from its principle. So this paper uses a charge amplifier circuit as a preamplifier to measure the electrostatic signal. An equivalent circuit to the electrostatic sensor and charge amplifier is shown in Fig. 2 [21].

The relationship between the output voltage U_o and the inducted charge Q can be expressed as

$$U_o = -\frac{j\omega K Q}{[(1/R_a)+(1/R_i)+(1+K)/R_f]+j\omega(C_a+C_e+C_i+C_f(1+K))} \quad (1)$$

In the above formula, U_o is the output voltage of the preamplifier, C_e is the cable capacitance, C_a and R_a are the equivalent capacitance and insulation resistance of the sensor, respectively, and C_i and R_i are the equivalent input capacitance and input resistance of the preamplifier, respectively.

When $K \gg 1$, $KC_f \gg C_a + C_e + C_i + C_f$. Equation (1) can be simplified as

$$U_o \approx -\frac{Q}{C_f} \quad (2)$$

So it can be seen that the output voltage U_o is approximately linear to the induced charge Q.

2.3 Signal processing

Huang et al. [13] introduced a general signal-analysis technique to efficiently extract the information in both time and frequency domains directly from the data, called Hilbert-Huang Transform (HHT). It is a two-step algorithm, combining Empirical Mode Decomposition (EMD) and Hilbert spectral analysis, to accommodate the nonlinear and non-stationary processes. This method is not based on a priori selection of kernel functions, but instead it decomposes the signal into intrinsic oscillation modes (represented by Intrinsic Mode Functions (IMF)) derived from the succession of extrema. It is adaptive, efficient, and without any prior assumptions [22].

Once the time series $x(t)$ has been decomposed into a finite number, n, of IMFs, $c_i(t), i = 1, 2, \ldots, n$ associated with various time scales and the residual $r(t)$, the time series can be reconstructed by the superposition of the IMFs and the residual [23], [24]:

$$x(t) = \sum_{i=1}^{n} c_i(t) + r(t) \quad (3)$$

Then the Hilbert transform is applied to each IMF to obtain a complex representation of IMF, $d_i(t)$

$$d_i(t) = \frac{1}{\pi} \int_{-\infty}^{\infty} \frac{c_i(t)}{t-\tau} d\tau \quad (4)$$

According to that, $c_i(t)$ and $d_i(t)$ form a complex conjugate pair that define an analytic signal, $z_i(t)$ as

$$z_i(t) = c_i(t) + j d_i(t) = a_i(t) e^{j\theta_i(t)} \quad (5)$$

where $a_i(t)$ is the amplitude, and $y_i(t)$ is the phase angle. They can be calculated by

$$a_i(t) = \sqrt{(c_i(t))^2 + (d_i(t))^2} \quad (6)$$

$$\theta_i(t) = \arctan\left(\frac{d_i(t)}{c_i(t)}\right) \quad (7)$$

Finally, the instantaneous frequency of each IMF, is defined as

$$\omega_i(t) = \frac{d\theta_i(t)}{dt} \quad (8)$$

At a given time t, the instantaneous frequency and the amplitude $a_i(t)$ are calculated simultaneously so that these values are assigned to the Hilbert spectrum, $H(\omega, t)$:

$$H(\omega,t) = \begin{cases} \sum \operatorname{Re} a_i(t) e^{j\int \omega_i(t)dt} & \omega_i(t) = \omega \\ 0 & \text{others} \end{cases} \quad (9)$$

The Hilbert marginal spectrum is a measure of total energy contributed from each frequency over the entire data span in a probabilistic sense. It provides a quantitative way to describe the time-frequency-energy representation by integrating the Hilbert spectrum over the entire time span [22],

$$h(\omega) = \int_0^T H(\omega, t) dt \quad (10)$$

where T is the total data length.

3 THE EXPERIMENTAL SYSTEM

The experimental system is shown in Fig. 3. It has been designed and developed to facilitate the electrostatic measurement of a charged solid phase in a pneumatic conveying line [20]. It comprises the following five major subsystems.

3.1 Carrier air supply system

In order to facilitate the experiment and to prevent the electronic circuit from being polluted in the field calibration, an induced draft, supplied to the system through an air supply line, is used as a carrier. Air supply rate is controlled finely with a control valve (2). The rate and pressure of the supply air is measured by means of a Pitot tube before entering the T-mixer (T).

3.2 The feeding system

The feeding system consists of a uniform speed motor, a continuously adjustable gearbox, and a fine linear screw feeder. Feeding ranges from 30 g/min to 3 kg/min. The feeding system is more stable and continuous than that adjusted by the SCR (Silicon Controlled Rectifier). The amount of discharging is adjusted and calibrated by the feeder and a mesh dust collector, respectively.

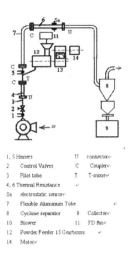

1, 5 Heaters   U connector
2 Control Valves   C Coupler
3 Pitot tube   T T-mixer
4, 6 Thermal Resistance
Sa electrostatic sensor
7 Flexible Aluminum Tube
8 Cyclone separator   9 Collector
10 Blower   11 FD fan
12 Powder Feeder 13 Gearboxes
14 Motor

Figure 3. A schematic diagram of pneumatic conveying system.

3.3 The test section

The test section is a horizontal ceramic pipe of length 1200 mm and inside diameter 100 mm with an upward flow direction. As shown in Fig. 3, the test section is placed in the downstream of the T mixer, while the reference section is in the upstream. In the test section, an electrostatic sensor is mounted to the distance of 800 mm from the pipe's lower terminal. A dummy sensor is fixed on the reference section. The electrostatic sensors are connected to the measuring circuit by a shielded twisted-pair cable.

3.4 The flow generator

Because of the objective of this experiment, a throttle is fixed in the pipeline as the flow regime generator. A transition flow regime is generated at the downstream of the throttle at a short distance. In this experiment, the throttling has generated three flow regimes, which are an annular flow, a roping flow, and a stratified flow. A stratified flow can be generated behind a gate valve, a roping flow can be generated behind a venturi tube or an orifice plate, and an annular flow can be generated behind an impact plate. The upstream edge of the throttle is especially designed to be streamlined to avoid powder accumulation, and the equivalent pore diameter of the throttle is about one tenth of its length.

3.5 The start-up and operation

Conveying air is generated by turning on the FD fan before the air flow rate can be supervised by Pitot tube (3) and controlled to desired level by regulating the valve (2) at the same time. The feeder is turned on before the feeding rate can be controlled to desired level by regulating the gearbox. Then, the solids particles flow into the T-mixer and establish gas-solid two-phase flow.

Table 1. Experimental condition in three different cases.

	Flow regime	Solid medium	velocity (m/s)
Case 1	Annular flow	Sand	15.09
Case 2	Roping flow	Sand	14.98
Case 3	Stratified flow	Sand	15.16

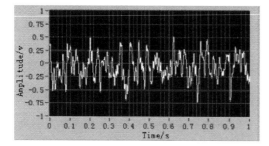

Figure 4. Electrostatic signal of annular flow.

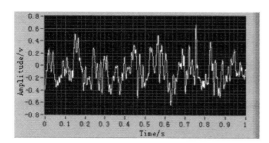

Figure 5. Electrostatic signal of roping flow.

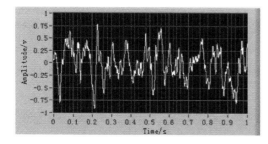

Figure 6. Electrostatic signal of stratified flow.

4 EXPERIMENT RESULTS AND DISCUSSION

Three different cases are shown in Table 1. In each case, an electrostatic fluctuation signal was measured many times.

The electrostatic fluctuation signals recorded by labview 8.0 in the pneumatic conveying system, Fig. 3, in the three cases are presented in Fig. 4, Fig. 5 and Fig. 6 for a short interval of time.

It is very difficult to identify their differences. So the Hilbert–Huang transform is applied to process the electrostatic signals in order to identify a flow regime.

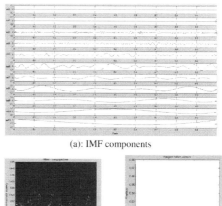

(a): IMF components

 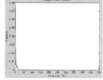

(b): Hilbert spectrum   (c): Marginal Hilbert spectrum

Figure 7. The electrostatic fluctuation signal in Case 1.

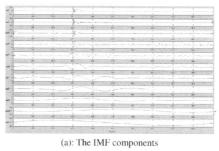

(a): The IMF components

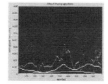

(b): Hilbert spectrum   (c): Marginal Hilbert spectrum

Figure 8. The electrostatic fluctuation signal in Case 2.

The IMFs, Hilbert spectrums and the Hilbert marginal spectrums of the electrostatic fluctuation signals of the three flow regimes are shown in Fig. 7, Fig. 8, and Fig. 9, respectively.

As stated in the theoretical section, the Hilbert–Huang transform separates a signal into IMFs with different scales. And the IMFs are plotted in the rows, shifting from the fine scale IMF 1 in the first row to the coarse scale (IMF 10/IMF 11) in the bottom two rows.

It can be seen that the fine scale features (corresponding to high frequency oscillations) of the electrostatic fluctuations are mainly captured by the detail components IMF1 and IMF2, while the coarse scale components (IMF10–IMF11) correspond to lower frequency oscillations. The result shows that the electrostatic fluctuation signals are concentrated in a low frequency. From the Hilbert-Huang spectrum we can

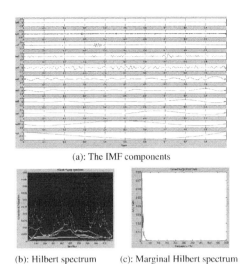

(a): The IMF components

(b): Hilbert spectrum (c): Marginal Hilbert spectrum

Figure 9. The electrostatic fluctuation signal in Case 3.

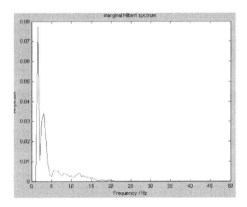

Figure 10. The partial enlargement of Hilbert marginal spectrums of Fig. 7 (Annular flow).

also see that the instantaneous frequency oscillates in the filed which is lower than 0.05 – a normalized frequency.

Furthermore, the Hilbert marginal spectrum produces an easily interpretable visual two-dimensional contour of the fluctuation, which depicts the amplitude dependent frequency of the fluctuation. From the Hilbert marginal spectrums shown in Fig. 7, 8 and 9 the frequency of the electrostatic fluctuation signals are concentrated in the field which is lower than 50 Hz. So the partial enlargements of Hilbert marginal spectrums of Fig. 7, 8 and 9 are show in Fig. 10, 11 and 12.

It can be seen more clearly that the frequencies of electrostatic fluctuation signals are lower than 10 Hz from Fig. 10, Fig. 11 and Fig. 12. From Fig. 10, we notice that the dominant frequency of Hilbert marginal spectrum of an annular flow in case 1 is 2 Hz, and its sub frequency is 3 Hz.

From Fig. 11, this paper also find that the dominant frequency of Hilbert marginal spectrum of roping flow of case 2 is 6 Hz, and its sub frequency is 8 Hz.

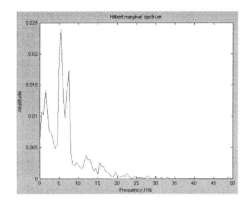

Figure 11. The partial enlargement of Hilbert marginal spectrums of Fig. 8 (Roping flow).

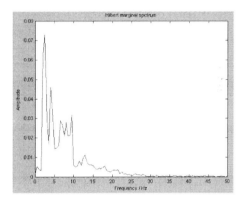

Figure 12. The partial enlargement of Hilbert marginal spectrums of Fig. 9 (Stratified flow).

It is also shown that the dominant frequency of the Hilbert marginal spectrum of stratified flow in case 3 is 3 Hz, and its sub frequency is 4 Hz from Fig. 12, which seems like Fig. 10. But it's very easy to find that there are three peaks between 5 Hz and 10 Hz in Fig. 12.

Comparing Fig. 10, Fig. 11 and Fig. 12, it can be found that the frequency of electrostatic signals in the annual flow is lower than the frequency of the roping flow, while its peak value is larger than the roping flow. The dominant frequency of the annual flow is between 0 Hz and 5 Hz while the dominant frequency of the roping flow is between 5 Hz and 10 Hz. The peak value of the annual flow is between 0.07 and 0.08 while the roping flow is between 0.02 and 0.025. Furthermore the Hilbert marginal spectrum of a stratified flow is just like the superposition of Hilbert marginal spectrums of annual flow and roping flow. The result shows that there are many obvious differences between the Hilbert marginal spectrums of annular flow, roping flow, and stratified flow and that they all have their own features. Thus, HHT can be used to identify the flow regime of a gas-solid two-phase flow.

5 CONCLUSIONS

The Hilbert–Huang transform shows how the instantaneous frequencies and amplitudes of the electrostatic fluctuation vary with time, while the Fourier transform cannot do this [25]. Thus, this method enables us to obtain the frequency content of local complex flow behaviours, which are associated with gas–solid flow. This paper used the HHT to analyse the electrostatic signal to identify gas-solid two-phase flow regimes. The result shows that the method of the HHT does well. Consequently, the Hilbert–Huang transform can be useful for analysing the electrostatic fluctuation signals to identify flow regimes of gas-solid two-phase flows in pneumatic conveying systems. In future research, the methods above will be approached and tested in pneumatically conveying pulverized coal system in a power plant.

ACKNOWLEDGEMENT

In this paper, the research was sponsored by the National Science and Technology Major Project (Project No. 2013ZX03005007).

REFERENCES

[1] Y. Yan 2001 Guide to the flow measurement of particulate solids in pipelines—Part 1: Fundamentals and principles *Powder Handl. & Process* **13** 343–52
[2] Y. Yan 2002 Guide to the flow measurement of particulate solids in pipelines—Part 2: Utilization in pneumatic conveying and emission monitoring *Powder Handl. & Process* **14** 12–21
[3] L.J. Xu, A.P. Weber, and G. Kasper 2000 Capacitance-based concentration measurement for gas-particle system with low particles loading *Flow Meas. Instrum.* **11** 185–94
[4] Y. Yan 1996 Mass flow measurement of bulk solids in pneumatic pipelines *Meas. Sci. Technol* **7** 1680–706
[5] Juliusz B. Gajewski 2006 Non-contact electrostatic flow probes for measuring the flow rate and charge in the two-phase gas–solids flow *Chemical Engineering Science* **61** 2262–70
[6] Y. Yan, L.J. Xu and Peter Lee 2006 Mass Flow Measurement of Fine Particles in a Pneumatic Suspension Using Electrostatic Sensing and Neural Network Techniques *IEEE Transactions on Instrumentation and Measurement* **55** 2330–4
[7] Juliusz B. Gajewski 2008 Electrostatic Nonintrusive Method for Measuring the Electric Charge, Mass Flow Rate, and Velocity of Particulates in the Two-Phase Gas–Solid Pipe Flows—It's only or as Many as 50 Years of Historical Evolution *IEEE Transactions on Industry Applications* **44** 1418–30
[8] Y. Yan, B. Byrne, S. Woodhead and J. Coulthard 1995 Velocity measurement of pneumatically conveyed solids using electrodynamic sensors *Meas. Sci. Technol* **6** 515–37
[9] J.B. Gajewski 1996 Monitoring electrostatic flow noise for mass flow and mean velocity measurement in pneumatic transport *J. Electrostat* **37** 261–76
[10] J.B. Gajewski, B. Glod and W. Kala 1990 Electrostatic method for measuring the two-phase pipe-flow parameters *IEEE Transactions on Industry Applications* **9** 650–5
[11] J.B. Gajewski 1994 Measuring probes, head, and system for the non-contact, electrostatic measurements of the two-phase flow parameters in pneumatic transport of solids *J. Electrostat* **32** 297–303
[12] R. Cheng 1996 A study of electrostatic pulverized fuel meters *Ph.D. Thesis* University of Teesside
[13] N.E Huang et al. 1998 The empirical mode decomposition and the Hilbert spectrum for nonlinear and non-stationary time series analysis *Proceedings of the Royal Society of London* **454** 903–95
[14] Hao Ding, Zhiyao Huang, Zhihuan Song and Yong Yan 2007 Hilbert–Huang transform based signal analysis for the characterization of gas–liquid two-phase flow *Flow Measurement and Instrumentation* **18** 37–46
[15] Chuanlong Xu et al. 2010 HHT analysis of electrostatic fluctuation signals in dense-phase pneumatic conveying of pulverized coal at high pressure *Chemical Engineering Science* **65** 1334–1344
[16] Lu Peng, Chen Xiaoping, Liang Cai, Pu Wenhao, Zhou Yun, Xu Pan and Zhao Changsui 2009 Study on Flow Regimes of High-pressure and Dense-phase Pneumatic Conveying *Journal of Physics: Conference Series* **147** 012078
[17] Sun Bin, Zhang Hongjian, Cheng Lu and Zhao Yuxiao 2006 Flow Regime Identification of Gas-liquid Two-phase Flow Based on HHT *Chinese J. Chem. Eng* **14**(1) 24–30
[18] Y. Yan, B. Byrne, S. R. Woodhead, and J. Coulthard 1995 Velocity measurement of pneumatically conveyed solids using electrodynamic sensors *Meas. Sci. Technol.* vol. 6, pp. 515–537
[19] Jan Krabicka, Y. Yan 2009 Optimised Design of Intrusive Electrostatic Sensors for the Velocity Measurement of Pneumatically Conveyed Particles *Technology Conference Singapore*, 5–7
[20] Hu H.L., Xu, T.M., Hui, S.E. 2006 A Novel Capacitive System for Concentration Measurement of Pneumatically Conveyed Pulverized Fuel at Power Station. *Flow Measurement and Instrumentation* 17(1) 87–92
[21] Chuanlong Xu et al. 2007 Sensing characteristics of electrostatic inductive sensor for flow parameters measurement of pneumatically conveyed particles *Journal of Electrostatics* **65** 582–592
[22] Rongjun Lu, Bin Zhou and Wei Gao, "Signal Analysis of Electrostatic Gas-Solid Two-Phase Flow Probe with Hilbert-Huang Transform", *The Ninth International Conference on Electronic Measurement & Instruments*. **2** 486–490
[23] G.G.S. Pegram, M.C. Peel and T.A. McMahon 2008 Empirical mode decomposition using rational splines: an application to rainfall time series *Proceeding of the Royal Society A-Mathematical Physical and Engineering Sciences* **464** 1483–501
[24] C.L. Xu et al. 2010 HHT analysis of electrostatic fluctuation signals in dense-phase pneumatic conveying of pulverized coal at high pressure *Chemical Engineering Science* **65** 1334–44
[25] Chen, C.H., Wang, C.H., Liu, J.Y., et al. 2010 Identification of earthquake signals from groundwater level records using the HHT method *Geophysical Journal International* 180 (2) 1231–1241

A fast multilevel reconstruction method of depth maps based on Block Compressive Sensing

Tao Fan & Guozhong Wang
School of Communication and Information Engineering, Shanghai University, Shanghai, China

ABSTRACT: The quality of synthesized view highly depends on the depth maps in the Free viewpoint TV (FTV), which is a popular system in 3D applications. In order to offer both good performance and low complexity, a fast multilevel reconstruction method of depth maps based on Block Compressive Sensing (BCS) is presented in this work. Different from other state-of-the-art CS algorithms, the proposed method mainly considers the special sparsity and some features of depth maps. Firstly, the wavelet domain as one of the sparsest expressions for depth is utilized to reduce the dimension. Then, an adaptive multilevel subsample strategy is proposed. Since different subbands of wavelet decomposition have different importance for depth maps, the unequal measurement is obtained with a first level strategy. For each subband, the BCS approach is used for reducing the memory space and computing complexity. For the sake of protecting the sharp edges and improving efficiency further, the second level adopts the Gini index strategy to judge adaptively the sample rate. To ensure the smoothness of the recovered depth maps, which can decrease the number of iterations and preserve the discontinuities at the edges, and which, in turn, can ensure the quality of the virtual viewpoint, the the projection landweber and an adaptive bilateral filter are incorporated at the reconstruction process. The experimental results show that the proposed method performs well, both in terms of subjective quality and as an objective measure towards the virtual views using the reconstructed depth maps. Moreover, the consequences of computation complexity also show good performances which have a high value in realistic application.

1 INTRODUCTION

With the continuous development of free viewpoint TV (FTV) system, the Multiview plus Depth (MVD) format has received increased attention in recent research [1]. Depth maps as key components of the system are not viewed by an end-user, but used as an aid for view rendering. According to the Depth-image-based Rendering (DIBR) technology, a minor error in the depth maps may cause serious distortion in the rendered map. In order to achieve a more realistic effect in application, the efficient compression and the high reconstruction quality for depth data should be considered simultaneously.

Commonly used depth maps processed frameworks include the down/up-sampling scheme, the typical image compression method, and the state-of-the-art Compressive Sensing (CS) strategy. The down/up-sampling scheme has been widely applied before [2]. However, this method inevitably introduced artefacts in view synthesis result. Another typical way to carry out these entities is to use directly standard image or video compression techniques like JPEG 2000 or High Efficiency Video Coding (HEVC) [3]. However, such schemes process the images in such a way as to maintain the visual quality of the result. Therefore, this may lead to a high amount of error in the reconstructed values of the depth map. In recent years, a new framework called Compressive Sensing (CS) has been developed for simultaneous sampling and compression. Due to the good robustness for transmitting, this scheme is well suited for the processing of the depth maps, which has been used in [4]. However, computation and storage complexity is the biggest bottleneck in application. A block compressive sensing (BCS) is proposed in [5] to address these issues. Nevertheless, the block-based image acquisition in these methods is implemented with the same measurement operator, ignoring the diversified characteristics in various blocks that would eventually influence their construction quality. Moreover, the blocking artefact is the defect in this method which may decrease the convergence speed of the algorithm.

In order to offer both good performance and low complexity, a fast multilevel reconstruction method of depth maps, based on block compressive sensing, is presented in this work. Different from other state-of-the-art CS algorithms, the proposed method mainly considers the special sparsity and some features of depth maps. Firstly, the wavelet domain as one of the sparsest expressions for depth is utilized to reduce the dimension. Then, an adaptive multilevel subsample strategy is proposed. Since different subbands of wavelet decomposition have different importance for depth maps, the unequal measurement is obtained with a first level strategy. For each subband, the BCS approach is used for reducing the memory space and computing complexity. For the sake of protecting the sharp edges and improving efficiency further, the second level adopts the Gini index strategy to judge adaptively the sample rate. To ensure the smoothness of the recovered depth maps, which can decrease the number of iterations and preserve the discontinuities at the edges, and which, in turn, can ensure the quality of the virtual viewpoint, the projection landweber and an adaptive bilateral filter are incorporated at the reconstruction process.

This paper is organized as follows. Section 2 gives a brief background of the article, including the BCS theory and the sparse analysis of a depth map. Section 3 proposes a fast multilevel scheme for depth map representation using CS. The proposed reconstruction algorithm is discussed in detail. The experimental results are presented in Section 4, and Section 5 concludes this paper.

2 BACKGROUND

2.1 Blocked-based compressed sensing

The BCS method for images or video signals supposes that an original image has N pixels with M measurements taken. The image is partitioned into $B \times B$ blocks and each block is sampled with the same operator. Let us assume that each image block x_i is taken by M_B measurements:

$$y_i = \Phi_B \cdot x_i \qquad (1)$$

where Φ_B is a $M_B \times B^2$ measurement matrix with $M_B = \left[\frac{M}{N} \cdot B^2\right]$. The equivalent measurement matrix Φ for the entire image is thus a block-wise diagonal one:

$$\Phi = \begin{bmatrix} \Phi_B & & & \\ & \Phi_B & & \\ & & \ddots & \\ & & & \Phi_B \end{bmatrix} \qquad (2)$$

It is clear that only a small measurement matrix with size $M_B \times B^2$ is needed to store in the BCS method, suggesting a huge memory saving.

2.2 Gini index

The Gini Index (GI) is considered the best criterion of signal sparsity which is defined below [6].

Given a vector $S = [S(1), \ldots, S(N)]$, with its elements re-ordered and represented by $S_{[k]}$ for $k = 1, 2, \ldots, N$, where $|S_{[1]}| \le |S_{[2]}|, \ldots, |S_{[N]}|$, then

$$GI(S) = 1 - 2\sum_{k=1}^{N} \frac{|S_{[k]}|}{\|S\|_1} \left(\frac{N-k+\frac{1}{2}}{N}\right) \qquad (3)$$

Figure 1. Comparing the GI of different transform domains for the depth maps of bookarrival.

where $\|S\|_1$ is the l_1 norm of S. We show in the Appendix that $GI(S)$ is a quasi-convex function in $|S|$.

2.3 The option of the sparse domain for depth maps

As Equation 3 described, the GI value range is from 0 to 1. That is mean 0 for the least spares and 1 for the sparest one. In terms of depth maps, many parts contain less detail than others. This means that the energy of depth maps may be concentrated in a few of coefficients if it adopt some sparse representation. Many transforms can be considered as the spare representation of the depth maps, but which one is the sparsest is our highest priority.

Figure 1 shows the comparison of GIs of different transform domains for the depth maps of BookArrival. It is obvious that the wavelet domain is the sparsest for the depth maps.

3 PROPOSED ALGORITHM

3.1 Mutilevel sampling strategy

As discussed above, it is significant to prevent the distortion of the structure and boundaries from the process of reconstructing depth maps. The most CS strategies existed force mainly on the quality of the final viewer. However, depth maps are not viewed by end-users, but are used as an aid for view rendering. Therefore, how to find the method of CS strategy to improve the quality of depth maps which suits for view rendering by exploiting the a priori information, is the main work in this paper.

In the case of depth maps, since the different levels of wavelet decomposition have different importance for the final image reconstruction quality, different blocks in each level may have different contributions to the virtual view. A multilevel sample strategy is designed to adjust the measurement so as to yield the different subrate, $B_{k,j}$, at each block in a different level. Assume that Ψ produces K levels of wavelet decomposition and each level is divided C block, where C equals $\frac{M \times N}{blocksize}$. Then we let the subrate for level K be:

$$B = \sum_{k=0}^{K} \omega_k \cdot B_k \quad \{k \in N_+, 0 \le k < K\} \qquad (4)$$

where B represent a target rate of the depth maps, and the ω is a set of weights toward each decomposition.

The second hierarchical design is focused on different blocks at each level. The subrate can be:

$$B_k = \sum_{j=0}^{C} \gamma_j \cdot B_{k,j} \quad \{j \in N_+, 0 \leq j < C\} \quad (5)$$

such that the overall subrate based on (4) and (5) becomes:

$$B = \sum_{k=0}^{K} \omega_k \cdot (\sum_{j=0}^{C} \gamma_j \cdot B_{k,j}) \quad \{j,k \in N_+ \mid 0 \leq j < C, 0 \leq k < K\} \quad (6)$$

As we all know, the energy of the wavelet coefficients is concentrated in the Discrete Wavelet Transform (DWT) baseband. The full measurement is set in this framework, i.e., $B_0 = 1$, such that the overall subrate becomes:

$$B = \frac{1}{4^K} \omega_0 \cdot (\sum_{j=0}^{C} \gamma_j \cdot B_{0,j}) + \sum_{k=1}^{K} \frac{3}{4^{K-k+1}} \omega_k \cdot (\sum_{j=0}^{C} \gamma_j \cdot B_{k,j}) \quad (7)$$

According to the formula above, the ω and γ need to set up adaptively so as to satisfy different depth maps. Firstly, we use level weights, $\omega_k = 16^{K-k+1}$ which we have found to perform well in practice. The γ stand for the block weights in each level. The Gini index of each block is used to confirm the value of it. Motivated by the property of the wavelet transform which the relevant coefficients are discontinuities the particular sampling positions are precisely those depth maps lying at the discontinuities. Based on the statistics of the Gini index, each block is analysed and then classified as FLAT, SHARP, and OTHER. Given the uniform each of decomposition, different measurement rates are assigned based on disparate characteristics of each block. Accordingly, the γ of each block is classified as:

$$\gamma_j \in \begin{cases} FLAT & if \ T_1 < G_j \leq 1 \\ SHARP & if \ 0 \leq G_j \leq T_2 \\ OTHER & if \ T_2 < G_j \leq T_1 \end{cases} \quad (8)$$

where, G_j stand for the jth block value of the Gini index. T_1 and T_2 are determined by the experiments. According to the discussion above, the block is sparse if $T_1 < G_j \leq 1$ and the flat block is considered. If $0 \leq G_j \leq T_2$, the block is sharp type. The remainder is regarded as other type. The γ_j is set as follows:

$$\gamma_j = \begin{cases} 0.5 & if \ \gamma_j \in FLAT \\ 1 & if \ \gamma_j \in SHARP \\ 0.75 & if \ \gamma_j \in OTHER \end{cases} \quad (9)$$

3.2 Fast reconstruction arithmetic

The projection-based CS-reconstruction techniques are adopted for depth reconstruction with the sampled data mentioned above. The reconstruction starts from some initial approximation $\hat{x}_{(0)}$, which is further refined in an iterative manner, as in the following:

$$\hat{x} = \hat{x}_{(i)} + \frac{\Psi \Phi^T}{\lambda}(y - \Phi \Psi^{-1} \hat{x}_{(i)}) \quad (10)$$

$$\hat{x}_{(i+1)} = \begin{cases} \hat{x}_{(i)}, & |\hat{x}_{(i)}| \geq \tau_{(i)} \\ 0, & otherwise \end{cases} \quad (11)$$

where λ a scaling is factor, and $\tau_{(i)}$ is the threshold used at ith iteration.

The BCS method may in some cases generate the blocking effects. To address this, some post processing must be designed during the process of reconstruction. Many methods have been proposed to handle this issue such as the Wiener filter and the median filter. These processes can remove the blocking artefacts and impose smoothness of sharp edges simultaneously. However, they may seriously affect the quality of the virtual view. On this basis, an Adaptive Bilateral Filter (ABF) is proposed as an image sharpening technique which can both sharpen edges and reduce blocking artefacts. In can be shown as follows:

$$s_a(p,q) = \exp(-\frac{1}{2} \cdot (x_p - x_q - \Delta_p)^2 / \sigma_p^2) \quad (12)$$

where, Δ_p and σ_p are adaptation parameters dependent on p, and they are used to control the centre and standard deviation of the Gaussian kernel that implements s_a. As opposed to the similarity filter kernel s, which is centred on x_p, the similarity filter kernel of ABF is centred at $x_p - \Delta_p$. The ABF has a very good sharpening ability if the adaptation parameters Δ_p and σ_p are calculated appropriately. In our approach, the ABF is incorporated into the basic Projected Landweber (PL) Framework in order to reduce blocking artefacts. In essence, this operation imposes smoothness in addition to the sparsity inherent to PL and protects the sharp edges so as to ensure the quality of virtual view. It is particularly important to the reconstructed process that the smoothness of the filter also can improve the convergence speed. The computational complexity may remarkably decrease. Specifically, the direction diffusion step is interleaved with the PL projection of Equation 10 and Equation 11.

4 RESULTS AND DISCUSSION

The proposed algorithms were evaluated using three test sequences with different resolution (1024 × 768, 1280 × 960, and 1920 × 1088) and characteristics. For each sequence, two views with texture and depth maps are selected for the experimental tests. The detailed information of the test sequences is provided in Table 1. In this section, we verify the merits of our proposed from three aspects. Firstly, the objective performances and subjective comparisons are revealed in terms of the views synthesized. Then the complexity analysis of this part will be discussed.

Table 1. Test sequence information.

Sequence	Resolution	Frame rate	Views
Poznan_Street	1920 × 1088	30	3-4-5
champagne	1280 × 960	30	37-38-39
BookArrival	1024 × 768	25	1-3-5

Table 2. The PSNR comparisons of the bicubic, BCS, TV, and proposed method.

Algorithm	Subrate				
	0.1	0.2	0.3	0.4	0.5
Ponan_Street					
Bicubic	31.093	31.818	32.026	32.104	32.112
BCS	31.743	32.076	32.133	32.102	32.123
TV	31.824	32.053	**32.148**	**32.173**	32.136
FMRBCS	**31.950**	**32.107**	32.130	32.154	**32.161**
Champagne					
Bicubic	32.208	32.920	33.503	33.740	33.914
BCS	33.818	34.176	34.191	34.098	34.034
TV	33.947	34.231	**34.235**	34.138	34.103
FMRBCS	**33.999**	**34.233**	34.222	**34.165**	**34.152**
BookArrival					
Bicubic	34.854	35.310	35.077	35.402	35.539
BCS	34.435	35.319	35.417	35.263	35.389
TV	35.044	35.478	35.553	35.628	**35.602**
FMRBCS	**35.399**	**35.574**	**35.618**	**35.670**	35.590

4.1 Performance

4.1.1 Objective performance

We now present experimental results using PSNR as objective quality metrics. Due to the depth maps don't use as display but as an aid for view rendering, it is meaningless to compare the performance of the depth maps. The quality of the synthesized views should be of concern. The Bicubic which is a classical depth up-sampling approach, and the BCS and TV which adopt an advanced CS theory reconstructed approach, are chosen to contrast with the proposed method. Each depth sequence, which has 100 frames, is processed frame by frame. In this paper, five subrates (which are from 0.1 to 0.5) for each depth sequence pair are processed and reconstructed via the Bicubic, BCS, TV and proposed methods. After the depth maps are processed, the reconstructed depth maps and their textures are used for view synthesis via the platform of View Synthesis Reference Software (VSRS) extracted from MPEG [7]. A summary of average PSNR results obtained for synthesized views are shown in Table 2. From Table 2 we can see that the proposed method performs well, mostly in terms of PSNR at different subrates.

4.1.2 Subjective performance

In order to confirm the effectiveness of our algorithm subjective quality should be analysed with care. The synthesized views with original depth and the reconstructed depth maps by different algorithms are

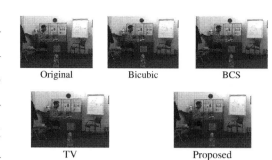

Figure 2. Subjective quality assessment.

Figure 3. The partial frames for visual comparison with different methods (A: Bicubic B: BCS C: TV D: Proposed).

provided in Figure 2, and the partial frames for visual comparison in Figure. 3. The "BookArrival" sequence is selected as shown in this paper. As shown in the figures, the results of using the proposed method not only obtain good performance on the whole image, but also surpass other methods in some details. It can be observed that the bicubic approach has obviously blurring result. The BCS strategy removes this effect, but generates jagged artefacts. The TV method provides better results, but the performance is achieved at the cost of a high computational complexity. The proposed method has fewer artefacts, while avoiding the computational complexity cost of the TV method, which will be analysed in the next section.

4.2 Complexity analysis

As we all know, there are many factors that affect the complexity of one algorithm, such as number of memory assessment, memory usage, etc. It is very difficult to balance all the factors in the practical applications. Good parallelism is a very important feature to reduce the implementation. In CS application, the total variation (TV) reconstruction is considered as one of the best retrieval algorithms for images or video signals. It is solved using a second-order-cone program that accommodates the TV-based norm. In essence, the TV approach to CS reconstruction adopts sparsity in the domain of a discretized gradient with the whole image, so it will be rather slow, perhaps even to point of being infeasible if the image size is large. In addition, good convergency is another pivotal factor for evaluating the algorithm. The number of iterations can measure this performance. Three methods have been tested in this part, each running for about half a minute on a dual-core 2.5-GHZ machine. As can be seen in Table 3, in term of execution times, the 10th frame of the

Table 3. Reconstruction time and iterations for the 10th frame of the 10th view of "BookArrival" depth maps at subrate 0.1.

Algorithm	Time (s)	Iterations	PSNR (depth)
AMRBCS	95.83	21	40.746
BCS	248.074	160	27.179
TV	1958.17	603	38.046

"BookArrival" sequence reconstruction with the proposed method at subrate 0.1, is nearly three times faster than the classical BCS because it has less iteration. The TV method requires over half an hour to reconstruct a single frame of depth sequence, despite the use of a fast structurally random matrices (SRM).

5 CONCLUSION

In this paper, "A Fast Multilevel Reconstruction Method of Depth Maps, Based on Block Compressive Sensing" has been proposed. In terms of synthesized views, the algorithm delivers high objective and subjective quality. It produces significantly fewer artefacts than classical up-sampling methods and the BCS method, and is computationally simpler than other high performance algorithms such as the TV method. More importantly, it has better robustness than classical compression or dimensionality reduction method about the transmission of depth maps in the 3D application. Given the growth in 3D multimedia systems requiring transmission and storage, the proposed method may have a wide range of possible applications.

REFERENCES

[1] Schwarz Heiko, Bartnik Christian, Bosse Sebastian *et al.*, "3D video coding using advanced prediction, depth modeling, and encoder control methods" [C], The 29th Picture Coding Symposium (PCS), Krakow Poland, May 2012, pp. 1–4.

[2] Min-Koo Kang, Dae-Young Kim, Kuk-jin Yoon, Adaptive Support of Spatial-Temporal Neighbors for Depth Map Sequence Up-sampling, IEEE Signal Processing Letters, V(21), 2, 2014, pp. 150–154.

[3] Qiuwen Zhang, Liang Tian, Lixun Huang, *et al.*, "Rendering Distortion Estimation Model for 3D High Efficiency Depth Coding", Mathematical Problems in Engineering, V(2014), 2014, pp. 1–7.

[4] Lee Sungwon, Ortega Antonio, "Adaptive compressed sensing for depth map compression using graph-based transform", 2012 IEEE International Conference on Image Processing (ICIP), 2012, Lake Buena Vista, FL, United states, pp. 929–932.

[5] S. Mun and J. E. Fowler, "Block compressed sensing of images using directional transforms," *Proc. Int. Conf. Image Process.*, Cairo, Egypt, pp. 3021–3024, Nov. 2009.

[6] N. Hurley and S. Rickard, "Comparing measures of sparsity", *IEEE Trans. Inf. Theory*, vol. 55, no. 10, pp. 4723–4741, Oct. 2009.

[7] ISO/IEC JTC1/SC29/WG11, View synthesis Algorithm in View Synthesis Reference Software 3.0 (VSRS3.0), Document M16090.

Dynamic modelling with validation for PEM fuel cell systems

Y. Chen, H. Wang, B. Huang & Y. Zhou
Electrical Technologies and Systems, GE Global Research, Shanghai, China

ABSTRACT: This paper studies modeling for 1KW water cooled Proton Exchange Membrane (PEM) fuel cell test bench. The presented model involves fluid dynamics and electrochemistry, addressing mainly the fast dynamics associated with flow and pressure. Semi-empirical models are employed for these descriptions, which combine empirical formulations with theoretical equations. The validity of the model was evaluated by determining the model parameters and its accuracy was verified against the experimental data.

Keywords: Proton exchange membrane; fuel cell; modeling

1 INTRODUCTION

Proton exchange membrane (PEM) fuel cell systems offer clean and efficient energy production, but the applications suffer from the short life of stacks due to transient violation to operational boundaries, which usually results in membrane electrode assembly (MEA) degradations. For example, fuel starvation due to instant load change makes higher voltage at anode and carbon oxidation; oxygen starvation leads to instant lower voltage and local hot spots due to reverse polarity; large unbalanced pressure leads to cross leakage or cross break of membrane; large cell voltage (e.g. >0.85V) during open-loop or at low-load condition erodes the carbon carrier and even oxides Pt electrodes. Beside limitations in material and stack design, the challenges in preventing MEA degradation and thereby extending the life of stack are of control problems. Actually, using system controls to extend fuel cell stack's life has been investigated in the last decade (Pukrushpan et al. 2004, Yu et al. 2010, Yi et al. 2011). The methods include increasing fuel velocity through recirculation to avoid fuel starvation, deliberately designed gas delivery sequence at start-up/shutdown to avoid transiently high voltage, installing auxiliary loads to restrict high voltage and etc. However, all these control strategies are predetermined and cannot adapt to process variation. For example, excess oxygen with a constant excess ratio is usually supplied to prevent oxygen starvation, but a variable excess ratio is preferred for varying load condition. At part loads, a higher excess ratio may be desired for fast loading up; but at full loads, a lower excess ratio may be preferred for a balanced cross-electrode pressure even though the loading capability may decrease. It is difficult for traditional control strategies to address these interactive constraints dynamically, and so the violation of operational constraints is inevitable to some extent.

As a kind of constrained optimization control method, model predictive control (MPC) was utilized to address these operation constraints (Golber et al. 2004, 2007, Panos et al. 2012). As MPC relies on a mathematical model that can predict process behavior, the first step in designing the control system is the derivation of the model. The model should be as accurate but simple as possible, so that a real-time controller can be implemented with repeated calculations for optimization problem solving.

This paper is the first part of research on PEMFC life extension through MPC controls. We studies PEMFC modeling techniques for a fuel cell test bench for MPC design purpose. The fuel cell system is firstly described, followed by the details of the model and the method to determine model parameters. The model is verified against the experimental data and conclusions are drawn afterwards.

2 PEM FUEL CELL SYSTEM AND MODEL DESCRIPTION

The fuel cell system to be studied in this paper is shown in Figure 1. It consists of a PEM fuel cell stack, a water-cooling system, gas delivery systems and exhaust discharge systems for fuel and oxidant flows respectively. The gas delivery systems include the components such as compressed hydrogen and air tanks, mass flow regulating valves and heated humidifiers. The exhaust discharge systems are equipped with back pressure valves. The mass flow regulating valves are used to achieve the desired hydrogen and air flow rate; the heated humidifiers to achieve the proper humidity and temperatures before the gases are channeled in stack. The back pressure valves can be used to adjust the working pressure of stack.

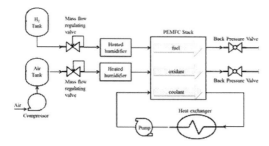

Figure 1. Overall scheme of the fuel cell test rig.

Figure 2. PEMFC test bench.

Figure 2 shows the photo of the associated PEM fuel cell test bench. It was upgraded from a fuel cell tester with software enhancement. Fuel cell power is delivered to a 1.5 kW electronic load. The adopted stack is a commercial 1.2 kW Sunrise PEMFC product, which is composed of 6 cells, each with a 270 cm^2 Nafion 212 membrane. LabVIEW was chosen as real-time controller which runs upon an industrial PC. It acquires the process measurements from sensors and delivers instructions to actuators. The process measurements include the mass flow rate of fuel and oxidant, the gas temperature and pressure at the both inlet and outlet of cathode and anode channels, the coolant flow rate and temperatures along the cooling passage, the cell voltages and etc.

The proposed model of the above fuel cell system involves three main modules that describe: fluid dynamics, thermodynamics and electrochemistry. Semi-empirical models are employed for these models, which combine empirical formulations with theoretical equations. Fluid dynamics model is composed of three interconnected modules: fuel flow stream, oxidant flow stream and the membrane. The thermodynamics model is employed to determine the homogeneous temperature of the stack. The electrochemistry model is a static model that determines the stack voltage. Since the fuel cell's dynamical behavior associated with the reactant pressure and flow is our focus, the slower dynamics associated with temperature regulation and heat dissipation is neglected.

To simplify the analysis, the following assumptions for the model have been made:

1) Zero-dimensional treatment: mass and energy transport is lumped using volume-average conservation equations; multiple cathode and anode volumes of fuel cells in the stack are lumped together as a single stack cathode and anode volume
2) Ideal and uniformly distributed gas
3) Consistent and instant humidification of inlet reactant flows
4) Uniform pressure of the stack and anode and cathode
5) Uniform stack temperature
6) Constant overall specific heat capacity of stack
7) Constant thermal properties if not specified
8) Constant pressure of hydrogen and air tank
9) Neglected effect of condensed water on gaseous flows.

2.1 *Nomenclature*

We denote the lumped variables associated with anodic gaseous flow by the subscript "*an*", and the cathodic gaseous flow by the subscript "*ca*". The lumped variables associated with supply manifold are denoted with the subscript "*sm*", return manifold with the subscript "*rm*" and heated humidifier with "*hh*". Other nomenclatures are listed below. Masses (kg) are denoted with m, mass flow (kg/s) with W, molar masses (kg/mol) with M, pressure (Pa) with P, temperature (K) with T, vapor saturation pressure at temperature T_x with $P^{sat}(T_x)$, relative humidity with Φ, humidity ratio with Ω, current (A) with I, current density (A/cm^2) with j, area (cm^2) with A, volume (m^3) with V, voltage with v. The variables associated with vapor are denoted with a subscript "*v*", water with "*w*", oxygen with "*O_2*", nitrogen with "*N_2*", hydrogen with "*H_2*", air with "*a*" and "dry air" with "*da*". The variables in specific volumes has a second subscript as the volume identifier (*sm*, *rm*, *ca*, *an*). The variables associated with the electrochemical reactions are denoted with "*rct*" as subscript. Similarly, the stack variables use "*st*", the individual fuel cell variables use "*fc*", the ambient conditions use "*amb*", the membrane variables use "*mbr*" and the back pressure valves use "bpv".

2.2 *Fluid dynamics model*

Figure 3 shows the fluid model scheme along reactant streaming channels. The inlet mass flow rates are the model inputs. Ambient condition is represented by T_{amb}, P_{amb} and Φ_{amb}, constituting the boundary conditions of the system. Compressed air and hydrogen are assumed of the same temperature as the ambient. As shown in Figure 3, the anodic gaseous flow has a similar fluid dynamics as the cathodic flow. Considering

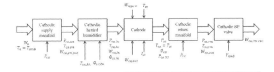
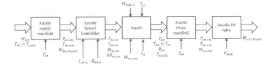

Figure 3. Model of fluid dynamics.

the space limitation, we omit the anodic fluid model in this paper.

2.2.1 Cathodic supply manifold

The rate of change of mass $m_{ca,sm}$ inside the supply manifold is governed by mass conservation equation, and the rate of change $P_{ca,sm}$ of supply manifold pressure is governed by energy conservation equation

$$\frac{dm_{ca,sm}}{dt} = W_a - W_{ca,sm,out} \quad (1)$$

$$\frac{dP_{ca,sm}}{dt} = \frac{\gamma R_a}{V_{ca,sm}}\left(W_a T_a - W_{ca,sm,out} T_{ca,sm}\right) \quad (2)$$

where R_a is the air gas constant, it can be calculated from $R_a = R/M_a^{amb}$, R is the universal gas constant, and M_a^{amb} is the molar mass of atmospheric air at ambient condition. $V_{ca,sm}$ is the supply manifold volume, $T_{ca,sm}$ is the supply manifold gas temperature, and $W_{ca,sm,out}$ is the outlet mass flow rate of supply manifold. $T_{ca,sm}$ and $W_{ca,sm,out}$ are determined with following equations.

$$T_{ca,sm} = \frac{P_{ca,sm} V_{ca,sm}}{R_a m_{ca,sm}} \quad (3)$$

$$W_{ca,sm,out} = K_{ca,sm,out}\left(P_{ca\_sm} - P_{ca}\right) \quad (4)$$

$K_{ca,sm,out}$ is the throttle coefficient of the interface that connects air supply manifold to the cathode channels of stack. It is a parameter to be determined.

2.2.2 Cathodic heated humidifier

The heating and humidification is approximated as an idea process, and so the outlet stream (also the inlet stream of cathode channels) has the desired temperature $T_{ca,hh}$ and the desired relative humidity $\Phi_{ca,hh}$.

$$T_{ca,in} = T_{ca,hh} \quad (5)$$

$$\Phi_{ca,in} = \Phi_{ca,hh} \quad (6)$$

The pressure of the stream is calculated by

$$P_{ca,in} = P_{ca\_sm} - P_{ca,sm} \cdot \frac{P^{sat}(T_{amb})\Phi_{amb}}{P_{amb}} + P^{sat}(T_{ca,hh})\Phi_{ca,hh} \quad (7)$$

The inlet mass flow of cathode is the supply manifold outlet flow added by the injected water in the heated humidifier, as calculated by

$$W_{ca,in} = W_{ca,sm,out} + \frac{M_v \Phi_{ca,hh} P^{sat}(T_{ca,hh})}{M_{da} P_{ca,hh,da}} \cdot W_{ca,sm,da} \quad (8)$$

where $P_{ca,hh,da}$ is the partial pressure for the dry air inside the heated humidifier, and $W_{ca,hh,da}$ is the dry air flow rate associated with the heated stream.

$$P_{ca,hh,da} = P_{ca,sm} - P_{ca,sm}\frac{\Phi_{amb}P^{sat}(T_{amb})}{P_{amb}} \quad (9)$$

$$W_{ca,hh,da} = W_{ca,sm,da} = \frac{W_{ca,sm}}{1 + \frac{M_v \cdot P_{ca,sm} \cdot \frac{\Phi_{amb}P^{sat}(T_{amb})}{P_{amb}}}{M_{da} \cdot \left[P_{ca,sm} - P_{ca,sm}\frac{\Phi_{amb}P^{sat}(T_{amb})}{P_{amb}}\right]}} \quad (10)$$

2.2.3 Cathode

Mass conservation yields governing equations for oxygen, nitrogen and water mass inside the cathode volume, given by

$$\frac{dm_{ca,O_2}}{dt} = W_{ca,in,O_2} - W_{ca,out,O_2} - W_{ca,rct,O_2} \quad (11)$$

$$\frac{dm_{ca,N_2}}{dt} = W_{ca,in,N_2} - W_{ca,out,N_2} \quad (12)$$

$$\frac{dm_{ca,w}}{dt} = W_{ca,in,v} - W_{ca,out,v} - W_{ca,rct,v} + W_{mbr,v} \quad (13)$$

It is assumed that liquid water does not leave the stack and evaporates into the cathode gas if cathode humidity drops below 100%. The mass of water in vapor forms until the relative humidity of gas exceeds saturation, at which point vapor condenses into liquid water.

Based on the gas properties of cathode inlet flow, specifically, the mass flow rate $W_{ca,in}$, pressure $P_{ca,in}$, relative humidity $\Phi_{ca,in}$, and temperature $T_{ca,in}$, the individual species are calculated by

$$W_{ca,in,O_2} = y_{O_2} \frac{1}{1+\Omega_{ca,in}} W_{ca,in} \quad (14)$$

$$W_{ca,in,N_2} = y_{N_2} \frac{1}{1+\Omega_{ca,in}} W_{ca,in} \quad (15)$$

$$W_{ca,in,v} = W_{ca,in} - W_{ca,in,N_2} - W_{ca,in,O_2} \quad (16)$$

where y_{O_2} and y_{N_2} are the mass fraction of oxygen and nitrogen in the dry air, respectively. They are calculated with

$$y_{O_2} = x_{O_2}\frac{M_{O_2}}{M_{da}}$$
$$y_{N_2} = (1-x_{O_2})\frac{M_{N_2}}{M_{da}} \quad (17)$$
$$M_{da} = x_{O_2}M_{O_2} + (1-x_{O_2})M_{N_2}$$

where $x_{O_2} = 0.21$ is the oxygen molar fraction in ambient dry air. The inlet humidity ratio $\Omega_{ca,in}$ in equation (14–15) is calculated by

$$\Omega_{ca,in} = \frac{M_v}{M_{da}} \cdot \frac{\Phi_{ca,in} P^{sat}(T_{ca,in})/P_{ca,in}}{1 - \Phi_{ca,in} P^{sat}(T_{ca,in})/P_{ca,in}} \quad (18)$$

The calculation of cathode outlet flow rate for these species is similar with (14)–(17). The formulations are omitted just for unnecessary repetition. It should be noted that the inlet flow variables in (14)–(18) must be replaced by the outlet flow variables for calculation of cathode outlet flow rates, and the associated oxygen molar fraction in cathode outlet dry air must be calculated with

$$x_{ca,out,O_2} = x_{ca,O_2} = \frac{P_{ca,O_2}}{P_{ca,O_2} + P_{ca,N_2}} \quad (19)$$

where the partial pressures of the oxygen $P_{ca,O2}$ and nitrogen $P_{ca,N2}$ inside the cathode volume are obtained through the ideal gas law based on the assumption that the cathode temperature is uniform and equal to the assumed uniform stack temperature T_{st}.

$$P_{ca,O_2} = \frac{m_{ca,O_2} R_{O_2} T_{st}}{V_{ca}}$$
$$P_{ca,N_2} = \frac{m_{ca,N_2} R_{N_2} T_{st}}{V_{ca}} \quad (20)$$

The partial pressure of vapor and the vapor mass inside the cathode volume is determined by

$$P_{ca,v} = \frac{m_{ca,v} R_v T_{st}}{V_{ca}} \quad (21)$$

$$m_{ca,v} = \min\left(m_{ca,w}, \frac{P^{sat}(T_{st}) * V_{ca}}{R_v T_{st}}\right) \quad (22)$$

With the partial pressures in (20) and (21), we can further calculate the pressure, the constant and the density of cathode gas

$$P_{ca} = P_{ca,O_2} + P_{ca,N_2} + P_{ca,v} \quad (22)$$

$$M_{ca,outgas} = M_{ca,gas} = (M_{O_2} P_{ca,O_2} + M_{N_2} P_{ca,N_2} + M_v P_{ca,v})/P_{ca} \quad (23)$$

$$\rho_{ca,outgas} = \rho_{ca,gas} = (m_{ca,O_2} + m_{ca,N_2} + m_{ca,v})/V_{ca} \quad (24)$$

The oxygen reaction rate $W_{ca,rct,O2}$ in equation (11), the water generation rate $W_{ca,rct,v}$ in equation (13) and hydrogen reaction rate $W_{ca,rct,H2}$ (used for anode calculation) are derived from the stack current I_{st} by using the electrochemical equations

$$W_{ca,rct,O_2} = M_{O_2} \cdot nI_{st}/(4F)$$
$$W_{ca,rct,H_2} = M_{H_2} \cdot nI_{st}/(2F) \quad (25)$$
$$W_{ca,rct,v} = M_v \cdot nI_{st}/(2F)$$

where n is the number of cells and F is the Faraday number.

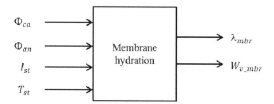

Figure 4. Model of membrane hydration.

2.2.4 *Membrane hydration*

The vapor mass flow $W_{mbr,v}$ across the membrane in equation (13) is calculated using mass transport principle

$$W_{mbr,v} = M_v n A_{fc} \cdot \left(n_d \frac{j}{F} - D_w \frac{C_{ca,v} - C_{an,v}}{t_m}\right) \quad (26)$$

where A_{fc} denotes the active area of the fuel cell, j the fuel cell current density and t_m the membrane thickness; D_w is the diffusion coefficient; n_d is the electro-osmotic coefficient; $C_{ca,v}$ and $C_{an,v}$ are the water concentration in cathode and anode volume. The determination of parameters n_d, D_w, $C_{ca,v}$ and $C_{an,v}$, can be referred in (Panos et al. 2012, Pukrusphan et al. 2005). Figure 4 shows the input-output model structure for membrane hydration.

2.2.5 *Cathodic return manifold*

The state equation of the return manifold pressure is

$$\frac{dP_{ca,rm}}{dt} = \frac{R_a T_{ca,rm}}{V_{ca,rm}}(W_{ca,out} - W_{ca,rm,out}) \quad (27)$$

where $V_{ca,rm}$ denotes the return manifold volume, $T_{ca,rm}$ the gas temperature that is assumed equal to T_{st}. Since the isothermal assumption is made for the return manifold, the state $m_{ca,rm}$ is omitted. $W_{ca,out}$ is the cathode outlet flow rate, which can be determined with the throttle coefficient $K_{ca,rm,in}$ of the interface that connects the cathode to the return manifold.

$$W_{ca,out} = K_{ca,rm,in} \cdot (P_{ca} - P_{ca,rm}) \quad (28)$$

The outlet flow rate of return manifold is governed by the back pressure valve. A simple formulation is used to describe the process

$$W_{ca,rm,out} = K_{ca,bpv} \cdot (P_{ca,rm} - P_{amb}) \quad (29)$$

where $K_{ca,bpv}$ denotes the throttle coefficient of the back pressure valve.

In summary, the fluid dynamics model based on the equation (1–2), (11–13), (28) and the omitted anode fluid models involves the eleven states

$$X = [m_{ca,sm}, P_{ca,sm}, P_{ca,rm}, m_{ca,O_2}, m_{ca,N_2},$$
$$m_{ca,v}, m_{an,sm}, P_{an,sm}, P_{an,rm}, m_{an,H_2}, m_{an,v}]^T$$

The parameters to be determined include $K_{ca,sm,out}$, $K_{ca,rm,in}$ and $K_{ca,bpv}$.

2.3 Electrochemistry

A static, semi-empirical mathematical model is used to characterize the fuel cell voltage, specifically, in the form of a polarization curve, which is the plot of fuel cell voltage versus current density. The current density, j, is defined as current per active area, $j = I_{st}/A_{fc}$. Since fuel cells are connected in series to form the stack, the total stack voltage is $V_{st} = n \cdot V_{fc}$. The polarization curve is described by Nernst equation with Nernst potential V_{Nernst}, also known as open-circuit voltage, subtracted by the activation voltage loss V_{act}, ohmic voltage loss V_{ohm} and concentration voltage loss V_{conc}

$$V_{fc} = E_{Nernst} - V_{act} - V_{ohm} - V_{conc} \tag{30}$$

Nerst potential is a function of stack temperature T_{st}, partial pressure of hydrogen $P_{an,H2}$ and oxygen $P_{ca,O2}$

$$E_{Nernst} = \frac{-\Delta G}{2F} + \frac{\Delta S}{2F}(T_{st} - T_{ref}) + \frac{RT}{2F}\ln\left[P_{an,H_2}\sqrt{P_{ca,O_2}}\right] \tag{31}$$

where ΔG is the Gibbs free energy change at the standard condition with $T_{ref} = 298.15$ K and ΔS the entropy change of the electrochemical reaction. $-\Delta G/2F = 1.229$ and $\Delta S/2F = -8.5 \times 10^{-4}$.

Activation voltage loss occurs because of the need to move electrons and to break chemical bonds in the anode and cathode. Part of available energy is lost in driving chemical reactions that transfers electrons to and from electrodes. The activation loss can be calculated with equation

$$V_{act} = \frac{RT}{\alpha \cdot 2F}\ln\left(\frac{j}{j_0}\right) \tag{32}$$

where α is the charge transfer coefficient and j_0 is the exchange current density (Hayre et al. 2006). They are two parameters to be determined.

Ohmic voltage loss is resulted from the resistance of the polymer membrane to protons transfer and resistance of the electrode to electrons transfer. It can be calculated with equation

$$V_{ohm} = \frac{t_{mbr}}{\sigma_{mbr}} \cdot j \tag{33}$$

where t_{mbr} is the thickness of membrane, and σ_{mbr} is the membrane conductivity. The conductivity model of Nafion 117 membrane is given below (Amphlett et al. 1995)

$$\sigma_{mbr} = b_1 e^{b_2\left(\frac{1}{303} - \frac{1}{T_{st}}\right)} \tag{34}$$

where b_2 is a constant ($=1268$) and b1 is heavily dependent on water content of membrane λ_{mbr}

$$b_1 = 0.005139\lambda_{mbr} - 0.00326 \tag{35}$$

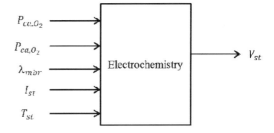

Figure 5. Model of stack voltage.

In consideration of the bias arising from lumped parametric model and accommodation for different types of Nafion membranes, we augment equation (34) with a correction factor b_0 that will be determined using experimental data.

$$\sigma_{mbr} = b_0 \cdot b_1 e^{b_2\left(\frac{1}{303} - \frac{1}{T_{st}}\right)} \tag{36}$$

Concentration voltage loss results from the drop in concentration of the reactants as they are consumed in the reaction. This explains the rapid voltage drop at high current density. The equation is

$$V_{conc} = -\beta\frac{RT}{2F}\ln\left(1 - \frac{j}{j_{max}}\right) \tag{37}$$

where $\beta = 1 + 1/\alpha$, and j_{max} is the limiting current density that will be determined.

In summary, the electrochemistry model is highly stack-specific. The parameters to be determined include α, j_0, b_0, and j_{max}. The input-output of electrochemistry model is illustrated in Figure 5.

3 MODEL VERIFICATION

The models in Figure 3, 4 and 5 are connected to form a complete model for the PEM fuel cell system. Since we concentrate on the dynamical fuel cell behavior associated with flow and pressure, a constant stack temperature is assumed. Table 1 gives constant coefficients used in the PEM fuel cell model.

As highlighted in the previous sections, some model parameters have to be determined for simulation and validation. We conducted offline experiments for this purpose. By injecting constant flow of hydrogen and air through the manifolds and the stack channels, we collected the data of mass flow rates and the associated inlet and outlet pressures; least square method is then employed to estimate the throttle coefficients in equation (28–29). The estimated parameters are shown in Table 2.

Polarization curve was further obtained through sweep current tests. The charge transfer coefficient, the exchange current density, the correction factor and the limiting current density were then fitted using some

Table 1. Constant coefficients used in the model.

Stack temperature, T_{st} (K)	333.15
Faraday's constant, F (C/mol)	96485
Universal gas constant, R (J/mol/K)	8.314
Air specific heat ratio, γ	1.4
Hydrogen specific heat ratio, γ_{H2}	1.4143
Molar mass of hydrogen, M_{H2} (kg/mol)	2×10^{-3}
Molar mass of oxygen, M_{O2} (kg/mol)	32×10^{-3}
Molar mass of nitrogen, M_{N2} (kg/mol)	28×10^{-3}
Molar mass of vapor, M_v (kg/mol)	18×10^{-3}
Fraction of oxygen in dry air, y_{O2} (kg/mol)	0.21
Cathode volume, V_{ca} (m^3)	0.083×10^{-3}
Cathode supply manifold volume, $V_{ca,sm}$ (m^3)	4.557×10^{-3}
Cathode return manifold volume, $V_{ca,rm}$ (m^3)	5.517×10^{-3}
Anode volume, V_{an} (m^3)	0.044×10^{-3}
Anode supply manifold volume, $V_{an,sm}$ (m^3)	4.543×10^{-3}
Anode return manifold volume, $V_{an,rm}$ (m^3)	4.864×10^{-3}
Fuel cell active area, A_{fc} (cm^2)	270
Number of cells in stack, n	6
Membrane thickness, t_{mbr} (cm)	0.005
Membrane dry density, ρ_{mbr} (kg/m^3)	0.002
Membrane dry equivalent weight, M_{mbr} (kg/mol)	1.1

Table 2. Estimated model parameters.

Cathode inlet throttle coeff., $K_{ca,sm,out}$ (kg/Pa/s)	9.78×10^{-8}
Cathode outlet throttle coeff., $K_{ca,rm,in}$ (kg/Pa/s)	9.78×10^{-8}
Cathode BP valve throttle coeff., $K_{ca,bpv}$ (kg/Pa/s)	6.44×10^{-8}
Anode inlet throttle coeff., $K_{an,sm,out}$ (kg/Pa/s)	1.02×10^{-8}
Anode outlet throttle coeff., $K_{an,rm,in}$ (kg/Pa/s)	1.02×10^{-8}
Anode BP valve throttle coeff., $K_{an,bpv}$ (kg/Pa/s)	5.37×10^{-9}
Charge transfer coeff. α	0.25
Exchange current density j_0 (A/cm^2)	0.0007
Correction factor b_0	2.57
Limiting current density j_{max} (A/cm^2)	2.5

optimization tools. The identified parameters are also shown in Table 2.

Figure 6 shows simulated polarization curve using the model parameters in Table 1 and Table 2. It has a very good match with the experimental one. Due to the current limit of electronic load (<240 A), the maximum current density tested is just 850 mA/cm^2.

We conducted experiments on the fuel cell test bench with a power type load. The loading path is shown in Figure 7. A stoichiometric ratio controller was implemented to provide excess fuel and oxidant flow. Specifically, the excess hydrogen ratio is 2 and the excess oxygen ratio is 2.5. The test case shown in Figure 7 was also simulated in Matlab/Simulink environment. The measured signals such as the current, the hydrogen flow rate, the air flow rate and the stack temperature from the test, were input to the mathematical model. The simulated responsive variables, such as voltages, inlet and outlet pressures, are then plotted and compared with the experimental data.

Figure 8 shows the comparison of inlet and out gas pressures. The dash lines denote the simulation results and solid lines the experimental data. Figure 9 compares the simulated stack voltage to the measured one. A high degree of conformity can be observed between simulation and experiment, especially in term of steady state. Large noises are observed in pressure measurements. We also noticed that a larger transient variation, for example, larger overshooting in stack voltage in the real plant. This can be ascribed to the errors of lumped parametric model.

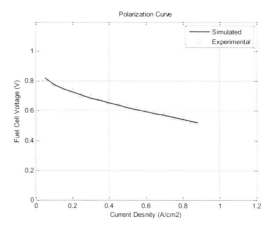

Figure 6. Comparison of the simulated polarization curve with the experimental one.

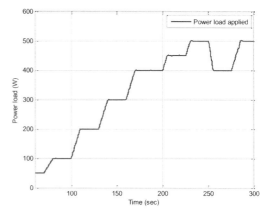

Figure 7. Load test with a stoichiometric controller.

4 CONCLUSIONS

A semi-empirical model was studied for a PEMFC test bench for MPC design purpose. The model is derived from theoretical equations combined with empirical

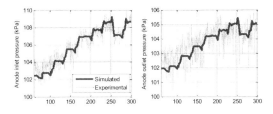

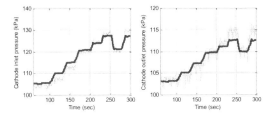

Figure 8. Gas pressure at cathode and anode.

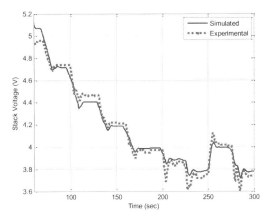

Figure 9. Stack voltage response.

formulations. Model parameters are determined with deliberately designed tests. The presented model was validated with a high degree of conformity in terms of polarization curve as well as power loading process. It offers a good model for MPC purpose which will be described in another paper.

ACKNOWLEDGEMENT

This research was supported by a grant from Science & Technology Commission of Shanghai Municipal (STCSM) Government Funds (12dz1200903).

REFERENCES

[1] Amphlett, J.C., Baumert, R.M. Mann, R.F., Peppley, B.A., Roberge, P.R. & Harris, T.J. 1995. Performance modeling of the Ballard Mark IV solid polymer electrolyte fuel cell. Journal of The Electrochemical Society, 142(1): 1–15.
[2] Golber, J. and Lewin, Daniel R. 2004. Model-based control of fuel cells: (1) Regulatory control. *Journal of Power Sources* 135: 135–151.
[3] Golber, J. and Lewin, Daniel R. 2007. Model-based control of fuel cells: (2) Optimal efficiency. *Journal of Power Sources* 173: 298–309.
[4] O'Hayre, R.P., Cha, S.W., Colella, W. & Prinz, F.B. 2006. Fuel Cell Fundamentals, John Wiley and Sons: New York.
[5] Panos, C., Kouramas, K.I., Georgiadis, M.C. & Pistikopoulos, E.N. 2012. Modeling and explicit model predictive control for PEM fuel cell systems, Chemical Engineering Science, 67: 15–25.
[6] Pukrushpan, J.T. Stefanopoulou, A.G. & Peng, H. 2004. Control of fuel cell breathing. IEEE Control System Magazine. 4: 30–46.
[7] Pukrushpan, J.T., Stephanopoulou, A.G. & Peng, H. 2005. Control of fuel cell power systems. In: Gremble, M.J. (Ed.), Principles, Modelling, Analysis and Feedback Design. Springer, London.
[8] Yi, B. & Hou, M. 2011. Solutions for the durability of fuel cells in vehicle application. Journal of Automotive Safety and Energy (Chinese), 2(2): 91–100.
[9] Yu, Y. & Pan, M. 2010. Research progress in system strategy of startup-shutdown for PEMFC. Chemical Industry and Engineering Progress (Chinese). 29(10): 1857–1862.

A highly sensitive new label-free bio-sensing platform using radio wave signal analysis, assisted by magnetic beads

Jae-Hoon Ji
Department of Mechanical Engineering, Yonsei University, Republic of Korea

Kyeong-Sik Shin & Yoon Kang Ji
Center of Biomicrosystems, Korea Institute of Science and Technology (KIST), Seoul, Republic of Korea

Seong Chan Jun
Department of Mechanical Engineering, Yonsei University, Republic of Korea

ABSTRACT: We introduce a new label and mediator-free detecting technique for biomaterial based on high frequency signal analysis, assisted by magnetic beads under a no reacting condition of enzyme or chemical. We chose OMSD as the first target material of our device. OMSD is a standard material for identifying the potential of biosensors. It is a chemically treated albumin complex. Thereby using OMSD, we can judge the sensitivity and possibility of the biosensor, not only its monomer target but also its oligomer or polymer target, due to the complex form of OMSD. Furthermore, albumin is a fundamental protein consisting of cell and most abundant protein. Thus, we can even identify the property of protein under a high frequency AC condition. The mechanism of detection is based on the sensitive reflective index change and dielectric constant dispersion in high frequency. It is widely known that the dielectric constant of liquid biopsy is dispersed inherently, depending on the reacting characteristics of solute and solvent under the AC voltage at different frequency ranges. In addition, well arranged beads that are coated with OMSD between the open signal electrodes, affect the reflection and transmission property of the device. The equivalent circuit is also established for obtaining the passive electrical property of OMSD. The components of equivalent circuits consist of the resistance (R), inductance (L), conductance (G), and capacitance (C). Additionally, the decomposing technique also provides Residual Signal (RS), and propagation constant (γ) from measured power signals. These various parameters, which are obtained by the analytical signal processing, exhibit more reliable results for instant *in-situ* sensing without any added supporters for reaction. Moreover, our device has some advantages of being reusable, having ream time sensing, and a high applicability.

1 INTRODUCTION

Recently, Radio Frequency RF-based bio-sensing techniques have been receiving attention for next generation sensing techniques. RF-based bio-sensors have some advantages compared to former sensing techniques. First, RF-based sensing techniques can make real time and continuous sensor that is a major demand of medical area for constructing automatic drug supply systems and remote medical services. By S-parameter, a fundamental signal in RF device, de-embedding technique, we can only extract the signal of the target material excluding other relative signal values, such as contact noises and signals of non-target materials. In addition, S-parameters are decomposed to multi-dimensional electrical properties using equivalent circuit analysis. Thus, by using an RF-based sensing technique, we can make a highly sensitive and selective bio-sensor. We identify the possibility of an RF platform by using OMSD that is a standard material in the bio-sensor area.

OMSD is a standard material for identifying the potential of a bio-sensor. It is a chemically treated albumin complex. Thereby using OMSD, we can judge the sensitivity and possibility of the bio-sensor, not only its monomer target but also its oligomer or polymer target, due to the complex form of OMSD. Furthermore, albumin is a fundamental protein consisting of cell and most abundant protein. Albumin that is non-glycosylated, multifunctional, and a negatively charged plasma protein is a fundamental protein in humans. Albumin that is 50% of the whole plasma protein is the most abundant protein and one of the smallest proteins in human plasma. Its main function in the human body is to regulate the osmotic pressure of blood. It has hydrophobic binding sites in its centre, and hydrophilic binding ligands in its outer parts. Albumin also acts as an ion and other molecule carrier. Due to the abundance and size issue in albumin, we choose albumin as the first target protein of sensing. In human serum, it is inevitable that OMSD and albumin exist as a mixture, unless we purify the serum to obtain the respective materials.

The mechanism of detection is based on the sensitive reflective index change, and the dielectric constant dispersion in high frequency. It is widely known that the dielectric constant of liquid biopsy is dispersed inherently, depending on the reacting characteristics of solute and solvent under the AC voltage at different frequency ranges. In addition, well arranged beads that are coated with OMSD between the open signal electrodes, affect the reflection and transmission property of the device when the interval of electrodes and the size of beads are well matched. This platform has some advantages of being reusable, having ream time sensing, and a high applicability. Above all, it has no need to calibrate the level of detection in the individual device.

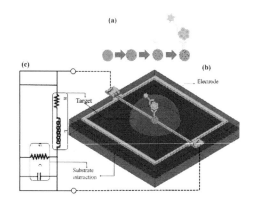

Figure 1. (a) Preparing the beads, (b) Schematic diagram of the device, and (c) Equivalent circuit of the device that consists of passive electrical components: R (resistance), L (inductance), G (Conductance), C (capacitance).

2 METHODS

2.1 Preparing the device

To minimize the signal loss, a highly resistive Si substrate was prepared, and then Ground-Signal-Ground (GSG) pattern was deposited using Ti/Au. This GSG pattern consists of two kinds of electrodes. One is a pair of transmission electrodes (S), and the other is a pair of ground electrodes (G). The S-parameter was measured by GSG probing with a network analyser (E5071C). The S-parameter, "S_{ij}" can be defined as the ratio between the incident voltage to port 'j' and the voltage measured at port 'i'. By matching equivalent circuits to the data, the passive electrical property (RLGC) of the standard material is extracted.

2.2 Preparing the material

The average particle size of silica-coated magnetic beads is 4.5 μm. That is the optimal size for attaching bio-molecules for research, purification, and functional studies. Above all, it is well matched to our electrode interval which is 3 μm. Because the silica magnetic beads are spherical and uniform, the detection of the material can be performed with outstanding efficiency. The beads were functionalized to react with the antibody. Then, the functionalized beads were incubated with the antibody to react with the target antigen, in this case OMSD. Magnetic beads are prepared as a non-treated condition, a functionalized condition, an antibody treated condition, and an OMSD target condition.

2.3 Signal measuring technique

The S-parameter was measured by GSG probing with a network analyser (E5071C). The S-parameter, "S_{ij}" can be defined as the ratio between the incident voltage to port 'j' and the voltage measured at port 'i'. It can be also separated into various components: RLGC. In Figure 1(b), these components can be expressed using the equivalent circuit.

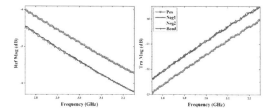

Figure 2. (a) Reflection parameter magnitude, (b) Imaginary parameter magnitude.

3 CONCLUSION

3.1 S-parameters of target material

The reflection parameter and the transmission parameter of magnetic beads, which have different kinds of surface conditions of OMSD, show a clear discrepancy depending on the existence of OMSD. Magnetic beads are prepared as a non-treated condition, a functionalized condition, an antibody treated condition, and an OMSD target condition. We can describe this result by the interaction of OMSD and electrodes. Both the OMSD and electrodes have a well matched size, and the radio frequency signal cannot pass through easily by the blocking of the signal due to the OMSD, that has a diameter of 4.5 μm which could block the signal between the gaps of electrodes (3 μm). Thus, in Figures 2(a) and (b), the transmission parameter shows that the signal cannot pass through easily to the other port, and the reflection parameter shows that the signal hardly reflected to its original port than other condition. The magnitude difference is at least 0.5 dB. It shows that our platform has a clear detection in the frequency range. This is because the reflective index is severely affected by the matched condition.

3.2 Multi-dimensional parameters of target material

The multi-dimensional parameters extracted from the S-parameters of OMSD also show discrimination with

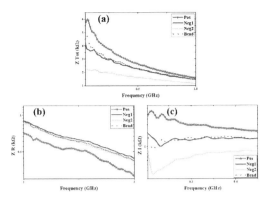

Figure 3. (a) Total impedance, (b) Real impedance, (c) imaginary impedance.

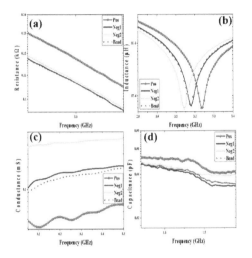

Figure 4. Extracted multi-dimensional parameters of OMSD in human serum, (a) R: resistance, (b) L: inductance, (c) G: shunt conductance, (d) C: shunt capacitance.

the concentration of OMSD in Figure 3. All of the data in Figures 3 and 4 shows quite good discrepancy of target material than other condition. For impedance, the target signal is shown to be at least 300 ohm higher than the other conditions. For resistance, the target signal is shown to be at least 100 ohm higher than the other conditions. All the other parameters also show a huge difference in the operating frequency. Especially, the inductance in Figure 4(b) exhibits that those are applied as an indicator of the concentration of OMSD, because the resonance dip points of each human serum that has a different concentration of OMSD, move to a higher frequency in almost the same intervals, depending on the concentration of OMSD.

4 CONCLUSION

In this study, we made an RF-based bio-sensor supported by magnetic beads, using the advantages of the RF technique. We detected the standard material: OMSD successfully *in situ* condition. The comparison of multi-dimensional electrical properties such as: R (resistance), L (inductance), G (conductance), C (capacitance) also make the accuracy of the sensor more precise and reliable. By applying our detection technique to the bio-sensor application, it will contribute to next generation sensor application by its unique advantages of being reusable, having ream time sensing, a high applicability, minimum operating energy, and ease of combining with mobile devices.

REFERENCES

Cristina Go'mez-Navarro, Jannik C. Meyer, Ravi S. Sundaram, Andrey Chuvilin, Simon Kurasch, Marko Burghard, Klaus Kern, and Ute Kaiser, "Atomic Structure of Reduced Graphene Oxide", Nano Lett., 10 (2010) 1144–1148.

Elizabeth A. Moschou, Bethel V. Sharma, Sapna K. Deo, and Sylvia Daunert, "Fluorescence Glucose Detection: Advances toward the Ideal in Vivo Biosensor", Journal of Fluorescence, 14 (2004) 535–547.

Faaizah Khan, Tania E. Saxl, John C. Pickup, "Fluorescence intensity- and lifetime-based glucose sensing using an engineered high-Kd mutant of glucose/galactose-binding protein", Biochemical and Biophysical Research Communications, 365, (2008) 102–106.

Ganhua Lu, Leonidas E. Ocola, and Junhong Chen, "Gas detection using low-temperature reduced graphene oxide sheets" Applied Physics Letters 94 (2009) 083111.

Hyung Goo Park, Sukju Hwang, Juhwan Lim, Duck-Hwan Kim, In Sang Song, Jae Hun Kim, Deok Ha Woo, Seok Lee, and Seong Chan Jun, "Comparison of Chemical Vapor Sensing Properties between Graphene and Carbon Nanotubes" JJAP 51 (2012) 045101.

Jing Luo, Sisi Jiang, Hongyan Zhang, Jinqiang Jiang, Xiaoya Liu, "A novel non-enzymatic glucose sensor based on Cu nanoparticle modified graphene sheets electrode", Analytica Chimica Acta 709 (2012) 47–53.

Joseph Wang, "Electrochemical Glucose Biosensors", Chem. Rev., 108 (2008) 814–825.

Karen E. Shafer-Peltier, Christy L. Haynes, Matthew R. Glucksberg, and Richard P. Van Duyne, "Toward a Glucose Biosensor Based on Surface-Enhanced Raman Scattering", J. Am. Chem. Soc., 125 (2003) 588–593.

Lakshmi N. Cella, Wilfred Chen, Nosang V. Myung, and Ashok Mulchandani, "Single-Walled Carbon Nanotube-based Chemiresistive Affinity Biosensors for Small Molecules: Ultrasensitive Glucose Detection", J. AM. CHEM. SOC., 132 (2010) 5024–5026.

Li-Min Lu, Hong-Bo Li, Fengli Qu, Xiao-Bing Zhanga, Guo-Li Shen, Ru-Qin Yu, "In situ synthesis of palladium nanoparticle–graphene nanohybrids and their application in nonenzymatic glucose biosensors", Biosensors and Bioelectronics 26 (2011) 3500–3504.

Morozov S.V., Novoselov K.S., Katsnelson M.I., Schedin F., Elias D.C., Jaszczak J.A., and Geim A.K., "Giant Intrinsic Carrier Mobilities in Graphene and Its Bilayer", PRL 100 (2008) 016602.

Novoselov K.S., Geim A.K., Morozov S.V., Jiang D., Katsnelson M.I., Grigorieval I.V., Dubonos S.V., Firsov A.A.,

"Two-Dimensional Gas of Massless Dirac Fermions in Graphene", Nature, 438 (2005) 197–200.

Wang S.G., Qing Zhang, Ruili Wang, and Yoon S.F., "A novel multi-walled carbon nanotube-based biosensor for glucose detection", Biochemical and Biophysical Research Communications, 311 (2003) 572–576.

Ying Mu, Dongling Jia, Yayun He, Yuqing Miao, Hai-Long Wu, "Nano nickel oxide modified non-enzymatic glucose sensors with enhanced sensitivity through an electrochemical process strategy at high potential", Biosensors and Bioelectronics 26 (2011) 2948–2952.

Noise analysis and suppression for an infrared focal plane array CMOS readout circuits

Pengyun Liu, Jiangliang Jiang & Chunfang Wang
School of Optoelectronics, Beijing Institute of Technology, Beijing, China

ABSTRACT: The purpose of this paper is to analyze noise generation and its suppression technology in readout circuits. The ideas are as follows: a Positive channel Metal Oxide Semiconductor (PMOS) transistor can reduce 1/f noise; Complementary Metal Oxide Semiconductor (CMOS) switches cancel KTC noise; and three circuit structures of correlated double sampling are studied, which can cancel most of noise, including 1/f, KTC, and fixed-pattern noises. The noise is also suppressed by using auto-zero adjustment technology. In addition, differential and chopping stabilization technology have been used successfully. These are described briefly in this paper.

1 INTRODUCTION

An Infrared Detector (IR) has several applications within different industries, particularly in the military, medical imaging systems, industrial exploration, forest fire prevention, and video surveillance industries (Razavi, 2005). Infrared detection is divided into the uncooled and cooled. Compared to the cooled quantum detectors, the uncooled infrared detector shows significant advantages in terms of being low cost and suitable for mass production. It has received much research attention. In many applications, the signal is very faint (usually nA, or even pA level). The readout integrated circuit (ROIC) design is a process seeking a trade-off between the various parameters and the optimization which can achieve the lowest noise is especially important.

2 NOISE IN THEROIC

The ROIC is one of the key components of an infrared focal plane array (IRFPA). It is used for processing and reading signal detected by an IRFPA (such as integrating, amplifying, filtering, sampling and holding). The readout circuit consists of tens or hundreds of thousands metallic oxide semiconductor field effect transistors (MOSFET). With the rapid progress in technology, the scale of an IRFPA becomes larger and the pixel pitch becomes smaller, which imposes more stringent requirements. Therefore, it is necessary to analysis the noise of the circuit.

There is 1/f, KTC, and Fixed-Pattern noise (FPN). Some of the noise is due to the circuit structure, materials or manufacturing processes, and others are due to the circuit structure or circuit control. Low noise, low power, and high speed are essential for an advanced IRFPA circuit while the pixel area is severely constrained by the detector. Therefore, it is important to reduce or restrain the noise and increase the signal–noise-ratio in the design of the readout circuit of CMOS imagers (Svard, 2012). Combined with the actual work experience, this paper gives several methods to reduce noise.

2.1 The intrinsic noise of MOSFET

The intrinsic noise of MOSFET is the result of all free fluctuations of voltage and current occurring. It is a consequence of the discrete nature of charge and matter. Unfortunately, the intrinsic noise cannot be eliminated but can only be suppressed. It includes thermal noise, 1/f noise and shot noise, etc.

Every resistor generates thermal noise. The physical origin of this noise is the thermal motion of free electrons inside a piece of conductive material, which is totally random. The voltage of thermal noise in MOSFET can be given by

$$\overline{V_{n,th}^2} = 4kT\gamma \frac{1}{g_m} \tag{1}$$

where k is a Boltzmann constant (k = 1.38 × 10^{-23} J/K), T denotes the temperature, γ is a technological constant which is about 2/3, and g_m is the transconductance. Due to a reduction in the level of thermal noise, we can reduce the length of the transistor or increase the transistor's width and bias current.

Unlike the thermal noise, shot noise does not depend on temperature. Since shot noise depends only on the DC current through the device, modifying the bias current represents an easy way to control the noise level. This property is particularly useful in the design of a pre-amplifier.

In electronic devices, 1/f Noise (Flicker Noise or Excess Noise) is always conditioned by the existence of a DC current in a discontinuous medium. It is assumed to be due either to defects affecting the semiconductor lattice (including unwanted impurity atoms), or to interactions between charge carriers and the surface energy states of the semiconductor. The 1/f noise can be modelled by an source cascaded with the gate. The model is given by:

$$\overline{V_{n,f}^2} = \frac{K}{C_{ox}WL}\frac{1}{f} \quad (2)$$

Here, K is a technological constant, C_{OX} is the gate oxide capacitance per unit area, and f is the frequency. Typically, Flicker Noise exists at low frequency. IR imaging generally works in low frequency, therefore, the 1/f noise is the main noise.

2.2 KTC noise

KTC noise is due to the circuit structure or circuit control. The infrared detector is composed of an array of pixels; a pixel detects a point. We should readout pixel signals gained in each of the columns in some ways. Firstly, we must eliminate the signal which sampled on pre-amplifier last time. So, we need to reset the input capacitance. The reset action brings KTC noise. It is given by:

$$\overline{V_{n,KTC}^2} = KT/C_p \quad (3)$$

Here, C_p is the input capacitance of the pre-amplifier.

2.3 Fixed-pattern noise

The fixed-pattern noise (FPN) is caused by the mismatch in pixels or the columns circuit. This mismatch is caused by semiconductor material and manufacturing process. It leads to a different geometry and threshold voltage of MOSFET. The geometry influences on the noise is small for high precision integrated circuit processing technology at present. However, the effect of deviation of threshold voltage is serious for circuit performance, especially in IRFPA, whose analogue signal is weak. The reason is that different threshold voltage can influence the Voltage Gate Source (VGS), and the channel current I_D changes with V_{GS} ($I_D = g_m \times V_{GS}$). So its influence is more serious for circuit performance. It severely limits the quality of signals (Roxhed, 2010).

3 SEVERAL METHODS TO REDUCE NOISE

3.1 Using PMOS instead of NMOS

According to Equation (2), we can reduce the filter noise by increasing the area of the devices. A PMOS transistor conveys hole in the "buried channel" which is a certain distance from the oxide to the silicon interface, so the 1/f noise of PMOS is lower. With the

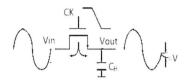

Figure 1. Charge injection.

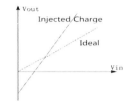

Figure 2. The input/output characteristics of the sampling circuit.

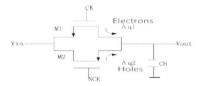

Figure 3. Complementary switch.

readout speed allowing, we can reduce 1/f noise by using PMOS to replace the NMOS in the circuit.

3.2 Complementary switch

In the readout circuit, we often use MOSFET as the switch. There must be a conduction channel between the silicon dioxide and silicon when the MOSFET is in a conducting state. A phenomenon called "channel charge injection" that allows electrons at the source to enter the channel and flow to the drain exists when the MOSFET is cutting off (Jin, 2010).

From the Figure 1, the charge in the left side is absorbed. However, the charge in the right is stored in C_H, so this introduces error to the voltage stored in the sample capacitor.

The distribution ratio of electricity is a complex function determined by many parameters, like electrical impedance to the earth in each end. The worst case is that the channel charge is stored all into the sampling capacitor. The input/output characteristics are shown in Figure 2.

According to the MOSFET structure and channel formation, one way to reduce the charge injection effect is by combining PMOS and NMOS. Then we can get

$$\Delta q_1 = W_1 L_1 C_{ox}(V_{CK} - V_{in} - V_{THN})$$
$$\Delta q_2 = W_2 L_2 C_{ox}(V_{in} - |V_{THP}|) \quad (3)$$

If $q_1 = q_2$, there is no noise. Another advantage is that the equivalent resistance value is much smaller than one.

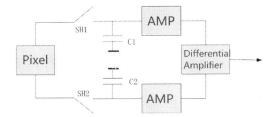

Figure 4. The first structure of CDS.

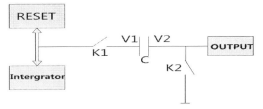

Figure 5. The second structure of CDS.

Of course, the complementary switch need the double chip area. With the rapid progress in technology, the scale of the IRFPA becomes larger and the pixel pitch becomes smaller. So the area for the ROIC is also an important aspect which we need to consider.

3.3 Correlated Double Sampling (CDS)

CDS is a widely adopted noise reduction method for discrete-time output signals. The basic idea of CDS is the elimination of the KTC noise and the reduction of the 1/f noise generated in the front-end of the amplifier. The use of a CDS integrator often requires voltage sampling at two or more different time points and records the differences. However, they can be mostly used in low frequency. The correlation of KTC and FPN in CDS can be given by:

$$P_R(\tau) = e^{-\Delta\tau/R_{eff}C} \qquad (4)$$

where $\nabla\tau$ is the sampling time interval, and R_{eff} is the resistance in the capacitance node. It has shown that $P_R(\tau)$ increases with decreasing $\nabla\tau$, and increasing $R_{eff}C$. The output voltage of the differential amplifier can be expressed by:

$$V_o = \frac{kT}{C}\left[\left(1 - e^{-\Delta\tau/R_{eff}C}\right)\right] \qquad (5)$$

In this equation, the KTC noise cannot be completely eliminated with $\nabla\tau$. However, it can eliminate most of the KTC noise with a small $\nabla\tau$ ($\nabla\tau \leq 100$ us) and suitable $R_{eff}C$. The CDS has three basic structures, respectively shown in Figure 4, 6 and 8.

The first structure of CDS is shown in Figure 4. It eliminates most noise with integral voltage difference, and it also can increase the dynamic range of the ROIC.

The second structure of CDS is shown in Figure 5. It is sampled by integrating on the sampling capacitors. It can eliminate or reduce the KTC noise, the flicker noise and the fixed-Pattern noise (Xu, 2002).

In addition, there is a simple CDS structure shown in Figure 6, and its control signal is shown in Figure 7.

It eliminates the low-frequency noise like a high pass filtering for the noise. When S_2 is connected with an independent voltage signal, the value of the noise is fixed; we can eliminate it in the digital signal processing.

Compared with the first CDS, the second and the third have fewer components, but the control signals

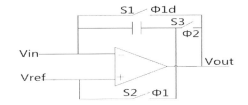

Figure 6. The third structure of CDS.

Figure 7. Time sequence of CDS.

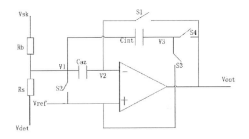

Figure 8. The ROIC architecture.

of them are more complex. The readout circuit needs better frequency characteristics. The most significant advantage is that there is no need for a subsequent differential amplifier after difference disposal is applied to the output signal, which makes it have a better integration level.

3.4 Auto-zero calibration technique

The structure of the readout circuit which uses auto-zero calibration technique is shown in Figure 8, and its control signal is shown in Figure 9. The circuit includes an amplifier, four switches, an auto-zero capacitor C_{AZ}, and an integrating capacitor C_{int}.

The four switches can control the capacitor integrating or resetting (Lu, 2010). High level means the switch is off, and low level means on (Nagate, 2004).

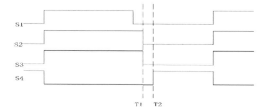

Figure 9. Time sequence of the ROIC.

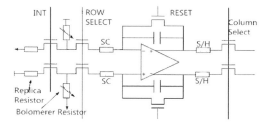

Figure 10. The structure of readout circuit of pixel.

C_{AZ} is the auto-zero capacitance storing the offset voltage and noise, and C_{int} is used for integration. If the gain and bandwidth of the amplifier can meet the requirements, the voltage of V_1 can be kept at V_{ref} in the integration process. At the same time, the amplifier requires node V_2 remaining unchanged, and the charge stored in the capacitor must be conserved, so the voltage of V_1 is kept at V_{ref}.

Therefore, the auto-zero technique ensures the voltage of the node V_1 maintained at V_{ref}. It can reduce the input offset voltage and noise of the amplifier.

3.5 Differential input stage

The differential input stage plays a key role in reducing the offset voltage and process deviation, suppressing noise, and improving sensitivity (Joon, 2010). The structure is shown in Figure 10.

3.6 Chopping stabilization

CHS is a continuous-time method which modulates the offset and the 1/f noise onto the high frequency to filter them out, while modulating the useful signals onto the high frequency and demodulating them to the baseband, based on digital modulation and demodulation technology. This kind of technology is very useful in the readout circuit of a continuous time microbolometer because it avoids the defect of white noise aliasing. The working principle is shown in Figure 11 (Alberto, 2002).

4 CONCLUSION

This paper analyses the noise of the ROIC, and gives several methods to reduce the noise. Noise in a CMOS readout circuit, especially in the pre-amplifier, greatly affects the performance of the IRFPA. The CMOS

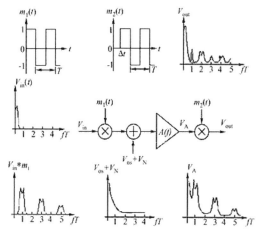

Figure 11. Working principle of CHS.

readout circuit structure should not only meet the low noise equirement to cater for large-scale focal plane array, it should also be content with other performance requirements, such as conversion speed, scale of circuit, and frame frequency. Several methods to reduce noise can be used in ROIC designing and they are explained in this paper.

REFERENCES

Alberto Bilotti, "Chopper-Stabilized Amplifiers with a Track-and-Hold Signal Demodulator," IEEE, Gerardo Monreal, 2002.

Behzad Razavi, "Design of Analog CMOS Integrated Circuits," USA. 2005.

Chen Xu and Shen Chao, "A New Correlated Double Sampling (CDS) Technique for Low Voltage Design Environment in Advanced CMOS Technology," ESSCIRC, 2002 (117–120).

Daniel Svard, Christer Jansson and Atila Alvand pour, "A Readout Circuit for an Uncooled IR Camera With Mismatch and Self-Heating Compensation," IEEE, 2012, 12(2): 102–105.

Jian Lu and Yun Zhou, "Novel Infrared Focal Array Readout IC with High Uniformity and Noise," Microelectronics, 2010, 40(2): 240–241.

Nagate H, Shtbai H and Hiraot "Cryogenic capacitive transimpedance amplifier for astro-nomical infrared detector," IEEE Trans Elec Dev, 2004, 51(2): 270–278.

N. Roxhed, F. Niklaus and A. C. Fischer, "Low-cost uncooled micro-bolometer for thermal imaging," Optical Sensing and Detection, SPIE, 2010.

Sang Joon Hwang, Ho Hyun Shin and Man Young Sung, "High performance read-out IC design for IR image sensor applications," Analog Integr Circ Sig Process, 2010 (64): 147–152.

Xiangliang Jin, "Principle of simple correlated double sampling and its reduced-area low-noise low-power circuit realization," Analog Integr Circ Sig Process, 2010(65): 209–215.

Speaker recognition performance improvement by enhanced feature extraction of vocal source signals

J. Kang, Y. Kim & S. Jeong
Electronics Engineering Department, ERI, Gyeongsang National University, Jinju, Republic of Korea

ABSTRACT: In this paper, an enhanced method for the feature extraction of vocal source signals is proposed for the performance improvement of speaker recognition systems. The proposed feature vector is composed of mel-frequency cepstral coefficients, skewness, and kurtosis extracted with low-pass filtered glottal flow signals. The proposed feature is utilized to improve the conventional speaker recognition system utilizing mel-frequency cepstral coefficients and Gaussian mixture models. That is, the scores evaluated from the convectional vocal tract and the proposed feature are fused linearly. Experimental results show that the proposed recognition system shows a better performance than the conventional one, especially in low Gaussian mixture cases.

1 INTRODUCTION

As society is developed, people want to seek convenience in their lives. Accordingly, the needs of e-commerce and digital government services using information technologies have increased rapidly. However, impersonation in order to access other people's databases happens frequently. Therefore, information security technology to protect personal information has been drastically on the rise. As a way of information security, biometric data has become a topical issue for several decades (T. Kinnunen et al., 2010). Among much bio-information, speech has been known to be the most convenient for human-computer interfaces. In conventional speaker recognition systems, Mel-frequency Cepstral Coefficients (MFCCs) or Linear Predictive Cepstral Coefficients (LPCCs) are extracted from input speech signals. Then, Gaussian Mixture Models (GMMs) are used to model the feature parameters of each speaker probabilistically. However, these methods have disadvantage in that they cannot model the morphological characteristics. Thus, in this paper, MFCCs, skewness, and kurtosis are extracted using the glottal flow signals in addition to conventional MFCCs. Also, low-pass filtering is performed to glottal flow before feature extraction because the high frequency range of the frequency spectrum of the glottal flow is relatively flat, that is, meaningless.

The structure of the paper is as follows. In Section 2, the proposed speaker recognition system is described. In Section 3, experimental results of the conventional, and the proposed system are presented comparatively. Section 4 concludes this study.

2 PROPOSED SPEAKER RECOGNITION SYSTEM

The proposed speaker recognition system pursues the performance improvement by using extracted features of glottal flow in addition to the MFCCs usually extracted in the vocal tracts. The overall structure of the proposed speaker recognition system is shown in Figure 1. Detailed explanations are as follows.

For short time input signals, the MFCCs to be utilized for the speaker recognition system are extracted basically. To extract feature parameters in glottal flow signals which are generated by vocal fold vibration, Voiced/Unvoiced (V/UV) speech classification should be preceded. The V/UV classification is performed by the scores evaluated by the MFCCs extracted using the input speech and their GMM-based probability functions. If a voiced interval is declared, Iterative Adaptive Inverse Filtering (IAIF) is applied to the input speech to generate the corresponding glottal flow signal and the estimation of MFCCs, skewness, and

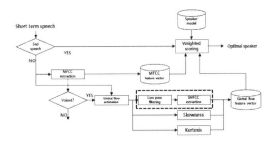

Figure 1. Proposed speaker recognition system.

kurtosis are followed using the glottal flow (T. Kinnunen et al. 2009). Hereafter, the MFCCs extracted using the glottal flow signal are denoted by the Source MFCC (SMFCC). Before the SMFCC extraction, low-pass filtering is performed because the high frequency range in the frequency spectrum of glottal flow is somewhat flat. When the analysis of the detected speech is finished, scores are evaluated by using extracted feature vectors and GMMs of the registered speakers. Finally, the optimal speaker is identified by the speaker model with the highest score.

2.1 GMM-based V/UV classification

For more precise classification of voiced and unvoiced speech, GMM-based probability models are constructed for the two classes. Generally, voiced speech is composed of harmonic signals by the periodical vibration of vocal folds and includes vocal source information (H. Kobatake, 2003). However, unvoiced speech has little vocal source information because it is produced while the vocal folds are open. Frequency spectral characteristics are different from each other and they are reflected to the MFCCs of input speech. Thus, the probabilistic modelling of voiced and unvoiced MFCCs can well discriminate the input short time speech. For the extracted MFCC vector, the V/UV classification is performed as:

$$\log(\mathbf{c}\mid\lambda_V) \overset{voiced}{\underset{unvoiced}{\gtrless}} \log(\mathbf{c}\mid\lambda_U) \quad (1)$$

where $\mathbf{c}$ = MFCC; λ_V = GMM of voiced speech; and λ_U = GMM of unvoiced speech.

2.2 Skewness and kurtosis of glottal flow

Human speech can be classified into voiced and unvoiced speech; whether the vocal folds are vibrated or not. For unvoiced speech, the sounds are produced by the air flow through the vocal organs while vocal folds are open and have no vibration. So, unvoiced speech and its vocal source signal are known to have approximately normal distribution in time domain. For voiced speech, vocal fold vibration makes the sounds basically. Thus, because the shape of its source signal, that is, the glottal flow is expected to be periodic, the distribution is somewhat peaky and asymmetric compared with that of the unvoiced speech Figure 2 shows an example of the sample distribution estimated using voiced and unvoiced source signals. As shown in Figure 2, the distribution of unvoiced speech is almost symmetric and similar to a Gaussian function with a variance.

Generally, the variance information is related to the scale of the input signal. Therefore, it cannot be an important feature for speaker recognition. The distribution of voiced glottal flow is peakier than that of unvoiced and skewed left as expected previously.

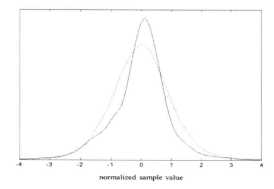

Figure 2. Distribution characteristics of the estimated glottal flow from voiced and unvoiced (solid line: voiced (/ah/), dashed line: un-voiced(/sh/)).

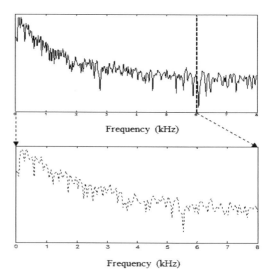

Figure 3. Original log power spectrum (upper) and stretched spectrum of the glottal flow (lower) (low-pass cut-off frequency: 6 kHz).

These characteristics are tightly related to those of the vocal fold vibration. Thus, they can be usefully applied to discriminate speakers. As mentioned earlier, the skewness and the kurtosis of the sample value are estimated by the 3rd and the 4th moments (C. Nikias et al., 1993). Equation 2 shows how to measure them:

$$S = \frac{\sum_{n=0}^{N-1}(x(n)-\mu)^3}{(N-1)\sigma^3}, K = \frac{\sum_{n=0}^{N-1}(x(n)-\mu)^4}{(N-1)\sigma^4} - 3 \quad (2)$$

where $x(n)$ = nth sample of input speech; N = number of samples; μ = mean of $x(n)$; and σ = variance of $x(n)$ The skewness becomes zero for symmetric distributions, positive for skewed-left ones, and negative for skewed-right ones. The kurtosis represents the sharpness of the distribution. For normal distributions, the value is zero. As the distribution becomes sharper, the value of its kurtosis becomes higher.

2.3 Low-pass filtering of glottal flow and spectral stretching

Usually, the high frequency range of the frequency spectrum of glottal flow becomes flat. Low-pass filtering of glottal flow is applied because its flatness is meaningless for feature extraction. Then, spectral stretching is performed to directly apply the spectrum to the MFCC extraction process (C.H. Shadle et al. 1978). At this time, if the cut-off frequency for low-pass filtering is too low, the loss of the useful information for speaker recognition can occur. Therefore, the optimal cut-off frequency should be searched for exhaustively. Finally, the filtered spectrum is stretched linearly and the MFCCs are extracted. Figure 3 shows an example of the spectral stretching of a glottal flow spectrum.

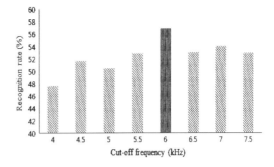

Figure 4. Recognition results for the conventional feature vector (number of mixture Gaussians: 16).

2.4 Score evaluation of speaker recognition

For each speaker the GMM is constructed by training feature vectors. The score for an utterance during test period is evaluated as:

$$Score(i) = \sum_{t=0}^{T-1} \log P(\mathbf{x}_t | \lambda_i) \quad (3)$$

where $\mathbf{x}_t = t$th MFCC vector of the input signal; $\lambda_i = i$th speaker GMM; and $T =$ number of feature vectors. Based on Equation 3, the fused score shown in Equation 4 is evaluated in this paper:

$$S_{tot} = (1-\alpha) \cdot S_{MFCC} + \alpha \cdot S_{SMFCC} \quad (4)$$

where S_{MFCC} = score evaluated by the MFCC of input speech; S_{SMFCC} = score evaluated by the SMFCC of input speech; and α = positive fusing factor less than one.

3 EXPERIMENTS AND RESULTS

As a training and test database for experiments, Texas Instruments and Massachusetts Institute of Technology (TIMIT) database production by the US National Institute of Standards and Technology (NIST) is used. The TIMIT database is composed of a total of 6,300 sentence files uttered by 438 males and 192 females from eight major regions in the United States. Each sentence is 2–4 seconds long and each speaker utters 10 different sentences with standard or dialectical styles. To train speaker models, eight sentences per speaker were selected. The other two sentences were utilized as test data. The analysis frame size was 30 ms and the Discrete Fourier Transform (DFT) was performed every 10 ms. The DFT size was 512. Basically, a twelve-order MFCC was extracted for input speech and its delta and accelerative components were concatenated to make a 36-order feature vector. Also, an SMFCC was extracted using glottal flow by the same process of the MFCC extraction, and skewness and kurtosis were additionally extracted. Thus, the

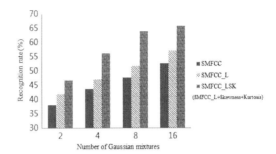

Figure 5. Comparative recognition performance of the proposed feature parameters (SMFCC_L: SMFCC extracted from low-pass filtered and spectrally stretched glottal flow).

dimension of the proposed feature parameter was 38. The number of mixtures for GMMs varied from 2 to 64 and the performance for each mixture was evaluated. α in Equation 4 was exhaustively searched from 0 to 1 with 0.01 step to find the best recognition performance.

3.1 Speaker recognition performance for conventional MFCCs

When the speaker recognition performances were measured, the accuracy started to decreased for Gaussian mixtures exceeding 16 due to the overfitting to small training database. Thus, the performance evaluation was measured up to 16.

3.2 Speaker recognition performance for proposed features

Figure 4 shows the performance of the SMFCC of low-pass filtered glottal flow according to the cut-off frequency. As shown in Figure 4, 6 kHz shows the best performance. Figure 5 shows the comparative performance for the proposed feature parameters. By concatenating the skewness and the kurtosis (SMFCC_LSK), more improved recognition performance was achieved, regardless of the number of mixtures.

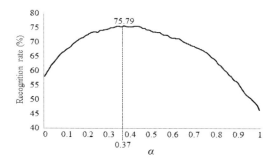

Figure 6. Recognition rates for fusing factor α.

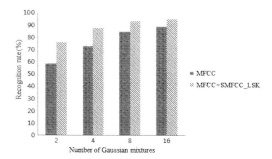

Figure 7. Recognition performance for score fusion.

3.3 *Performance improvement by fusing scores*

Figure 6 shows the recognition performance by varying α in Equation 4. At 0.37, the best performance is obtained. Figure 7 shows the recognition performance for score fusion. Consistently, the recognition rates by fusing the two heterogeneous scores are improved. The performance improvements were 17.6%, 14.6%, 8.7%, and 6.0% for mixture 2, 4, 8, and 16.

4 CONCLUSIONS

In this paper, a new feature parameter extraction method by low-pass filtering and spectral stretching of glottal flow was proposed. Additionally the skewness and the kurtosis were extracted to improve the recognition performance. The scores by conventional MFCCs and the proposed features were optimally fused to obtain higher recognition rates. The proposed speaker recognition system showed the better performance in all test cases. The performance improvement especially, was more noticeable in low Gaussian mixture conditions.

ACKNOWLEDGEMENTS

This research is supported by the Ministry of Culture, Sports and Tourism (MCST) and the Korea Creative Content Agency (KOCCA) in the Culture Technology (CT) Research & Development Program, 2014.

REFERENCES

[1] C.H. Shadle and B.S. Atal, "Speech synthesis by linear interpolation of spectral parameters between dyad boundaries," Speech, and Signal Pro., Vol. 3, pp. 577–580, 1978.
[2] C. Nikias and A. Petropulu, Higher-Order Spectra Analysis, Prentice Hall, 1993.
[3] H. Kobatake, "Optimization of voiced/Unvoiced decisions in nonstationary noise environments," IEEE Trans. Speech and Signal Proc., Vol. 35, No. 1, pp. 8–18, 2003.
[4] T. Kinnunen and P. Alku, "On separation glottal source and vocal tract information in telephony speaker verification," ICASSP, pp. 4545–4548, 2009.
[5] T. Kinnunen and H. Li, "An overview of text-independent speaker recognition: From features to supervectors," Speech Communication Vol. 52, No. 1, pp. 12–40, 2010.

… *Electronics and Electrical Engineering – Zhao (ed.)*
© 2015 Taylor & Francis Group, London, ISBN 978-1-138-02809-8

Online detection and disturbance source location of low frequency oscillation

Jun Luo, Fangzong Wang & Chongwen Zhou
College of Electrical Engineering & New Energy, China Three Gorges University, Yichang, Hubei Province, China

Baijian Wen
Guangdong Power Grid Co. Ltd., Power Dispatching Control Center, Guangzhou, Guangdong Province, China

ABSTRACT: Recently, low frequency oscillation has occasionally taken place in power girds. There are two steps to control low frequency oscillation. Firstly, low frequency oscillation needs to be detected quickly. Secondly, the parameters of low frequency oscillation as well as the earliest time of power oscillation, relevant equipment, or lines to lock the disturbance source, need to be identified. To this end, this paper used normalized kurtosis to judge whether the grid disturbance occurred, then applied a fast power method to identify the parameters of low frequency oscillation. The normalized kurtosis is used to determine the earliest time point of power oscillation, and to lock the disturbance source via a disturbance-time correlation analysis method. The above methods have been put into trial operation in the Guangdong Power Grid Corporation Dispatching and Communication Center. The result shows that the method has advantages of fast calculation, locating the disturbance source effectively, and it possesses good practicability.

1 INTRODUCTION

With the continuous expansion of the power grid scale, low frequency oscillation has occurred (Liang, Xiao, Zhang, Zhou & Wu, 2011). Quick identification and effective control are of great significance in ensuring the safe operation of the power grid. At present, a Wide Area Measurement System (WAMS) consisting of Phasor Measurement Units (PMUs) has been installed in the power system; it provides a platform for online detection, identification, and control of low frequency oscillation.

Researchers have put forward many kinds of identification methods of low frequency oscillation so far (Wang & Su, 2011; Li, Xie, Zhang & Li, 2007; Li & Wang, 2009; Wang & Li, 2010), but the detection method and disturbance source location method have not yet received much study (Dong, Liang, Yan & Yang, 2012). The existing sampling frequency of PMUs is 50 Hz; this means the sampling time is 20 ms. The number of PMUs installed in the power grid is very large, therefore, the amount of data collected by the WAMS is very large too. So, to detect low frequency oscillation quickly from the massive data is the basic link between security analysis and control of the power system (Liang et al., 2011).

A fundamental solution for low frequency oscillation detection online is to identify the parameters of the PMU data continually. Obviously, it requires or even wastes a large amount of computing resources, and it is difficult to achieve real-time calculation. In fact, low frequency oscillation is often occurred after a series of power grid disturbances. Therefore, in order to solve the problem of low frequency oscillation detection online, firstly, the power grid disturbances need to be detected. So far, the common disturbance signal detection methods mainly include the difference time domain method (Wei, Zhang, Geng, Zhang, Li & Liu, 2004), the wavelet analysis method (Pan & Li, 2013), and the mathematical morphology method (Wang, He & Zhao, 2008), etc. The difference time domain method is simple, with rapid calculation, but the singularity of the difference result is not enough, and cannot detect the grid disturbance accurately under the case of large load fluctuations. The wavelet analysis method using wavelet transform modulus maxima theory, can accurately get the break point of the signal, but the computational complexity leads to its limitations in practice. The mathematical morphology method has been used in the field of power quality disturbance detection, but it is difficult to set the disturbance threshold. In this paper, the normalized kurtosis of a PMU signal has been used as the index to detect disturbance (Zhang, 2002, p. 282; Di & Xu, 2013). This method calculates the normalized kurtosis of the standardization PMU signal, which is real-time updated by the sliding window, and compares the result with the disturbance threshold, then determines whether the disturbance exists in the sliding window or not.

After detecting grid disturbances, it needs to determine whether there is an occurrence of low frequency oscillation by identifying the parameters of the PMU data. Researchers have presented many identification methods including the Prony method (Wang & Su,

2011), the Hilbert-Huang Transform (HHT) method (Li, Xie, Zhang & Li, 2007), and the subspace tracking method (Li & Wang, 2009; Wang & Li, 2010), etc. A Prony algorithm is appropriate for processing stationary signals without noise, but it requires a large amount of calculations. The HHT algorithm has the ability of filtering high-frequency noise and processing non-stationary signals, but it has the end effect and the frequency aliasing effect. Subspace tracking methods such as PASTd (Wang & Li, 2010) belong to modern spectrum estimation which has not only high-resolution ability, but also fast calculation and high stability for tracking time-varying signals. So, a method called the "Fast Power Method" (FPM) subspace tracking algorithm (Hua, Xiang, Chen, Meraim & Miao, 1999; Wei & Bu, 2012) has been used in this paper to identify the parameters of low frequency oscillation.

In addition, it needs to lock the disturbance source of low frequency oscillation for dispatchers to take effective control measures. This paper used normalized kurtosis to determine the earliest time point of power oscillation, and locked the disturbance source by comparing the time point and the corresponding time interval of the position change information of remote signalling.

2 GRID DISTURBANCE ONLINE DETECTION METHOD, BASED ON NORMALIZED KURTOSIS

The PMU signal can be divided into the stationary signal and the dynamic signal in the power system. The stationary signal is caused by some random small disturbance such as load switching, and the dynamic signal is caused by some large disturbance. These two kinds of signals have random characteristic, but both of them obey a certain statistical regularity, which can be described by probability density distribution. The signal which satisfies normal distribution is called a Gauss signal, otherwise it is called a non-Gauss signal. Zhang (2002) defined an index called "normalized kurtosis" to distinguish Gauss signals and non-Gauss signals, and Di & Xu (2013) undertook some experiments which show that the stationary signal or the dynamic signal's normalized kurtosis is in the small fluctuations at 3; the normalized kurtosis is much larger than 3 when the sliding window includes a stationary signal and a dynamic signal simultaneously; normalized kurtosis instantaneous transitions indicate the disturbance has just happened.

Consider the data $x(t), (t=1,2,\ldots,N)$, the formula of its normalized kurtosis is given by:

$$K_x = \frac{E\{x^4(t)\}}{E^2\{x^2(t)\}} \quad (1)$$

where E means average calculation, $E\{x^4(t)\}$ means average calculation of the sum of the fourth power of $x(t)$, and $E^2\{x^2(t)\}$ means square the average calculation of the sum of square of $x(t)$.

The step of grid disturbance detection online is as follows:

(1) Calculate the current time point of normalized kurtosis of a standardized PMU signal in the sliding window.
(2) Compare the normalized kurtosis and the disturbance threshold to judge whether disturbance exists in the grid.
(3) Analyse the next time point data in the sliding window according to the sliding step.

3 IDENTIFICATION METHOD OF LOW FREQUENCY OSCILLATION, BASED ON A FAST POWER METHOD SUBSPACE TRACKING ALGORITHM

3.1 Introduction of a power method subspace tracking algorithm

The modern spectrum estimation method has been widely applied in the signal processing field. The subspace tracking method belongs to the modern spectrum estimation method, and its essence is a fast solution and update of the signal subspace. Due to its small amount of calculation, high resolution, and good tracking performance of time-varying signals, the subspace tracking method had been applied to identify the parameters of low frequency oscillation in recent years (Li & Wang, 2009; Wang & Li, 2010). Hua, Xiang, Chen, Meraim, & Miao (1999), put forward a method called "the nature power method" for subspace tracking, and the relevant experiment shows that the nature power method has better convergence than some common subspace tracking methods such as PAST, NIC (Miao & Hua, 1998), etc.

The data under consideration are a $m \times 1$ sequence: $\boldsymbol{x}(n) = [x(n), x(n+1), \ldots, x(n+m-1)]^T$. The covariance matrix of the sequence is denoted by $\boldsymbol{C}_{xx} = E(\boldsymbol{xx}^H)$. The principal subspace $Range(\boldsymbol{W})$ spanned by the sequence, of dimension $p < m$, is defined to be the span of the p principal eigenvectors of the covariance matrix. The weight matrix $\boldsymbol{W}$ of the signal subspace can be obtained by the following iterative formula:

$$\boldsymbol{W}(i) = \boldsymbol{C}_{xx}(i)\boldsymbol{W}(i-1)\left[\boldsymbol{W}(i-1)^H \boldsymbol{C}_{xx}(i)^2 \boldsymbol{W}(i-1)\right]^{-1/2} \quad (2)$$

The covariance matrix is updated by the following formula:

$$\boldsymbol{C}_{xx}(i) = \beta \boldsymbol{C}_{xx}(i-1) + \boldsymbol{x}(i)\boldsymbol{x}(i)^H \quad (3)$$

where β is a forgetting factor chosen between 0 and 1.

Multiplying $\boldsymbol{W}(i-1)$ on both right sides of (3) yields:

$$\boldsymbol{C}_{xx}(i)\boldsymbol{W}(i-1) = \beta \boldsymbol{C}_{xx}(i-1)\boldsymbol{W}(i-1) + \boldsymbol{x}(i)\boldsymbol{x}(i)^H \boldsymbol{W}(i-1) \quad (4)$$

Table 1. The FPM algorithm.

Initialization
$W(0) = [I_p; O_{(m-p)\times p}]$, $P(0) = I_p$
For $i = 1, 2 \ldots DO$
$y(i) = W(i-1)^H x(i)$
$h(i) = P(i-1) y(i)$
$g(i) = (\beta + y(i)^H h(i))^{-1} h(i)$
$\varepsilon^2 = \|x(i)\|^2 - \|y(i)\|^2$
$\tau(i) = \dfrac{\varepsilon^2(i)}{1 + \varepsilon^2(i)\|g(i)\|^2 + \sqrt{1 + \varepsilon^2(i)\|g(i)\|^2}}$
$\eta(i) = 1 - \tau(i)\|g(i)\|^2$
$y'(i) = \eta(i) y(i) + \tau(i) g(i)$
$h'(i) = P(i-1)^H y'(i)$
$P(i) = \beta^{-1}(P(i-1) - g(i) h'(i)^H)$
$e'(i) = \eta(i) x(i) - W(i-1) y'(i)$
$W(i) = W(i-1) + e'(i) g(i)^H$

A simple choice of approximation is:

$$C_{xx}(i) W(i-1) = \beta C_{xx}(i-1) W(i-2) + x(i) x(i)^H W(i-1) \quad (5)$$

Note:

$$C_{xy}(i) = C_{xx}(i) W(i-1) \quad (6)$$

$$y(i) = W(i-1)^H x(i) \quad (7)$$

Upon substituting (6) and (7) into (5) yields:

$$C_{xy}(i) = \beta C_{xy}(i-1) + x(i) y(i)^H \quad (8)$$

Then, with (6), (2) becomes:

$$W(i) = C_{xy}(i) [C_{xy}(i)^H C_{xy}(i)]^{-1/2} \quad (9)$$

The iterative method uses equations (7), (8), and (9) to calculate W called the "nature power method subspace tracking algorithm", and it requires $O(mp^2) + O(p^3)$ flops to compute. Clearly, the complexity of the algorithm is too high and not suitable for real-time computing. Therefore, Wei & Bu (2012) derived a new, fast subspace tracking algorithm called the "Fast Power Method subspace tracking algorithm" (FPM), based on matrix inversion lemma, which only requires $O(mp)$ flops to compute W. The recurrent process of the weight matrix W is mentioned in Table 1.

After i times of iteration, the weight matrix W will be convergent, and it equals to the signal feature vector matrix.

3.2 Parameter calculation of low frequency oscillation, based on the FPM

After detecting grid disturbances, the occurrence of low frequency oscillation needs to be determined by identifying the parameters of the PMU data.

Consider the PMU data $x(0), x(1), \ldots, x(N-1)$ that is to be modelled by:

$$\hat{x}(n) = \sum_{i=1}^{p} b_i z_i^n, \ (n = 0, 1, \cdots N-1) \quad (10)$$

$$b_i = A_i e^{j\varphi_i} \quad (11)$$

$$z_i = e^{(\alpha_i + j2\pi f_i) T_s} \quad (12)$$

where T_s is the sampling interval, and p is the rank of signal covariance matrix. A_i, f_i, φ_i, and α_i are amplitude, frequency, phase, and attenuation factor.

Use a fast power method subspace tracking algorithm to calculate the signal feature vector matrix W and compute the eigenvalues of signal subspace $S = (W_\downarrow)^+ W_\uparrow$. $W_\downarrow$ means remove the last row of W, and $W_\uparrow$ means remove the first row of W. $(\cdot)^+$ denotes pseudo-inverse.

Note that λ_i is the eigenvalue of signal subspace S; the frequency and the attenuation factor are given by:

$$f_i = angle(\lambda_i)/(2\pi T_s) \quad (13)$$

$$\alpha_i = \ln|\lambda_i|/T_s \quad (14)$$

Finally, use the least squares method to compute the relevant amplitude and phase as follows:

$$Z = \begin{bmatrix} 1 & 1 & \cdots & 1 \\ z_1 & z_2 & \cdots & z_p \\ \cdots & \cdots & \cdots & \cdots \\ z_1^{N-1} & z_2^{N-1} & \cdots & z_p^{N-1} \end{bmatrix} \quad (15)$$

$$B = [Z^H Z]^{-1} Z^H x(n) = (b_1, b_2 \cdots, b_p)^T \quad (16)$$

$$A_i = 2|b_i| \quad (17)$$

$$\varphi_i = angle(b_i) \quad (18)$$

Detection and control of low frequency oscillation is usually concerned about the dominant oscillation frequency only in the power system, thus, the order p of the signal, set to 3 or 4, is enough. Obviously, with the small value of p, the amount of calculation of the algorithm is very small, so it can calculate fast.

4 DISTURBANCE SOURCE LOCATIONS

Effective control measures need to be taken to quell power oscillation quickly after low frequency oscillation occurs, in order to avoid faults expanding further. According to the dispatching regulation (Liang et al.,

2011), the treatment of low frequency oscillation control can be summarized as direct control and indirect control. Direct control could eliminate oscillation quickly, accurately, and effectively, and influences only a narrow area, but it requires locking the disturbance source quickly. Therefore, a disturbance-time correlation analysis method, which analyses the relationship between the disturbance and its start time, has been provided in this paper.

Theoretically, the disturbance source location problem should be solved via finding the cause of low frequency oscillation. It is known that low frequency oscillation usually accompanies more than one power oscillation of a generator unit in the power system, and the generator occurred oscillation earliest is the source of power oscillation. So, the earliest time point (denoted by t_k) is an important index. Use normalized kurtosis to determine the earliest time point of power oscillation, and compare the time point and the position change information of remote signalling included in the corresponding time interval $[t_k - T, t_k]$ (T express the time domain length of searching forward). The comparison result could determine that the relevant disturbance of the position change information of remote signalling causes the low frequency oscillation.

In practice, due to using the PMU in WAMS, the time resolution of Sequence Of Events (SOE) is on a microsecond level, thus, it is feasible to use the above method to locate the disturbance source with PMU data.

5 SUMMARY OF ONLINE DETECTION AND DISTURBANCE SOURCE LOCATION OF LOW FREQUENCY OSCILLATION

Low frequency oscillation online detection and the disturbance source location method provided in this paper mainly includes the following three steps:

(1) Combine normalized kurtosis and the sliding window to detect grid disturbance online;
(2) Use the FPM to identify the parameters of the PMU data of the generator unit or tie line where occurred disturbance earliest;
(3) Use the disturbance-time correlation analysis method to lock the disturbance source. The whole flow is shown in Figure 1.

6 PRACTICAL APPLICATION CASES AND ANALYSIS

The above low frequency oscillation detection and the disturbance source location method have been put into trial operation in the Guangdong Power Grid Corporation Dispatching and Communication Center. Recently, two low frequency oscillation faults (denoted by Event 1 and Event 2) have been detected.

Figure 2 and Figure 3 show a part of the field data of a certain line which detected disturbance earliest in

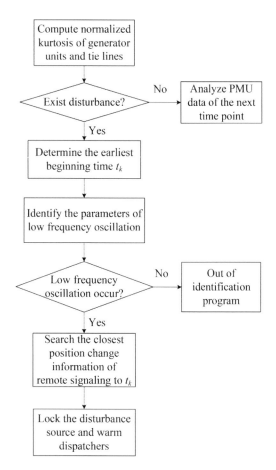

Figure 1. System operation flowchart of online detection and identification of low frequency oscillation.

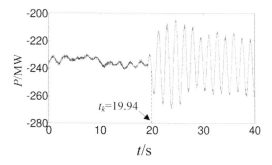

Figure 2. Active power field data in Event 1.

Event 1 and Event 2, respectively. Figure 4 and Figure 5 show the online calculation results of normalized kurtosis in Event 1 and Event 2, respectively.

It is clear from Figure 4 and Figure 5 that the stationary signal and the dynamic signal's normalized kurtosis fluctuates around 3; normalized kurtosis instantaneous transitions indicate the disturbance has just happened. In Event 1, the earliest time is equals to 19.94, and in Event 2, the earliest time is equal to 21.12.

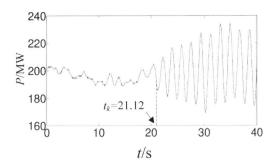

Figure 3. Active power field data in Event 2.

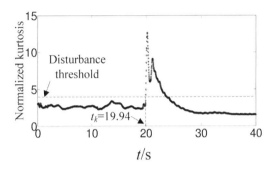

Figure 4. Calculation result of normalized kurtosis in Event 1.

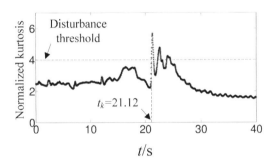

Figure 5. Calculation result of normalized kurtosis in Event 2.

Table 2. Parameter computational results of active power in Event 1.

Frequency [Hz]	Amplitude [MW]	Phase [°]	Attenuation factor
0.5262	11.4451	−177.8117	−0.0897
0.6150	22.8709	52.9916	0.0003

Table 3. Parameter computational results of active power in Event 2.

Frequency [Hz]	Amplitude [MW]	Phase [°]	Attenuation factor
0.5557	12.6568	151.8074	0.0886
0.6816	3.2184	−1.6609	−0.1582

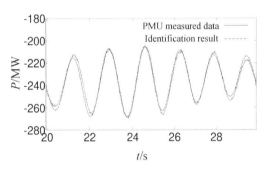

Figure 6. Identification results comparison chart in Event 1.

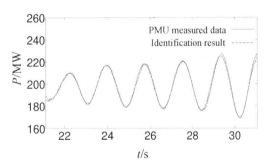

Figure 7. Identification results comparison chart in Event 2.

The online identification results of low frequency oscillation in Event 1 and Event 2 are shown in Table 2 and Table 3, respectively. Use the parameters to draw a fitting curve and compare with the measured data, as shown in Figure 6 and Figure 7. It is clear from Figure 6 and Figure 7 that the fitting curve is coordinated with the measured data. In other words, the identification results based on the FPM is reliable.

The disturbance-time correlation analysis method locked the disturbance sources exactly in the above two low frequency oscillation faults, and warned the dispatchers in time. In the future, this method would be carried out in further tests.

7 CONCLUSIONS

(1) Combine normalized kurtosis and the sliding window to compute the power grid disturbance; use the fast power method subspace tracking algorithm to identify the parameters of low frequency oscillation. This paper provides an online detection and identification method of low frequency oscillation which has characters of fast calculation and reliable identification.

(2) This paper provides a disturbance-time correlation analysis method to locate the disturbance source

which has the advantages of being simple and practical.

(3) This online detection and disturbance source location method had been put into trial operation in the Guangdong Power Grid Corporation Dispatching and Communication Center. The preliminary operation result shows that this method has good practicability.

REFERENCES

[1] Di, B. & Xu, Y.T. 2013. On-line power grid disturbance signal detect based on WAMS and normalization kurtosis. *Power System Protection and Control* 41(5): 140–145.

[2] Dong, Q., Liang, J., Yan, X.W. & Yang, R.J. 2012. Locating method of disturbance source of low frequency oscillation in large scale power grid. *Proceeding of the CSEE* 31(1): 78–83.

[3] Hua, Y.B., Xiang, Y., Chen, T.P., Meraim, K.A. & Miao, Y.F. 1999. A new look at the power method for fast subspace tracking. *Digital Signal Process* 9: 297–314.

[4] Li, C.C. & Wang, F.Z. 2009. *Online Parameters Identification of Low Frequency Oscillation by Neural Computation*. International Conference on Intelligent Computing and Intelligent Systems.

[5] Li, T.Y., Xie, J.A., Zhang, F.Y., & Li, X.C. 2007. Application of HHT for extracting model parameters of low frequency oscillations in power system. *Proceeding of the CSEE* 27(28): 79–83.

[6] Liang, Z.F., Xiao, M., Zhang, K., Zhou, J. & Wu, J. 2011. Discussion on control strategy for low frequency oscillation in China Southern Power Grid. *Automation of Electric Power Systems* 35(15): 54–58.

[7] Miao, Y.F. & Hua, Y.B. 1998. Fast subspace tracking and neural network learning by a novel information criterion. *IEEE Trans on Signal Processing* 46(7): 1967–1979.

[8] Pan, C.M. & Li, F.T. 2013. The detection and identification of transient power quality based on wavelet transform. *Electrical Measurement and Instrumentation* 50(10): 69–72.

[9] Wang, F.Z. & Li, C.C. 2010. *Online Identification of Low-Frequency Oscillation Based on Principal Component Analysis Subspace Tracking Algorithm*. Asia-Pacific Power and Energy Engineering Conference 2010.

[10] Wang, H. & Su, X.L. 2011. Several improvements of Prony algorithm and its application in monitoring low-frequency oscillations in power system. *Power System Protection and Control* 39(11): 140–145.

[11] Wang, L.X., He, Z.Y. & Zhao, J. 2008. Detection and location of power quality disturbance based on mathematical morphology. *Power System Technology* 32(10): 63–68+88.

[12] Wei, L., Zhang, F.S., Geng, Z.X., Zhang, B.L., Li, N. & Liu P.J. 2004. Detection, localization and identification of power quality disturbance based on instantaneous reactive power theory. *Power System Technology* 28(6): 53–58.

[13] Wei, Z.Q. & Bu, C.X. 2012. Fast power method for subspace tracking. Mathematics in Practice and Theory 42(10): 153–159.

[14] Zhang, X.D. 2002. *Modern Signal Processing (Second Edition)*. Beijing: Tsinghua University Press.

A soft-start Pulse Frequency Modulation-controlled boost converter for low-power applications

Min-Chin Lee & Ming-Chia Hsieh
Department of Electronic Engineering, Oriental Institute of Technology, New Taipei City, Taiwan

Tin-I. Tsai
Department of Communication Engineering, Oriental Institute of Technology, New Taipei City, Taiwan

ABSTRACT: This paper uses the Pulse Frequency Modulation (PFM) and current feedback control mechanism to implement a boost converter, which can stabilize the output voltage when the input voltage and load change. The converter, operating in Discontinuous Conduction Mode (DCM), is designed and simulated using a standard 0.35 um Complementary Metal-Oxide Semiconductor (CMOS) process. The converter has a 1421×1396 um^2 chip size and power dissipation of about 9.5 mW. This chip can operate with an input supply voltage from 3.0 V to 3.3 V, and the maximum output current is 95 mA. The converter's output voltage can be stabilized at 5.0 V with a soft-start time of 0.2 ms when the converter is started (power-on) to prevent inrush current and output voltage overshoot. The converter with a maximum output ripple voltage of 56 mV, has a load regulation that is less than 0.2 mV/mA and a line voltage regulation that is less than 4 mV/V With lower power consumption and smaller output ripple, the proposed soft-start PFM-controlled boost converter is suitable for portable or low-power devices.

1 INTRODUCTION

In recent years, the rapid development of portable electronic products with compact size, diversified application and prolonged usage time require high efficiency power management modules. Although fixed-frequency Pulse-width Modulation (PWM) is the mainstream design for heavy load and high efficiency power converters, it does have the characteristics of a high quiescent current and switching loss. Portable electronic products often work in the light load condition of standby or shot-down mode. Therefore, the varied-frequency PFM in discontinuous conduction mode is more preferable for such light load conditions. Also the inrush current of the boot of a PFM power converter often causes the effects of an overshoot of the output current, an impulse voltage drop of the output voltage, the degradation of circuit operating performance, and a shortening of the lifetime of the lithium batteries. Therefore it requires a soft-start circuit to protect the beginning of the turn-on stage for voltage regulators [1]. In this paper, a soft-start PFM-controlled boost power converter is proposed by a standard 0.35 um CMOS technology. The converter can be used in the backlight driver circuit and in various applications of the low-power power management module.

2 PRINCIPLE AND STRUCTURE OF A PFM-CONTROLLED BOOST POWER CONVERTER

The schematic diagram of a PFM-controlled boost power converter is shown below in Figure 1. The input range of this circuit is from 3.0 to 3.3 V and the output can be stabilized at 5.0V. The converter consists of the power stage and the PFM control stage. The power stage is composed of the power MOS transistors, the inductor, and the filtering capacitor. The PFM control stage is composed of a hysteresis comparator (three-stage source-coupled differential pair comparator), a current-sensing limiting circuit, an off-time (control pulse) generator, a soft-start circuit, a voltage-level shifter, and a non-overlap clock generator.

The voltage control loop uses the hysteresis comparator to detect the output voltage. When the output voltage, which is fed back through a voltage divider to one input of a comparator, is lower than the reference voltage (soft-start circuit output), it will let the current control loop produce a square wave of higher frequency to boost the power converter circuit.

Meanwhile, the current control loop also limits the inductor's current to let the output voltage of the power converter rise to 5V. Then it changes to voltage-mode control. This is the basic principle of the

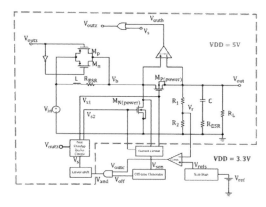

Figure 1. PFM-controlled CMOS boost power converter.

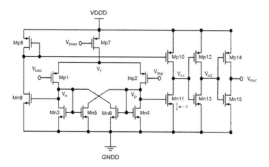

Figure 2. Source-coupled differential pair.

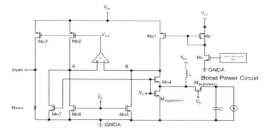

Figure 3. Current sensing limiter circuit.

switching power converter in the current-mode control. In this paper we focus on the analysis and design of component units of the current-mode PFM control loop [2].

3 ANALYSIS AND DESIGN OF CIRCUITS OF THE PFM CONTROL STAGE

3.1 Comparator

There are five comparators required in this soft-start PFM-controlled boost power converter. The main comparator controls the modulator which is the control pulse frequency generator. Another one is in the soft-start circuit, and a further one is used as a current control loop to limit the inductor's current (zero-current detector circuit). The off-time generator takes the final two comparators. In this article, the positive-feedback source-coupled differential pair is used as the structure in all comparators. This converter circuit needs to select the P-type or N-type input differential pair for different purposes. The following is an example using the P-type input differential pair. The structure is shown in Figure 2 [3][4]. This hysteretic comparator is composed of the PMOS differential pair stage, the Common-source (CS) amplifier stage, and the two inverter stages as buffers to improve the transient response of the comparator.

3.2 Current sensing limiter

The PFM current-control boost power converter requires the sense and also the limit magnitude of current flowing through the inductor. The structure of the circuit is shown in Figure 3 [5]. This circuit operation principle is to sense the current through the inductor by the V_{DS} voltage in power NMOS transistor is transmitted to Mn7 transistor through switch Mn4 transistor and error amplifier (virtual short). And the Mn7 is 1/k size of power NMOS. Its purpose is to facilitate the control of the following stage and reduce the power consumption. The error amplifier is an operational amplifier (OPA) with high gain and low drift voltage [6]. To avoid sensing error, the circuit senses the waveform of the rising inductor current when the inductor stores energy. By transmitting to MP3 transistor through error amplifier and current mirror, the sensing inductor current through Rsen is transferred into Vsen. Then the Vsen further transmits to the next stage (Off-time Pulse Generator) of the PFM control loop.

3.3 Off-time pulse generator

This circuit is the core of the PFM-controlled boost power converter. Its main function is a frequency generating circuit to generate the PFM control pulse in the circuit of the power converter, as shown in Figure 4 [2][7]. The circuit is divided into two parts. The first part uses the signal Vsen, generated by the previous stage (current sensing circuit), to do the control. First, it confines the inductor current loop through a comparator. Then it is further transmitted to a digital circuit to generate the desired square wave. Be aware that the inverter INVS is an inverter with a delay function. It is very important because its amount of delay time affects the operation of the whole circuit. The second loop is a frequency generating circuit. It transmits the square wave generated by charging and discharging the capacitor. The signal of these two parts is controlled by SR flip-flop and then transmitted to the control loop of the circuit.

3.4 Non-overlap buffer

A non-overlap buffer, as its name suggests, is a buffer circuit used to select and drive the power MOS circuit with synchronous rectifier technology. Because

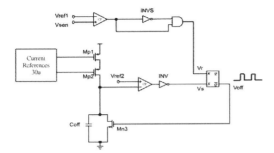

Figure 4. Off-Time pulse generator.

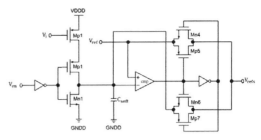

Figure 6. Soft-start circuit.

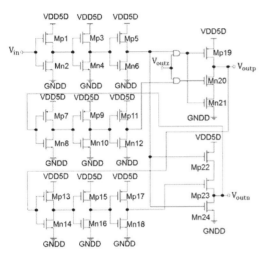

Figure 5. Non-overlap buffer.

the amount of current flowing through the power MOS is quite large, the size of the power MOS is also quite large respectively. Thus it requires a very strong capability to fully drive the power MOS. Non-overlap is a choice to drive the power MOS because the power NMOS and PMOS transistors of synchronous rectifier technology can-not be opened simultaneously. If both transistors are turned on at the same time, it will cause a large current to flow to the ground. Then the power MOS components will be burned out. Therefore, a non-overlap buffer with switched capability is used to select power NMOS or PMOS transistor.

In this paper, two digital logic AND gates are added into the traditional structure [8]. The Voutz, output of the zero-current detecting circuit is connected to one of the AND gate inputs. This will make the power PMOS close early enough to avoid the extra power consumption, as shown in Figure 5.

3.5 Soft-start circuit

With power-on on a PFM-control boost power converter which has no soft-start circuit, its output voltage will be zero. The negative input of the P-type differential input pair will be connected to this output. Therefore, the difference between this output voltage and the reference voltage will be too large, resulting in a power PMOS continuous conduction. It forces the output voltage of this boost power converter to rise to normal value very quickly. During this period, there will be excessive current flowing through this boost power converter, which causes damage to the power stage circuit in the power converter.

The structure of the soft-start circuit is shown in Figure 6 [1]. When the converter is started (Ven = 1 at this moment), the converter enters soft-start operation first. The soft-start circuit generates an increasing reference voltage and the comparator first compares the Vsoft of the plug-in capacitor with the reference voltage, Vref. Then the output voltage of the comparator selects the Vsoft of the multiplexer. The output voltage. Vrefs (=Vsoft) affected by the current source, rises slowly. This mechanism can avoid to a large input current. Until Vsoft is higher than Vref, the comparator will output, another voltage and the multiplexer sends Verf as output voltage Vrefs. For normal operation, the output voltage of this boost power converter sets up a negative feedback to regulate the output voltage by the Verf adjusted.

The references of the current reference circuit, the zero-current detector, the anti-ringing circuit, and the level shifter circuit of the soft-start PFM-controlled boost converter are in [4]–[9].

4 SIMULATION RESULTS

4.1 Output ripple voltage

Figure 7 shows that the output voltage and its ripple voltage of the soft-start PFM-controlled boost converter can stabilize at 5 V steadily, under an input source voltage of 3 V, when the circuit is at full load condition (Iload = 95 mA). The results of post-sim and pre-sim show that the output ripple voltages are 52 mV and 51 mV respectively. The steady-state time of both are about 200 us, and the output voltages are all at 5 V steadily.

Figure 8 shows the output voltage, and the ripple voltage of the boost converter can stabilize at 5 V steadily, under an input source voltage of 3.3 V, when the circuit at full load condition (Iload = 95 mA). The results of post-sim and pre-sim show that the output ripple voltages are 56 mV and 55 mV respectively. The

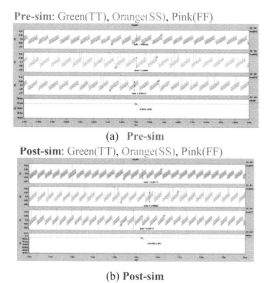

(a) Pre-sim

(b) Post-sim

Figure 7. Ripple voltage [Vin = 3V, Iload = 95 mA].

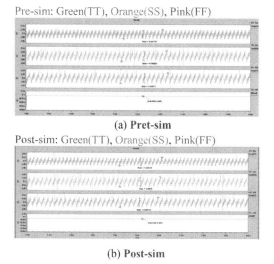

(a) Pret-sim

(b) Post-sim

Figure 8. Ripple Voltage [Vin = 3.3V, Iload = 95 mA].

steady-state time of both are about 200 us, and the output voltages are all at 5 V steadily.

4.2 Load regulation

Figure 9 shows that when the input voltage is fixed at 3 V the step variations of the load current (5 mA to 95 mA) are from light to heavy load at time 1 ms and current (95 mA to 5 mA) are from heavy to light load at time 2 ms, and the output voltage of the circuit is observed. From pre-sim and post-sim the output voltage eventually is stabilized at 5 V and its steady time is about 200 us. The variation of the output voltage of this boost power converter is about 10.4 mV. In

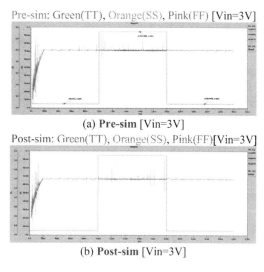

(a) Pre-sim [Vin=3V]

(b) Post-sim [Vin=3V]

Figure 9. Load Regulation [ILoad = 5 mA → 95 mA].

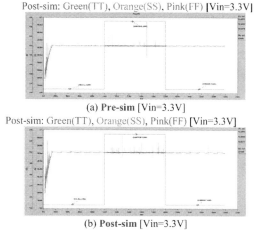

(a) Pre-sim [Vin=3.3V]

(b) Post-sim [Vin=3.3V]

Figure 10. Load Regulation [ILoad = 5 mA → 95 mA].

other words, post-sim and pre-sim show that the load regulation of the circuit is less than 0.2 mV/mA.

Figures 10(a) and (b) show the output voltage variation of the converter is about 10.5 mV (pre-sim) and 10.2 mV (post-sim) when the load current step variations are from 5 mA to 95 mA. The output voltage can stabilize at 5 V and its steady time is 200 us under an input source voltage operated at 3.3 V; the load regulation of the circuit is less than 0.2 mV/mA.

4.3 Line regulation

Figure 11 shows that when the output of the converter is fixed at light load condition (Iload = 5 mA), the step variation of the input voltage (3 V to 3.3 V) at time 1 ms and the input voltage (3.3 V to 3 V) at time 2 ms and the effects given by the variation of output voltage of the circuit are observed. From pre-sim and post-sim the output voltage eventually is stabilized at 5 V and

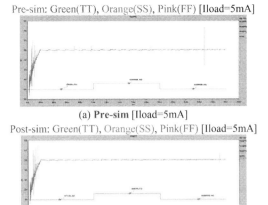

(a) **Pre-sim** [Iload=5mA]

(b) **Post-sim**[Iload=5mA]

Figure 11. Line Regulation [Vin = 3V → 3.3 V].

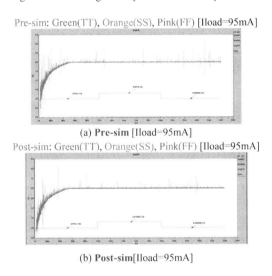

(a) **Pre-sim** [Iload=95mA]

(b) **Post-sim**[Iload=95mA]

Figure 12. Line Regulation [Vin = 3V → 3.3V].

its steady time is about 200 us. The variation of the output voltage of the converter is about 0.3 mV and 0.4 mV. In other words, pre-sim and post-sim show the line regulation of this boost power converter is within 3 mV/V.

Figures 12(a) and (b) show the output voltage variation of the converter is about 0.6 mV (pre-sim) and 0.3 mV (post-sim) when the input voltage step varies from 3 V to 3.3 V. The output voltage can stabilize at 5 V and its steady time is 500 us under heavy load condition (Iload = 95 mA); the line regulation of the circuit is less than 4 mV/V.

4.4 Switching frequency

Figures 13(a) and (b) show the switching frequency variation range of the proposed PFM-controlled boost converter versus the input voltage when the output

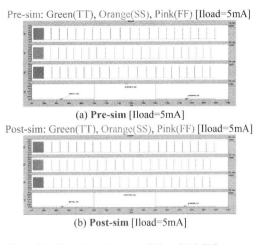

(a) **Pre-sim** [Iload=5mA]

(b) **Post-sim** [Iload=5mA]

Figure 13. Switching Frequency [Vin = 3V, 3.3V].

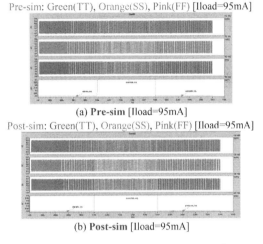

(a) **Pre-sim** [Iload=95mA]

(b) **Post-sim** [Iload=95mA]

Figure 14. Switching Frequency [Vin = 3V, 3.3V].

load is fixed at light load condition (Iload = 5 mA). Ppost-sim and pre-sim show the switching frequency is 7.8 kHz and 7.9 kHz respectively when the input voltage is fixed at 3V. The switching frequency is 6.6 kHz and 6.7 kHz respectively when the input voltage is fixed at 3.3V.

Figures 14(a) and (b) show the switching frequency variation range of the converter versus the input voltage when the output load is fixed at full load condition (Iload = 95 mA). Post-sim and pre-sim show the switching frequency is 13.6 kHz and 26.3 kHz respectively when the input voltage is fixed at 3V. The switching frequency is 50 kHz and 52.6 kHz respectively when the input voltage is fixed at 3.3V.

5 CONCLUSIONS

The integrated circuit (IC) physical layout of this soft-start PFM-controlled CMOS power converter is shown in Figure 15 [10].

Chip Size: 1421*1396 um²
Power Dissipation: 9.5 mW
Max. Frequency: 400 kHz

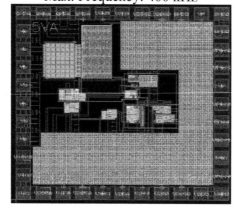

Figure 15. Integrated circuit layout of a soft-start PFM-controlled boost converter.

Table 1. Corner case simulation.

Parameter		TT	SS	FF
Input Voltage Range		3 V to 3.3 V	3 V to 3.3 V	3 V to 3.3 V
Power Dissipation		7.9 mW	9.5 mW	8.1 mW
Iout (max.)		95 mA	95 mA	95 mA
Output Voltage		5 V	5 V	5 V
Operating Frequency (max.)		350 kHz	360 kHz	320 kHz
		350 kHz	360 kHz	320 kHz
Ripple Voltage		55 mV	56 mV	50 mV
Load Regulation (mV/mA)	5 to 95 mA	0.116	0.117	0.12
	95 to 5 mA	0.114	0.115	0.105
Line Regulation (mV/V)	3 V to 3.3 V	1.3	1.3	3
	3.3 V to 3 V	2	3.3	2.6
Switching Frequency	ILoad = 5 mA	6.6 kHz~	6.8 kHz~	6.4 kHz~
	Vin = 3 V to 3.3 V	7.9 kHz	8.1 kHz	7.5 kHz
	ILoad = 95 mA	13.6 kHz~	26.3 kHz~	26.3 kHz~
	Vin = 3 V to 3.3 V	52.0 kHz	52.6 kHz	62.5 kHz

By using the simulation and design of a standard 0.35 um CMOS technology, the characteristic parameters of different corners from the previous section are shown in Table 1 and 2 for pre-simulation, post-simulation and required specification. Based on the aforementioned discussions, we can conclude that the proposed soft-started PFM-controlled boost converter has a chip size of $1421 \times 1396\,um^2$, and power dissipation of about 9.5 mW. The chip supply voltage can be from 3.0 V to 3.3 V, and the switching frequency can be from 6.6 kHz (with a high input-voltage and light load) to 52.6 kHz (with a low input-voltage and heavy load), and its output voltage can stabilize at 5.0 V, with less than a 60 mV ripple voltage at the maximum loading current 95 mA. The converter with a soft-start time of about 0.2 ms effectively suppresses the inrush current and the overshooting of the output voltage during startup period. The load regulation and line regulation of the converter are about 0.12 mV/mA and 3.3 mV/V, respectively. Finally, the simulated result shows that the soft-start PFM-controlled boost converter is suitable for a light power system, for example handy terminals or low-power devices.

Table 2. Comparison between expected specification and pre-sim and post-sim results.

Parameter	Spec.	Pre-sim	Post-sim
Input Voltage Range	3V to 3.3V	3V to 3.3V	3V to 3.3V
Power Dissipation	<10 mW	9.5 mW	9.5 mW
Iout (max.)	95 mA	95 mA	95 mA
Output Voltage	5 V	5 V	5 V
Operating Frequency (max.)	350 kHz	350 kHz	400 kHz
Ripple Voltage	<60 mV	56 mV	56 mV

ACKNOWLEDGEMENT

The authors would like to thank the Chip Implementation Center (CIC) of the National Science Council, Taiwan, for supporting chip manufacturing.

REFERENCES

[1] Wan-Rone Lio, Mei-Ling Yeh and Y. L. Kuo. "A High Efficiency Dual-Mode Buck Converter IC for Portable Application", IEEE Trans. Power Electronics, Vol. 23, No. 2, pp. 667–677, 2008.
[2] Yu-Ming Chen, "Design and Implementation of PFM Boost Converter for White LED Driver", Master Thesis, 2012.
[3] Phillip E. Allen and Douglas R. Holberg, "Analog Circuit Design", Saunders College Publishing HBJ, 2002.
[4] Chih Liang Huang, "Low Voltage, Zero Quiescent Current PFM Boost Converter for Battery-operated Devices", Master Thesis, 2006.
[5] C. F. Lee and P. K. T. Mok, "A monolithic current-mode DC-DC converter with on-chip current-sensing technique", IEEE J. Solid-State Circuits, No. 39, pp. 3–14, Jan. 2004
[6] Behazad Razavi, "Design of Analog COMS Integrated Circuits", International edition, McGraw-Hill, 2001.
[7] You-Hsiang Lu, "Design of the White LED Boost Chip with PWM for TFT LCD", Master Thesis, 2008.
[8] Chun-Chang Tseng, "Design of a Low EMI Synchronous Dual-Mode Boost Regulator", Master Thesis, 2010.
[9] Hong-Yi Huang, "Lab for Analog Integrated Circuits Design", 2003.
[10] Hong-Yi Huang, "Analysis of Mixed-Signal IC Layout", 2006.

Thermal analysis of phase change processes in aquifer soils

D. Enescu, H.G. Coanda, O. Nedelcu & C.I. Salisteanu
Department of Electronics, Telecommunications and Energy, Valahia University of Targoviste, Romania

E.O. Virjoghe
Department of Automatics, Informatics and Electrical Engineering, Valahia University of Targoviste, Romania

ABSTRACT: Thermal computation of the soil freeze supposes the knowledge of thermophysical properties of rocks in order to dig the underground tunnels. In this paper some ground-freezing techniques are analyzed using the total freezing wells on a linear gallery with two parallel paths. The model enables to solve by numerical methods the equations needed in order to obtain the Neumann constant as well as the velocity and acceleration of freezing front propagation and their average values. In many cases, the approximation approaches can provide some simple and fast solutions, with acceptable accuracy for practical engineering purposes. The authors illustrated a dedicated procedure for computing these parameters, implemented by them in the Visual C platform.

1 INTRODUCTION

The artificial ground freezing (AFG) was practiced in order to form a mineshaft being patented by Poetsch in Germany. Recent technological advances, such as computational analysis, drilling technologies as well as mobile freeze units have increased the number and types of applications.

To protect excavations, the ground freezing was used to stabilize sample weak ground, construct temporary access roads, and also to maintain permafrost below overhead pipeline foundations. Execution of some civil engineering in the unsteady soils with water meets some difficulties which are insurmountable at the big depths. A method which solves this problem is the aquifer rock consolidation by freeze. Therefore, the soil is artificially frozen on the whole digging area. Mechanical strength of soil increases and soil permeability decreases with freezing. Artificially, the frozen soil is also utilized for underground tunnel construction in urban areas and for soft ground construction in caves. First of all, this paper is focused on the ground-freezing technique, using as example the total freezing wells on a linear gallery of 100 m, with two parallel paths. For this study which is made on linear meters, the wells with a length of 10 m are used. External radius of the external wall of pipe has the value of 127 mm according to the standard STAS 404/1–80. An important condition is that the interactions between wells must be neglected and the temperature field is strongly affected by the well placement within the working area. In order to solve its complexity, the problem can be approached using some approximations, among which it is possible to set up a formulation depending on a parameter called the Neumann constant. The aim of this paper is to compute the Neumann constant by means of a software implemented by the authors in the Visual C platform. The heat load charts are obtained by using the freezing time and the well number. Application of numerical methods in order to investigate some processes has to be justified by some kind of verification. Therefore, these verifications for thermal models are confined to comparison with analytical solutions for some cases with constant initial and boundary conditions (e.g. Neumann solution). Mathematical modeling of the thermal interaction between the wells and frozen soil under steady-state conditions has been carried out in a number of works. Most of these studies present solutions obtained under numerical methods in order to solve boundary value problems.

2 PROBLEM FORMULATION

Ground freezing is based on withdrawing heat from soil. This method is very safe but it is necessary to be designed in detail. The process converts in-situ pore water into ice. To create a frozen soil body either a row of vertical, horizontal or inclined freeze wells have to be drilled. An opened inner pipe refers to as a down pipe, is inserted into the centre of the closed-end freeze pipe (Fig. 1). The down pipe is used in order to supply a cooling medium (brine or liquid nitrogen). The inner pipe is connected to a supply line and outer pipe to a return line (when brine is used) or exhaust line (when liquid nitrogen is used). The coolant flows through the inner pipe. Brine (usually, CaCl2 calcium chloride) freezing requires a closed circulation system and the use of refrigerating plant. This flows back through a manifold system before returning to the freeze plant. Generally, the brine supply temperature is from −20°C

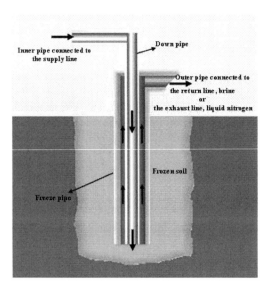

Figure 1. Ground freezing technique.

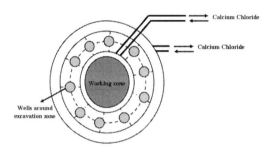

Figure 2. Wells placed on the excavation area.

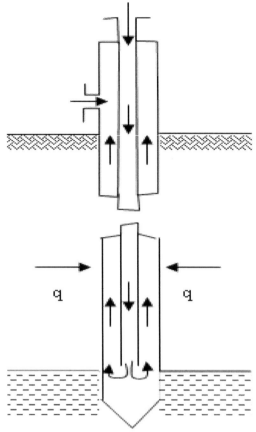

Figure 3. Well necessary for the soil freeze area.

to −37°C. Liquid nitrogen (LN2) starts to vaporize at a temperature −196°C in the annulus between freeze and inner pipe, picking up heat on its way up. The cold nitrogen gas is directly vented into atmosphere. A frozen soil body can be formed within a few days, whereas it takes weeks for the brine system.

A classical compression refrigerating plant to freeze the soil is used for small depths (30–40 m). The evaporator is composed of special wells introduced in soil and placed in circle, in line, on the perimeter of a rectangle or ellipse depending on the shape of final construction. As cooling agent, ammonia in direct circuit for small depths or calcium salt solution for big depths is used. Figure 2 shows a freezing well necessary for freezing the aquifer soil. It is compulsory for the wells to be vertically inserted, otherwise unfrozen zones could appear and water could enter the working area of interest for the excavation. A well has an inner pipe by which cooling agent flows in soil and an external pipe coaxial with the first one by which the heated cooling agent flows towards the refrigeration plant (Fig. 3).

Pipes have the ratio of diameters given by the condition of the Reynolds criterion:

$$\text{Re}_i > \text{Re}_e \qquad (1)$$

where: $Re = \frac{w \cdot d}{v}$ is the Reynolds number; w [m/s] is the velocity; d [m] is the diameter; v [m²/s] is kinematic viscosity; Re_i is the Reynolds number corresponding to the inner diameter of the well; Re_e is the Reynolds number corresponding to the external diameter of the well.

The well tip has to be built in order to withstand mechanical stresses which can occur during their introduction in soil. The wells are inserted in the soil till an impermeable layer is reached. If there is no such a layer, the soil is artificially waterproof at the bottom of the well. The wells with little length are inserted in the soil by vibration, while the long wells are inserted in soil by an orifice drilled. It is compulsory for the wells to be vertically inserted, otherwise unfrozen zones could appear and water could enter the working area of interest for the excavation. The wells are fixed on the working area at distances of about 1 m. As the distance between wells is lower, as the freezing time is less. The well connection to refrigerant distributor and refrigerant collector is made taking into account that

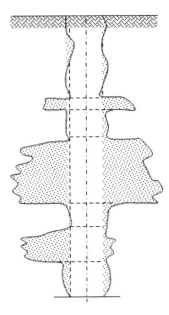

Figure 4. The rock body consolidated by freeze around a well area.

during freezing the aquifer soil increases its volume, thus leading to bigger mechanical efforts of the pipes, which could finally damage the wells.

Therefore, it is imposed that the refrigerant distribution and refrigerant storage is made by rubber pipes, on which there is the hoar-frost during their operation. The hoar-frost melting is a signal that the well does not work.

The cooling agent receives heat from the rocks as well as from the water which freezes. Around each well, there is a frozen area and after a time interval called closed time all frozen areas are connected each other forming a rock layer strengthened by freeze which does not allow within the working area to be water or running ground (Fig. 4). In order to determine ground monitoring as well as the growth of the freeze wall, it was necessary to install thermo-couple string at several locations around the freezing region. The thermo-couples were lowered into a cased hole. Each string was comprised of sensors located at interval down any given temperature monitoring hole.

Closure of the rock wall consolidated by freeze can be controlled by means of the control wells placed inside or outside of working area. Inside of the perimeter where the wells are placed, all rocks are consolidated by the water freezing. Therefore, the excavation works are made by explosion or by means of a hammer breaker in order to break the rocks. Working area is insulated on perimeter as well as is temporary consolidated with timber sets. Final consolidation is made by gallery concreting on galleries.

Therefore, the insulation is compulsory in order to avoid the occurrence of the rock thawing from inside.

Figure 5 presents the rock wall formation of the consolidated by freeze, where the control well are denoted as Sc inside the working zone and outside of it.

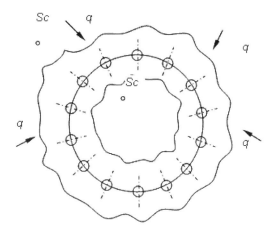

Figure 5. External rock wall consolidated by freeze zone.

For this study which is made on linear meters, the wells with a length of 10 m are used. External radius of the external wall of pipe has the value of 127 mm according to the standard STAS 404/1-80.

3 MATHEMATICAL MODEL

In analysis of this problem, we consider the soil having two different areas: a frozen zone and an unfrozen zone. The soil is considered as a homogeneous material having constant thermal properties, even if in reality the properties of the frozen and thawing zones may be different. In Figure 6 is presented the temperature field along two zones. The computing parameters are: T_f is the heat carrier temperature in well axis, T_0 is the freezing temperature of water, T_1 is the soil temperature in frozen zone 1, T_2 is the soil temperature in unfrozen zone 2, h is the convection heat transfer coefficient by surface $\left[\frac{W}{(m^2 \cdot K)}\right]$, l_s is the latent heat of solidification $\left[\frac{J}{kg}\right]$, k is the thermal conductivity in other two zones $\left[\frac{W}{(m \cdot K)}\right]$, ρ is density in other two zones $\left[\frac{kg}{m^3}\right]$, T_i is the initial soil temperature, α is thermal diffusivity $\left[\frac{m^2}{s}\right]$ in other two zones, Ts is the temperature of contact surface between soil and external pipe of well. To solve this problem a few important hypothesis are specified: soil temperature Ts is constant in time, temperature in well axis is considered constant, heat source in soil zero and there are not considered the reciprocal influences between two neighbor wells. The heat transfer equations for other two zones are given by:

$$r<R; \frac{\partial T_1}{\partial r} = \alpha_1 \nabla^2 T_1 \qquad (2)$$

$$r>R; \frac{\partial T_2}{\partial r} = \alpha_2 \nabla^2 T_2 \qquad (3)$$

The initial and boundary conditions are:

$$T_1(r_1,t) = T_s \qquad (4)$$

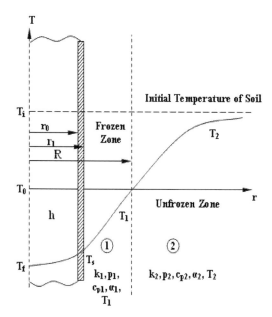

Figure 6. Temperature field in frozen soil and unfrozen soil.

$T_2(r,0) = T_i$ (5)

$T_2(\infty,t) = T_i$ (6)

$T_1(R,t) = T_2(R,t) = T_0$ (7)

Equation of the heat balance at the freezing boundary for $r = R$:

$$k_1 \nabla T_1 = k_2 \nabla T_2 + \rho_1 l_s W \qquad (8)$$

that means the rate of heat transfer which passes through the frozen zone is equal to the rate of heat transfer which passes through the unfrozen zone at which the rate of heat transfer from the latent heat of solidification is added.

The temperature field is given by the following general equation:

$$T = A \cdot E_i(-u) + B \qquad (9)$$

where:

$$E_i(-u) = \int \frac{e^{-t}}{t} \cdot dt \qquad (10)$$

is called the exponential integral. Using the general form of the temperature field given by equation (9), the following functions of the temperature fields in other two zones can be proposed:

$$T_1(r,t) = A\left[E_i\left(\frac{-r^2}{4\alpha_1 t}\right) - E_i\left(\frac{-r_1^2}{4\alpha_1 t}\right)\right] + B \qquad (11)$$

$$T_2(r,t) = C \cdot E_i\left(\frac{-r^2}{4\alpha_2 t}\right) + D \qquad (12)$$

From the first intitial condition (4) we obtain the unknown quantity noted with B:

$$T_1(r_1,t) = T_s = B \qquad (13)$$

The initial conditions given by equations (5) and (6) indicate in variable $u = \frac{r^2}{(4\alpha \cdot t)}$ just one condition of uniqueness: $r = \infty$ or $t = 0$ require as $u = \infty$. Therefore:

$$T_2(\infty,t) = T_2(r,0) = T_2(\infty) = T_i = D \qquad (14)$$

The initial condition (7) which represents the continuity of temperature fields in the freezing front is given by:

$$A \cdot \left[E_i\left(\frac{-R^2}{4\alpha_1 t}\right) - E_i\left(\frac{-R_1^2}{4\alpha_1 t}\right)\right] + T_s =$$
$$= C \cdot E_i\left(\frac{-R^2}{4\alpha_2 t}\right) + T_i = T_0 \qquad (15)$$

This equation is verified only if the following condition called the Neumann hypothesis is verified:

$$R = n\sqrt{t} \qquad (16)$$

where n represents the Neumann constant which can be determined by approximation calculus, R represents the freezing radius and t is the time.

From equation (11) and (12) we obtain:

$$A = \frac{T_0 - T_s}{E_i\left(\frac{-r^2}{4\alpha_1 t}\right) - E_i\left(\frac{-r_1^2}{4\alpha_1 t}\right)} \qquad (17)$$

$$C = \frac{T_0 - T_i}{E_i\left(\frac{-R^2}{4\alpha_2 t}\right)} \qquad (18)$$

The solutions of temperature field in other two zones of soil, in the case of the propagation processes of a freezing front in a liquid, homogeneous and isotropic medium are:

$$T_1(r,t) = \frac{T_0 - T_s}{E_i\left(\frac{-n^2}{4\alpha_1}\right) - E_i\left(\frac{-r_1^2}{4\alpha_1 t}\right)} \left[E_i\left(\frac{-r^2}{4\alpha_1 t}\right) - E_i\left(\frac{-r_1^2}{4\alpha_1 t}\right)\right] + T_s \qquad (19)$$

$$T_2(r,t) = \frac{T_0 - T_i}{E_i\left(\frac{-n^2}{4\alpha_2}\right)} E_i\left(\frac{-r^2}{4\alpha_2 t}\right) + T_i \qquad (20)$$

By means of the condition concerning the rate of heat transfer storage at the freezing limit (8):

$$k_1 \frac{dT_1}{dR} = k_2 \frac{dT_2}{dR} + \rho_1 l_s \frac{dR}{dt} \qquad (21)$$

the following equation is obtained:

$$\frac{k_1(T_0 - T_s)}{E_i\left(\frac{-n^2}{4\alpha_1}\right) - E_i\left(\frac{-r_1^2}{4\alpha_1 t}\right)} \exp\left(\frac{-n^2}{4\alpha_1}\right) - \frac{k_2(T_0 - T_i)}{E_i\left(\frac{-n^2}{4\alpha_2}\right)} \exp\left(\frac{-n^2}{4\alpha_2}\right) = \rho_1 l_s \frac{n^2}{4} \quad (22)$$

from where the Neumann constant depending on the time can be obtained. Equation (22) can be rewritten as:

$$\frac{k_1(T_0 - T_s)}{E_i\left(\frac{-n^2}{4\alpha_1}\right) - E_i\left(\frac{-n^2 \cdot r_1^2}{4\alpha_1 \cdot R^2}\right)} \exp\left(\frac{-n^2}{4\alpha_1}\right) - \frac{k_2(T_0 - T_i)}{E_i\left(\frac{-n^2}{4\alpha_2}\right)} \exp\left(\frac{-n^2}{4\alpha_2}\right) = \rho_1 l_s \frac{n^2}{4} \quad (23)$$

where the implicit function from which the Neumann constant n depends on the freezing radius [10]–[12].

Determination of the Neumann constant on another time interval: 0,..., t is given by the following equation:

$$\bar{n} = \left(\frac{1}{t}\right)\int_0^t n \cdot dt \quad (24)$$

The velocity of freezing front propagation is given by:

$$w = \frac{dR}{dt} = \frac{n}{2\sqrt{t}} = \frac{n}{2} t^{-0.5} = \frac{1}{2} \cdot \frac{n}{\frac{R}{n}} = \frac{n^2}{2R}, \quad [m/s] \quad (25)$$

depending from equation (16) with: $\sqrt{t} = \frac{R}{n}$.

The average velocity of freezing front propagation in a period of time t is clasically defined by:

$$\bar{w} = \frac{1}{R - r_1} \int_{r_1}^R \frac{n^2}{2R} dR = \frac{n^2}{2(R - r_1)} \ln \frac{R}{r_1} \Rightarrow$$

$$\bar{w} = \frac{n^2}{2(R - r_1)} \ln \frac{R}{r_1}, \quad [m/s] \quad (26)$$

The acceleration of the freezing front propagation is computed in the same manner:

$$a = \frac{dW}{dt} = -\frac{n}{4} t^{-1.5}, \quad [m/s^2] \quad (27)$$

Because the values of acceleration are negative this process is a damped process. In the range (0,1) the average acceleration is:

$$\bar{a} = \frac{1}{t_2 - t_1} \int_{t_1}^{t_2} a \cdot dt \quad (28)$$

or:

$$\bar{a} = \frac{n^2}{2 \cdot R \cdot r_1} \cdot \frac{1}{R - r_1} < 0, \quad [m/s^2] \quad (29)$$

where:

$$t_2 = \frac{R^2}{n^2}, \quad [m \cdot s^{-0.5}] \quad (30)$$

and:

$$t_1 = \frac{r_1^2}{n^2}, \quad [m \cdot s^{-0.5}] \quad (31)$$

From relation (27), the average acceleration has the following form:

$$\bar{a} = \frac{1}{t_2 - t_1}\int_{t_1}^{t_2} a\, dt = \frac{1}{t_2 - t_1}\int_{t_1}^{t_2} \frac{-k}{4} t^{-1.5} dt = \frac{-k}{4} \cdot \frac{1}{t_2 - t_1} \cdot \left.\frac{t^{-1.5+1}}{-1.5+1}\right|_{t_1}^{t_2}$$

$$= \frac{-k}{4} \cdot \frac{1}{\frac{R^2}{k^2} - \frac{r_1^2}{k^2}} \cdot \left.\frac{t^{-0.5}}{(-0.5)}\right|_{t_1}^{t_2} = \frac{k^3}{2} \cdot \frac{1}{(R - r_1)(R + r_1)} \cdot \left.\frac{t^{-0.5}}{(-0.5)}\right|_{t_1}^{t_2} \quad (32)$$

where $t^{-0.5}|_{t_1}^{t_2}$ can be written as:

$$t^{-0.5}|_{t_1}^{t_2} = \left(\frac{1}{\sqrt{t_2}} - \frac{1}{\sqrt{t_1}}\right) = \left(\frac{k}{R} - \frac{k}{r_1}\right) = -\frac{(R - r_1)}{R r_1} \quad (33)$$

Therefore, the average acceleration is:

$$\bar{a} = \frac{k^4}{2R r_1} \cdot \frac{-(R - r_1)}{(R - r_1)(R + r_1)} = -\frac{k^4}{2R r_1} \cdot \frac{1}{(R + r_1)} < 0 \quad (34)$$

4 THERMAL PROPERTIES AND PARAMETERS

For purposes of illustration, the theoretical model was solved and analyzed for a set of values necessary in numerical computation. Therefore, to compute the constants A, B, C, D from equations (13), (14), (17), (18), we consider the thermophysical parameters in the frozen zone according to the temperature $t_s = -15°C$. The values for the temperatures at $t_s = 0°C$ and $t_s = -20°C$ and the temperature at $t_s = -15°C$ determined by interpolation are presented in Table 1.

Thermophysical parameters of water for values of 0,10,15,20,25,30 °C are given in Table 2.

We consider the mass latent heat of solidification of water: $l_s = 334.84\ [kJ/kg]$

5 NUMERICAL SOLUTION

To solve equation (21) in order to obtain the Neumann constant by approximation calculus we need to bring

Table 1. Thermal properties used in analyses for $t_s = 0°C$ and $t_s = -20°C$.

t_s, [°C]	ρ_1, [kg/m^3]	k_1, [W/(mK)]	c_{p1}, [kJ/(kgK)]
0°C	917	2.1	2.26
−20°C	920	2.44	1.94
−15°C	919.25	2.38	2.02

Table 2. Thermal properties of water at different temperatures.

t_s, [°C]	ρ_2, [kg/m^3]	k_2, [W/(mK)]	$c_{p2} \cdot 10^3$, [kJ/(kgK)]
0	999.9	0.567	$4.218 \cdot 10^3$
5	920	0.571	$4.205 \cdot 10^3$
10	919.25	0.579	$4.193 \cdot 10^3$
15	999	0.580	$4.187 \cdot 10^3$
20	998.2	0.584	$4.182 \cdot 10^3$
25	997	0.588	$4.180 \cdot 10^3$
30	995.6	0.592	$4.178 \cdot 10^3$

this equation to a simple form. Therefore, we divide by $\rho_1 l_s \alpha_1$ equation (21) making the following notations:

$$x = \frac{n^2}{4\alpha_1} \quad (35)$$

$$p = \frac{k_1(T_0 - T_s)}{\rho_1 l_s \alpha_1} \quad (36)$$

$$m = \frac{k_2(T_0 - T_i)}{\rho_1 l_s \alpha_1} \quad (37)$$

$$s = \sqrt{\frac{\alpha_1}{\alpha_2}} \quad (38)$$

$$q = \frac{r_1^2}{R^2} \quad (39)$$

By means of these notations the following equation is obtained:

$$p \cdot \frac{\exp(-x)}{E_i(-x) - E_i(-x \cdot q)} - b \cdot \frac{\exp(-x \cdot s)}{E_i(-x \cdot s)} = X \quad (40)$$

This equation can be written as a function having the following form:

$$F(x) = f(x) - \overline{X} \quad (41)$$

where:

$$F(x) = b \cdot \frac{\exp(-x \cdot s)}{E_i(-x \cdot s)} \quad (42)$$

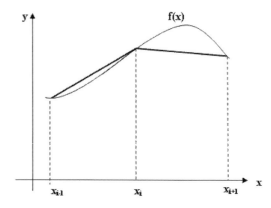

Figure 7. Trapezium method.

and:

$$f(x) = p \cdot \frac{\exp(-x)}{E_i(-x) - E_i(-x \cdot q)} \quad (43)$$

To solve equation (43), the exponential integral from (10) is used:

$$E_i(-x) = -\int_x^\infty \frac{e^{-t}}{t} dt = \int_\infty^x \frac{e^{-t}}{t} dt = \int_{-\infty}^{-x} \frac{e^{-t}}{t} dt \quad (44)$$

$$E_i(-x) = C + \ln x + \sum_{n=1}^\infty (-1)^n \frac{x^n}{n!n} \quad (45)$$

where C is the Euler-Mascheroni constant being equal with:

$$C = \lim_{n \to \infty} \left(\sum_{k=1}^\infty \frac{1}{k} - \ln n \right) = 0.577215665 \quad (46)$$

Function Ei is approximated by trapezium method. The trapezium method is the simplest to understand and implement. An important thing about the trapezium method is that this only gives approximate results if the y-values are derived from a continuous function. The area of the trapezium will either overestimate or underestimate the true value by some amount. Integration is often described as the calculation of the area under the curve. It the area under the curve is divided into a number of strips, each strip can be approximated to a trapezium and the sum of the areas of all the trapezia will be an approximation of the area under the curve (Figure 7):

$$\int_a^b f(x)dx = \sum_{i=1}^n h_i \frac{f(x_{i-1}) + f(x_i)}{2} + R_n \quad (47)$$

In order to carry out the integral computation with preset approximation, in the case when function f(x) is analytically defined, the more suitable approach is to divide the integral interval in n equal segments.

In this case the integration was made by 20,000 small intervals and solution will be search in the

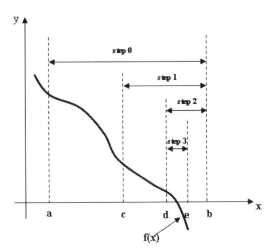

Figure 8. Bisection method.

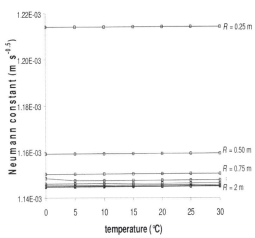

Figure 9. Neumann constant values versus temperature.

opened interval (0,1,1) by the bisection method with 20 iterations:

$$E_i(x) = \frac{1}{2(i+1)} \left[\exp\left(\frac{-xi}{n}\right) - \exp\left(\frac{-x(i-1)}{n}\right) \right] \quad (48)$$

Bisection method is used to find roots of an equation given that points a and b are f(a) < 0 < f(b). Thus, we assume that the function together with points at which its values are negative and positive are initially given. The essence of the half-interval method lies in the fact that the sign of the function changes on opposite sides of a root. Suppose the function f(x) has one root in the interval between x = a and x = b, or [a,b] as shown in Figure 8.

This method is based on the fact that when an interval [a,b] contains a root, the sign of the function at the two ends f(a) and f(b) are opposite each other, namely:

$$f(a) \cdot f(b) < 0 \quad (49)$$

The first step in the bisection method is to bisect the interval [a,b] in two halves, namely [a,c] and [c,b] where:

$$c = \frac{a+b}{2} \quad (50)$$

By checking the sign of $f(a) \cdot f(c)$ the half interval containing the root can be identified. If $f(a) \cdot f(c) < 0$, then the interval [a,c] has the root, otherwise the root is in interval [c,b]. Therefore, the new interval containing the root is bisected again. As the procedure is repeated the interval becomes smaller and smaller. At each step, the midpoint of interval containing the root is taken as the best approximation of the root. Therefore, if we study the sign of the function from equation (41) it is identified the interval where this function has zero value (F(x) = 0).

For this, the following interval is considered:

$$(a, b) = (x_{min}, x_{max})$$

where:

$$f(x_{min}) > 0$$
$$f(x_{max}) > 0$$

Step 0 has:

$$x_{med}^{(0)} = \frac{x_{min} + x_{max}}{2} = \frac{a+b}{2} = c \quad (51)$$

if $f(x_{med}^{(0)}) > 0$.
The new interval is $(x_{med}^{(0)}, x_{max}) = (c, b)$.
Next, step 1 has:

$$x_{med}^{(1)} = \frac{x_{med}^{(0)} + x_{max}}{2} = \frac{c+b}{2} = d \quad (52)$$

with which is computed F(x) and if:

$$f(x_{med}^{(1)}) > 0 \quad (53)$$

step 2 will follow, and so forth.

The iterative procedure is stopped when the half interval size is less then a prespecified size. After the value x is determined, the Neumann constant is obtained from equation (35). For all variants the constants p, r, and s (with p = 0.090, m = −0.014 and s = 9.275) are the same, being modified at each temperature only the constant q (corresponding to each radius).

After running the Visual C program, the Neumann constants obtained for different values of the radius R at various temperatures $t_i = 0 \div 30°C$ are represented in Figure 9. As it can be seen, the Neumann constant has a very low variation with temperature. Considering the variation of the Neumann constant with the radius R.

Fig. 10 shows the values of the Neumann constant obtained numerically from equation (21), and compares them with the time average values calculated numerically from equation (23). The points referred to those temperatures from 0 to 30°C are almost superposed for each radius on the corresponding curves. This happens because of the almost negligible variation of the Neumann constant with temperature in the range considered (0–30°C).

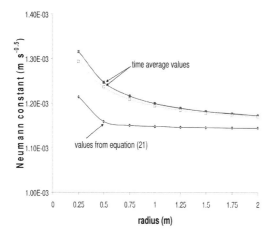

Figure 10. Neumann constant values obtained numerically from equation (21) and time average values calculated numerically from equation (23).

Comparison between the two curves shows that the behaviour of the curves is consistent, and the values are relatively close (the per cent difference ranges from about 6.5% for R = 0.25 m to about 2.2% for R = 2 m). The values of Neumann constant obtained in Visual C belong to the following range:

$$0.0014407 \leq \bar{n} \leq 0.001148, [m \cdot s^{-0.5}] \quad (54)$$

The velocities and accelerations of freezing front propagation are computed by means of some programs dedicated in Visual C, taking into consideration the equations (25), (26), (27) and (34). The values obtained are presented in Table 3. Velocities and accelerations of freezing front propagation depend on time, their equations being correlated with the solvable problem of determining the Neumann constant. As accelerations are negative, freezing front propagation is a damped process. It is considered that the average velocity of freezing front propagation is about 2 cm/day, having a technological charater. This velocity can be introduced in engineering computations, even if the problem is reduced at computation of Neumann constant. The drifting velocity of construction works is about 2 m/day.

Results obtained by means of this model are juxtaposed with of Leonachescu and Mateescu. These

Table 3. Thermal properties of water at different temperatures.

	R [m]	0.25	0.50	0.775	1.00	1.25	1.50	1.75	2.00
$t_1 = 0°C$	w [m/s]	2.9484	1.3435	0.8821	0.6595	0.5254	0.4373	0.3745	0.3275
	$\bar{w}$ [m/s]	4.0582	2.4680	1.8859	1.5589	1.3374	1.1796	1.0592	0.9642
	a [m/s^2]	−3.4764	−0.3610	−0.1032	−0.0434	−0.0220	−0.0127	−0.0080	−0.0053
	$\bar{a}$ [m/s^2]	−9.0761	−2.2667	−1.0479	−0.6078	−0.3946	−0.2776	−0.2059	−0.1589
$t_1 = 5°C$	w [m/s]	2.9486	1.3435	0.8821	0.6583	0.5254	0.4373	0.3745	0.3275
	$\bar{w}$ [m/s]	4.0589	7.09093	1.8859	1.5562	1.3373	1.4423	1.0592	0.9642
	a [m/s^2]	−3.4776	−0.3610	−0.1037	−0.0433	−0.0220	−0.0127	−0.0080	−0.0053
	$\bar{a}$ [m/s^2]	−9.0791	−2.2668	−1.0479	−0.6056	−0.3946	−0.2776	−0.2059	−0.1589
$t_1 = 10°C$	w [m/s]	2.9487	1.3437	0.8822	0.6584	0.5254	0.4373	0.3745	0.3276
	$\bar{w}$ [m/s]	4.0589	2.4684	1.8862	1.5565	1.2901	1.1798	1.0594	0.9644
	a [m/s^2]	−3.4776	−0.3611	−0.1037	−0.0433	−0.0220	−0.0127	−0.0080	−0.0053
	$\bar{a}$ [m/s^2]	−9.0791	−2.2675	−1.0483	−0.6059	−0.3946	−0.2775	−0.2060	−0.1589
$t_1 = 15°C$	w [m/s]	2.9488	1.3438	0.8824	0.6585	0.5255	0.4374	0.3746	0.3276
	$\bar{w}$ [m/s]	4.0593	2.4686	1.8865	1.55666	1.3376	1.1799	1.0596	0.9645
	a [m/s^2]	−3.4782	−0.3611	−0.1038	−0.0433	−0.0220	−0.0127	−0.0080	−0.0053
	$\bar{a}$ [m/s^2]	−9.0809	−2.2679	−1 0486	−0.6059	−0.3748	−0.2784	−0.2060	−0.1590
$t_1 = 20°C$	w [m/s]	3.0896	1.3439	0.8825	0.6585	0.5255	0.4374	0.3746	0.3276
	$\bar{w}$ [m/s]	4.2530	2.4688	1.8868	1.5567	1.3377	1.1800	1.0596	0.9646
	a [m/s^2]	−3.4182	−0.3612	−0.1038	−0.0433	−0.0220	−0.0127	0.0080	−0.0053
	$\bar{a}$ [m/s^2]	−9.9684	−2.2682	−1.0490	−0.6060	−0.3948	−0.2778	0.2060	−0.1590
$t_1 = 25°C$	w [m/s]	2.9471	1.3440	0.8825	0.5738	0.5258	0.4374	0.3746	0.3277
	$\bar{w}$ [m/s]	4.0597	2.4690	1.8868	1.5569	1.3377	1.1801	1.0598	0.9647
	a [m/s^2]	−3.4790	−0.3612	−0.1038	−0.0433	−0.0220	−0.0127	−0.0080	−0.0053
	$\bar{a}$ [m/s^2]	−9.0830	−2.2685	−1.0489	−0.6062	−0.3948	−0.2778	−0.2061	−0.1590
$t_1 = 30°C$	w [m/s]	2.9493	1.3441	0.8826	0.6587	0.5256	0.4375	0.3747	0.3277
	$\bar{w}$ [m/s]	4.0600	2.4695	1.8869	1.5570	1.3380	1.1802	1.0599	0.9647
	a [m/s^2]	−3.4793	−0.3613	−0.1038	−0.0433	−0.0221	−0.0127	−0.0080	−0.0053
	$\bar{a}$ [m/s^2]	−9.0842	−2.2689	−1.0491	−0. 6063	−0.3950	−0.2779	−1.2061	−0. 1590

results confirm the theoretical and practical development in the case of heat transfer in the bodies with the mobile boundaries.

6 CONCLUSIONS

In this paper, the case of using total freezing on a portion of straight gallery is analyzed. Some analytical equations for estimation of the Neumann constant and the velocity and acceleration of freezing front propagation have been analyzed. These equations are based on an analysis of the numerical solution for temperature versus Neumann constant and radius versus Neumann constant, for different soil temperatures and the same temperature of contact surface between soils. The software product gives a calculus support for evaluating of the freezing/thawing processes, providing a simple and convenient tool in calculations of real temperature fields. This comparative analysis was carried out for a number of values of the parameters as the Neumann constant, the velocity and acceleration of freezing front propagation and their average values. The present approach simplifies the mathematical description of the solution while assuring a satisfactory level of precision.

REFERENCES

[1] J.S. Harris, "Ground Freezing in Practice", Thomas Telford Services Ltd., London, 1995.

[2] Lu, T. Wang, K.-S., "Numerical analysis of the heat transfer associated with freezing solidifying phase changes for a pipeline filled with crude oil in soil saturated with water during pipeline shutdown in winter", Journal of Petroleum and Science Engineering, 2008 (52–58).

[3] M. Schultz, M. Gilbert, H. Hass, "Ground freezing – principles, applications and practices", Tunnels & Tunneling International, September, 2008 (http://www.cdm-ag.de/eu/Ground _Freezing).

[4] L. Bronfenbrener, E. Korin, "Thawing and refreezing around a buried pipe", Journal of Chemical Engineering and Processing, Elsevier, 1998, pp. 239–247.

[5] D. Enescu, "Heat power study of consolidation processes by freeze in aquifer soils", Master dissertation, Politehnica University of Bucharest, 1996.

[6] N. P. Leonăchescu, "Heat Engineering", Didactic and Pedagogic Publishing House, Bucharest, 1980.

[7] N. P. Leonăchescu, "Heat Transfer between constructions and soil", Technical Publishing House, Bucharest, 1989.

[8] D. Enescu, "Interference of the temperatures field during the freezing/thawing processes", Doctoral Dissertation, Bucharest, 2004.

[9] Mateescu, "Contributions to heat transfer study by bodies with mobile boundaries", Doctoral Thesis, Bucharest, 1998.

[10] V. E. Romanovsky, T. E. Osterkamp, N. S.Duxbury, "An evaluation of three numerical models used in simulations of the active layer and permafrost temperature regimes", Cold Regions Science and Technology, 26(1997), 195–203, Elsevier.

[11] I. Caraus, I. Necoara, "A cutting plane method for solving convex optimization problems over the cone nonnegative polynomials", WSEAS Transaction on Mathematics, Issue 7, Volume 8, July 2009, pp. 269–278.

[12] K.T. Chiang, G.C. Kuo, K.J. Wang, Y.F. Hsiao, K.Y. Kung, "Transient temperature analysis of a cylindrical heat equation", WSEAS Transaction on Mathematics, Issue 7, Volume 8, January 2009, pp. 309–319.

[13] M. Shala, K. Pericleous, M. Patel, "Unstructured staggered mesh methods for fluid flow, heat transfer and phase change", WSEAS Transactions on Fluid Mechanics Issue 5, Volume 1, May 2006, pp. 431–438.

[14] N.E. Mastorakis, "An Extended Crank-Nicholson Method and its Applications in the Solution of Partial Differential Equations: 1-D and 3-D Conduction Equations", WSEAS Transactions on Mathematics, Vol. 6, No. 1, 200, pp. 215–225.

[15] G.P. Newman, "Artificial ground freezing of the Mc Arthur River Uranium Ore Deposit" (http://www.cubex.net/Library/Media/Artificial_Ground_Freezing.pdf).

[16] D. Enescu, "Freezing/Thawing Phenomena in homogeneous mediums", Bibliotheca Publishing House, Targoviste, 2005.

[17] D. Enescu, H.G. Coandă, I. Caciula, "Experimental considerations concerning the thermal field around the pipes placed in a medium with phase change", International Conference Energy-Environment, "Politehnica" University of Bucharest, October, the 20th–22nd, 2005.

[18] http://www.coilgun.eclipse.co.uk/math_1.html.

The development of a slotted waveguide array antenna and a pulse generator for air surveillance radar

Mashury Wahab, Daday Ruhiyat, Indra Wijaya, Fajri Darwis & Yussi Perdana Saputera
Research Centre for Electronics and Telecommunications of the Indonesian Institute of Sciences (RCET-LIPI), Bandung, Indonesia

ABSTRACT: This paper presents our research on the development of S-band frequency air surveillance radar with a frequency centre of 3 GHz. The design and development of the slot waveguide antenna array and pulse generator are described in this paper. The waveguide antenna was designed using aluminium material with a dimension of 3581.6 × 114.12 mm (length × height). The slot V method was implemented to achieve a narrow beamwidth of the radiation pattern. From the simulation, the obtained bandwidth was about 14 MHz, Voltage Standing Wave Ratio (VSWR) was 1.15, and gain was about 27 dBi. A pulse generator was designed using a Radio Frequency (RF) switch to form Frequency Shift Keying (FSK) configuration. A Pulse Width Modulation (PWM) control and user interface were implemented using microcontroller, Atmel ATmega 328P.

Keywords: slot waveguide antenna array, pulse generator, air surveillance radar

1 INTRODUCTION

1.1 Research background

Indonesia is geographically located between the continents of Asia and Australia. There are more than 17,000 islands in Indonesia. For such a large area, a large amount of radar is needed for monitoring the air and water against illegal activities carried out by foreign and domestic entities. The need for the radar was announced by users from the government, the military, and the authorities, during the National Radar Seminar held annually [Wahab et al., 2012, 2013].

Figure 1 shows the combined coverage of military and civilian radar belonging to the armed forces and the airport authorities. There are many spots that are still uncovered by the radar.

Figure 2 depicts the spread of airports in Indonesia, including middle and large-sized airports. There are more than fifty operational airports with high traffic. Most of the airports do not have active radar to detect incoming and outgoing aircraft, so there is a problem of transportation safety at Indonesian airports. According to the ICAO (International Civil Aviation Organization), the use of airport radar is vital to increase the security and safety of civil aviation [World Aero Data, 2012, ICAO, 2012].

Air surveillance radar is commonly used by the air force and airport authorities to monitor and control the air space inside, and at the border of, a country. The price of air surveillance radar is expensive (millions of USDs), so it will cost a lot to install radar in all airports in Indonesia. Due to this reason, a national capability is being encouraged to produce such radar locally. Therefore, the price of air surveillance radar

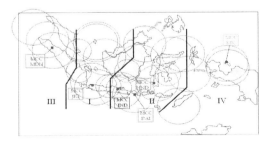

Figure 1. The combined coverage of military and civilian Indonesian Radar.

Figure 2. Airport locations in Indonesia.

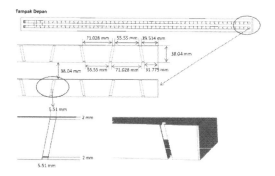

Figure 3. Front view of the slotted waveguide antenna array.

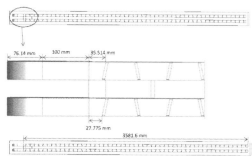

Figure 4. Dimension of the left side of the slot antenna.

can be reduced. In addition, the maintenance of the radar can also be supported by local expertise.

Other benefits can be achieved by producing the radar locally, such as initiating the growth of the local supporting industries, the employment of a large number of workers, enhancing data security, ensuring the ease of operations due to the manuals and menus of the user interface being written in the Indonesian language, and maintenance support can be given by local companies.

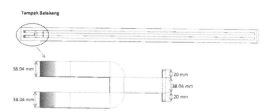

Figure 5. Dimension of the back panel of the antenna.

1.2 Research goals

The goal of this research is to enhance national capabilities for producing air surveillance radar and other types of radar locally, to grow the supporting industries, to open up the job market in the radar industries, to increase the capabilities of human resources, to produce innovations, and to equip the laboratories with state-of-the-art measuring equipment and tools. National and international publications, and patent submissions are the scientific outputs of this research.

1.3 Research method

This research is performed, based on the defined specifications. Designs and simulations for the hardware, software, and antenna were carried out. After the components were delivered to the laboratories, assembly and testing were performed. Improvement will be done, based on the test results. Reports will be provided at the end of the research project year.

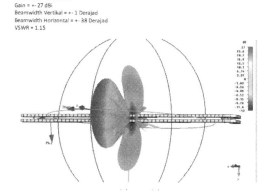

Figure 6. Simulation results of the antenna radiation pattern.

2 DESIGN AND SIMULATION

2.1 Slotted waveguide antenna array

The antenna type for this air surveillance radar is a slotted waveguide antenna array. The design of this antenna is depicted in Figure 3. The slot method is the V method as shown in Figure 3 [Skolnik 1990, 2002].

It can be seen from Figure 4 that the length of the antenna is 3581.6 mm. The back view of the antenna is shown in Figure 5.

For this antenna module, the simulation result of the antenna radiation pattern is depicted in Figure 6.

Figure 6 shows the simulation results of the slot antenna. A gain of about 27 dBi, a vertical beamwidth of 38°, and a horizontal beamwidth of 1° were achieved. The frequency bandwidth of this antenna module is about 14 MHz.

The simulated VSWR of this antenna module is depicted in Figure 7. We obtain VSWR of 1.15.

2.2 Pulse generator

The function of this module is to modulate the carrier signal, f_s, with pulse signals of certain values of Pulse Width (PW) and Pulse Repetition Frequency (PRF).

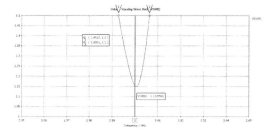

Figure 7. Simulated VSWR.

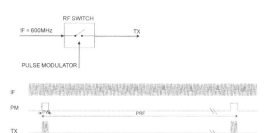

Figure 8. FSK configuration.

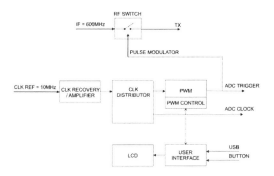

Figure 9. Block diagram of the pulse generator.

The modulation process is performed at an intermediate frequency (IF) IF frequency of 600MHz using a RF Switch, so that a FSK (Frequency Shift Keying) configuration is constructed.

To maintain coherency, this module also generates clock and trigger signals for data acquisition. These signals must be derived from the same source. The block diagram of the pulse generator is shown in Figure 9.

3 FABRICATION

The design of the slot waveguide antenna array with a centre frequency of 3 GHz is fabricated using aluminium. At the time this paper was written, the fabrication was not completed yet, so the antenna cannot be measured. Figures 10 and 11 below show the fabrication processes.

Fabrication of the pulse generator is described as follows. The PWM control and user interface were

Figure 10. Fabrication of the slotted antenna.

Figure 11. Fabrication of the slotted antenna (another view).

Figure 12. Pulse generator module.

implemented using an Atmel ATmega 328P microcontroller with a system clock derived from a reference clock of 10 MHz. This PWM was generated using 16 bit features on the microcontroller so that a PW minimum of 0.1 of a microsecond can be generated. The configuration of PW and PRF was set using a PC via a universal serial bus (USB) protocol. A liquid crystal display (LCD) and Push Button were used to be able to control manually without using the PC. This pulse generator is shown in Figure 12.

The outputs of this pulse generator were measured by an oscilloscope and the results are shown in Figures 13–15.

4 DISCUSSION

The design, simulation, and implementation of a slot waveguide antenna array are relatively new research

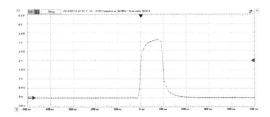

Figure 13. Pulse signal with PW = 100 ns.

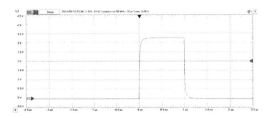

Figure 14. Pulse signal with PW = 1 us.

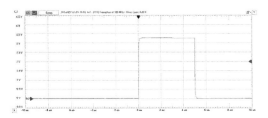

Figure 15. Pulse signal with PW = 5 us.

fields in Indonesia. This is due to the fact that the implementation of such an antenna is limited to certain applications, such as for high power radar and transmission systems. It is expected that there will be little difference between simulated and fabricated results. The fabrication of this antenna requires special tools and equipment. High precision is required to obtain an accurate antenna system. Therefore, it takes several months for a local workshop to manufacture this slot antenna. The specifications of the antenna are: gain ∼27 dBi, vertical beamwidth of 38°, and horizontal beamwidth of 1°. The frequency bandwidth of this antenna module is about 14 MHz.

The pulse generator is able to generate a pulse signal with a variety of pulse widths. Thus, this pulse generator is ready to be used for the pulse radar system. The coherence/synchronization is also controlled by this generator as the timing is derived from the same clock. Thus, the video data that will be converted by the ADC (Analog to Digital Converter) is also synchronized.

5 CONCLUSION

A slot waveguide antenna array has been designed and fabricated for air surveillance radar. In addition, a pulse generator has also been designed, assembled, and tested. The antenna and pulse generator will later be integrated with the transceiver modules. The complete system will be tested in 2015.

REFERENCES

[1] Wahab, Mashury, et al, 'Proposal Litbang Radar 2012–2026 (Indonesia): Proposal for Radar R&D 2012–2026', Internal Report, 2012.
[2] Wahab, Mashury, et al., 'Desain dan Implementasi Antena Generasi I untuk Litbang Konsorsium Radar (Indonesian): Design and Implementation of 1st Generation Antenna for Radar Consortium R&D', Seminar INSINAS 2012.
[3] World Aero Data, 'Airport In Indonesia', http://www.worldaerodata.com/countries/Indonesia.php.
[4] ICAO, http://www2.icao.int/en/home/default.aspx.
[5] Wahab, Mashury, et al., 'Desain dan Implementasi Antena Generasi II dan Pembangkit Frekuensi untuk Litbang Radar Pengawas Udara (Indonesian): Design and Implementation of 2nd Generation Antenna for Air Surveillance R&D', Seminar INSINAS 2013.
[6] Skolnik, M.I., 'Radar Handbook', McGraw-Hill, 1990.
[7] Skolnik, M.I., 'Introduction to Radar Systems', McGraw-Hill, 2002.

A harmonic model of an orthogonal core controllable reactor by magnetic circuit method

Weisi Gu
Zhejiang University of Technology, Hangzhou, China

Hui Wang
Ningbo Ningbian Power Technology Co. Ltd., Ningbo, China

ABSTRACT: With a special orthogonal core structure, the Ferrite Orthogonal Core Controller Reactor (FOC-CR) has the advantages of simple controlling and low harmonic content. This paper establishes its equivalent magnetic circuit model and the basic mathematical equations. Through the assumption of an Magnetic Motive Force (MMF)-flux relation in a form of an odd polynomial, the relationship of the FOC-CR's input voltage and flux is derived. Thus, the simulation model is established in Matlab/Simulink by using S-function. Good agreement is obtained by contrasting the test results of the simulation model and the experimental prototype.

Keywords: Orthogonal core controllable reactor; Magnetic equivalent circuit; S-function

1 INTRODUCTION

In recent years, power development and its effective utilization have been elevated to a crucial position. Controllable reactors are widely used in power grids. They can regulate the network voltage and be used as reactive compensation. Ferrite orthogonal core controllable reactors, have approximate linear control characteristics and relatively low harmonic content [1,2]. Recently, researchers have tried to achieve flexible and dynamic adjustment of reactive power [3,4]. Thus, establishing a simple and highly effective FOC-CR model for computer simulation is very significant. Some approaches for FOC-CR modelling, such as 3-dimensional magnetic circuit modelling and numerical modelling, have been presented in certain references [5–7]. Among them, 3-dimensional magnetic circuit modelling is much more accurate but complicated, needing massive calculation which results in low effectiveness. For the reason mentioned above, this paper describes the structure of the FOC-CR and proposes a new simple method of FOC-CR modelling in Matlab/Simulink, based on the analysis of the FOC-CR's equivalent magnetic circuit.

2 CONFIGURATION AND OPERATION PRINCIPLE

The FOC-CR is composed of two U-cores rotated at 90° in respect to each other. The material of the two cores is grain oriented silicon steel. The control and

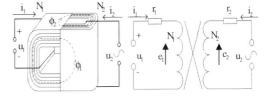

Figure 1. Configuration of the FOC-CR.

working windings are fixed on the two cores respectively. Figure 1 gives a schematic diagram of the FOC-CR. N_1 and N_2 are the control and working winding turns. The control and working currents are i_1 and i_2, respectively. The dashed curves, ϕ_1 and ϕ_2, illustrate the control and working fluxes. When the current i_1 is zero in the control coil, only alternating flux exists in the core. Because of its symmetric structure, there will be no coupling flux in the control coil. The fluxes, which are in the interfacing sections of two orthogonal cores, consist of two parts; one is constant flux ϕ_1, generated by the DC control source, the other is alternating flux ϕ_2, generated by AC power. Therefore, the ϕ_2 can be impacted with the changing of ϕ_1, which influences the degree of flux saturation in the interfacing sections. Due to the nonlinear characteristic of the B-H curve, when the magnitude of DC flux ϕ_1 is enough to enable the junction regions to go into magnetization saturation zone, the equivalent reactance of the working winding will be reduced, with the current increase in the control winding.

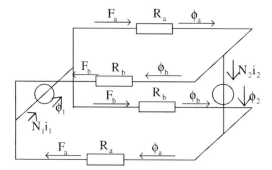

Figure 2. Magnetic equivalent circuit of the FOC-CR.

3 MATHEMATICAL MODEL

3.1 Equivalent magnetic circuit and basic equation

Ignoring the leakage magnetic reluctance and reluctance in the linear region of the orthogonal core, the equivalent magnetic circuit of the FOC-CR can be regarded as is shown in Figure 2.

F_a and F_b are MMF's in common regions, ϕ_a and ϕ_b the fluxes, and R_a and R_b the reluctances.

According to the basic principle of the magnetic circuit, the relation can be derived as:

$$\begin{cases} N_1 i_1 = F_a + F_b \\ N_2 i_2 = F_a - F_b \\ \varphi_1 = \varphi_a + \varphi_b \\ \varphi_2 = \varphi_a - \varphi_b \end{cases} \quad (1)$$

From Equation (1), we can obtain:

$$\begin{cases} \varphi_a = \frac{1}{2}(\varphi_1 + \varphi_2) \\ \varphi_b = \frac{1}{2}(\varphi_1 - \varphi_2) \end{cases} \quad (2)$$

Assuming the relationship between the MMF and flux is known, that is:

$$F = F(\varphi)$$

MMF can be derived as follows:

$$\begin{cases} N_1 i_1 = F(\frac{\varphi_1 + \varphi_2}{2}) + F(\frac{\varphi_1 - \varphi_2}{2}) \\ N_2 i_2 = F(\frac{\varphi_1 + \varphi_2}{2}) - F(\frac{\varphi_1 - \varphi_2}{2}) \end{cases} \quad (3)$$

As the working winding is connected to the network, its port voltage is the network voltage, thus ϕ_2 is known. In this case, the relationship between i_2 and i_1 can be given as (4), with respect to ϕ_2:

$$i_2 = f(i_1)\big|_{\varphi_2} \quad (4)$$

Equation (4) reveals the control characteristics of the working winding under a DC bised condition.

3.2 Relationship of input voltage and fluxes

To solve Equation (3), the MFF-flux relationship of the core must be known, namely, Equation (4). However, due to the saturation characteristics of the magnetic circuit of FOC-CR being nonlinear, it is hard to determine the MMF-flux relationship. For a real manufactured reactor, the relationship can be measured by an experiment. For simplification, we assume the nonlinear MMF-flux relationship as an odd polynomial, as follows:

$$F(\phi) = a_1\phi + a_2\phi^3 + a_3\phi^5 \quad (5)$$

a_1, a_2, and a_3 are constants related to the size and material characteristics of the FOC-CR.

From (5), F_a and F_b can be given by:

$$\begin{cases} F_a = F_a(\phi) = a_1\phi_a + a_2\phi_a^3 + a_3\phi_a^5 & (6) \\ F_b = F_b(\phi) = a_1\phi_b + a_2\phi_b^3 + a_3\phi_b^5 & (7) \end{cases}$$

Then we have:

$$\begin{cases} N_1 i_1 = a_1\phi_1 + (a_2/4)\phi_1^3 + (3a_2/4)\phi_1\phi_2^2 + (a_3/16)\phi_1^5 \\ + (5a_3/8)\phi_1^3\phi_2^2 + (5a_3/16)\phi_1\phi_2^4 \end{cases} \quad (8)$$

$$\begin{cases} N_1 i_1 = a_1\phi_2 + (a_2/4)\phi_2^3 + (3a_2/4)\phi_1^2\phi_2 + (a_3/16)\phi_2^5 \\ + (5a_3/8)\phi_2^3\phi_1^2 + (5a_3/16)\phi_1^4\phi_2 \end{cases} \quad (9)$$

According to Figure 2, we have:

$$\begin{cases} u_1 = r_1 i_1 + e_1 & (10) \\ u_2 = r_2 i_2 + e_2 & (11) \\ e_1 = N_1 \frac{d\phi_1}{dt} & (12) \\ e_2 = N_2 \frac{d\phi_2}{dt} & (13) \end{cases}$$

By adding (10), (11), (12), and (13) to (8) and (9), (14) and (15) can be derived:

$$\frac{d\phi_1}{dt} = \frac{u_1}{N_1} - \frac{r_1}{N_1^2}(a_1\phi_1 + (a_2/4)\phi_1^3 + (3a_2/4)\phi_1\phi_2^2$$
$$+ (a_3/16)\phi_1^5 + (5a_3/8)\phi_1^3\phi_2^2 + (5a_3/16)\phi_1\phi_2^4) \quad (14)$$

$$\frac{d\phi_2}{dt} = \frac{u_2}{N_2} - \frac{r_2}{N_2^2}(a_1\phi_2 + (a_2/4)\phi_2^3 + (3a_2/4)\phi_1^2\phi_2$$
$$+ (a_3/16)\phi_2^5 + (5a_3/8)\phi_2^3\phi_1^2 + (5a_3/16)\phi_1^4\phi_2) \quad (15)$$

(14) and (15) show the relationship between the input voltage and the causing fluxes of the FOC-CR.

4 SIMULATION IN MATLAB

4.1 Simulation modelling

According to (14) and (15), we build an S-function model of voltage-flux. By combining (10), (11), (12),

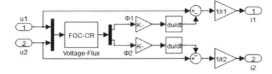

Figure 3. Voltage-Flux S-function model of the FOC-CR.

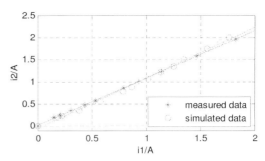

Figure 5. Control characteristics of the FOC-CR.

winding current waveform and control characteristics is provided. Figure 4 gives the contrastive current waveforms when i_1 was 0.28 A and 1.77 A.

Figure 5 shows the contrastive control characteristics. From Figure 4, it is not difficult to find that the harmonic content of the working winding current is greatly reduced with the increase of the DC current in the control winding, which leads the common regions of cores into the deep depth of magnetic saturation. The simulation current waveforms are similar with the real measured and basically reflect the harmonic characteristics of the FOC-CR. By contrasting the control characteristics, we can see both agree well, and the FOC-CR has approximate linear control characteristics in its adjustment range.

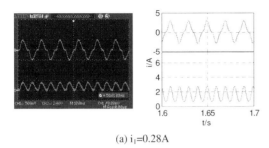

(a) i_1=0.28 A

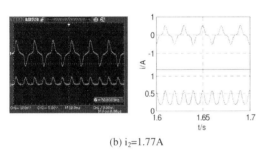

(b) i_2=1.77 A

Figure 4. The measured and simulated current of the FOC-CR.

Table 1. Parameters of the FOC-CR.

N_1/turns	N_2/turns	r_1/Ω	r_2/Ω	ϕ_1/mm	ϕ_2/mm
245	240	2.76	2.73	0.7	0.7

and (13), the final FOC-CR simulation model is obtained in Matlab/Simulink, as Figure 3 shows.

On the Matlab analysis, it is necessary to provide the constants a_1, a_2, and a_3. In this paper, we design an experimental prototype, according to Table 1, to determine the coefficients by the simple circuit shown in Figure 4. The core material is grain oriented silicon steel with a lamination thickness of 0.35 mm. The rated voltage and rated current are 50 V and 0.82 A, respectively.

4.2 Simulation results

From the experimental data, we get the constants: $a_1 = 1.02 \times 10^5$ A/Wb, $a_2 = -1.25 \times 10^2$ A/Wb^3, and $a_3 = 6.13 \times 10^{17}$ A/Wb^5. For verifying the feasibility of this modelling method, the contrastive analysis of the

5 CONCLUSION

We proposed a modelling method for a FOC-CR in Matlab and gained the winding waveforms and control characteristics. The simulation results agree well with the measured ones. Thus, the modelling method presented in this paper is effective, and useful for the analysis and design of a power reactor with an orthogonal core.

ACKNOWLEDGEMENT

This work was financially supported by the Natural Science Foundation of Zhejiang Province (No. LY13E070004).

REFERENCES

[1] Nakamura K., Ichinokura O. & Maeda M. 2000. Analysis of orthogonal-core type linear variable inductor and application to VAR compensator, *IEEE Transactions on Magnetics* 36(4): 3565–3567.

[2] Ichinokura O., Tajima T. & Jinzenji T. 1993. A new variable inductor for VAR compensation. *IEEE Transactions on Magnetics* 29(6): 3225–3227.

[3] Yao Y., Chen B.C. & Tian C.H. 2008. EHV magnetically controlled reactor for suppressing interior over-voltage and secondary arc current. *Journal of Jilin University* (Engineering and Technology Edition) 38: 201–208.

[4] Hong H., Zhang J.Y., Song M., et al. 2013. Magnetization study on a new type of orthogonally configured magnetic core structure and its potential application to superconducting controllable reactors. *IEEE Transactions on Applied Superconductivity* 23(3).

[5] Tajitna K., Kaga A., Anazawa Y., Ichinokura O. 1993. One method for calculating flux-mmf relationship of orthogonal-core. *IEEE Transactions on Magnetics.* 29(6): 3219–3221.

[6] Nakamura K., Yoshida H., Ichinokura O. 2004. Electromagnetic and thermal coupled analysis of ferrite orthogonal-core based on three-dimensional reluctance and thermal-resistance network model. *IEEE Transactions on Magnetics* 40(4): 2050–2052.

[7] Ichinokura O., Sato K. & Jinzenji T. 1991. A SPICE model of orthogonal-core transformers. *American Institute of Physics* 69(8): 4920–4930.

A risk assessment model of power system cascading failure, considering the impact of ambient temperature

Baorong Zhou, Rongrong Li & Lanfen Cheng
Electric Power Research Institute, China Southern Power Grid, Guangzhou, China

Pengyu Di, Lin Guan, Suyun Wang & Xiaocan Chen
School of Electric Power, South China University of Technology, Guangzhou, China

ABSTRACT: In recent years, cascading failure has become an important cause leading to large-scale blackouts of power systems. It is of great value to predict the possible chains of cascading failure effectively and make a scientific and rational cascading failure risk assessment of power systems. Based on the analysis of the cascading failure developing process, a cascading failure risk assessment model of a power system, considering the impact of ambient temperature, is proposed, which divides the cascading process into two stages, i.e., a slow dynamic stage and a fast dynamics stage. The probability models of the two stages are designed respectively according to their own characteristics. In terms of risk assessment, the impact factor of failure depth and slow dynamic stages are introduced in order to fully consider the influence of dispatchers' adjustments. Finally, the rationality of this model is verified through a sample analysis based on the 10-machine New-England Power System.

1 INTRODUCTION

With the expansion of power systems, a series of large-scale blackouts have occurred in many countries over recent years [1,2], such as the blackout of interconnected North America power grid on Aug. 14th, 2003, the blackout of Italy power grid on Sep. 28th, 2003, the blackout of Western Europe power grid on Nov. 4th, 2006, etc. Many scholars have found that the relationship between the blackout probability and the blackout size presents a power law distribution, which also explains the reason why the large-scale cascading failure has happened more frequently in recent years. Therefore, the mechanism of the occurrence and development for cascading failure has become a subject of much debate.

Based on the complex system of self-organization characteristics, the "OPA" model [4], which is short for three research institutes, the "CASCADE" model [5], and the branching process model [6] are introduced successively. These models either try to search out the critical elements of a given power system using the DC flow model, or design simple physical models to simulate the basic characteristic of cascading failure evolution. Therefore, it is difficult to accurately assess the cascading failure risk of a practical system. The cascading chains are searched for in current literature [8], based on the analysis of the occurrence and development of cascading failure process. However, the models designed in these papers are not precise enough, which will result in relatively great evaluation errors, especially when not considering the impact of ambient temperature.

This paper proposes an innovative risk assessment model of power system cascading failure, considering the impact of ambient temperature. Based on the detailed analysis on the developing process of a real cascading failure case, the model proposed takes the impact of ambient temperature into consideration, so that the assessment results may be more accordant with engineering practice.

The paper is organized in eight sections. Section 2 describes the detailed developing process of the 8.14 blackouts in North America, and divides the cascading process into two stages, i.e. the slow dynamic stage and the fast dynamic stage. Section 3 proposes the rated capacity model considering the temperature effect, on which the method about stage identification is presented. Section 4 gives a detailed introduction to the probability models for both the slow dynamic stage and the fast dynamic stage. Section 5 describes the risk assessment model of cascading failure, considering the impact of dispatchers' adjustments. Section 6 introduces the risk assessment process systematically. Section 7 discusses the rationality of the model proposed based on the 10-machine New-England Power System. Finally, Section 8 concludes the proposed method and its advantages.

2 THE ANALYSIS OF THE CASCADING FAILURE PROCESS

2.1 Cascading failure case

The 8.14 blackouts in North America, which occurred on Aug. 14th, 2003, are one of the most serious blackout accidents in history, causing power outages for fifty million people, which lasted for twenty-nine hours. This paper selects the North American 8.14 blackouts as a typical cascading failure case for analysis, and the evolution of the blackout is as follows:

(1) At 14:02-15:42, four 345 kV lines including Stuart-Atlanta, Harding-Chamberlin, Hanna-Juniper, and Star-South Canton tripped because of a ground short circuit.
(2) After the outage of the Star-South Canton line, the power flow of the 138kV power grid, which supplies electricity for Cleveland, extraordinarily increased. From 15:39 to 16:05, sixteen lines tripped successively and 600MW load was shed. At 16:06:03, the load rate of the 345kV Sammis-Star line surged to more than 120%, following the outages of many other 345kV lines.
(3) Until then, the system was confronted with voltage collapse, which caused outages of many lines in 5 minutes, and led to a systematic collapse.

2.2 Partition of cascading failure stages

Through the above analysis of the North American 8.14 blackout, the cascading failure process could be clearly divided into two stages:

1) the slow dynamic stage between 14:02-15:42. Generally, the fault reasons for this stage were short circuit faults because of the thermal balance destruction after the power flow transferring. The destruction of thermal balance brought about enlargement of the lines' sag, which increased the fault probability greatly.
2) the fast dynamic process after 15:42. Generally, the fault reasons for this stage were the operation of relays, which resulted in seriously exceeding the protective relays' setting values.

The essential cause of the slow dynamic process is the impact of ambient temperature on the lines' thermally stable capacity, i.e., the rated capacity. Since the rated capacity and the ambient temperature are negatively correlated, the rated capacity will decline as the ambient temperature keeps increasing. However, the load rates of lines monitored by the protective relays are based on the rated capacity in some specific temperature, such as 20°C. Therefore, when the ambient temperature increases to some extremely high temperature, like 40°C, the situation will occur that the load rates monitored by the protective relays are still under protective settings while the thermal balances of some power lines have already been broken in fact. As the relays fail to recognize the actual condition of the overload, the lines will continue heating until the line sag is large enough to touch the trees, which will cause the next level of ground short circuit. Since the duration of this stage is relatively long, the situation is called the slow dynamics stage.

When the load rate monitored by the relays exceeds the present settings, the protective relays will operate and trip the related lines in a short time delay. Since

Table 1. Comparisons between the load rate monitored by the relays and the actual load rate considering ambient temperature

The short circuit line	voltage classes	load rate (20°C)	load rate (40°C)
Harding-Chamberlin	345 kV	43.5%	60.4%
Hanna-Juniper	345 kV	87.5%	121.4%
Star-South Canton	34 5kV	93.2%	129.4%

the duration of this stage is relatively short, it is called the fast dynamic stage.

The load rates under the different ambient temperature of the three fault transmission lines are given in Table 1. From the table, it can be seen that when the ambient temperature is up to 40°C, the actual load rate of the Hanna-Juniper line and the Star-South Canton line exceed 120%, while the load rate monitored by the relays is still less than the settings, and the relays will not respond. Therefore, the two lines will make contact with the trees and be finally tripped by the relays because of the ground short circuit.

Generally, the slow dynamic stage and the fast dynamic stage have the following characteristics:

1) The slow dynamic stage reflects a process of the heat accumulation when the thermal stability balance is broken, while the fast dynamic stage reflects the operation of relays as the load rate exceeds the protection setting values.
2) The slow dynamic stage develops more slowly, i.e., 0.5 to 1.5 hours, while the fast dynamic stage develops rapidly, usually in 5 to 10 minutes.

3 CASCADING FAILURE STAGE IDENTIFICATION

3.1 Rated capacity model of power lines considering the temperature's effect

It can be learned from the thermal stability characteristics of power lines that with a rise of the ambient temperature, the line's thermally stable capacity will gradually decline. Based on the analysis of a dynamic equation of the line thermal stability, literature [8] derives the relationship between actual rated capacity (thermally stable capacity) and ambient temperature, as Formula (1) shows:

$$S'_{max} = S_{max}\sqrt{1+(T_0-T'_0)\frac{v}{\alpha}\left(\frac{V}{P_{max}}\right)^2} \qquad (1)$$

where S_{max} is the rated capacity of overhead lines at initial temperature T_0, P_{max} is the rated active power of overhead lines at initial temperature T_0, S'_{max} is the rated capacity of overhead lines at actual temperature T'_0, and V is the voltage between nodes of the line end. The related parameters can be obtained by Formula (2):

$$\frac{v}{\alpha} = \frac{Hp/\rho c\omega}{0.239/\rho c\omega^2\sigma} = \frac{Hp\omega}{0.239\sigma} = \frac{Hp}{0.239R} \qquad (2)$$

where R is the resistance per unit length of lines, p is Line's perimeter, and H is the coefficient of thermal effects.

3.2 Stage identification

Steps identifying the cascading failure's stage are as follows:

1) Use Equation (4) and (5) respectively to calculate the load rate monitored by the relays, and the actual load rate at the current temperature:

$$B_i = \frac{S_i}{S_{i,\max}} \quad (4)$$

$$B_{i,Real} = \frac{S_i}{S_{i,MaxReal}} \quad (5)$$

where B_i is the line's load rate monitored by the relays, which is based on the rated capacity at normal temperature. $B_{i,Real}$ is the line's actual load rate, which is based on the rated capacity at the current temperature, S_i is the line's transmission capacity, $S_{i,\max}$ is the rated capacity of the line at normal temperature, and $S_{i,MaxReal}$ is the rated capacity of the line at the current temperature.

2) Determine the stage of cascading failures according to $B_{i,Real}$ and $B_{i,Real}$ and B_i. If $B_{i,Real} \geq 1$ but $B_i < 1$, the concerned line is under a slow dynamic process. If $B_{i,Real} \geq 1$ and $B_i \geq 1$, the concerned line is under fast dynamic stage. If $B_{i,Real} < 1$ and $B_i < 1$, the concerned line is in a normal state.

4 PROBABILITY MODEL OF EACH STAGE

Since the development stage of cascading failures has been identified, based on the methods described above, the line outage probability for each cascading failure stage can be obtained by using the corresponding probability model, and thus the cascading chains can be searched out. This section will propose the cascading failure probability model considering the characteristics of each stage.

4.1 Slow dynamic probability model

Since the actual overload situation of the lines cannot be monitored by the relays when the system enters the slow dynamic stage, some lines will be continuously overloaded while the relays will have no response. Taking into account the differences in the load rate, the lines with maximum load rate usually heat up most quickly and are most prone to ground short circuit caused by contact with the trees. The slow dynamics probability model is elaborated in Formulae (6) and (7):

$$B_{i,Real} = \frac{S_i}{S_{i,MaxReal}} \quad (6)$$

$$P_{i,S}(B_{i,Real}) = \begin{cases} \dfrac{B_{i,Real}}{Max\{B_{i,Real}\}} & (B_{i,Real} > 1) \\ 0 & (B_{i,Real} \leq 1) \end{cases} \quad (7)$$

where $P_{i,S}(B_{i,Real})$ represents the outage probability of line i with the actual load rate $B_{i,Real}$, and $Max\{B_{i,Real}\}$ represents the maximum actual load rate of all the power lines.

In particular, parts of the lines may continue exceeding the rate limits. By considering the cumulative effect of line heating in the process of cascading failure, the probability of a ground short circuit for continuous overload lines will be higher than for others. Therefore, it is necessary to take into consideration the effect of continuous overload when designing the probability models. The weighted value of α_i is introduced:

$$B^*_{i,Real} = \alpha_i \cdot B_{i,Real} \quad (8)$$

$$\alpha_{i,m} = \begin{cases} \alpha_1 & S_i(m) \geq S_{i,MaxReal} \text{ 且 } S_i(m-1) < S_{i,MaxReal} \\ \alpha_2 & S_i(m) \geq S_{i,MaxReal} \text{ 且 } S_i(m-1) \geq S_{i,MaxReal} \end{cases} \quad (9)$$

where $B^*_{i,Real}$ is the comprehensive load rate considering the continuous over-limit of line i, $S_i(m-1)$ is the actual line transmission capacity for line i when cascading failure develops to $m-1$ order failure, $S_i(m)$ is the actual line transmission capacity for line i when cascading failure develops to m order failure, $S_{i,MaxReal}$ is the actual upper limit of the normal power flow, and $\alpha_{i,m}$ is the weighted value considering continuous overload.

The line outage probability $P_{i,S}(B^*_{i,Real})$ which considers the continuous over-limit can be obtained by substituting $B^*_{i,Real}$ into Formula (7).

4.2 Fast dynamic probability model

The system will enter the fast dynamic process when the load rate monitored by the relays exceeds the setting values. When the transmission line capacity stays in the normal range, the statistical value $\overline{P}_l$ can be chosen as the line outage probability $P_{i,F}(B_i)$, which is 1.81×10^{-4} as usual. However, when the transmission line capacity exceeds the ultimate range $S_{i,LimitMax}$, the relays will operate normally and trip the line, so the outage probability will be as high as 1.0. When the transmission line capacity stays between the normal range and the ultimate value, the probability of line tripping will rise with an increase of power flow in this line. A linear equation can be taken as the outage probability model, which is shown in Formula (10):

$$P_{i,F}(B_i) = \begin{cases} \overline{P}_l & (B_{i,Min} \leq B_i \leq B_{i,Max}) \\ \dfrac{1-\overline{P}_l}{R_{i,LimitMax} - R_{i,Max}} B_i + \dfrac{\overline{P}_l \times R_{i,LimitMax} - R_{i,Max}}{R_{i,LimitMax} - R_{i,Max}} & (B_{i,Max} \leq B_i \leq B_{i,LimitMax}) \\ 1 & (B_i \geq B_{i,LimitMax}) \end{cases} \quad (10)$$

where B_i is the load rate monitored by the relays, which can be calculated by Formula (11):

$$B_i = \frac{S_i}{S_{i,\text{Max}}} \quad (11)$$

The fast dynamic probability model considering continuous overload is shown in Formulae (12) and (13):

$$B_i^* = \beta_i \cdot B_i \quad (12)$$

$$\beta_i = \begin{cases} \beta_1 & S_i(m) \geq S_{i,\text{Max}} \text{且} S_i(m-1) < S_{i,\text{Max}} \\ \beta_2 & S_i(m) \geq S_{i,\text{Max}} \text{且} S_i(m-1) \geq S_{i,\text{Max}} \end{cases} \quad (13)$$

In formulae (12) and (13), B_i^* is the comprehensive load rate considering the continuous overload, $S_{i,\text{Max}}$ is the upper limit of the normal power flow, and $\beta_{i,m}$ is the continuous overload weight for line i when cascade failure develops to m order failure.

The line outage probability $P_{i,F}(B_i^*)$, which considers the continuous overload, can be obtained by substituting B_i^* into Formula (10).

5 RISK ASSESSMENT MODEL OF CASCADING CHAINS

5.1 Risk assessment model without considering the adjustments of dispatchers

The probability model, described above, does not consider the adjustments of dispatchers. In this case, the probability of each level can be calculated using the relevant model after the stage identification. Assuming the failure probability for each level fault is $P_1, P_2, \ldots, P_N$, the probability of the whole cascading chain can be obtained by Formula (14):

$$P_{cas} = P_1 \cdot P_2 \cdots P_N \quad (14)$$

In this paper, the risk index of cascading failure chains can be obtained by Formula (15):

$$R_{cas}(k) = P_{cas}(k) \times C_{cas}(k) \quad (15)$$

where $P_{cas}(k)$ is the probability of the cascading chain k, $C_{cas}(k)$ is the loss of load caused by the cascading fault chain k, and $R_{cas}(k)$ is the risk index of the cascading fault chain k.

The loss of load $C_{cas}(k)$ originates from two aspects:

1) the unbalanced power when the power system splits into two electrical islands.
2) the load shed when the system power solution is divergent.

The termination condition of the search for a cascading chain is listed as follows:

1) The power system splits into two electrical islands.
2) All operating parameters are in an allowable range.

For a number of cascading chains caused by a certain initial failure, the index defined in Formula (16) can be used to assess the systematic risk level:

$$R_{cas} = \frac{1}{N} \sum_{k=1}^{N} P_{cas}(k) \times C_{cas}(k) \quad (16)$$

where R_{cas} is the systematic cascading failure risk index, and N is the number of cascading chains caused by the initial failure.

5.2 Risk assessment model considering the adjustments of dispatchers

In the practical operation, the dispatchers will take a series of emergency security measures to limit the expansion of the cascading failures.

Generally, if the depth of a cascading chain is too small, less time will be left for the dispatchers to adopt some measures, and thus the probability of this cascading chain is greater.

In addition, the duration of the fast dynamic stage is 2–10 min, while the duration of the slow dynamic stage is 6–7 times slower than the fast dynamic stage. Therefore, the more slow dynamic stages a cascading chain has, the more time dispatchers would have to take some security measures to reduce the risk of the cascading chain, and thus the probability of this cascading chain is smaller.

(1) Impact Factor of the Failure Depth (IFFD)

Ignoring the impact of the slow dynamic stages, which assumes that each step of the cascading chain is in a fast dynamic stage, the definition of the IFFD is shown in Formula (17):

$$IFFD = \frac{IFD_{SET}}{IFD} \quad (17)$$

In the formula, IFD is the failure depth of a cascading failure chain, and IFD_{SET} is the failure depth threshold which is set by operators according to the time required to take security measures. IFD_{SET} is set as 5 in this paper.

(2) Impact Factor of the Slow Dynamic (IFSD)

The definition of the IFSD is shown in Formula (18):

$$IFSD = \frac{1}{\lambda^n} \quad (18)$$

In the formula, λ is the ratio between the duration of the slow dynamic and fast dynamic; usually λ is set as 6 or 7 and it is set as 6 in this paper. n is the number of slow dynamic stages in the cascading failure chain.

(3) The composite risk

IFFD and IFSD can fully consider the impact of dispatcher adjustments on the development of cascading failure chains, so the risk assessment model considering the adjustments of dispatchers is defined as:

$$R'_{Cas} = IFFD \cdot IFSD \cdot R_{Cas} \quad (19)$$

6 PROCESS OF RISK ASSESSMENT CONSIDERING THE IMPACT OF AMBIENT TEMPERATURE

The process of risk assessment considering the impact of ambient temperature is as follows:

Step 1. Select the initial failure according to the operators' practical requirements.
Step 2. Make use of the BPA (Bonneville Power Administration) software to calculate the power flow after the last failure.
Step 3. Identify the cascading failure stage according to the result of the power flow calculations.
Step 4. Use the probability model of the corresponding stage to obtain the fault probability of each line, and get the optional fault lines filtered by the probability threshold.
Step 5. Analyse the power flow after the outage of each selected line respectively, and get the correlation coefficient of each line. The composite search index obtained can be used to screen out the next fault lines. The methods to obtain the corresponding coefficient can be found in [9].
Step 6. If the power flow calculation converges, jump back to Step 3 and continue searching for the next potential failure lines. If the power flow calculation does not converge, utilize the load shed model [10] to calculate the load curtailment until the power flow calculation converges.
Step 7. If there are some electrical islands or all the operating parameters are in the allowable range, record the cascading chain, and begin to search the next fault line.
Step 8. Make a detailed analysis of the searching results and calculate the impact factor of the failure depth and impact factor of the slow dynamic of each cascading chain respectively, so as to obtain a reasonable systematic cascading failure risk.

The process of risk assessment is shown in Figure 1.

7 SAMPLE ANALYSIS

The sample analysis is based on the 10-machine New-England Power System.

7.1 The impact of the slow dynamic stage on the cascading chain developing process

The influence of the slow dynamic stage on the cascading chain developing process is shown in Table 2. The numbers of the cascading failure lines are described by the buses' number. For example, line (3-4) in Table 2 refers to the line connecting the bus 3 and bus 4.

Without considering the slow dynamic stage, the initial failure, i.e., the failure of line (3-4), will result in the failure of line (18-3). After that, the system will stay at the normal stage. In this case, the depth of the cascading chain is 2, and the load shedding amount is 0.

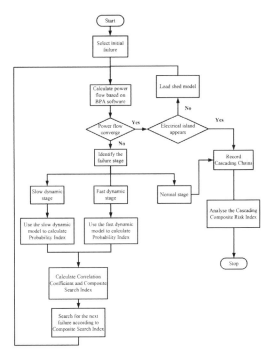

Figure 1. The process of risk assessment.

Table 2. The influence of the slow dynamic stage on the cascading chain developing process

	Cascading Failure Chain	Load Shed MW	Probability Index	Risk Index
Considering slow dynamic stage	3-4 → 18-3 → 1-2 → 17-27	121.8	0.0286	3.489
Without considering slow dynamic stage	3-4 → 18-3	0	0.0286	0

When considering the slow dynamic stage, the early developing process of the cascading chain is the same as without considering it. After the tripping of line (18-3), although the load rates of all the lines are under the protective setting values, the transmitted power of some lines has exceeded the thermal stable capacity at the current temperature. After a period of time, line (1-2) will be tripped due to the ground short circuit caused by contact with the trees, which will cause new cascading failure. The process will not stop until line (17-27) is tripped and the system will split into two electrical islands. In this case, the depth of the cascading chain is 4, and the load shedding amount is 121.8 MW.

7.2 Risk assessment of system's cascading failure under different load rates

The composite risk index of the system's cascading failure under different load rates is shown in Figure 2.

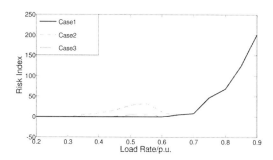

Figure 2. The composite risk index of the system's cascading failure under different load rates.

Where:

1) Case 1: Ignore the slow dynamics stage
2) Case 2: Consider the influences of the slow dynamic stage, but ignore the dispatcher adjustments
3) Case 3: Consider both the influences of the slow dynamic stage and dispatcher adjustments

(1) Case 1

For Case 1, when the load rate is lower than 0.58, the composite risk of the system remains 0, but when the load rate is greater than 0.58, the composite risk index rises with the increase of the load rate. This is because without considering the slow dynamic process, when the load rate is lower than 0.58, many risky lines have not been fully considered. So, the assessment results of the cascading chains are relatively optimistic.

(2) Case 2

For Case 2, because it has considered the impact of ambient temperature on the lines' thermal stable capacity, the model can find out those lines whose thermal stability has been broken, so the composite risk index rises gradually from 0.3, reaching its peak at 0.56, and then declining gradually until the load rate equals 0.62. Then the curve changes almost the same as in Case 1.

Within the range of 0.5 to 0.6, the slow dynamic stage and the fast dynamic stage coexist. Therefore, the system risk index will experience a small peak, which is the result of interaction between those two stages. And the development of system cascading failure will not totally enter the fast dynamic stage, until the load rate increases to 0.62.

(3) Case 3

For Case 3, because there is enough time for the dispatchers to take some security measures during the slow dynamic process, the composite risk index of the system declines significantly during the slow dynamic stage.

8 CONCLUSION

In this paper, a cascading failure risk assessment model is proposed, which considers the impact of ambient temperature. This model divides the cascading failure process into a slow dynamic stage and a fast dynamic stage, which are in accordance with engineering practice. By defining the failure depth impact factor and the slow dynamic impact factor, it is much easier to consider the adjustments of the dispatchers. The analysis of numerical examples verifies the rationality of the model proposed, which shows that the risk assessment results will be more optimistic when the slow dynamic stage is not considered, and the monitoring of the slow dynamic stage is the key to reducing the systematic cascading failure risk.

REFERENCES

[1] Gan Deqiang, Hu Jiangyi, Han Zhenxiang, "A Pondering Over Some International Blackouts in 2003," Automation of Electric Power Systems, Vol. 28, No. 3, 2004, pp. 1–4.

[2] Chen Xiangyi, Chen Yunping, Li Chunyan and Deng Changhong, "Constructing Wide-area Security Defensive System in Bulk Power Grid—A Pondering over the Large-scale Blackout in the European Power Grid on November 4", Automation of Electric Power Systems, Vol. 31, No. 1, 2007, pp. 4–8.

[3] B. A. Carreras, D. E. Newman, I. Dobson, Senior Member, IEEE and A. B. Poole. Benjamin A. Carreras, David E. Newman, Ian Dobson, Senior Member, IEEE, and A. Bruce Poole. "Evidence for self-organized criticality in a time series of electric power system blackouts," IEEE Transactions on Circuits and Systems—I: Regular Papers.

[4] Dobson, B. A. Carreras, V. E. Lynch and D. E. Newman, "An initial model for complex dynamics in electric power system blackouts," Proceedings of the 34th Hawaii International Conference on System Sciences, 3–6 January 2001.

[5] Dobson, B. A. Carreras, D. E. Newman, "A probabilistic loading-dependent model of cascading failure and possible implications for blackouts," Proceedings of the 36th Hawaii International Conference on System Sciences, 6–9 January 2003.

[6] Dobson, B. A. Carreras, D. E. Newman, "A branching process approximation to cascading load-dependent system failure," Proceedings of the 37th Hawaii International Conference on System Sciences, 5–8 January 2004.

[7] P. A. Parrilo, S. Lall, F. Paganini, G. C. Verghese, B. C. Lesieutre and J. E. Marsdenl, "Model reduction for analysis of cascading failures in power systems," Proceedings of the American Control Conference, San Diego, California, 2–4 June 1999.

[8] M. Anghel, K. A. Werley and A. E. Motter, "Stochastic Model for Power Grid Dynamics," Proceedings of the 40th Hawaii International Conference on System Sciences, January 2007.

[9] M. S. Rodrigues, J. C. S. Souza, M. B. Do Coutto Filho and M.Th. Schilling, "Automatic contingency selection based on a pattern analysis approach," International Conference on Electric Power Engineering, 29 August–2 September 1999.

[10] Chen Lin, Sun Yuanzhang, Zheng Wanqi and Chao Jian, "Large-Scale Composite Generation and Transmission System Adequacy Evaluation and It's Application," Automation of Electric Power Systems, Vol.28, No.11, 2004, pp. 75–78.

Target speech detection using Gaussian mixture modeling of frequency bandwise power ratio for GSC-based beamforming

J. Lim, H. Jang, S. Jeong & Y. Kim
Electronics Engineering Department/ERI, Gyeongsang National University, Jinju, Republic of Korea

ABSTRACT: Among many noise reduction techniques using microphone array, Generalized Sidelobe Canceller (GSC)-based one has been one of the strong candidates. The performance of GSC is directly affected by its Adaptation Mode Controller (AMC). Namely, accurate target signal detection is essential to guarantee the sufficient noise reduction in pure noise intervals and the less distortion in target speech intervals. This paper proposes an improved AMC design technique based on the frequency bandwise power ratio of the outputs of fixed beam-forming and blocking matrix. Then, the ratios are probabilistically modeled by mixture of Gaussians for pure noise and target speech, respectively. Finally, the log-likelihood is evaluated to discriminate the noise and target speech. Experimental results show that the proposed algorithm outperforms conventional AMCs in output SNR and receiver operating characteristics curves.

1 INTRODUCTION

Of various human-machine interface technologies, speech interface is the most convenient and researches on speech-based interfaces have been carried out for decades. Recently, the speech interfaces have been applied to smartphones, car navigations, intelligent robots, and electronic devices. However, the performance of the speech interfaces would inevitably and severely be degraded by diverse types of noises. Therefore, speech enhancement techniques have been widely studied in order to achieve notable performance improvement in real adverse noisy environment.

Noises can be largely categorized into stationary and nonstationary ones according to their long-term statistical characteristics. For instance, noises of air-conditioners and PC fans belong to stationary noises, whereas human voice and music belong to nonstationary noises. Stationary noises can be sufficiently removed by applying a single channel filter such as the Wiener or the Kalman filter (S. Jeong et al. 2001). However, nonstationary noise reduction with the single channel filters causes severe speech distortion. Thus, to cope with nonstationary noises, microphone array techniques including beamforming (BF) and blind source separation (BSS) have been studied (A. Hyvarinen et al. 2000). It is known that BF algorithms produce better performance than BSS in real applications because they additionally utilize the directional information of the target source. Thus, a microphone array-based BF algorithm is primarily implemented to reduce nonstationary noises in this paper.

Linearly constrained minimum variance (LCMV) and generalized sidelobe canceller (GSC) are widely used multichannel beamforming algorithms. In particular, GSC beamforming algorithm is theoretically equivalent to LCMV but can yield further improved noise reduction performance than LCMV when adaptation mode controller (AMC), works successfully.

In this paper, in order to detect accurate target signal intervals, a new AMC method is proposed. The AMC is based on a frequency bandwise power ratio of the outputs of fixed beamforming (FBF) and blocking matrix (BM). In addition, the ratios are probabilistically modeled by Gaussian mixtures of target speech and noise-only intervals.

The remainder of this paper is organized as follows. In Section 2, the proposed AMC is explained in detail. In Section 3, several experiment results of the conventional and proposed AMC methods will be compared and discussed. Finally, conclusion will be given in Section 4.

2 PROPOSED AMC ALGORITHM

In Figure 1, $B_i(k)$ and $Y_{FBF}(k)$ are the kth frequency responses of the ith BM and FBF outputs, respectively. A feature vector is composed of a frequency bandwise power ratio of the FBF and BM outputs. Log-likelihood ratio (LR) is estimated by comparing trained target speech and noise GMMs and the extracted feature vector. Finally, the LR is compared to an empirical threshold, LR_{TH}. The interval of higher LR than the threshold is decided as the target interval and vice versa.

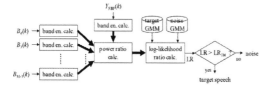

Figure 1. Block diagram of the proposed AMC algorithm.

2.1 Definition of frequency bandwise energy

Equation 1 denotes the BM output power of the *l*th frequency band,

$$P_{BM,m}^{(l)} = \sum_{k=k_s^{(l)}}^{k_e^{(l)}} |B_m(k)|^2, \quad l = 0,1,...,N_{B-1} \quad (1)$$

where l = frequency band index; m = microphone index; N_B = number of frequency bands; $k_s^{(l)}$ = minimum frequency index of the *l*th band; and $k_e^{(l)}$ = maximum frequency index of the *l*th band. In addition, the frequency bandwise FBF output power, $P_{FBF}^{(l)}$, can be similarly obtained by replacing $B_m(k)$ in equation 1 by $Y_{FBF}(k)$

2.2 Band-wise power ratio

The *l*th band power ratio is calculated as a ratio of the FBF output power to the average of the *M*-1 BM output power and can be represented by

$$PR^{(l)} = \frac{P_{FBF}^{(l)}}{\frac{1}{M-1}\sum_{m=2}^{M} P_{BM,m}^{(l)}} \quad (2)$$

where M = number of microphones.

2.3 Gaussian mixture model-based scoring

LR is estimated by GMM-based probability distributions with frequency bandwise power ratios. The ratios of target speech and noise intervals are separately calculated in the training process. The LR is evaluated by

$$LR = \log P(\mathbf{PR}|\lambda_T) - \log P(\mathbf{PR}|\lambda_N) \quad (3)$$

where $\mathbf{PR} = [PR(0), PR(1),\ldots, PR(N_B-1)]^T$; λ_T = estimated GMM parameter set of target speech intervals; and λ_N = the estimated GMM parameter set of noise intervals.

2.4 Target speech detection

The FBF yields a large output in the target speech interval and a small value in the noise-only interval, whereas the BM yields a small output in the target speech interval and a large value in the noise-only interval. Therefore, it is highly probable that the LRs of target speech intervals and noise-only intervals are large and small, respectively. A decision rule for target speech detection can be represented by

$$LR > LR_{TH}. \quad (4)$$

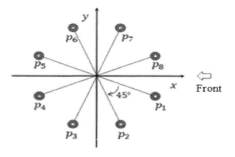

Figure 2. Arrangement of the microphone array.

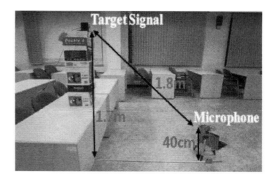

Figure 3. Environment for target signal acquisition.

3 EXPERIMENT AND RESULT

3.1 Multichannel database acquisition and experiment condition

The arrangement of the microphone array is described in Figure 2 Eight microphones were arranged by uniform circular array and the radius of the array was 8 cm. p_1 and p_8 were encountered to the look direction. The databases of target speech and noise were separately acquired by using on 8 channel recording device. Target speech is collected by locating the source at the front of the array. All the data were sampled with 16 kHz and the resolution was 16 bits. The chamber for DB acquisition was a lecture room and its size was 9 m × 11 m × 3 m. Target speech signals were around one second-long 1808 Korean isolated words and the speech signals were played and record by high-quality speaker and the recording device, respectively (see Fig. 3). In addition, multichannel noise was collected by playing babble noise of a restaurant through 6 speakers surrounding the microphone array (see Fig. 4). The noisy DB was constructed by artificially adding the noise to the target speech at −5, 0, 5 dB SNRs. The total number of noisy DB was 5424 (3 × 1808).

A half of the noisy DB was utilized for training and the other half was used for performance evaluation test.

3.2 GSC algorithm

GSC algorithm was implemented in the frequency domain. The sizes of window for short-time signal

Figure 4. Environment for directional noise acquisition.

analysis and discrete Fourier transform (DFT) were 512. Frequency analysis was performed per 5 ms. The reference of the BM output calculation was the first channel. The NC filter coefficients were updated by normalized LMS (NLMS) (M. Hayes, 1996).

3.3 Proposed AMC algorithm

To implement the proposed AMC in Figure 1, the frequency domain input signals decomposed into 1 kHz of 8 uniform frequency bands, i.e. a band is composed of 32 frequency bins. Thus, dimension of **PR** of the equation 3 was 8. The target speech and noise intervals of the noisy DB were manually marked. In equation 3, the numbers of Gaussian mixtures of target speech and noise were 4.

3.4 Performance evaluation

For performance evaluation, receiver operating characteristic (ROC) curves and output SNR of GSC were measured. Conventional target signal detection methods were normalized cross correlation (NCC) and power ratio (PR) methods (Y. Jung et al. 2005). The NCC method was calculated by the NCC of the two channels. The NCC can be represented by

$$\phi_t^{(NCC)} = \frac{\sum_{k=k_1}^{k_2} X_i(k) X_j^*(k)}{\sqrt{\sum_{k=k_1}^{k_2} |X_i(k)|^2} \sqrt{\sum_{k=k_1}^{k_2} |X_j(k)|^2}} \quad (5)$$

where t = frame index; $(\cdot)^*$ = complex conjugate; $X_i(k)$ = kth frequency response of the ith channel; k_1 = minimum frequency index; and k_2 = maximum frequency index In addition, PR method is a ratio of the FBF and BM outputs and can be represented by

$$\phi_t^{(PR)} = \frac{\sum_{k=k_1}^{k_2} |Y_{FBF}(k)|^2}{\sum_{k=k_1}^{k_2} \sum_{m=2}^{M} |B_m(k)|^2} \quad (6)$$

where $B_m(k)$ = kth frequency response of the mth BM outputs; and L = DFT size. In order to obtain the NCC in equation 5, the two selected channels were the 2nd

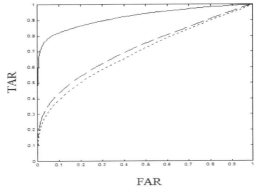

Figure 5. ROC curves calculated from training DB (dotted: NCC, dashed line: PR, solid: proposed).

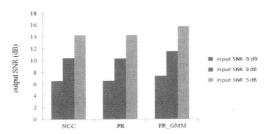

Figure 6. Output SNRs of the GSC with various AMC (NCC, PR, PR_GMM (proposed)).

and 7th of the array. The frequency band for calculating $\Phi_t^{(NCC)}$ and $\Phi_t^{(PR)}$ in equation 5 and 6 was 250 Hz–1.5 kHz, where target speech is highly distributed in general.

Figure 5 represents the ROC curves of the three method, i.e. NCC, PR, and the proposed method, calculated from all the training DB. In Figure 5, false acceptance rate (FAR) and true acceptance rate (TAR) indicate the misdetection rate considering noise as target speech and the correct detection rate of the target speech, respectively. In Figure 5, FARs of NCC, PR, and the proposed method in equal error rate (EER) were 0.38, 0.34, and 0.15, respectively. In addition, the thresholds of NCC, PR, and the proposed method in EER were 0.87, 16.7, and −1.7, respectively. As represented in Figure 6, the performance of the proposed target speech detection outperforms the performance of the conventional methods. The main reason for the performance improvement is analyzed as follows.

In case of NCC, the limitation of NCC was a multichannel feature obtained from only two channels and most of the energy distribution of acoustic signals was concentrated on the low frequency region. As the result, NCC of the noise-only interval was relatively not small. In addition, the PR method used all channel information Thus, the performance of the PR method was superior to that of the NCC. However, the performance improvement was not notable. On the other hand, it is considered that the performance of the proposed method was improved due

Table 1. FAR and TAR of the proposed method obtained from test DB.

Input SNR (dB)	FAR	TAR
−5	0.15	0.74
0	0.14	0.84
5	0.17	0.92

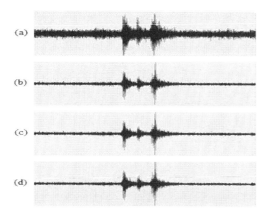

Figure 7. Waveform results of GSC beam-formin (target speech: /ie-kun-dae/) (a) noisy input (5 dB) (b) NCC-AMC (c) PR-AMC (d) proposed AMC.

to applying the probabilistic modelling technique to the bandwise power ratio. The proposed AMC uses relatively large information. The increased information for target signal detection affects the performance improvement.

FAR and TAR of the proposed method when the threshold obtained by EER of the training DB is applied to test DB are presented in Table 1. Figure 6 presents the output SNRs of the GSC with various AMC methods. The output SNR is measured by a ratio of the average target speech to noise energies. The obtained thresholds of EER in Figure 5 are applied to each AMC algorithm. As shown in Figure 6, the proposed method yields the best performance at all the input SNRs. Especially, the output SNRs of GSCs with NCC, PR, and the proposed method were 14.27 dB, 14.28 dB, 15.7 dB at 5 dB input SNR. Figure 7 presents waveform results of the GSC with the AMCs.

4 CONCLUSION

In this paper, a novel AMC method for GSC beamforming is proposed. The proposed AMC is based on the GMM of the frequency bandwise power ratio.

The experiment results of the proposed AMC presented the lowest EER and highest output SNR of GSC beamforming. As future works, an AMC working in each frequency band will be implemented by integrating the proposed AMC with additional feature parameters.

ACKNOWLEDGEMENT

This research was supported by Basic Science Research Program through the National Research Foundation of Korea (2011-0011800) funded by the Ministry of Education, Science and Technology.

REFERENCES

[1] Hyvarinen and E. Oja, Independent component analysis: Algorithms and applications, Neural Networks, vol. 13, no. 4, pp. 411–430, 2000.
[2] M. Hayes, Statistical Digital Signal Processing and Modeling, John Wiley & Sons, 1996.
[3] S. Jeong and M. Hahn, "Speech quality and recognition rate improvement in car noise environments", Electronics Letters, Vol. 37, No. 12, pp. 801–802, 2001.
[4] Y. Jung, H. Kang, C. Lee, D. Youn, C. Choi, and J. Kim, "Adaptive microphone array system with two-stage adaptation mode controller," IEICE Trans. Fund., vol. E88-A, no. 4, pp. 972–977, Apr. 2005.

ND
A compressive sampling method for signals with unknown spectral supports

Enpin Yang, Xiao Yan & Kaiyu Qin
School of Aeronautics and Astronautics, University of Electronic Science and Technology of China, Chengdu, Sichuan, China

Feng Li & Biao Chen
The State Administration of Press, Publication, Radio, Film and Television of the People's Republic of China, Beijing, China

ABSTRACT: Signals with unknown spectral support frequently appear in modern signal processing, such as blind frequency hopping signal detection and cognitive radio application. A representative characteristic of this kind of signals is that they are sparse in frequency domain. And because Nyquist sampling is inapplicable or highly inefficient for sparse signals, we propose a compressive sampling method to deal with this problem. The method uses a bank of periodical Run-Length-Limited sequences to modulate input signal. As a result, spectrum of the modulated signal is equivalent to the shifted weighted sum of input signal's spectrum. Obviously, in the baseband of modulated signal, the weighted sum contains all spectral slices of the original signal. Hence, the modulated signal can be filtered by a low-pass filter and sampled at a low rate. Then, signal reconstruction can be performed based on the Compressive Sensing theory.

1 INTRODUCTION

1.1 Signals with unknown spectral support

Shannon-Nyquist sampling theorem assumes that signals to be sampled are all derived from a single vector space, which means sampled signals are projected onto a vector space by sampling operator. For instance, the interpolating function set of the theorem $\{\text{sinc}(t - nT), n \in Z\}$ is an orthogonal basis for the vector space (Unser, 2000). However, this assumption makes the theorem inapplicable or highly inefficient in many cases, because signals to be sampled may actually reside in a union of subspace (Lu et al. 2008). For example, signals with unknown spectral support come from a union of subspace.

Features of signals with unknown spectral support can be summarized as follows: their Fourier transform $X(f)$ only occupy a few disjoint intervals of unknown locations on a spectral band. For a fixed set of locations, these signals reside in a subspace. Further, with all possible spectral locations, those signals live in a union of subspaces. And this class of signals is also called multiband signals (Eldar et al. 2009). Due to the diversity of the set of subspaces, one possible way for acquiring signal is to compress the set. In this case, for signals with unknown spectral support, compressing subspace means compressing spectral locations.

In Figure 1, the set $\{\phi 1, \phi 2, \phi 3\}$ spans a subspace where the signal $X(f)$ lives in. We can compress or project $X(f)$ onto a lower dimension subspace $\{\phi 1\}$ to simplify sampling operation.

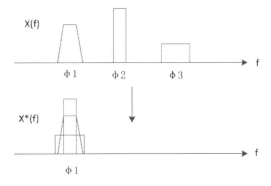

Figure 1. Spectrum compressing for one signal with unknown spectral support.

1.2 Related compressive sampling method

Since the compressive sensing theory (Candès et al. 2006) was proposed in 2006, several sub-Nyquist sampling strategies based on it have been carried out, including Random Filter (RF) (Tropp et al. 2006a, b), Random Demodulator (RD) (Kirolos et al. 2006, Laska et al. 2007), Constrained RD (CRD) (Harms et al. 2011a, b, 2013) and Modulated Wideband Converter (MWC) (Mishali et al. 2009, 2010a, b, 2011).

Signal processing flows of all strategies mentioned above are extremely similar, as shown in Figure 2. Firstly, spectrum of signal to be sampled is smeared across the entire Nyquist band by different means.

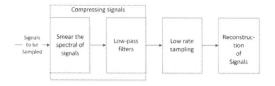

Figure 2. Signal processing flow of several sub-Nyquist sampling strategies.

And then, the non-bandlimited signal is appropriately filtered and sampled. Finally, original signal is reconstructed according to the compressive sensing theory.

Except those similarities, RF, RD and CRD require that spectral support of input signals should be a set of points rather than a set of intervals. Obviously, those three methods are invalid in real world; they can only be used in some specific situations. However, the MWC can process almost all kinds of signals with unknown spectral support.

Due to the merit of the MWC, we present a modified compressive sensing method based on it. We choose Run-length-limited sequences (RLL) as the modulation operator to smear spectra of input signals, which is more applicable than sequences used in the original MWC. Then, we revise the key parameter calculation formula of the MWC, so that RLL sequences can perfectly smear spectra of input signals.

The following part of this paper is organized as follows. Section 2 introduces the basic principle of Compressive Sensing theory. Section 3 illustrates all details of the proposed method. Section 4 shows the result of numerical simulation.

2 BACKGROUND: COMPRESSIVE SENSING

2.1 Sparse or compressible signal

First of all, the sparse or compressible signal should be introduced. A discrete-time signal vector x, whose length is L, can be called sparse if the amount of nonzero entries of the vector is much less than L. However, a considerable number of signals in the real world are not sparse. Fortunately, there is a way to decompose those signals. And the decomposed signals, that we call them compressible signals, have same properties with sparse signals.

If the signal x can be expressed as $x = \Psi\alpha$, in which Matrix Ψ is an orthogonal basis (each column ψ_i is basis vector) or dictionary (each column is called atom), and if length of α is greatly larger than the number of its nonzero entries, it is compressible. And for the matrix Ψ, some are commonly used such as DFT matrix, wavelet matrix, Curvelet, Ridgelet, Contourlet and Bandelet.

2.2 The projection matrix

Due to the sparsity of vector α, it can be approximated via a z-terms vector, where the z indicates the number

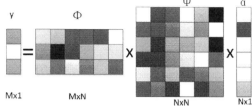

Figure 3. Principle of compressive sensing (white blocks indicate zero entries).

of nonzero entries. That is, we can project the vector α from L dimension to z dimension.

In compressive sensing theory, there is a matrix Φ which can realize the projection, as shown in Figure 3.

$$y = \Phi x = \Phi\Psi\alpha \quad (1)$$

where y is the outcome after compression operation. And its dimension has been greatly reduced. Under different circumstances, Φ should obey different conditions, such as Coherence-based Guarantee, RIP Guarantee, Statistical-RIP Guarantee and Extend-RIP Guarantee.

2.3 Signal reconstruction

If the sparest vector α can be obtained through (2), then the original signal can be calculated by solving $x = \Psi\alpha$. Rewrite (1) as

$$y = \Theta\alpha \quad (2)$$

where $\Theta = \Phi\Psi$, and Θ is an $M*N$ matrix. We can find that there is no solution for (2) due to $M \ll N$. It is ill-posed. However, under the compressive sensing framework, there are two kinds of algorithms can deal with this problem. One is Basis Pursuit algorithm which calculates the l_1-norm of α

$$\hat{\alpha} = \min \|\alpha\|_{l_1} \quad s.t. \Theta\alpha = y \quad (3)$$

Another one is Matching Pursuit algorithm, and it calculates the l_0-norm of α

$$\hat{\alpha} = \min \|\alpha\|_{l_0} \quad s.t. \|\Theta\alpha - y\| < \xi \quad (4)$$

where ξ is a tolerable error.

3 THE COMPRESSIVE SENSING METHOD

As addressed before, MWC can handle a large class of signals with unknown spectral support. Therefore, we propose a modified compressive sensing method based on it. This method conquers a limitation of MWC—waveforms of sequences used in modulation stages should alternate at the Nyquist rate of input signal—by utilizing Run-Length-limited sequences. This limitation makes MWC inapplicable when input signals have large Nyquist frequencies.

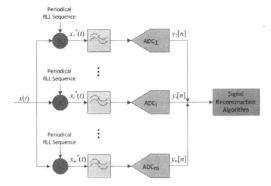

Figure 4. Block diagram of the proposed method.

3.1 Signal model

The proposed method is designed for processing signals with unknown spectral support, which can be formulated as:

$$X(f) = \sum_i X(f_i) \quad i \in Z \quad f \in [-f_{max}, f_{max}] \quad (5)$$

$X(f)$ is the Fourier transform of a bandlimited complex signal $x(t)$, and its FT consists of several disjoint spectral slices (including the conjugate symmetry), while location of each slice (sub-band) is unknown.

3.2 Run-Length-Limited sequence

Value set of RLL sequence is $\{-1, +1\}$. There are two parameters (d, k) which control the alternating rate of the sequence, and those two parameters indicate the minimum run length and the maximum run length of symbols respectively. Minimum run length of RLL sequence restricts the maximal alternating rate. For example, in unit time T, alternating rate of a pseudo-random sequence with c symbols is c/T; alternating rate of a $(2, k)$ RLL sequence with same symbols is $c/2T$. The rate of RLL sequence is much less than that of the pseudo-random sequence. Therefore, we can use this property to break the limitation of MWC.

3.3 The compressive sampling model

3.3.1 Block diagram of the method

The block diagram of the proposed method is shown in Figure 4. The model consists of M channels, and in each channel the input signal $x(t)$ is modulated by a RLL sequence with period T_{rll}. Then, the modulation is followed by a low-pass filter and a ADC operating at T_s ($1/T_s \geq B$, where B denotes the biggest bandwidth of all $X(f_i)$; B is much less than the Nyquist band).

3.3.2 Fourier domain analysis of the method

Because the input signal $x(t)$ is sparse, we analyse it in the Fourier domain. It is easy to derive that the Fourier series of the Run-Length-Limited sequence $p_i(t)$ is

$$p_i(t) = \sum_{l=-\infty}^{+\infty} c_{il} e^{j\frac{2\pi}{T_{rll}}lt}, \quad 1 \leq i \leq M \quad (6)$$

In which

$$c_{il} = \frac{1}{T_{rll}} \int_0^{T_{rll}} p_i(t) e^{-j\frac{2\pi}{T_{rll}}lt} dt, \quad 1 \leq i \leq M \quad (7)$$

The Fourier transform $X_i^*(f)$ of the modulated signal $x_i * (t) = x(t) \times p_i(t)$ is

$$\begin{aligned}
X_i^*(f) &= \int_{-\infty}^{+\infty} x_i^*(t) e^{-j2\pi ft} dt \\
&= \int_{-\infty}^{+\infty} x(t) \left(\sum_{l=-\infty}^{+\infty} c_{il} e^{j\frac{2\pi}{T_{rll}}lt} \right) e^{-j2\pi ft} dt \\
&= \sum_{l=-\infty}^{+\infty} c_{il} \int_{-\infty}^{+\infty} x(t) e^{-j2\pi \left(f - \frac{l}{T_{rll}}\right)t} dt \\
&= \sum_{l=-\infty}^{+\infty} c_{il} X(f - lf_{rll}), f \in [-f_{max}, f_{max}]
\end{aligned} \quad (8)$$

Note that $f_{rll} = 1/T_{rll}$; the f in $X_i^*(f)$ does not belong to the interval $[-f_{max}, f_{max}]$, because the modulated signal is non-bandlimited.

Filter the non-bandlimited signal via a low-pass filter with cutoff frequency f_{lp}

$$f_{lp} = \frac{1}{2T_s} = \frac{f_{rll}}{2} = \frac{B}{2} \quad (9)$$

And then, the filtered signal can be sampled at a low rate. Let $y_i[n]$ denotes the sample sequence of i-th channel, thus, the discrete-time Fourier transform (DTFT) of $y_i[n]$ is

$$Y_i(e^{j2\pi fT_s}) = \sum_{n=-\infty}^{+\infty} y_i[n] e^{-j2\pi fnT_s} = \sum_{l=-L_0}^{+L_0} c_{il} X(f - lf_{rll}) \quad (10)$$

where L_0 is a positive integer, which ensures all spectral slices of the Nyquist band can be moved to the baseband.

(10) shows that the DTFT of $y_i[n]$ is the weighted sum of f_{rll}-shifted $X(f)$, and we can rewrite it in matrix form

$$\mathbf{Y}(f) = \mathbf{A}\mathbf{Z}(f) \quad (11)$$

where $\mathbf{Y}(f)$ is a column vector composed of the DTFT of the samples from each channel; $\mathbf{Z}(f)$ is a column vector derived by equally dividing the $X(f)$ into $2L_0 + 1$ slices; $\mathbf{A}$ is an $M*N$ matrix whose i-th row is the Fourier series of RLL sequence.

Comparing (2) and (11), we can find that they have the same structure. Therefore, solution of (11) can be derived within the Compressive Sensing framework.

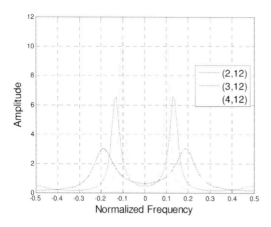

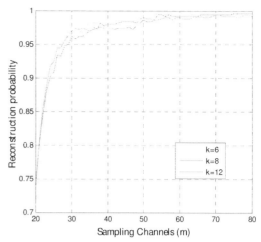

Figure 5. Normalized power spectrum of Markov chain generated Run-Length-Limited sequences (those spectra are normalized according to the maximal frequency f_{max} of the input signal).

Figure 6. Reconstruction probability with different RLL sequence (SNR is set to 10 dB).

3.3.3 Key parameter of the matrix A

We know that the matrix A is an $M*N$ matrix, and M is equal to the number of channels and N equals $2L_0 + 1$. For one specific row, the ith entry indicates amplitude of ith harmonic of the RLL sequence. So it is natural for us to think of analyzing spectrum of the RLL sequence.

Figure 5 shows several normalized power spectra of different RLL sequences, and we can find that amplitude decays very fast in their decreasing intervals. Consequently, a lot of columns of A are almost zero, which leads to terrible signal reconstruction. Therefore, we should find a way to change this situation. We can cut out the power spectrum of the RLL sequence to eliminate those zero amplitude. It is easy to see all spectra in Figure 5 have nonzero value in the original point, so we can use the origin to cut out the spectrum. That is to say, we should find a point f on the abscissa (for the (2, k) class of RLL sequences), which satisfies the following equation

$$\sup(P_{RLL}(e^{j2\pi f})) = P_{RLL}(e^0), f \in (0.25, 0.5) \quad (12)$$

where $P_{RLL}(\cdot)$ is the power spectrum dense of Markov chain generated RLL sequence (Bilardi et al. 1983).

Since power spectra are normalized according to f_{max}, we can derive this equation

$$f_{RLL} = \frac{1}{2} \times \frac{f_{max}}{2f} \quad (13)$$

where f_{RLL} is the alternating rate of RLL sequences.

The alternating rate also can be calculated by

$$f_{RLL} = \frac{1}{2} \times \frac{c}{T_{rll}} \quad (14)$$

where c is the number of symbols in one single period T_{rll}. Since T_{rll} is decided by the biggest bandwidth B, c is the key parameter which has a direct influence on the matrix A. This is because c totally determines the amplitude of the ith harmonic.

There is another condition (Mishali et al. 2010b) (14) about c,

$$c \geq c_{min} \square 2\left\lceil f_{rate} \times T_{period} + \frac{1}{2} \right\rceil - 1 \quad (15)$$

where f_{rate} equals the Nyquist rate of input signal, and also equals the alternating rate of modulation sequence (Mishali et al. 2010b). T_{period} is equivalent to the T_{rll} in this paper.

Based on (13), (14) and (15), the key parameter calculation equation can be formulated as

$$c \geq c_{min} = 2\left\lceil \frac{f_{max}}{4 f\!f_{rll}} + \frac{1}{2} \right\rceil - 1 \quad (16)$$

And (16) is the key stone of the proposed method, which guarantees the method can reconstruct signals with high probability.

4 NUMERICAL SIMULATION

The numerical simulation can certify performance of the proposed method. We apply our method to signals whose Nyquist frequencies are fixed at 5 GHz. And there are six sub-bands (including the conjugate symmetry; bandwidth of each sub-band is less than 50 MHz) existing in the Nyquist band with unknown locations.

Figure 6 shows the reconstruction probability with different RLL sequences. It is obvious that if the number of channels is big enough, the original signal can be perfectly reconstructed. Although it is impossible to build so many ADC channels in hardware, there is a solution, which is elaborated in those MWC literatures (Mishali et al. 2010a, b, 2011), to solve this problem.

5 CONCLUSION

In this paper, we have presented a compressive sampling method to process signals with unknown spectral support. Our method is based on the novel Modulated Wideband Converter (MWC). We revise the modulation stage of the MWC and propose a new key parameter calculation formula. Compared with the Nyquist sampling strategy, the proposed method can greatly reduce the sampling rate; meanwhile it is capable to reconstruct signals with high probability.

ACKNOWLEDGEMENT

This work was supported by the Specialized Research Fund for the Doctoral Program of Higher Education of China under Grant 20130185110023.

REFERENCES

[1] Bilardi, G. & Padovani, R. 1983. Spectral analysis of functions of Markov chains with applications. *IEEE Trans. Commun.*: 853–861.

[2] Candès, E. & Romberg, J. & Tao, T. 2006. Robust uncertainty principles: Exact signal reconstruction from highly incomplete frequency information. *IEEE Trans. Inf. Theory* 52(3): 489–509.

[3] Eldar, Y.C. & Michaeli, T. 2009. Beyond bandlimited sampling. *IEEE Mag. Signal Process.* 26(4): 48–68.

[4] Harms, A. & Bajwa, W. U. 2011a. Faster than Nyquist, slower than Tropp. in *Proc. IEEE CAMSAP 2011*: 345–348.

[5] Harms, A. & Bajwa, W. U. 2011b. Beating Nyquist through correlations: A constrained random demodulator for sampling of sparse bandlimited signals. in *Proc. IEEE ICASSP 2011*: 5968–5971.

[6] Harms, A. & Bajwa, W. U. 2013. A Constrained Random Demodulator for Sub-Nyquist Sampling. *IEEE Trans. Signal Process.* 61(5): 707–723

[7] Kirolos, S. & Laska, J. 2006. Analog-to-information conversion via random demodulation. in *Proc. of the IEEE DCAS 2006*: 71–74.

[8] Laska, J. N. & Kirolos, S. 2007. Theory and implementation of an analog-to-information converter using random demodulation. in *Proc. IEEE Int. Symp. Circuits Syst. 2007*: 1959–1962.

[9] Lu, Y.M. & Do, M.N. 2008. A Theory for Sampling Signals From a Union of Subspaces. *IEEE Trans. Signal Process.* 56(7): 2334–2345.

[10] Mishali, M. & Eldar, Y. C. 2009. Blind multiband signal reconstruction: Compressed sensing for analog signals. *IEEE Trans. Signal Process.* 57(4): 993–1009.

[11] Mishali, M. & Elron, A. 2010a. Sub-Nyquist Processing with the Modulated Wideband Converter. in *Proc. IEEE ICASSP 2010*: 3626–3629.

[12] Mishali, M. & Eldar, Y. C. 2010b. From Theory to Practice: Sub-Nyquist Sampling of Sparse Wideband Analog Signals. *IEEE J. Sel. Topics Signal Process.* 4(3): 375–391.

[13] Mishali, M. & Eldar, Y. C. 2011. Xampling: Signal Acquisition and Processing in Union of Subspaces. *IEEE Trans. Signal Process.* 59(10): 4719–4734.

[14] Tropp, J.A. & Wakin, M.B. 2006a. Random Filters for Compressive Sampling and Reconstruction. in *Proc. ICASSP 2006*, Toulouse: 872–875.

[15] Tropp, J.A. 2006b. Random Filters for Compressive Sampling. in *Proc. ISS 2006*: 216–217.

[16] Unser, M. 2000. Sampling–50 years after Shannon. *Proc. IEEE* 88: 569–587.

Design and analysis of SSDC (Subsynchronous Damping Controller) for the Hulun Buir coal base plant transmission system

Guangshan Li
Huaneng Yimin Coal and Electricity Co. Ltd., Hulun Buir, China

Shumei Han, Xiaodong Yu & Shiwu Xiao
North China Electric Power University, Beijing, China

Xiaohuan Xian
Rice University, Texas, USA

ABSTRACT: A wide-band SSDC based on the PLL (Phase Locked Loop) subsynchronous signal extraction method was designed to control the SSO (Subsynchronous Oscillation) risk of the Hulun Buir coal base plant transmission system, and the effects of the SSDC was analysed with PSCAD/EMTDC electromagnetic transient simulation. The subsynchronous component of the HVDC (High Voltage Direct Current Transmission) rectifier AC bus voltage was extracted based on the PLL signal extraction method as the input of the SSDC control system. The transfer function was designed based on the frequency correction algorithm to meet the amplitude-frequency and phase-frequency requirements of controlling the SSO in a wide range of subsynchronous frequencies. The result of the time-domain simulation indicated that SSDC controls the SSO effectively under both large and small disturbances.

1 INTRODUCTION

When the generators transmit power through HVDC, subsynchronous oscillation problems will be caused in the shafting and threaten the shafting security. In order to control SSO caused by DC transmission, SSDC as the current repressive measures in the grid side, will be equipped in the HVDC rectifier side in China. However, in some practical engineering, SSDC has a problem in that it is unable to start and cannot produce an inhibition effect, so it is necessary to research the design and inhibiting ability of SSDC.

2 DESIGN AND INHIBITION PRINCIPLE OF SSDC

The SSDC was first proposed by the Electric Power Research Institute to inhibit SSO caused by HVDC. The basic principle is to provide additional positive electrical damping for the steam turbine generator shaft at the subsynchronous point. A modal frequency signal is extracted by the SSDC, added to the loop of the rectifier constant current controller, and thereby provides additional electromagnetic torque ΔT_{SSDC} to the generators. If the parameters of the SSDC are appropriate, the additional torque will provide positive electrical damping to the generator at the subsynchronous frequency point. Therefore, the SSO risk will be inhibited.

The design of the SSDC includes structure selection, selection and extraction of the input signal, and transfer function design. The signal extraction and transfer function design are the focus of this study.

2.1 Structure selection

The existing SSDC structure mainly includes two categories: narrowband SSDC and broadband SSDC. Narrowband SSDC needs to extract and restrain each modal frequency and this makes the control of the system very complex, so broadband SSDC is often used in practical engineering.

Broadband SSDC needs to provide appropriate phase correction and gain for each subsynchronous frequency mode in the subsynchronous frequency band, which SSDC plays a part in, so broadband SSDC design requires more detailed optimization algorithms. A broadband SSDC structure diagram is shown in Figure 1.

2.2 Selection and extraction of the input signal

SSDC input signal information should include frequency, usually the shaft slip, the converter station bus voltage, the bus current, power, and so on. At the same time, SSDC inhibits SSO problems caused by HVDC,

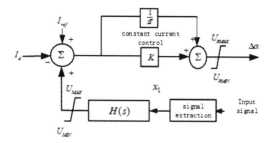

Figure 1. Broadband SSDC structure diagram.

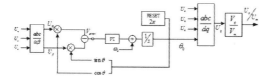

Figure 2. The design of the method based on the PLL diagram.

and the input signal should have characteristics and effectiveness, so the HVDC rectifier AC bus voltage is taken as the input signal.

The SSDC signal extraction sectors need to meet the following three basic conditions:

1) While extracting the subsynchronization signal, in order to achieve good inhibition effect, try to eliminate noise;
2) After extracting the signal it should be able to accurately reflect the state of the system;
3) Since the HVDC rectifier stations frequency conversion will occur on both sides of the DC and the AC, so converting the extracted signal frequency should be realized in the process of extraction.

From the above analysis, the frequency deviation measurement has become the key to this link. When comparing DFT (Discrete Fourier Transform) and PLL, the PLL measuring accuracy is slightly lower, but it can realize the conversion of the extracted signal frequency in the process of extraction, eliminate noise, and achieve better effects from signal extraction. Therefore, the PLL as the SSDC signal extraction method is selected. Figure 2 shows the design of the method based on the PLL diagram.

2.3 Transfer function design

The purpose of the SSDC transfer function is to mathematically process the extracting signal in order to achieve inhibition of subsynchronous oscillation. The frequency domain correction method is fitting for this phase. First of all, the different frequency compensation phases need to be calculated, then the transfer function is designed according to the obtained phase frequency characteristics.

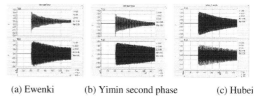

(a) Ewenki   (b) Yimin second phase   (c) Hubei

Figure 3. The modal figure of units when SSDC is not put into use.

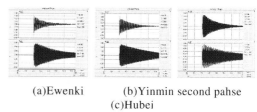

(a) Ewenki   (b) Yinmin second pahse
(c) Hubei

Figure 4. The modal figure of units when the gain of SSDC is 1/2.

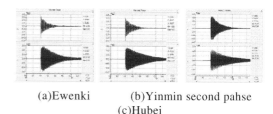

(a) Ewenki   (b) Yinmin second pahse
(c) Hubei

Figure 5. The modal figure of units when the gain of SSDC is 1.

The SSDC transfer function is:

$$H(s) = (3.745e-007s^4 + 0.0001628s^3 + 0.0177s^2) / (2.083e-011s^6 + 2.882e-009s^5 + 7.62e-007s^4 + 6.646e-005s^3 + 0.004042s^2 + 0.1097s + 1)$$

3 SIMULATION VERIFICATION OF THE SSDC DESIGN EFFECT

In order to verify the validation of SSDC design, under large-disturbance conditions, the inhibitory effects of SSDC with different gains of SSDC are analysed, using the PSCAD/EMTDC tool to simulation the inhibitory effect of SSDC.

Simulation time is 1 second, and there are three simulation conditions, one is without SSDC, another is that the gain of SSDC is 1/2 unit, and the last one is that the gain of SSDC is 1 unit. The simulation curve of units in the Ewenki power plant, the Yimin second phase power plant, and the Hubei power plant, are shown in Figure 3, Figure 4, and Figure 5, respectively. The modal decay ratios of the units in different power plants are listed in Table 1.

Table 1. The modal decay ratio of units before and after the SSDC is put into use.

	Ewenki	Yimin second phase	Hubei
No SSDC	−0.1104	−0.1925	−0.1457
	−0.0513	−0.0816	−0.0376
The gain of SSDC is 1/2	−0.2234	−0.3134	−0.2799
	−0.0913	−0.1378	−0.0801
The gain of SSDC is 1	−0.4548	−0.6582	−0.5557
	−0.1698	−0.2588	−0.1949

According to the comparison between Figure 3 and Figure 4, it can be found that the convergence speed of the unit speed difference increases when SSDC is put into use; according to the comparison between Figure 4 and Figure 5, it can be found that the convergence speed of the unit speed difference increases when the gain of SSDC increases too.

According to Table 1, the decay ratios of the unit first modal greatly increase in the Ewenki power plant, when SSDC is put into use, or the gain of SSDC increases. The conclusion is that SSO is greatly inhibited. The trend is similar in the other power plants, which shows that SSDCs designed under large disturbance have good inhibitory effects on units in other power plants.

Moreover, since the larger the gain of SSDC, the better the inhibitory effect of SSDC for the SSO excited by HVDC, the scale factor of SSDC cannot be too large when the stability problem of the system and the output limiter are taken into account.

4 CONCLUSION

Because of the SSO risk existing in the Hulun Buir coal base power plant, the SSDC in Huliao HVDC rectifier serving as inhibitory measures at power network side, are designed in detail to well inhibit the SSO risk of units in the Hulun Buir coal base power plant transmission system.

The inhibitory effects of SSDC for SSO, with different gain, are compared under large disturbance conditions, with time domain simulation results in PSCAD/EMTDC. The results show that the designed SSDC has a good inhibitory effect on units in the Hulun Buir coal base power plant transmission system, under large disturbance conditions. In reasonable scale, the larger the gain of the SSDC, the greater the inhibitory effects.

REFERENCES

[1] Subsynchronous Resonance Working Group of the System Dynamic Performance Subcommittee. Reader's guide to subsynchronous resonance [J]. Transactions on Power Systems, 1992, 7(1): 150–157.
[2] Piwko R J, Larsen E V. HVDC System Control for Damping of Subsynchronous Oscillations [J]. IEEE Trans on Power Apparatus and Systems, 1982, 101(7): 2203–2211.
[3] Linie Li, Chao Hong. The Analysis of SSO Problem on Gui-GuangII Line HVDC Project [J]. Automation of Electric Power Systems, 2007, 31(7): 90–93.
[4] Zheng Xu. The Analysis of AC/DC Power System Dynamic Behavior [M] Beijing: China Machine Press, 2004: 157–202.
[5] Rauhala T, Järventausta P. On Feasibility of SSDC to Improve the Effect of HVDC on Subsynchronous Damping on Several Lower Range Torsional Oscillation Modes [C]. Power and Energy Society General Meeting, 2010 IEEE. IEEE, 2010: 1–8.
[6] Benfeng Gao, Chengyong Zhao, Xiangning Xiao, Weiyang Yin, Chunlin Guo, Yanan Li. Design and Implementation of SSDC for HVDC [J]. High Voltage Engineering, 2010. 36(2): 501–506.
[7] Wang Zhang, Sen Cao, Hongguang Guo, Tianfeng Li. Application of SSR Damping Controller in Gaoling Back to Back HVDC Project [J]. Power System Protection and Control, 2009, 37(15): 27–32.
[8] EPRI EL-2708. HVDC System Control for Damping Subsynchronous Oscillation. Project 1425-1 Final Report, 1982.
[9] Fan Zhang, Zheng Xu. A Method to Design a Subsynchronous Damping Controller for HVDC Transmission System [J]. Power System Technology, 2008(11): 13–17.
[10] Yuyao Feng, Chao Hong, Yu Yang, Xitian Wang, Chen Chen. Synthesis of Supplementary Subsynchronous Damping Controllers for HVDC Systems [J]. East China Electric Power, 2008, (05): 53–56.
[11] Shousun Hu. The Principle of Automatic Control [M], Beijing: Science Press, 2007.

Stage division and damage degree of cascading failure

Xiangqian Yan
Hohai University, Nanjing, China

Feng Xue, Ye Zhou & Xiaofang Song
Nari Technology Development Limited Company, Nanjing, China

ABSTRACT: Aiming at the cascading failure, this paper gives a description of each process of cascading failure after researching and analysing the process of blackouts all over the world, and expounds the characteristics of different stages. It then puts forward a method of stage division, whereby the process of cascading failure has been divided into five stages: the pre-fault stage, the initial stage, the slow evolution stage, the fast evolution stage, and the collapse stage. What is more, based on the stage division and considering the damage of cascading failure to the power system, this paper proposes a quantitative index to measure the damage, which gives several suggestions for preventing the extension of cascading failure. Finally, the viability of the quantization index is demonstrated through the simulation of an IEEE-118 power system.

1 INTRODUCTION

In recent years, several blackouts have happened around the world, China has also undergone some outages, such as the "7.1" blackout in the Central China Power Grid in 2006. Although the probability of blackout is very small, once blackout happens, it will lead to huge economic loss and affects society. As Xue said in 2011, if events might happen, they would happen under certain conditions, according to Murphy's Law. Similarly, Cao et al. (2010) pointed out that statistical data showed the probability distribution of the size of a blackout followed the power-law tail characteristics, which make the blackout inevitable under certain conditions.

Cascading failure is an important cause leading to blackout; the research on cascading failure has received widespread attention. The analysis of cascading failure can be based on the conventional simulation method on account of various mathematical models, or based on complexity theory. The physical concept of the former is clear and the conclusion is visual; the latter can describe the whole behaviour from an overall perspective. In the analysis, the process is divided into several stages by most scholars, based on some important events in the process of cascading failure. These methods of stage division are intuitive, but the description of each stage is not enough, and these methods may fall into the complicated narration easily only according to the time order. The process of cascading failure has been divided into an initial stage and a cascading failure stage by Zhu (2008), which is not detailed enough to describe each stage. In another paper, Xue (2006) has divided the August 14 Northeast blackout into five stages: the slow breaking stage, the fast breaking stage, the oscillation stage, the crash stage, and the final blackout, which describe the process of blackout clearly. However, the method put forward was aimed directly at the Northeast blackout which does not apply to other events, for example, the blackout occurring in South America and North Mexico on September 8, 2011 did not include the oscillation stage. In a different way, another scholar, Hu (2004), divided the Northeast blackout into an initial stage and a collapse stage. We can see that the process of cascading failure can be divided into several stages according to different methods, which is not convenient to study and take measures to block the cascading failures because the stages are different according to different methods.

Currently, the damage caused by cascading failure is measure by several aspects: the economic losses, the loss of the elements and the load. Money, calculated from all aspects, is the most intuitive and comprehensive way to measure the damage caused by cascading failure, but the losses are manifold and a distinction is needed between direct and indirect losses, such as the difference, and it is difficult to convert all forms of loss into money. Load loss is suitable for measuring the size of cascading failures relatively, the loss of total electric quantity is equal to the load loss multiplied by the outage time, but the outage load is scattered and each outage time is not the same. Furthermore, the outage time in simulation can only rely on the experience, therefore, this parameter is also very difficult to be accurate. The number of opened branches can also measure the damage which can be obtained easily, and the value is a determined constant However, all the opened branches are the same after using this index, which ignores the different voltage levels of each power line, as do the different effects caused by different opened lines on the power grid and users. This limitation is obvious. On the other hand, all kinds of indicators to measure the loss of cascading failures are independent of each other. There is a lack of an effective statistics index to assess the damage of cascading failure in the different stages.

According to Xue (2012), it is feasible to switch control modes automatically after mastering the mechanism and rules of cascading failure, based on the characteristics of each stage. Cascading failure was divided into three stages by Wang in 2010. Then the scholar studied the methods suitable for different stages based on different characteristics. There is an idea pointed by Xue (2012) that we can adopt ordered measure actively which is under control, in order to avoid uncontrolled loss, so that the cost can be as small as possible after the active control. Therefore, it is necessary to calculate the cost of active control and the damage of cascading failure without the control, and make comparisons between them. Furthermore, some study should been carried out on the degree of damage in each stage of cascading failure.

This paper puts forward a method of dividing the process of cascading failure into five stages: the pre-fault stage, the initial stage, the slow evolution stage, the fast evolution stage, and the collapse stage, based on the analysis of different characteristics in the process of cascading failure. Then a quantitative index is proposed which has a practical application value in the mastery of the process of cascading failure and the selection of control measures, to prevent the event from expending.

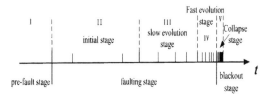

Figure 1. Schematic diagram of cascading failure stage.

2 STAGE DIVISION OF CASCADING FAILURE

2.1 Description of each stage

Cascading failure relates to the operation state of the power system closely before any failure happens, therefore, the operation state of the power grid before the fault should also be considered. In this paper, the process of cascading failures are divided into three stages: the pre-fault stage, the faulting stage, and the blackout stage. Then the faulting stage is divided further into another three stages: the initial stage, the slow evolution stage, and the fast evolution stage, because the faulting stage is the main process in cascading failure. After the fast evolution stage, the collapse stage comes which belongs to the blackout stage. Figure 1 illustrates the division of processes in cascading failure.

The power system may be in a unhealthy state, although there is no fault happening. For example, the summer temperature was high and the load on the line was heavy before the August 14 Northeast blackout in 2003, and the load levels of the transmission section between Italy and Europe were close to the limit of safety before the accident. On the other hand, when a lot of components cease operating, this may also lead to the power system into a sub-healthy state, such as the blackout in India in 2012. As can be seen, the unhealthy situation before the fault offers the green house to cascading failure.

After the initial fault occurs, the faulting stage is realized, and the power system transforms from unhealthy to morbid. Other faults may occur at this stage if the initial fault has not been removed in time, due to some certain or uncertain reasons. In the initial stage these faults build up a set in which the causation between each fault is not obvious. It is the set that makes a great effect on the running state rather than a single fault. If these faults have not been eliminated on time, the health of the power system becomes poor, which leads it into the slow evolution stage because of the load transfer and the infelicity of the setting value, malfunction, or hidden failure of relay protection device. After the transmission line of Stuart-Atlanta tripped on August 14, 2003 in the Northeast blackout, another transmission line on the same transmission section tripped because of load transfer. At this stage, the fault coverages increase gradually and the damage becomes more and more serious. If the faults have not been eliminated in the slow evolution stage, the fast evolution stage will come after a serious fault. This serious fault has great damage to the power system which is called a crucial fault. Then the situation of the power system will become increasingly serious. With the development of cascading failure, the operation status of the power system becomes worse and worse. When all the measures are out of action, faults occur like an avalanche, so the cascading failure transfers into the collapse stage, and large blackouts become inevitable.

A typical complete blackout caused by cascading failure will experience these stages, but the form of cascading failure will also have many differences because the structure of the power grid and the operation state is different between each other, and the causes of each fault are perplexing. According to the main characteristic of each blackout, cascading failure is divided into three forms: overload-leading type, structure-leading type and match-leading type, by Liu et al. (2012). The three types of cascading failure show different patterns because of different characteristics, so the evolution process of cascading failures range from a few minutes to several hours. The difference in

Table 1. The characteristics in each stage and the division basis of cascading failure.

Stages		Basis of division	Characteristic of stage
Pre-fault stage		No fault occurs	The system operates normal or be in the alert.
Faulting stage	Initial stage	I–II: fault occurs but have no causality	Faults occur but have no causality, the system is damaged preliminary.
	Slow evolution stage	II–III: causality between faults	The causality is obvious, the faults increase but slowly.
	Fast evolution stage	III–IV: key fault occurs	Important fault happens and the number of faults increases significantly.
Blackout stage	Collapse stage	IV–V: Controls have no use.	Faults occur intently, control methods cannot prevent the cascading failure.

the time of the process is due to the loss of key components early on, which transfers the cascading failure into the fast evolution stage.

2.2 Characteristics and the division basis of each phase

In the development of cascading failure, different characteristics show in each stage. Each characteristic and reference of division are shown in Table 1.

Because of the complexity of cascading failure in time and space, what should been paid attention to is that an overlap region may occur on the intersection of each stage. In other words, the edge between each stage is fuzzy.

3 DAMAGE DEGREE OF CASCADING FAILURE

After quantifying the damage caused by cascading failure, it is convenient to determine the stage of cascading failure, based on the value and growth trends of the damage degree, which can provide reference for active control. With the evolution of cascading failures, the generation capacity continues to shrink, and the structure of the power grid become weaker, along with the gradual loss of load, so the damage can be calculated from multiple aspects.

3.1 The index of damage degree

By considering the effect on the operation state and structure of the power system, this paper gives the definition of damage degree of cascading failure: the damage degree is an index to calculate the loss caused by cascading failure, which includes two aspects. The first is the destruction of the grid topology, reflecting the structural vulnerability of the power system, and the second, is the loss of load and the degeneration of the running status of the components, reflecting the state vulnerability. This paper gives a formula for calculating the damage degree of cascading failures, as shown in Equation (1):

$$D = \frac{1}{\varepsilon}(1 - e^{-\sum_{i=1}^{n}\frac{\Delta p}{P_N}}) \bullet e^{\alpha(1-\frac{n}{N})\sum_{j=1}^{m}\gamma_j u_j} \bullet e^{\beta(1-\frac{l}{L})} \quad (1)$$

In this equation, D is the damage degree of cascading failure, ΔP is the variable quantity of active power on each branch after the fault, P_N is the rated power of the corresponding branch, i is the number of branches of the power grid, n is the number of remaining branches after the fault, N is the number of branches at the initial time, u_j means different voltage level, γ_j is the weight of each voltage level, l is the remaining load of the power system, L is the total load at the beginning, α and β are content, and so is the ε, which is related to α and β.

3.2 The steps to calculate the damage degree

The following shows the steps needed to calculate the damage degree:

1) Obtain the initial data, such as the initial power flow;
2) Get a sequence of cascading failure which includes all the faults in order of time;
3) Set the initial set of fault, the number of elements in this set is not too much because the previous blackouts usually caused by few failures;
4) If any isolated network occurs, handle the isolated network, and calculate the loss. Otherwise go to Step 3;
5) Calculate the power flow, and end the calculation if the power flow is not convergence;
6) If the power flow is convergence, calculate the damage degree based on the formula shown above;
7) Set the next fault components according to the fault sequence, and then go to Step 4.

The calculation process of the damage degree is shown in Figure 2.

4 ANALYSIS OF CASES

4.1 Analysis of the index of damage degree

The damage degree is composed of three parts:

$$D_1 = 1 - e^{-\sum_{i=1}^{n}\frac{\Delta p}{P_N}} \quad (2)$$

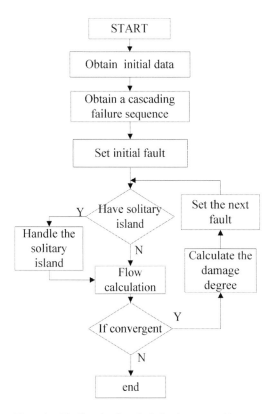

Figure 2. The flow chat for calculating the extent of damage.

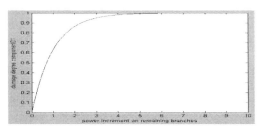

Figure 3. The growth curve of the first part of the damage extent.

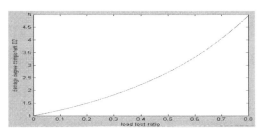

Figure 4. The growth curve of the second part of the damage extent.

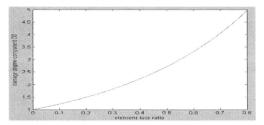

Figure 5. The growth curve of the third part of the damage extent.

$$D_2 = e^{\beta(1-\frac{l}{L})} \quad (3)$$

$$D_3 = e^{\alpha(1-\frac{n}{N})\sum_{j=1}^{m}\gamma_j u_j} \quad (4)$$

Here, D_1 shows the effect on components by cascading failure, D_2 indicates the loss of cascading failure directly, and D_3 is the damage to the power network structure.

Figure 3 is used to illustrate the tendency of the effect on components in the process of cascading failure. The horizontal coordinate is the power increment on the remaining branches after the faults; the ordinate is the component of the damage. The power increment on the remaining branches is a fluctuating value, but the value will increase with the development of cascading failure. D_1 begins to increase from zero after the initial fault. Usually, in the initial stage the load is not lost, and the damage to the structure of the power grid is not serious; the damage degree is mainly composed of D_1 which grows fast. As time goes by, the loss of the load increases gradually and the damage to the network structure becomes serious; D_1 and D_2 account for the major part in the damage degree. At this point, the proportion of D_1 is small, so the growth curve tends to be gentle.

The growth trend of the load loss is shown in Figure 4. The abscissa is the load lost ratio, and the ordinate D_2 is the component of the damage. In the initial phase, the load loss is zero or very small, so the component of the load loss accounting for the damage index is very small; D_2 increases slowly from 1. With the evolution of cascading failure, the load loss increases gradually, and the D_2 scores a higher ratio of D, which shows in the curve.

The damage to the power network structure caused by the loss of components is shown in Figure 5. The abscissa is the loss of components with different voltage grades, and the ordinate is the damage component D_3. Similar to D_2, the number of fault elements is small in the initial stage. The effect on the damage degree D caused by D_3 is not serious, so D_3 increases slowly from 1. After a period of time, the number of fault elements increases, and the damage to the structure of the power grid tends to be more serious. The increase of D_3 has great contribution on D, so D_3 leads to the increase of the proportion of D_3.

From the analysis above, the damage degree is mainly characterized by the power increment at the beginning of cascading failure. After a period, the cascading failure become serious, the index is

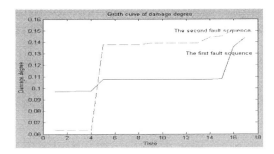

Figure 6. The growth curve of damage extent.

Table 2. The division of the 118 node system.

Areas	Buses in each area
A	Bus 1–19, 30, 113, 117
B	Bus 20–23, 25–29, 31, 32, 114, 115
C1	Bus 33–61, 63, 64
C2	Other buses

Table 3. The growth of the damage degree in each stage of the first cascading failure.

The first sequence	Damage degree	Stages
L10	0.0970	Initial stage
L169	0.0975	
L48	0.1076	
L103	0.1082	Slow evolution
L97	0.1358	stage
L96, L98, L99	0.1444	Fastevolution
L100, L104, L105, L106, L107, L108	Not convergent	stage

Table 4. The growth of damage degree in each stage of the second cascading failure.

The second sequence	Damage degree	Stages
L142	0.0632	Initial stage
L35, L141	0.1375	Slow
L139	0.1390	evolution
L138, L137	0.1444	stage
L107, L129, L131, L136, L143, L149, L185	0.1451	Fast evolution
L119	0.1507	stage
L104, L105, L106, L108, L116, L120, L140	Not convergent	

reflected mainly by load loss and the structure failure of the power grid.

4.2 *The simulation on IEEE 118 bus system*

Here MATPOWER is used to analysis the reasonableness of the damage degree in the IEEE 118 bus system. In the simulation, the faults are set to be permanent, and the faults are mainly on branch. As the process of cascading failure is fast, the load fluctuation has not been taken into consideration.

The number of branches in the IEEE118 bus system is set according to the results of the power flow. Two cascading failure sequences are obtained. The curve of the damage degree is shown in Figure 6.

For convenient analysis, the IEEE118 bus system is divided into four regions, as shown in Table 2.

The damage degree of the first fault sequence in each stage of the extent of damage index increase, is shown in Table 3.

The damage degree of the first fault sequence in each stage of the extent of damage index increase, is shown in Table 4.

Through the analysis above we can see the effect of the faults in the initial stage is localized in Case 1 and Case 2, which is small on the whole power system and the damage degree increases slowly. The reason why the damage degree of Case 1 is larger in the initial stage, is that the specific faults are different and the seriousness is also different, so that the value of damage degree may different to a great extent in the initial stage. The causality between the faults in the slow evolution stage is remarkable, and the event spreads in the local. The number of faults in this stage is not large; the damage degree increases slowly. However, the loss of the important branch in Case 2 leads to a rapid increase of the index in this stage. According to Liu (2012), Case 2 is the structure leading the cascading failure. The loss of an important element will greatly accelerate the process of cascading failures and even cause mutations of the index. In the fast evolution stage, the loss of load increases and the damage of the structure of the power grid is serious; the event is developing rapidly, and faults occur frequently. The event often extends from local to the whole power system and the index increases rapidly.

5 CONCLUSION

This paper puts forward a method of dividing the process of cascading failure into five stages. According to the characteristics of each stage, it is feasible to switch control modes automatically, to realize self-organized in terms of functionality, and to optimize the measures in each stage after master the mechanism and rules of cascading failure. According to the division of the process, a quantitative index is proposed, which has a practical application value in the mastery of the process of cascading failure and the selection of control measures to prevent the event from expending This method is mainly applicable to ordinary cascading failure, and whether the method is suited or not to successive faults caused by extreme weather or other external factors, will be studied next.

REFERENCES

Cao Yijia, Wang Guangzeng. The complexity of power system and related problems [J]. Electric Power Automation Equipment, 2010, 30 (2).

Hu Jin. The human factors of "8.14" the United States and Canada blackout [J]. North China Electric Power, 2004.

Li Chunyan, Sun Yuanzhang, Chen Xiangyi, Deng Guiping. The preliminary analysis of Western Europe blackout and prevention measures of blackout accident in China [J]. Power System Technology, 2006, 30 (24).

Liu Youbo, Hu Bin, Liu Junyong, et al. The analysis and application of the power system cascading failure (a) related theory method and application of [J]. Power System Protection and Control, 2013, 41 (9).

Liu Youbo, Hu Bin, Liu Junyong. Analysis of the theory and application of power system cascading failure (two), key features and edification [J]. Automation Of Electric Power Systems, 2013, 41 (10).

Wang Si. Vulnerability assessment and stability control based on fault chains [D]. Hubei Province, Wuhan, Huazhong University of Science and Technology, 2010, 71–73.

Xue Yusheng, Xiao Shijie. Comprehensive prevention of small probability but high risk event—the thinking of blackout by natural disaster in Japan and nuclear leak [J]. Automation of Electric Power Systems, 2011, 35 (8).

Xiao Shijie, Wu Weining, Xue Yusheng. Power system security and stability and control application technology [M]. China Electric Power Press 2011, 1–2.

Xue Yusheng. Framework for defending blackouts cooperatively in space and time (three) optimization of each line and coordinate between different lines [J]. Automation, 2006, 30 (3).

Xue Yusheng, Xiao Shijie. The generalized block of power system in India [J]. Power System Automation, 2012, 36 (16).

Yin Yonghua, Guo Jianbo, Zhao Jianjun, Bu Guangquan. Preliminary analysis and lessons of the United States and Canada blackout [J]. Power System Technology, 2003, 27 (10).

The design of a highly reliable management algorithm for a space-borne solid state recorder

S. Li, Q. Song, Y. Zhu & J.S. An
National Space Science Center, Chinese Academy of Sciences, Beijing, China
University of Chinese Academy of Sciences, Beijing, China

ABSTRACT: There are always bad blocks in NAND Flash, which puts the data reliability at risk, and comes up with a higher requirement for the storage system management. In this paper, a new bad block management algorithm for a spacecraft solid state recorder has been presented with higher reliability and less response time. Software is responsible for the normal address assignment, while hardware allocates the substituted address immediately a bad block appears. The Triple Modular Redundancy (TMR) is adopted to strengthen the reliability of the Bad Block Table (BBT). There are three copies of the bad block table in the Central Processing Unit (CPU) memory Synchronous Dynamic Random Access Memory (SDRAM), and a simplified copy in Field Programmable Gate Array (FPGA). One simplified copy in the FPGA ram decreased the response time. To further improve system reliability, the address allocation algorithm is optimized and simplified. The algorithm has been tested in a four-pipe-line and 32-bit parallel extended solid state recorder.

Keywords: FPGA, TMR, SSR, NAND Flash, management algorithm

1 INTRODUCTION

1.1 Solid state recorder

A Solid State Recorder (SSR) is a key component in modern electronic systems, and it has been widely used in communications, radar, aerospace, and other fields [1]. Due to the extreme space environment, a SSR in spacecraft has its own characteristics [2]. Space mission data can be transferred to ground receiving stations when satellites are in a certain monitoring scope. Therefore, most high-speed large amounts of data need to be stored locally, and then replayed when the connection is established. As so much invaluable data is stored in the SSR, even a small error may put the mission at risk. The single event upset (SEU) effect, strong vibration, and instability of the power supply all threaten the data. A competent spaceborne storage system needs to be high-speed, highly reliable, have a high storage density, low-power, non-volatile, and low-cost [3].

1.2 NAND Flash

Spaceborne NAND Flash uses 3-D packaging technology against radiation and to improve storage density, which means that memory die are stacked using an adhesive to form a cube. A NAND-type flash memory may be written and read in pages while being erased in blocks [4].

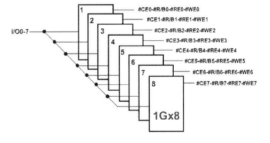

Figure 1. The signal of the packaged chip.

3-D packaging technology [9] is usually employed to form one high integration chip with many dies. Take the Zhuhai Orbita Company's NAND Flash chip, VDNF64G08-F, as an example to show how the 3-D packaging technology affects the chip in both shape and electronic characteristics. It is packed with eight pieces of 8Gbit NAND Flash chips, which are made by the SUNSANG Company to attain the capacity of 64Gbit for one chip. It looks like a golden cube. The eight pieces in one cube share the I/O bus power, but have independent CE (Chip Enable), CLE (Command Latch Enable), ALE (Address Latch Enable), RE (Read Enable), and WE (Write Enable).

The internal structure of the 8Gbit chip, which is called a die in a packaged chip, is shown in Figure 2. The die is composed of 4,096 blocks. One block is composed of 64 pages. There are two areas, called the

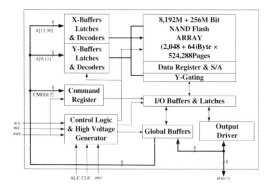

Figure 2. Block diagram of one die.

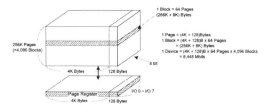

Figure 3. The array organization of one die.

data area and the spare area, in a page. In this chip, the data area is 4K bytes while the spare area is 128 bytes. The array organization of one die is shown in Figure 3.

1.3 Bad block

To make the NAND Flash serve as a low cost, solid state, mass storage medium, the standard specification for the NAND allows for the existence of bad blocks in a certain percentage. These kind of bad blocks are called inherent bad blocks. All good blocks will be marked FFH in the sixth word in the spare area of the first page, while where of the bad block is a word other than FFH which is usually 00H. NAND Flash devices can degrade over time and develop bad blocks. Bad blocks may not be able to be fully erased, or may have stuck bits that will not toggle correctly, but the data that have been programed successfully are kept. Since the block has gone bad, there is no guarantee that writing a bad block marker will succeed. The block will then fail to be detected as bad on subsequent scans, which may lead to data corruption if used later. So, there must be a bad block table (BBT) in the storage system to keep the system always aware of the block status. Therefore, maintenance should be done as long as SSR is in use.

2 TRADITIONAL ALGORITHMS [3, 5, 6, 7, 8]

Because all flash memory will eventually wear out and no longer be useable, a BBT needs to be maintained to track blocks that fail in use. The BBT records all known bad blocks that are present on a device. The main content of all bad block management algorithms are:

1) Identify the inherent bad blocks.
2) Identify a bad block and write data effectively to a good block when encountering a bad one.
3) Establish and dynamically refresh a BBT.

The traditional BBT management algorithm working process is shown below. Firstly, sequentially scan all the blocks to create the primary BBT. Secondly, good blocks are subdivided into two categories: the exchange block area and the user addressable block area. The address allocation is in the addressable block area. Thirdly, if attempts to program or erase a block give errors in the status register, the block is marked as bad. Therefore, a new address in the exchange block area will replace it and the BBT would be updated.

The evaluation of the capacity of the exchange block area is too difficult. If the area is too little, there may be not enough exchanging blocks. If it is too large, many blocks will be left unuse.

Usually the BBT is stored in the memory which is SDRAM. CPU manages this table directly, while all the information is collected by FPGA. Moreover, in space, the environment is rather harsh so that the SDRAM is prone to being affected by Total Ionizing Dose (TID), Single Event Upset (SEU), Single Event Latch-up (SEL), Single Event Functional Interrupt (SEFI), Single Event Burnout (SEB), Single Event Transient (SET), and so on. Therefore, the data in the SDRAM can be changed easily, and it has many disadvantages as shown below:

1) The table is volatile; it must be rebuilt when power up by scanning of NAND flash.
2) The updating process is slow and easily disturbed.
3) The table is too big to quickly search.
4) The management occupies a lot of CPU resources.
5) The BBT is not very reliable.

There is also a new presented management approach which uses other storage media to store BBT. However, this new media has a limited capacity and the cost is high.

3 NEW ALGORITHM DESIGN AND IMPLEMENTATION

3.1 Introduction to the new algorithm

The management of SSR is mainly about how to keep a reliable BBT and how to use it reliably. To achieve this goal, we improve the traditional method in two aspects: the storage of BBT, and the bad block replace strategy.

3.2 BBT storage reliability design

At first, FPGA scans all the blocks and recognizes all the initial bad blocks. The initialization process is only started when it is power up. The initialization process is shown in Figure 4.

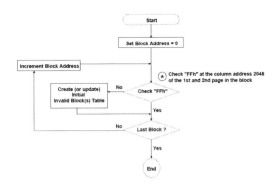

Figure 4. The bad block replace process.

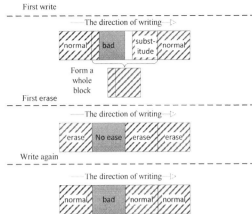

Figure 5. The bad block replace process.

After initialization, the BBT table will be stored in SDRAM and will help the SSR management software to allocate the next write or erase address. Additionally, when a bad block occurs, CPU will amend the BBT. Therefore, BBT will be read and wrote a lot when using SSR.

To improve the reliability of storage of the BBT, a TMR method is employed in the process. While the operations to the chips are done by FPGA directly, so we also set a copy of BBT in the buffer of FPGA to minimize the waiting time when a bad block occurs. Therefore, the table is stored in four places: three of them are in the SDRAM controlled by the CPU, the fourth is in the buffer set by the FPGA. The fourth copy is simplified for the limited resources in the FPGA, which just contains the addresses of the current blocks and the substituted block and the page number when a current block turns bad.

3.3 *Bad block replace strategy design*

Because of the extreme environment and the low computing power, space missions need algorism as simple as possible to ensure reliability. After initialization, the organized BBT will be stored in five places. The software picks out the bad blocks addresses and prepares the remaining address for data sequentially. Usually, the SSR has plenty of time to do other operations besides writing. The FPGA will start the erasing operation to guarantee there are enough empty blocks to write on when free. The FPGA will give out interruptions when the Flash program or erase fails, and use the next block to take the place of the bad one directly. The data will be rewritten on the same page as the bad block, and the data written before this page will remain in the bad block, which will never be transferred to the substituted block. The page number will be written in the BBT to make the data read continuously. When the data needs updating, all the blocks will be erased except the worn-out bad blocks. When writing again, the bad blocks will be rejected and the substituted block will be treated as a normal one. Then, the new data will be written into the substituted block. The whole process is explained in Figure 5.

4 EXPERIMENT AND RESULTS

4.1 *Prototype implementation*

We evaluate our algorithm in a typical SSR used in spacecraft, and most key components are domestic. The platform structure is CPU+FPGA+NAND Flash. Longson, is the CPU responsible for sending commands, allocating addresses, and communicating with external devices. Its external frequency is 50 MHz and internal frequency is 100 MHz. Its average execution speed of fixed-point instructions is 300 MIPS, while the average execution speed of floating-point instructions is 50 MIPS. The FPGA is flash FPGA, ACTEL A3PE3000, which is responsible for managing the BBT and controlling all the hardware directly. Flash FPGA has a lower power consumption than the other two kinds of FPGA. A3PE3000 has a 66 MHz 64-bit PCI core and a true dual-port SRAM, which could enhance the operating speed substantially.

The NAND Flash chips are produced by Orbita. A single chip storage capacity is 8G *8 bit which is composed of eight memory die. The writing operations are composed of the data input process and the programming process. Usually, the data input time for a single page is 132us (the system clock is 32M), while the programming time ranges from 200us to 700us. Before programming succeeds, all the data written must be stored in the buffer, in case of rewriting. The four-pipeline is employed for balancing the writing speed and the buffer capacity. The system adopt pipe-line and 32-bit parallel extended to achieve 350 Mbps input data rate.

The system block diagram is shown in Figure 6.

4.2 *Results*

The experiment shows that SSR using the proposed algorithm could understand nearly two times error rate in SDRAM than the one using tradition method in a normal operation. Also, the waiting time in the proposed method can be decreased. Moreover, all the

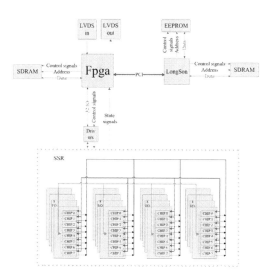

Figure 6. System block diagram.

improvement do not need much extra software or hardware cost.

5 CONCLUSION

In this work, how to improve the reliability of a management algorithm for a spaceborne solid state recorder has been investigated. A novel algorithm to store BBT in five places is proposed.

The algorithm is based on the thorough consideration of the characteristics of NAND Flash and the environment of the space mission. In this method, the BBT has four copies. Three out of four copies of the BBT of the whole system are stored in SDRAM, which will improve the reliability of the BBT. The remaining copy in FPGA is intended to reduce the waiting time when a bad block occurs. The results show that the algorithm improves the reliability of SSR management substantially.

ACKNOWLEDGMENT

This paper is supported by the "Strategic Priority Research Program of the Chinese Academy of Sciences" under Grant No.XDA04060300.

REFERENCES

[1] Lei Lei, High performance memory based on NAND flash design, Hunan University. (2008).
[2] Zhu Yan, High-speed spacecraft solid state storage design, Chinese Academy of Sciences (Space Science and Applied Research Center). (2006).
[3] Lang Fei, Sun Yue, Development on high-speed digital image acquisition, Infrared and Laser Engineering. 34 (2005) 227–231.
[4] Said F. Al-sarawi, Derek Abbott, and Paul D. Franzon. A review of 3-D packaging technology, IEEE Transactions on Components, Packaging and Manufacturing Technology. 21(1998). 2–14
[5] Kang Jeong-Uk, Kim Jin-Soo, Chanik Park, et al. A multichannel architecture for high-performance NAND flash-based storage system [J]. Journal of Systems Architecture, 2007, 53: 644–658.
[6] Luo Xiao, Liu Hao, Method of NAND flash memory bad block management based on FAT File System. Chinese Journal of Electron Devices 31(2008) 716–719.
[7] Lim S, Park K. An efficient NAND Flash file system for flash memory storage [J]. IEEE Transactions on Computers, 2006, 55(7): 906–912.
[8] Kim J, Kim J M, Noh S H, et al. A space efficient flash translation layer for compact flash systems [J]. IEEE Transaction on Consumer Electronics, 2002, 48(2): 366–375.
[9] Said F. Al-sarawi, Derek Abbott, and Paul D. Franzon. A review of 3-D packaging technology, IEEE Transactions on Components, Packaging and Manufacturing Technology 21(1998). 2–14.

A patrol scheme improvement for disconnectors based on a logistic regression analysis

Jieshan Li, Yonghu Zhu & Zhiqin Zhao
IT & Communication center, EHV Power Transmission of CSG, Guangzhou, China

ABSTRACT: In main power grids the disconnector switch is one of the most important electric pieces of power equipment. A wide range of production and performance parameters has resulted in the difficulty of implementing patrol schemes. Therefore, it is necessary to put forward a new method to make the patrol scheme more scientific and reasonable. This paper introduces a patrol scheme improvement scheme, based on a logistic regression analysis. The analysis process is based on equipment defect data and product information. Through the results it is responsible for collecting valid messages for evaluating the quality of the disconnector in running time, and improving the patrol scheme to adapt the equipment's running status.

1 INTRODUCTION

In a main power grid, the disconnector switch is one of the most important electric power components. On account of its wide usage and frequent operation, the disconnector has a great influence on the transformer substations and the power grid. However, a wide range of production and performance parameters results in the difficulty of implementing a patrol scheme. Therefore, it is necessary to put forward a new method to make the patrol scheme more scientific and reasonable.

Because the operating data which is shared on a unified platform is the basis of a condition assessment, so there is a need to accumulate equipment defect data and produce information in order to realize any improvement. Coincidentally, the development of smart power girds makes this possible.

According to the mathematical model and method, the failure of the distribution and the performance of equipment, in different transformer substations and belonging to different manufacturers, will be classified intelligently. And then, traditional practices of manual patrol schemes should be transformed into a rational allocation which is based on panoramic data and failure analysis data.

2 LOGISTIC REGRESSION ANALYSIS

A logistic regression algorithm is one kind of common method which is used to describe the relationship between multi-type response variables and predictive variables. This is shown in Formula (1), where there is a vector quantity x' whose number of autonomous variables is n.

$$x' = (x_1, x_2, \cdots x_n) \quad (1)$$

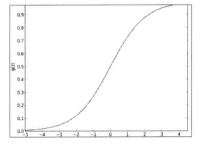

Figure 1. Weighted value of different response variables are used to establish the fitted curve.

At the same time, the expression of contingent probability P is shown as Formula (2).

$$P(Y=1\,|\,x) = p \quad (2)$$

It means the possibility of a predictive variable relative to response variables. Then, the curve fitting can be implemented among the discrete point by means of Formula (3).

$$P(Y=1\,|\,x) = \pi(x) = \frac{1}{1+e^{-g(x)}} \quad (3)$$

Afterwards, a weighted value of different response variables is used to establish the fitted curve, as shown in Figure 1. This curve can be used to realize a trend analysis later.

This function is an effective method of logistic regression conversion, and the expression to deduce this conversion is shown in Formula (4).

$$g(x) = \ln\frac{\pi(x)}{1-\pi(x)} = \beta_0 + \beta_1 x_1 + \cdots \beta_n x_n \quad (4)$$

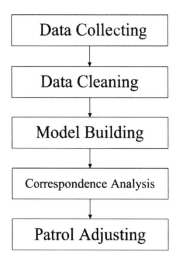

Figure 2. Five main steps in logistic regression model building.

The contingent probability of an event not having occurred is shown in Formula (5).

$$P(Y=0 \mid x) = 1 - P(Y=1 \mid x) = 1 - \frac{e^{g(x)}}{1+e^{g(x)}} = \frac{1}{1+e^{g(x)}} \tag{5}$$

Then, it would seem the ratio of two kinds of contingent probabilities, including those that occurred and those that did not.

$$\frac{P(x=1 \mid x)}{P(x=0 \mid x)} = \frac{p}{1-p} = e^{g(x)} \tag{6}$$

This ratio means the odds of experiencing an event. For short, it is called the 'odds'. Because the value of p belongs to [0, 1], the value of the odds has to be greater than 0. The linear function is built by evaluating the logarithm of added value. Inference process is shown is Formula (7).

$$\log\left(\frac{p}{1-p}\right) = \beta_0 + \beta_1 x_1 + \cdots \beta_n x_n \tag{7}$$

3 MODEL BUILDING AND ALGORITHM IMPLEMENTATION

As the operating data reflects the running condition of a disconnector switch, it is useful to analyse this data by a logistic regression model. Using the above method, it is possible to discover a regular lurk in discrete data. The regular lurk will be taken as the practical evidence for a patrol scheme, and then the patrol work will be more reasonable and effective than now.

The specific process is shown as Figure 2. There are five main steps in logistic regression model building: data collecting, data cleaning, model building, correspondence analysis, and patrol adjusting.

Table 1. Source parameters collected from ED-MIS.

Year	Equipment	Department Nanme	Failure Class	Found Time of Defects	Product Factory
2013	Disconnector	500kV罗LII-#2M50432-A	Urgent	2009/2/24 12:48	SYTBDG
2013	Disconnector	#1-500kV-50111-A	Urgent	2009/8/22 18:40	SYTBDG
2013	Disconnector	#2-500kV-50232-A	Urgent	2011/7/6 9:50	SYTBDG
2013	Disconnector	500kVLHII-50332-A	Urgent	2012/6/17 22:32	XAXD
2013	Disconnector	#2-500kV-50232-A	Urgent	2013/6/4 14:04	CZDZ
2013	Disconnector	#2-500kV-50232-A	Common	2001/1/29	SYBYQ
2013	Disconnector	500kVLHI-50132-A	Serious	2001/3/12	MG
2013	Disconnector	220kV-#1-22121-B	Common	2001/7/16	XAXD
2013	Disconnector	500kV-50421	Common	2001/8/12	XAXD
2013	Disconnector	35kV313-3132	Serious	2001/8/25	XAXD
2013	Disconnector	缓I-66160	Common	2001/9/19	XAXD
2013	Disconnector	220kVHDI-20542-A	Common	2001/9/25	SYBYQ
2013	Disconnector	#2-35kV-3026	Common	2001/10/30	CZDZ
2013	Disconnector	500kVSLII-50411-C	Serious	2001/10/31	CZDZ
2013	Disconnector	500kVLHII-52012-A	Common	2001/12/21	SYBYQ

Relation: shebeiquexian

No.	shengchanchangjia (Nominal)	touyunshijian (Numeric)	class (Nominal)
1	HNCGGYDQ	2.8	yiban
2	SYGYKG	20.5	yiban
3	AESTGYDQ	0.3	yiban
4	XDGYKGC	5.7	yiban
5	HNPGJT	5.2	yiban
6	XDGYKGC	4.2	yiban
7	MG	20.4	yiban
8	BJABBGYKG	3.0	yiban
9	MG	20.4	yiban
10	MG	20.4	yiban
11	MG	20.4	yiban
12	MG	20.4	yiban
13	PDSGYKGC	14.5	yiban
14	HNPGJT	0.[illegible]	[illegible]
15	PDSGYKGC	0[illegible]	[illegible]

Right click

Undo OK Cancel

Figure 3. Reserve data.

3.1 *Data collecting*

Data collecting is the base of logistic regression analysis. This work will provide practical evidences to make the patrol scheme more reasonable. The source parameters collected from an Equipment Data Manage Information System (ED-MIS) contains the main elements about the disconnector, including: date, Found Time of Defects, the equipment, department name, failure class, and the product factory.

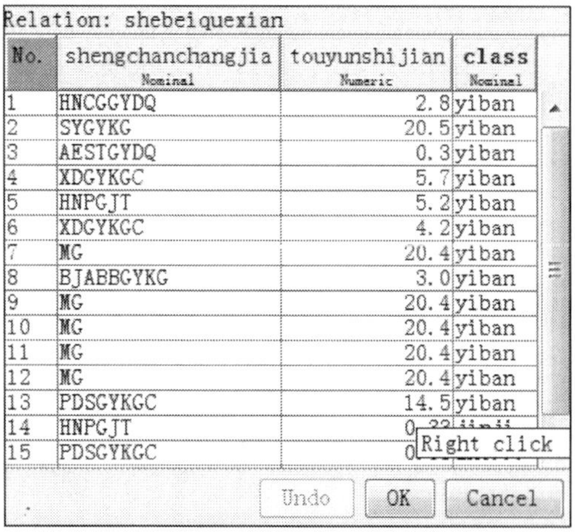

3.2 *Data cleaning*

As the source parameters, collected from management information system, are used to describe the running equipment's status, so they are inevitably miscellaneous. In order to obtain significant information, it is necessary to clean the source parameters and evaluate them before modelling. The cleaning instrument in this paper is an 'SAS Data Integration Studio.' Afterwards, the reserve data is shown in Figure 3. The first column is the product data set F_1, the second is using time data set T_1, and the last one is the failure class data set C_1.

The reserve data is more pertinent and reasonable than the source data. So it is possible to build a logistic regression model using reserve data directly.

3.3 Model building

A logistic regression algorithm is one kind of common method which is used to describe the relationship between multi-type response variables and predictive variables.

The product data set F_1 is defined as vector quantity x′ which consists of a response variable, and failure class set C_1 is defined as vector y′ quantity which consists of predictive variable. Then it would seem that the expressions of x′ and y′ which is shown in Formula (8) and (9).

$$x' = F_1 = \{HNCGGYDQ, SYGYKG, AESTGYDQ, DGYKGC, HNPGJT, MG, BJABBGYKG, PDSGYKGC\} \quad (8)$$

$$y' = C_j = \{common, serious, urgent\} \quad (9)$$

The probability of failure that occurred for a disconnector produced by a different factory is:

$$P(c_j = 1 | f_i) = \pi(f_i) = \frac{1}{(1+e^{-g(f_i)})} \quad (10)$$

Probability of failure that did not occur for a disconnector produced by a different factory is:

$$P(c_j = 0 | f_i) = 1 - P(c_j = 1 | f_i) = 1 - \pi(f_i) = \frac{1}{e^{g(f_i)}} \quad (11)$$

Afterwards, the odds of experiencing an event could be got:

$$\frac{P(c_j=1|f_i)}{P(c_j=0|f_i)} = e^{g(f_i)} \quad (12)$$

Because of:

$$g(f_i) = \beta_0 + \beta_1 * f_1 + \beta_2 * f_2 + \beta_3 * f_3 + \cdots \beta_8 * f_8 \quad (13)$$

the linear function is built by evaluating the logarithm of added value. The inference process is shown is Formula (14).

$$\beta_0 + \beta_1 * f_1 + \beta_2 * f_2 + \beta_3 * f_3 + \cdots \beta_8 * f_8 \quad (14)$$

Therefore, the value of β_1 to β_8 means the relevance between the product collection element and the failure class collection element.

3.4 Relevance analysis

According to the above, it is easy to obtain the relevant coefficient of different products. According to the characteristic of the logistic regression model, the number "0" will be the cutting line to differentiate 'safety' and 'danger'. When the value of a relevant coefficient exceeds the cutting line, the higher the safer it is; when the value is below the cutting line, the lower it is the more dangerous it is. The details are shown in Table 1.

Here, the order of severity for serious class is lower than 'Urgent', and higher than 'Common'.

Table 2. Relevance coefficient of different product.

Product	Num	Relevance coefficient		
		Urgent	Serious	Common
AEST	67	−14.137f4	34.1769	4.7615
SYGYKGC	8	13.6376	−9.1677	−16.939
XAXDGYKG	762	−9.2086	−11.7245	19.1861
HNPGJT	1936	26.3853	−23.3885	13.7279
MG	210	18.4668	−5.1653	−31.9812
BJABBGYKG	472	−20.8318	−13.8057	33.9078
HNCGGYDQ	220	−23.3308	−14.4043	37.3996
PDSGYKGC	461	−22.9607	31.106	−3.9812

3.5 Task adjusting

Taking the first and second line as an example, although the relevance coefficient of AEST is higher than SYGYKGC in the serious column, the former's relevance coefficient is −14.1374 which is lower than the latter. So according to the analysis and comparison, the possibility of breakdown for the disconnectors produced by SYGYKGC is higher than those produced by AEST.

As well, because of the relevance coefficient of HNPGJT is 26.3853. which is the highest in urgent column, the possibility of breakdown for the disconnectors produced by HNPGJT is higher than others. So it is necessary to pay close attention to them, and change the patrol scheme to increase the number of examinations.

In this way, it would be possible to found the problems, and solve it without delay.

4 SUMMARY

This paper introduces a patrol scheme improvement formula, based on a logistic regression analysis. It is realized on the basis of equipment defect data, product information, and a logistic regression model. The steps of this analysis include: data collecting, data cleaning, model building, correspondence analysis, and task adjusting. This method is also applied to other equipment. Also, it can be extended for actual situations to make it more widely used.

REFERENCES

Daniel T. Laose, "Data Mining Methods and Models", Higher Education Press.
De Heng Zhu, Zhang Yan, Kexiong Tan, "Electric Equipment Condition Monitoring", China Electric Power Press.
Mei Yu Yuan, "Data Mining and Machine Learning", Tsinghua University Press.
Ping Ju, "Power System Modeling Theory and Method", Science Press.
Wen Jian Yu, "Numerical analysis and algorithm", Tsinghua University Press.

Progress on an energy storage system based on a modular multilevel converter

Bin Ren, Chang Liu, Yonghai Xu & Chang Yuan
North China Electric Power University, Changping District, Beijing, China

Shanying Li & Tao Wu
North China Electric Power Research Institute, Xicheng District, Beijing, China

ABSTRACT: Energy storage system and multilevel converters are important components in new energy power system. Applications to energy storage systems will promote the optimization of the grid structure, and can achieve the goal of new energy friendly access and coordination of control. The Modular Multilevel Converter (MMC) is a member of a multilevel converter family, and has been widely applied in the medium- or high-voltage high power applications. Applications of energy storage system based on modular multilevel converters are concluded in this paper. The developing progress of different energy storage technologies is firstly addressed together, and the operational principles and the technical characteristics of an MMC are introduced briefly. Then the integrating methods of energy storage and the key technologies of the MMC-ES, especially the main circuit parameter design, modulation strategy, voltage control, and control are summarized and analysed. Finally, suggestions for the key issues of an energy storage system based on a modular multilevel converter are proposed.

1 INTRODUCTION

Energy storage technology mainly means using an Energy Storage System (ESS) to make electrical energy convert to other forms and store them, then to re-convert it to electrical energy release when you need. An ESS includes the storage unit and the Power Conversion System (PCS). In recent years, a multi-level converter with its unique structural features in the high-voltage and high-power applications attract broad interest (Abu-Rub H. 2010). Among them, Cascaded H-Bridge Converter (CHC) used as PCS in ESS has been widely studied (Maharjan L. et al. 2008a, b, 2009). A Modular Multilevel Converter (MMC) topology (Marquardt R. 2003) has been widely studied and applied in high voltage direct current transmission fields, and it also has broad application prospects (Franquelo L.G. 2008).

A MMC combined with an energy storage device (MMC-ES, Modular Multilevel Converter-Energy Storage) had been studied by scholars abroad (Teodorescu R. 2011, Baruschka L. 2011) and many advantages were proposed: access to the high-voltage grid directly without a transformer, improved efficiency and saving investment; the modular structure allowing the storage unit to be embed in a split manner, so that the energy storage unit operates at a lower voltage, improvement in efficiency and reliability; intermediate links giving AC and DC power grid interconnection, capability of storing energy from AC and DC power systems or the release of energy etc. Since, the study of MMC-ES has attracted the attention of researchers abroad, but the study of MMC-ES is still so far less in the domestic field (Zhang Z. 2012).

In this paper, an overview of energy storage technologies is concluded, and MMC topology and technical characteristics are introduced briefly. Then the integrating methods of energy storage are concluded and analysed, the progress on MMC-ES also is summarized and analysed. Finally a vision and recommendations for researching energy storage systems, based on MMC, are proposed.

2 AN OVERVIEW OF ENERGY STORAGE TECHNOLOGY

Present energy storage technologies include physical storage, magnetic storage, chemical storage, and phase-change energy storage, etc., (Zhang W. 2008). Physical storage includes pumped storage, flywheel energy storage and compressed air energy storage; electromagnetic energy storage includes super capacitor energy storage and superconducting magnetic energy storage, etc.; chemical energy storage includes lead-acid batteries, lithium-ion batteries, sodium sulphur batteries and flow batteries, etc.; phase-change energy storage, including molten salt storage and phase-change materials storage. Compared with other energy storage methods, phase-change does not release the stored energy in the form of electricity, and

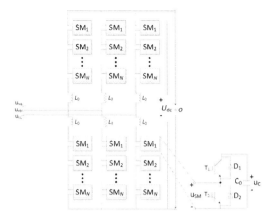
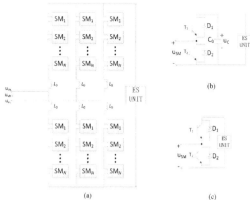

Figure 1. The configuration of an MMC and its sub-module.

Figure 2. (a) ES paralleling to the common DC link directly, (b) ES paralleling to DC side of SM directly, (c) DC capacitor of SM replaced by ES.

the power and energy levels involved very complex factors. It will play an important role in Demand-Side Management (DSM) with the development of the smart grid (Liu S. 2013).

Different energy storage technologies have different performances in energy density, power density, response speeds and the capacity scale of energy storage systems. While different technical requirements of energy storage systems, applied in power systems, are proposed, therefore, various needs must be taken into account and the appropriate storage methods used in power systems must be selected (Li J. 2013).

3 RESEARCH PROFILES OF MMC AND INTEGRATING METHODS OF ENERGY STORAGE UNIT

3.1 Topology and technical characteristics of MMC

The Topology of the three-phase MMC is illustrated in Figure 1, O represents the zero potential reference point. A converter is constituted of six arms, each arm consists of a reactor and the N sub-modules (SM) connected in series, the upper and lower arm forming a phase unit, and the AC output terminal of the corresponding phase is constituted by the connection point of the two arms reactors. A single SM structure is also shown in the figure, T_1 and T_2 are IGBTs, D_1 and D_2 are anti-parallel diodes, C_0 is the DC-side capacitor of sub-module; U_C represents the voltage of the capacitor, u_{SM} represents the voltage of the sub-module.

An MMC has many technical features compared to the conventional two- or three-level converter topology: a highly modular structure; excellent output characteristics; a common DC bus; unbalanced operational ability; advantages in its switch device; and fault ride-through and recovery capabilities.

3.2 Research on methods of energy storage unit integrated to MMC

Existing research on the parallel manner of energy storage unit integrated to the DC side of an MMC includes: direct parallels, through direct DC/DC parallels, and through isolated DC/DC parallels. The methods of energy storage unit integrated to the DC side of an MMC are related to a selection of energy storage unit, the complexity of the control strategy, and the power transmission efficiency, operational reliability, overall economy, and many other factors should be taken into account.

3.2.1 Energy storage unit paralleled on the DC side directly

The advantages of an energy storage unit paralleling to the DC side directly is that it is a simple structure, without a transformer in the DC side, and relatively low energy consumption. The main drawback is the lack of choice flexibility for capacity in the energy storage system, and this affects the life of the energy storage unit. The topology of the energy storage paralleling to the DC side directly is shown in Figure 2.

The capacitors of sub-model are replaced by super capacitor directly in (Zhang Z. 2012). The performance of large centralized battery paralleling on common DC bus of MMC directly are analysed in (Baruschka L. 2011, Soong T. 2014). The simulation results show that the performance of this topology is poor with respect to the spilt manner of battery embedding in sub-modules. The battery is paralleled to a sub-module DC side directly in (Ciccarelli F. 2013), and it is used in the electric vehicle charging station as an energy buffer. In (D'Arco S. 2012, 2014), an MMC with battery energy storage system is used in an electric vehicle, and considering that the added DC/DC link will reduce the efficiency of the entire device, the capacitor of sub-module is replaced by batteries. For the effects of a secondary low-frequency oscillating current on the battery, a resonant filter tuned at the second harmonic frequency combined with a low-pass filter is presented to reduce the effects of a second low-frequency oscillating current and prolong battery life (Vasiladiotis M. 2012).

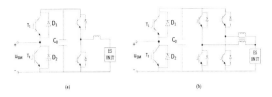

Figure 3. (a) ES paralleled on the DC side through non-isolated DC/DC, (b) Interleaved DC/DC.

3.2.2 *An energy storage unit paralleled on the DC side through non-isolated DC/DC*

A battery energy storage unit is used in (Soong T. 2014) and the advantages of DC/DC link are presented. It decouples the battery from the sub-module capacitor, reducing the DC filter required for the battery, increases the lifespan of the battery, and allows the sub-module capacitor to decrease. The major disadvantage of the introduction of DC/DC part is that the efficiency of the energy conversion of the whole system will be reduced. The topology which the storage unit paralleled to the DC side directly through DC/DC manner is shown in Figure 3.

In the power traction converters (Coppola M. 2012), a super capacitor energy storage unit is used, and a DC/DC converter is needed to improve the voltage of the storage energy unit. Considering the low-frequency component of the sub-module will flow to the battery, in order to avoid that low frequency component. A non-isolated DC/DC link is used to connect the battery and the DC side sub-module (Vasiladiotis M. 2015). And the DC/DC link offers an additional degree of freedom for sub-model capacitor voltage balance control. An interleaved bi-directional boost converter is proposed in (Teodorescu R. 2011), and the inductor ripple can be reduced and then the inductor' size can also be reduced. The oscillated current can't be used to charge the battery (Schroeder M. 2013), which leads to an additional DC/DC link for the bi-directional power flow.

3.2.3 *An energy storage unit paralleled on the DC side through isolated DC/DC*

An isolated DC/DC link is proposed in (Teodorescu R. 2011), and its main advantage is that it has the ability of using a relatively low voltage of energy storage unit for a high AC voltage output, and has electrical isolation, but its drawback is that it uses more switch devices and the control strategy is much more complex. Applications of MMC in photovoltaic cells integrated to a grid are described in (Perez M A. 2013, Rivera S. 2013), and it is proposed that photovoltaic cells should be paralleled on the sub-module DC side through an isolated DC/DC link when the photovoltaic cells need to be grounded. The topology of that energy storage unit paralleled on the DC side though an isolated DC/DC link is shown in Figure 4.

In summary, different access method should be considered for different energy storage units and different applications. The battery usually shunts on the sub-module DC side through a DC/DC link in order to

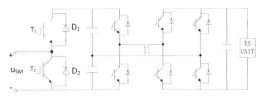

Figure 4. ES paralleled on the DC side by isolated DC/DC.

prevent any inflow of sub-module low-frequency component to the battery and, thus, extend the battery's lifetime. However, a battery can be used to replace the sub-module DC capacitor or parallels on the sub-module DC side directly for applications that require high efficiency. When using super-capacitor energy storage, it requires DC/DC link to increase the terminal voltage, considering its discharge characteristics. When the storage unit has special requirements, such as be grounded, it should parallel on the DC side through an isolated DC/DC link.

4 KEY TECHNOLOGY RESEARCH PROGRESS OF MMC-ES

Key issues of an MMC include the pulse modulation, the DC voltage control, pre-charge, circulation, harmonic, mathematical model, the main circuit parameters, and fault protection, all of which have been researched at home and abroad. As a combination of MMC and energy storage devices, an MMC-ES has a lot in common with an MMC, but also has differences. Since the MMC-ES is in emerging research, existing literatures include but are not limited to its voltage control, power balance control, its control of storage unit and its other aspects. This section focuses on the key technical research issues of an MMC-ES, which include main circuit parameter design, pulse modulation, the sub-module capacitor voltage balance control, and control strategy.

4.1 *Main circuit parameters design of MMC-ES*

The design of the main circuit parameters affects the dynamic and static performance of an MMC system. Research on the main circuit parameters of MMC-ES in existing literatures is not abundant. The optimal parameter design of MMC topology used in an energy storage system is introduced in (Hillers A. 2013). First, control variables are decoupled, and then the calculation of the bridge arm voltage is derived, the sub-module capacitor parameter is calculated from the energy view, and illustrates that arm reactor parameter does not depend on the value of the ripple of the output current, but depend on the voltage pre-control reaction time when a fault occurs in the grid side. It concludes that it is most critical to choose switch devices of the MMC in the energy storage systems.

4.2 *Pulse modulation technology of MMC-ES*

If the storage units are paralleled on an MMC directly, the modulation strategy of an MMC-ES is the same as an MMC. If the storage units are paralleled on an MMC with direct DC/DC access or isolated DC/DC access, the pulse modulation is divided into the pulse modulation of a sub-model and the modulation of a DC/DC link.

Three pulse modulation technologies, Space Vector Pulse Width Modulation (SVPWM), Level Shift Pulse Width Modulation (LS-PWM), and Carrier Phase-Shifted Pulse Width Modulation (CPS-SPWM) are described in (Teodorescu R. 2011). To improve the CPS-PWM, upper and lower arms of each bridge arm constituted by n sub-modules of the MMC-ES use a n/2 carrier to compare with the modulated wave to generate the duty cycle signal, and to simplify the capacitor voltage at the fundamental frequency balance, but reduces the number of signals at the output level. Two modulators are used in (Vasiladiotis M. 2015), for modulating the sub-module using the CPS-PWM, the modulated signal of the DC/DC link is obtained from the battery-side controller of the sub-module capacitor voltage balance.

4.3 Sub-module capacitor voltage balance control of MMC-ES

There are two basic ideas for DC capacitor voltage balance control currently: a balance is achieved by an external control circuit, the other is through its own balance control algorithm. The algorithm design of former can be simplified, but this requires additional hardware and control systems, increases the cost and complexity, and reduces the reliability. In contrast, the latter does not have these problems, the difficulty lies in the algorithm design (Yang X. 2013).

Since the energy storage units are paralleled on the DC-side of sub-module, sub-module capacitor voltage balance of MMC-ES and MMC are not identical. The sub-module capacitor voltage balance is controlled by using the DC side battery through the DC/DC link, while the inner current outer voltage controller is designed to control sub-module capacitor voltage balance (Vasiladiotis M. 2015). The applications of MMC in photovoltaic cells integrated to grid are analysed in (Perez M A. 2013, Rivera S. 2013), due to the use of the maximum power point tracking control for photovoltaic cells, so that each sub-module runs at different voltages, and no longer uses the original sub-module balancing strategy, but ensures that the voltage of each sub-module is kept in the vicinity of the reference values.

In summary, because the DC side paralleled the energy storage unit in MMC-ES, it provides an additional degree of freedom for sub-module capacitor voltage balance control, and simplifies the sub-module capacitor voltage balance control algorithms.

4.4 Control study of MMC-ES

The overall control of the MMC-ES system includes control of an MMC and a storage unit, and other aspects, while research on the control of the energy storage unit is mainly focused on the State Of Charge (SOC) balancing control.

The SOCs are modelled as first-order systems with integral behaviour in (Vasiladiotis M. 2015). An explicit transfer function of the closed-loop controller is derived and limitations on gain selection are analysed. In addition, the voltage unbalanced conditions caused by the failure of the grid side is likely to cause an imbalance of battery state of charge, and the use of common-mode voltage injection methods ensure that the unbalanced network side does not affect the power distribution of each phase of the upper and lower arm. In order to maintain the balance control of active power, the balance current is injected into each battery sub-modules (Schroeder M. 2013), the calculation method of balance current is given and the system's operational state under balanced and unbalanced battery voltage is discussed in simulation. Then it draws a conclusion that the sub-module capacitor voltage will be unbalanced under the unbalanced battery voltage, but the paper does not give a solution. In addition, a way is given in this paper that can balance the sub-module capacitor voltage by injecting reactive current when the battery is idle.

Therefore, the MMC-ES control should consider the energy storage unit control and MMC control. The overall control is divided into various sub-controls for coordinating the control links to achieve operational requirements.

5 PROSPECT AND RECOMMENDATIONS OF MMC-ES

The above analysis shows that, at present, a certain amount of research into theoretical analysis, modulation strategy, and voltage control of an MMC-ES has been done, but research still needs further improvement and development in the following aspects:

1) Select the storage unit considering the practical applications of grid and MMC topology. Single storage is difficult to meet a variety of applications in power system, and research on energy storage unit selection is to be done.
2) Study the way how energy storage unit connects to an MMC. It is of theoretical significance and practical value in engineering to study a more efficient and applicable way of access to the storage element.
3) Existing studies on the MMC-ES is more concentrated in the AC system equilibrium conditions. But when a failure occurs on the AC side and leads to unbalanced conditions, taking into account the impact on the MMC-ES, and circulation analysis, then harmonic analysis of MMC-ES needs to be made.
4) The existing literature has carried out certain research on circulation and harmonics on MMC topology, but there are few researches when ES is added. Considering the properties of MMC-ES, study on circulation and harmonics on MMC-ES should be made.

5) Study on the main circuit parameters on MMC-ES does not see more in the existing literatures, but the main circuit parameters affect the system's dynamic performance and static performance. Therefore, the main circuit parameter designs need to be studied further.
6) Expand the application of MMC-ES to make it relevant to the fields of new energy integrated into the grid, power quality regulation, peak and frequency control, promoting the stable operation of power systems and improving the power grid's reliability etc.

6 CONCLUSION

Energy storage technology, topological structure and technical features of an MMC are briefly introduced in this paper. Then the energy storage integrating methods and the key technologies of an MMC-ES were summarized and analysed. Finally, some suggestions about some aspects of MMC-ESs are put forward. These will provide some reference for the research and application of MMC-ES.

ACKNOWLEDGMENT

This paper is funded by the National High Technology Research and Development of China (863 Programme: 2011AA05A113).

REFERENCES

Abu-Rub H, Holtz J, Rodriguez J, "Medium-voltage multi-level converters-state of the art, challenges, and requirements in industrial applications," Industrial Electronics, IEEE Transactions on, vol. 57, no. 8, Aug. 2010, pp. 2581–2596.

Baruschka L, Mertens A, "Comparison of Cascaded H-Bridge and Modular Multilevel Converters for BESS application," Energy Conversion Congress and Exposition (ECCE), 2011.

Ciccarelli F, Del Pizzo A, Iannuzzi D, "An ultra-fast charging architecture based on modular multilevel converters integrated with energy storage buffers," Ecological Vehicles and Renewable Energies (EVER), 2013 8th International Conference and Exhibition on.

Coppola M, Del Pizzo A, Iannuzzi D, "A power traction converter based on Modular Multilevel architecture integrated with energy storage devices," Electrical Systems for Aircraft, Railway and Ship Propulsion (ESARS), 2012.

D'Arco S, Piegari L, Quraan M S, Tricoli P, "Battery charging for electric vehicles with modular multilevel traction drives," Power Electronics, Machines and Drives (PEMD 2014), 7th IET International Conference on, 2014.

D'Arco S, Piegari L, Tricoli P, "A modular converter with embedded battery cell balancing for electric vehicles," Electrical Systems for Aircraft, Railway and Ship Propulsion (ESARS), 2012.

Franquelo L G, Rodriguez J, Leon J I, Kouro S, Portillo R, Prats M AM, "The age of multilevel converters arrives," Industrial Electronics Magazine, IEEE, vol. 2, no. 2, June 2008, pp. 28–39.

Hillers A, Biela J, "Optimal design of the modular multi-level converter for an energy storage system based on split batteries," Power Electronics and Applications (EPE), 2013.

Lesnicar A, Marquardt R, "An innovative modular multilevel converter topology suitable for a wide power range," IEEE Power Tech Conference Proceedings, Bologna Italy: 23–26 June 2003.

Li Jianlin, "Key technologies of wind/photovoltaic/ storage in smart grid," Beijing: China Machine Press, 2013

Liu Shilin, Wen Jinyu, Sun Hai-shun, Cheng Shijie, "Progress on applications of energy storage technology in wind power integrated to the grid," Power System Protection and Control, vol. 41, no. 23, 2013, pp. 145–153.

Maharjan L, Inoue S, Akagi H, "A Transformer less Energy Storage System Based on a Cascade Multilevel PWM Converter With Star Configuration," Industry Applications, IEEE Transactions on, vol. 44, no. 5, Sept.–Oct. 2008, pp.1621–1630.

Maharjan L, Inoue S, Akagi H, Asakura J, "A transformer-less battery energy storage system based on a multilevel cascade PWM converter," Power Electronics Specialists Conference, 15–19 June 2008.

Maharjan L, Inoue S, Akagi H, Asakura J, "State-of-Charge (SOC)-Balancing Control of a Battery Energy Storage System Based on a Cascade PWM Converter," Power Electronics, IEEE Transactions on, vol. 24, no. 6, June 2009, pp. 1628–1636.

Perez M A, Arancibia D, Kouro S, Rodriguez J, "Modular multilevel converter with integrated storage for solar photovoltaic applications," Industrial Electronics Society, IECON 2013 – 39th Annual Conference of the IEEE, 2013.

Rivera S, Bin Wu, Lizana R, Kouro S, Perez M, Rodriguez J, "Modular multilevel converter for large-scale multi-string photovoltaic energy conversion system," Energy Conversion Congress and Exposition (ECCE), 2013.

Schroeder M, Henninger S, Jaeger J, Ras A, Rubenbauer H, Leu H, "Integration of batteries into a modular multilevel converter," Power Electronics and Applications (EPE), 2013 15th European Conference on.

Soong T, Lehn P W, "Evaluation of Emerging Modular Multilevel Converters for BESS Applications," Power Delivery, IEEE Transactions on, vol. 29, no. 5, Oct. 2014, pp. 2086–2094.

Trintis I, Munk-Nielsen S, Teodorescu R, "A new modular multilevel converter with integrated energy storage," IECON 2011 – 37th Annual Conference on IEEE Industrial Electronics Society, 7–10 Nov. 2011.

Vasiladiotis M, Rufer A, "Analysis and Control of Modular Multilevel Converters With Integrated Battery Energy Storage," Power Electronics, IEEE Transactions on, vol. 30, no. 1, Jan. 2015, pp. 163–175.

Vasiladiotis M, Rufer A, Beguin A, "Modular converter architecture for medium voltage ultra-fast EV charging stations: Global system considerations," Electric Vehicle Conference (IEVC), 2012.

Yang Xiaofeng, Lin Zhiqin, Zheng Trillion Q., You Xiaojie, "A Review of Modular Multilevel Converters," Proceedings of the CSEE, vol. 33, no. 6, 2013, pp. 1–14.

Zhang Wenliang, Qiu Ming, Lai Xiaokang, "Application of Energy Storage Technologies in Power Grids," Power System Technology, vol. 32, no. 7, 2008, pp. 1–9.

Zhang Zhenhua, Feng Tao, "Study of used modular multi-level storage units to improve the stability of wind farms integrated to grid," Sichuan Electric Power Technology, vol. 35, no. 1, 2012, pp. 6–8+53.

Robust fall detection based on particle flow for intelligent surveillance

Changqi Zhang & Yepeng Guan
School of Communication and Information Engineering, Shanghai University, Shanghai, China

ABSTRACT: A novel particle flow based method has been developed to detect fall events automatically without any pre-knowledge for the scene in advance in this paper. The motion description is achieved robustly by building a particle flow filed. The fall event is detected by utilizing the motion gradient of particle that presents significant difference between normal walking and falling. We also decompose the particle motion at any angle into two orthogonal directions (horizontal and vertical). A gradient ratio is employed to describe the gradient change in both directions to improve the capability of generalization in different view-angle. The method performs well in some challenging scenes. It does not require any object detection, tracking, or training in detecting fall events which may be very difficult in practice. The developed approach is highly robust to work well both with indoor and outdoor scene. Experimental results show that the proposed method can be applied to detect fall events effectively.

1 INTRODUCTION

Fall detection is a very important computer vision-based application that has been used for abnormal detection or healthcare of older adults. A fall occurs when a person accidentally falls/slips while walking or standing. It may be due to health and ageing-related issues, abnormality of walking surface or abnormal behavior. It has become very important to develop intelligent surveillance systems, especially vision-based systems, which can automatically monitor and detect falls.

There are numerous studies on the fall detection using different technologies and techniques (Mubashir et al. 2013). The existing fall detection approaches can be categorized roughly into two different categories: wearable device based and vision based. In the following we mainly summarize the different vision-based methods.

Computer vision provides a promising solution to analyze personal behavior and detect certain unusual events such as falls. Vision based approaches use one or several cameras to detect falls for the elderly or general population and convey multiple advantages over other sensor based systems. They do not require a device attached to the person as they are able to detect the human motion, using computer vision algorithms. Cameras can also be used to detect multiple events simultaneously with less intrusion. A mobile human airbag release system was designed for fall protection for the elderly in (Shi et al. 2009). A high speed camera is used for the analysis of falls. Gyro thresholding is applied to detect a lateral fall. The classification of falls is performed by using a support vector machine (SVM) classifier. A classification method for fall detection by analyzing human shape deformation was proposed in (Rougier et al. 2011). A Gaussian Mixture Model (GMM) classifier is implemented to detect falls. In (Liao et al. 2012) the Bayesian Belief Network (BBN) was used to model the causality of the fall event with other events, based on the motion activity measure and human silhouette shape variations. Depth cameras provide 3D information without requiring calibration of multiple cameras. Depth cameras for fall detection are used in (Rougier 2011, Diraco et al. 2010, Leone et al. 2011). Those three methods only use two features, namely distance from the floor and acceleration, and make a decision by applying hardcoded thresholds to individual features. Recently, Kinect sensor is used in human behavior analysis (Lee et al. 2013, Li et al. 2014), which uses three types of sensors: an RGB camera, an IR and an acoustic sensor. In (Mastorakis et al. 2012), a fall is detected by analyzing the 3D bounding box's width, height and depth and ignoring the global motion of 3D bounding box. A statistical method based on Kinect depth cameras was proposed, which makes a decision based on information about how the human moved during the last few frames (Zhang et al. 2012).

The existing vision based solution can work well in various scenarios. But they mainly shared the following shortage: some solutions cannot work with common camera. They have higher requirements to the camera and depend on some special cameras, such as Kinect, high speed camera. This limits the generality of the solution. Some other solutions needs to train a classifier with appropriate amount feature samples collected before training phase. In the real-world environment, collecting a large number of samples is not an easy job or even impracticable at sometimes. Moreover, it needs to re-train a new classifier, when the scene is changed.

In this paper, we propose a method that is capable of detecting fall events in the real-world scene with common camera. This method is based on three main ideas. First, we use the particle based motion estimation both in long-range and moderately dense to describe object motion effectively. Second, we decompose the motion of any angle in image plane into two orthogonal directions. Here, the direction of horizontal and vertical is selected. Third, we introduce the original and simple idea of gradient ratio (denoted GR in the remaining of this paper) to detect fall events. This ratio is obtained by dividing the vertical motion gradient by the horizontal. For people falling down, this ratio is high compared to when they are walking or standing up. This feature is less sensitive to noise than methods based on motion foreground detection or object tracking. Finally, the real-time implementation demonstrates the developed algorithm is effective.

The rest of the paper is organized as follows. In Section 2, particle based motion estimation is introduced, followed by the method of fall detection in Section 3. The experimental results are given in Section 4 followed by some conclusions in Section 5.

2 PARTICLE FLOW MOTION DESCRIBING

Motion describing is the process of determining motion vectors that describe the transformation from one image to another, usually from adjacent frames in a video sequence. Video motion estimation is often performed using optical flow (Beauchemin et al. 1995) or feature tracking (Shi et al. 1994), which corresponds to the perceived movement of pixels. Optical flow estimates a dense motion field from one frame to the next, whereas feature tracking follows a sparse set of salient image points over many frames. The optical flow representation is best suited to successive pairs of frames, not to long sequences. It is often disturbed with some noises caused by camera sensor or slight shake. Longer-range correspondences can be obtain by concatenating frame-to-frame flow fields, but the resulting multi-frame flow must be refined at each step to avoid drift. Feature tracking requires that the same feature can be detected reliably and consistently across many frames. A big disadvantage of this technique is that correspondence errors tend to be very large.

Our approach represents people motion using a set of particles that move through time to produce motion estimates that are both long-range and moderately dense. Particles can represent complicated geometry and motion because they are small; a particle's appearance will not change as rapidly as the appearance of a large feature patch and it is less likely to straddle an occlusion boundary. Particles represent motion in a nonparametric manner; they do not assume that the scene consists of planar or rigid components.

The particle flow algorithm uses frame-to-frame optical flow to provide an initial guess of particle motion. The algorithm treats flow estimation as a black box that can be replaced with an alternate flow algorithm. Before building the particle flow, we need to provide a stable optical flow. Let $u(t)$ and $v(t)$ denote the components of an optical flow field that maps image point $I_t(x,y)$ to an image point in the next frame $I_{t+1}(x+u(t), y+v(t))$.

We optimize the optical flow field using a variational objective function combines a data term E_D and smoothness term E_S.

$$E_{Flow}(u,v,t) = E_D(u,v,t) + \alpha E_S(u,v,t) \quad (1)$$

The data term measures the global deviations from the grey value constancy assumption, that the grey value of a pixel is not changed by the displacement.

$$E_D(u,v,t) = \sum_{x,y} \Psi\left(\left[I_{t+1}(x+u, y+v) - I_t(x,y)\right]^2\right) \quad (2)$$

$$\Psi(s^2) = \sqrt{s^2 + \varepsilon^2} \quad (3)$$

The function Ψ is a differentiable form of the absolute value function. Moreover, this choice of Ψ does not introduce any additional parameters, since ε is only for numerical reasons and can be set to a fixed value.

The smoothness term measures the variation of the flow field using the robust norm Ψ. This is achieved by penalizing the total variation of the flow field.

$$E_S(u,v,t) = \sum_{x,y} \Psi\left(u_x(t)^2 + u_y(t)^2 + v_x(t)^2 + v_y(t)^2\right) \quad (4)$$

By minimizing the objective function, we can find the functions u and v.

Particles are initialized on a uniform grid. The particle flow algorithm uses frame-to-frame optical flow to provide an initial guess for particle motion. Particle i has a time-varying position $(x_i(t), y_i(t))$ that is defined between the particle's start and end frames. Each particle has its own start time and end time. To propagate particle i from frame $t-1$ to t, we use the flow field $u(t-1)$, $v(t-1)$.

$$x_i(t) = x_i(t-1) + u(t-1) \quad (5)$$

$$y_i(t) = y_i(t-1) + v(t-1) \quad (6)$$

After propagating the particles, we prune particles that continue to have high distortion energy values. Distortion energy is defined between a pair of linked particles i and j.

$$E(i,j,t) = l_{ij}\psi\left(\left[u_i(t) - u_j(t)\right]^2 + \left[v_i(t) - v_j(t)\right]^2\right) \quad (7)$$

$$u_i(t) = x_i(t) - x_i(t-1) \quad (8)$$

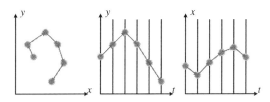

Figure 1. The particle trajectory in image space (left), spatio-temporal space for horizontal direction (middle) and vertical (right).

$$v_i(t) = y_i(t) - y_i(t-1) \qquad (9)$$

These particles with high energy indicate possible drift. To reduce the impact of a single bad frame, we filter each particle's energy values using a Gaussian. It filters the values for the given particle, which is moving through image space. If in any frame the filtered energy value is greater than d, the particle is deactivated in that frame.

After pruning, there may be gaps between existing particles, where we need to add new particles. We check the point on the grid. If there are particles, one new particle will be added.

3 GRADIENT ANALYSIS

For each one specific particle can generate a polynomial curve denoted as path line. These path lines describe a particle trajectory through a video cube, that a sequence of video frames stacked such that time forms a third dimension. We can view this trajectory in image coordinates or spatio-temporal coordinates, as shown in Figure 1.

We used the particle-gradient based method to describe the different between normal walk/standing and fall behavior. The gradient can be used to measure changes in the direction of greatest change. From Figure 1, we can find that the gradient will have a significant change when the particle alters the patterns of motion. To analyze the motion of the particles, the gradient is computed in the direction horizontal and vertical.

$$\nabla T_x = \frac{\partial T}{\partial x}, \ \nabla T_y = \frac{\partial T}{\partial y} \qquad (10)$$

where T is the trajectory of particle.

We consider the change in both directions by introducing the gradient ratio.

$$GR = \frac{\nabla T_x}{\nabla T_y} \qquad (11)$$

The gradient ratio can describe the motion of particle changes in both directions. It will be one great value and generate a peak when the particle moves quickly in horizontal or vertical direction. In contrast, the gradient will be constant approximately when the particle moves smoothly.

The motion of particle will represent one stable movement pattern, when people walk normally. If the particle moves stable, the gradient ratio will be a constant or changed slightly. The stable movement pattern will be replaced, when the fall occurs.

4 EXPERIMENTAL RESULTS

To evaluate the proposed method, we select three public datasets, the CAVIAR dataset (CAVIAR 2001), the MCFD dataset (Auvinet et al. 2010) and the CASIA dataset (CASIA 2007). In CAVIAR dataset, the illumination conditions are not controlled and backgrounds are complex and not static. People walk in an almost open area, while the image content is captured from above with low quality. Thus, humans are presented distortedly as they get close or far away from the camera. The MCFD dataset designs scenarios that were carried out by an actor who performed the falls in the laboratory with appropriate protection (mattress). The CASIA dataset is outdoor scene. People collapse suddenly while walking in the path. The image content is captured from a camera with angle against the walking direction, and therefore the motion of people is not only in horizontal or vertical direction. The motion in both direction needs to be considered simultaneously.

The original resolution of CAVIAR dataset is half-resolution PAL standard 384 × 288. The MCFD dataset is 720 × 480. The CASIA dataset is 320 × 240. The results of experiments are given at an Intel Dual-core processor 3.2 GHz, 4G RAM computer with Open source computer vision library (OpenCV).

In experiment, we compute the particle on a grid size 32 × 32. We test the proposed method on each video sequence, as shown in Figure 2. We use two measurements for describing the accuracy of fall event detection.

$$\text{Precision} = \frac{\#TP}{\#TP + \#FP} \times 100\%$$
$$\text{Recall} = \frac{\#TP}{\#TP + \#FN} \times 100\% \qquad (12)$$

where #TP (True Positive) is the number of detected fall events which actually happen, #FP (False Positive) is the number of detected fall events which do not occur (false alarms), and #FN (False Negative) is the number of fall events which do occur but are not detected. The experimental results of the test videos with frame rate 25 FPS are shown below in Table 1.

The real-time implementations of this method have been tested. Computation times for the proposed method on three datasets are reported in Table 2.

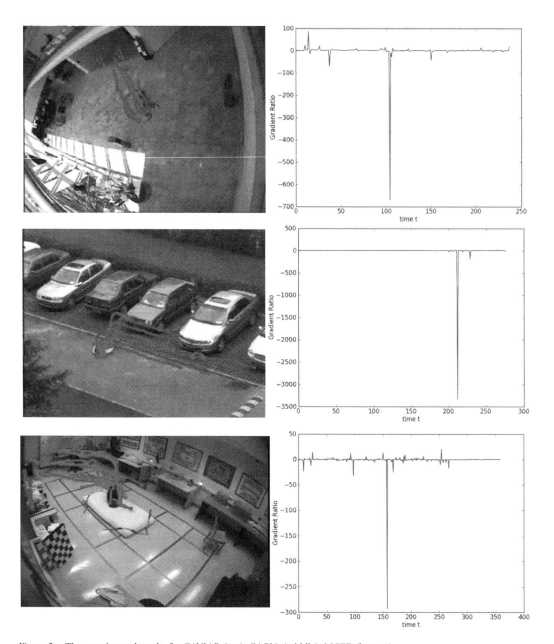

Figure 2. The experimental results for CAVIAR (top), CASIA (middle), MCFD (bottom).

Table 1. Fall event detection (%).

	CAVIAR	MCFD	CASIA
Ground truth	3	192	23
Precision	0.75	0.88	0.96
Recall	1	0.98	1

Table 2. Running time (unit: ms).

CAVIAR	MCFD	CASIA
14.5	15.3	43.1

5 CONCLUSIONS

In this paper, we propose a method to detect fall events automatically in video surveillance scenes based on the particle flow and the motion activity measure. The motion measure is obtained by computing motion gradient of the trajectories of particle flow. In image space, the motion is decomposed into two orthogonal directions (horizontal and vertical). The gradient ratio is defined to describe the significant differences between falling and walking. The proposed method does not need the object detection and tracking method, and can achieve the purpose of detecting fall events in practical applications. The algorithm is tested on real world video sequences, and experiments show that the proposed approach is flexible and computationally cost effective.

ACKNOWLEDGMENTS

This work is supported in part by the Natural Science Foundation of China (Grant no. 11176016, 60872117), and Specialized Research Fund for the Doctoral Program of Higher Education (Grant no. 2012108110014).

REFERENCES

[1] Leone, G. Diraco, and P. Siciliano, "Detecting falls with 3d range camera in ambient assisted living applications: A preliminary study," Medical Engineering & Physics, Vol. 33, No. 6, 2011, pp. 770–781.

[2] K. Lee, V. Y. Lee, "Fall detection system based on kinect sensor using novel detection and posture recognition algorithm," Proceedings of International Conference on Smart Homes and Health Telematics, 2013, pp. 238–244.

[3] Rougier, J. Meunier, A. St-Arnaud, J. Rousseau, "Robust Video Surveillance for Fall Detection Based on Human Shape Deformation," IEEE Transactions on Circuits and Systems for Video Technology, Vol. 21, No. 5, 2011, pp. 611–622.

[4] Rougier, "Fall detection from depth map video sequences," Proceedings of the 9th international conference on Toward useful services for elderly and people with disabilities: smart homes and health telematics, 2011, pp. 128–129.

[5] CAVIAR. CAVIAR project/IST 2001 37540. Available from: http://www.prima.inrialpes.fr/PETS04/caviar_data.html.

[6] CASIA. CASIA action database 2007. Available from: http://www.cbsr.ia.ac.cn/china/Action%20Databases%20CH.asp

[7] Auvinet, C. Rougier, J. Meunier, A. St-Arnaud, J. Rousseau, "Multiple cameras fall dataset," Technical report 1350, DIRO – Université de Montréal, 2010.

[8] Diraco, A. Leone, P. Siciliano, "An active vision system for fall detection and posture recognition in elderly healthcare," Design, Automation & Test in Europe Conference & Exhibition, 2010, 1536–1541.

[9] Shi, C.-S. Chan, W. J. Li, K.-S. Leung, Y. Zou, Y. Jin, "Mobile Human Airbag System for Fall Protection Using MEMS Sensors and Embedded SVM Classifier," IEEE Sensors Journal, Vol. 9, No. 5, 2009, pp. 495–503.

[10] Mastorakis, D. Makris, "Fall detection system using Kinect's infrared sensor," Journal of Real-Time Image Processing, 2012, pp. 1–12.

[11] Shi and C. Tomasi, "Good features to track," Proceedings of IEEE Conference on Computer Vision and Pattern Recognition, 1994, pp. 593–600.

[12] M. Mubashir, L. Shao and L. Seed, "A survey on fall detection Principles and approaches," Neurocomputing, Vol. 100, 2013, pp. 144–152.

[13] S. S. Beauchemin and J. L. Barron, "The computation of optical flow." ACM Computing Surveys, Vol. 27, No. 3, 1995, pp. 433–467.

[14] Y. Li, K.C. Ho, M. Popescu, "Efficient source separation algorithms for acoustic fall detection using a microsoft Kinect," IEEE Transactions on Biomedical Engineering, Vol. 61, No. 3, 2014, pp. 745–755.

[15] Y. Wang, K. Huang, T. Tan, "Human Activity Recognition Based on R Transform," Proceedings of IEEE Conference on Computer Vision and Pattern Recognition, 2007, pp. 1–8.

[16] Y. T. Liao, C.-L. Huang, S.-C. Hsu, "Slip and fall event detection using Bayesian Belief Network," Pattern Recognition, Vol. 45, No. 1, 2012, pp. 24–32.

[17] Z. Zhang, W. Liu, V. Metsis, V. Athitsos, "A Viewpoint-Independent Statistical Method for Fall Detection," Proceedings of International Conference on Pattern Recognition, 2012, pp. 3626–3630.

An IEC 61850 based coordinated control architecture for a PV-storage microgrid

Hongyuan Huang, Feijin Peng & Xiaoyun Huang
Foshan Power Supply Bureau, Guangdong Power Grid Co. Ltd., Foshan, Guangdong, China

Aidong Xu, Jinyong Lei, Lei Yu & Zhan Shen
Electric Power Research Institute of China Southern Power Grid, Guangzhou, China

ABSTRACT: In order to meet the requirements of the diversity of function and performance, of the different types of control equipment, of the non-uniformity of communication protocol, and of other issues for the microgrid applications, an IEC 61850 based layered comprehensive coordinate control architecture is proposed for PV-Storage microgrid systems. The concrete implementation method is explained in detail in combination with a PV-Storage integrated microgrid project. In the proposed architecture of a microgrid system, the intelligence, standardization, and digitization of information exchange in the control process, can be achieved. The decentralized arrangement and distributed implementation of advanced applications also can be implemented. The functions of utmost interoperability between IEDs, the plug and play of DER and its controller, and the stable and economical operation of microgrid systems are also designed in this architecture.

1 INTRODUCTION

A microgrid is often defined as a small generating and distributing system, consisting of Distributed Generations (DGs), energy storage, Power Conversion Systems (PCSs), loads, control system, and monitoring and protection devices. It is an autonomous system which can realize self-control, protection and management, and can run in both grid-connected mode and isolated mode. The existing research and practice have shown that connecting distributed generators to the public grid in the form of microgrids and supporting each other, is the most effective way to play the effectiveness of distributed generators (Lu 2007, Wang 2010, Pepermans 2005).

Large numbers of power electronics are widely used as interface equipment in microgrids, which have small or less inertia and poor overload capacity. The intermittency and fluctuation of distributed generations' output power increases the difficulty of instantaneous power balance in a microgrid. Therefore, real-time control of the microgrid is necessary in order to ensure the voltage and frequency stability of the system (Wang 2008 & 2012, Mou 2010). At the same time, in order to improve the schedulable ability of DGs and the economic performance of the entire microgrid, forecast of DG and load on large time scale is needed for the coordinated control and optimal dispatch of DGs (Guo 2012, Chen 2011, Gu 2012, Liu 2011). The existing research has focused on the principle research and software simulation of DG and microgrid control, and some of this verified control functions based on the established experimental platform (Pei 2010, Mao 2010). However, less consideration is found on controller implementation, communication system and system architecture. In view of the diversity of microgrid application functions and different requirements of real-time performance, control architecture with flexible functional layout and strong scalability is required to support the microgrid control and energy management.

For various kinds of microgrid control, technologies are still in a developmental phase, it also has many problems, such as lack of sufficient test in practice, products with low maturity, differences between control equipment in the actual project implementation process, utilization of non-standard process control protocol among equipment, and different communication interface types, which bring great difficulties to information integration, operational control and scheduling management of the microgrid system. In addition, because the communication specifications for different control equipment are not standard, it is very difficult to implement "plug and play" and "interoperability", and brings great challenges to the project implementation of microgrid. The IEC 61850 standard becomes the first choice to solve these problems for its open and interoperable characteristics. The current study for the application of IEC 61850 in microgrids mostly focuses on the protocol conversion

(Yoo 2011), logic modelling of distributed generators (Frank 2009, Apostolov 2009) and communication services (Colet-Subirachs 2012). The research has not yet formed a complete information system of IEC 61850 in microgrid application, also it did not yet find specific modelling and implementation methods. A microgrid hierarchical information system structure based on IEC 61850 standard, and the general design and implementation method of equipment information model at each level is given (Deng 2013), but the research is mainly based on the Manufacturing Message Specification (MMS) service of IEC 61850 standard, have not been involved in General Object-Oriented Substation Event (GOOSE) services and Sampling Value (SV) services, etc.

In view of the above requirements for microgrid application functions, real-time performance, and the need for standardization, this paper puts forward a kind of microgrid coordinated control architecture based on IEC 61850 standard from the perspective of system implementation and engineering practice, and tests it in a PV-storage microgrid project. The rest of the paper is organized as follows. An introduction of IEC 61850 and its extension in DG and microgrid is given in Section 2, and a coordinated control architecture based on IEC 61850 is put forward in Section 3. In Section 4 the implementation method is illustrated by a PV-storage microgrid project, and the paper is concluded in Section 5.

2 IEC 61850 AND ITS EXTENSION IN THE FIELD OF DISTRIBUTED GENERATION

The IEC 61850 standard constituted by International Electrotechnical Commission (IEC) is widely used in the field of substation automation. It utilizes hierarchical distributed architecture and object-oriented modelling methods, realizes the self-description of data objects, and provides different vendors of Intelligent Electronic Devices (IED), an effective way to achieve interoperability, and seamless integration (Tan 2001). As the IEC 61850 technology gradually matures and is widely used, its technology and method gradually extends to other application areas beyond the substation automation (Tong 2008, Wang 2007, Shi 2012, Ling 2012).

At present, the IEC 61850 standard is expanding its scope of application. The second version is under revision, and the title has been changed to "Communication networks and systems for power utility automation", which means that the scope of application of IEC 61850 will be no longer limited to substation, and may expand to the whole power system. Distributed power system monitoring and control communication system, new energy sources, such as wind power plants and solar power plant, all will be based on the IEC 61850 standard.

The IEC 61850 standard develops a service model satisfying the transmission requirements of real-time and other information according to the characteristics of the production process of the electric power system, using the abstract communication service interface and specific communication service mapping to adapt to the rapid development of network technology requirements (Tan 2001).

2.1 *MMS service model*

MMS is the main specific communication service mapping in the IEC 61850 standard, used for data exchange between IEDs and monitoring systems and the implementation of most of the services required for the monitoring functionality of a power system. For example, signal quantity data such as digital input, events, alarm etc. and analog quantity data such as analog input, protection measurement etc. can be sent upward through the implementation of the Report Control Block (RCB), mapping to the reading and writing service and report service of MMS. Setting management functions can be implemented by Setting Group Control Blocks (SGCB), mapping to the reading and writing service of MMS. Control functions, such as digital output, analog output, etc., can be implemented by control models, mapping to the reading and writing services and report services (IEC 61850-7-2 2003; IEC 61850-8-1 2004).

2.2 *GOOSE service model*

GOOSE is an important service model provided by IEC 61850, based on high speed point-to-point communication instead of the hard-wired communication between traditional intelligent electronic devices, which provides a fast, efficient and reliable method for communication between logical nodes. In order to ensure the real-time and reliability of GOOSE service, GOOSE message uses ASN.1 syntax coding associated with basic encoding rules, jumps over the TCP/IP protocol, and transmits in the link layer of Ethernet directly, and adopts a combination mechanism of heart-beat messages and rapid retransmission of variable-bit messages (IEC 61850-8-1 2004).

2.3 *Information model of DG for microgrid*

In the second edition of IEC 61850, the standard of communication systems for distributed generations, has been incorporated into the IEC 61850-7, named IEC 61850-7-420 for publication. Object models for the vast majority of distributed generations are defined in the IEC 61850-7-420. Except for wind power, various Logical Devices (LDs) and Logical Nodes (LNs) of distributed power systems are also included. These distributed power systems include photovoltaic, energy storage, Combined Cooling Heating and Power (CCHP), fuel cell, electric vehicle charging systems, etc. This standard also includes the logical node application of power conversion, energy storage, and commutation devices in the distributed power system (IEC 61850-7-420 2009).

Table 1. Common functional requirements of a microgrid.

Functions	real-time	info scope
Emergency frequency regulation	ms	local
Dynamic voltage adjustment	ms	local
Island detection	10s ms	regional
Power smooth	s & m	regional
Coordinated control	s	global
Power forecast	m & h	global
Load forecast	m & h	global
Economic dispatch	m & h	global

3 A COMPREHENSIVE COORDINATED CONTROL ARCHITECTURE BASED ON IEC 61850

3.1 Functional requirements of microgrid control system

The control system of microgrids aims to ensure stable, reliable, and economic operations, and the common functions and requirements of real-time performance and information scope is shown in Table 1 where ms, 10s ms, s, s&m, m&h stands for millisecond scale, several tens of millisecond scale, second scale, second or minute scale, minute or hour scale, respectively.

3.2 Architecture of microgrid control system

The microgrid control system monitors the real-time operation information and evaluates the current running state. It can automatically generate the control instruction and send up the executed results to the public power grid monitoring system, and can also receive scheduling or control instructions from the public power grid. Through statistical analysis and optimization calculation with historical, real-time, and prediction data, the comprehensive operation control instructions in microgrids are generated. As shown in Table 1, the requirements of real-time performance and information scope of all kinds of control functions differ greatly. Hence, the functions of real-time control, coordinated control and optimal operation should be deployed at different levels from the reliability point of view.

This paper presents a kind of coordinated control architecture for microgrids, which is divided into three layers according to functionality. The first layer is that of local controllers which control locally distributed generations, energy storage units, and controllable loads and can be specified as Micro-Source Controllers (MSC) and Load Controllers (LC). The second layer is made up of coordinated controllers which implements the coordinated control of MSCs and LCs both in grid connected and isolated modes of operation and can be specified as Microgrid Central Controllers (MGCC) and Micro-grid Mode Controllers (MGMC). The third layer comprises a Microgrid Energy Management System (MEMS) which realizes comprehensive

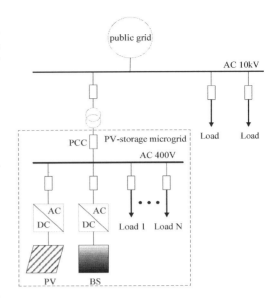

Figure 1. Schematic diagram of the PV-storage microgrid.

optimization and a schedule of various kinds of energy such as cold, heat, oil, gas and electricity.

3.3 Communication architecture for microgrid

The establishment of an information model for DGs and other devices in microgrid conforms to the IEC61850 standard. For micro-source controllers, modelling follows IEC 61850-7-420 and IEC 61400-25, and follows IEC61850-7-4 for other devices.

According to the real-time performance requirements for information exchange demanded by various control functions listed in Table 1, the communication network is divided into a monitoring network and a control network.

The monitoring network adopts an IEC 61850/MMS services model for the slow monitoring information exchange between control equipment in the third and first layers and the second layers. The control network adopts IEC 61850/GOOSE service model to achieve rapid control information exchange between control equipment in the first and second layer. The monitoring network and control network are separated to achieve physical isolation.

4 AN INSTANCE OF A PV-STORAGE MICROGRID

A practical PV-storage microgrid system is shown in Figure 1. The system is composed of PV arrays (150 kept in total), battery storage (500 kWh in total) with a PCS (250 kW), and four controllable load branches.

The coordinated control system of the PV-storage microgrid adopts the three-layer control architecture described above. Three layers are connected

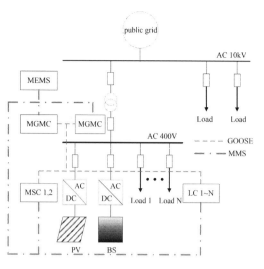

Figure 2. Illustration of deployment of proposed control architecture and communication networks.

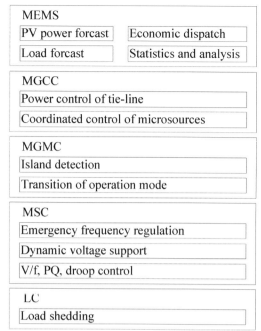

Figure 3. Deployment of control functions of microgrid.

by communication networks for effective information exchange. Controller deployment and communication architecture is shown in Figure 2. MGCC, MGMC, MSCs and LCs are connected together by IEC 61850/GOOSE network for rapid transmission of important and real-time information. MEMS, MGCC, MGMC, MSCs and LCs are linked together through an IEC 61850/MMS network for transmission of comprehensive operational information.

The deployment of control functions of the microgrid is shown in Figure 3. In this PV-storage microgrid project, the functions of each controller are described below.

The MGCC is the core of system control. It connects micro-source controllers of PV and battery storage (BS), microgrid mode controller and load controllers. By coordination with MSCs and LCs, it can enable stable operation of the microgrid especially when in isolated mode.

MGMC is applied to island detection and smooth transition between grid-connected and islanded modes. MGMC sends the instruction of mode transition to MSC of BS, opens the breaker at point of common coupling immediately after islanded status is detected or closes the breaker when the public grid has returned to normal.

The MSC controls all kinds of distributed energy resources, such as PV and energy storage in this microgrid project. It can perform emergency frequency regulation and dynamic voltage support to keep stable operation.

The LC can collect information of key loads inside the microgrid and realize a low frequency/undervoltage load shedding function when in isolated mode or in the process of transition from the grid-connected to the isolated.

The field test results show that the delay time between the two controllers through GOOSE is about 6.12 ms and the time interval from issuing the command to response of energy storage system is about 22.81 ms. The test data indicates that the proposed coordinated control architecture based on IEC 61850 can satisfy the real-time requirement of microgrid control.

5 CONCLUSIONS

The content of IEC61850 standard has covered the distribution network and distributed energy resource field, which will become the main communication standard of microgrid coordinated control architecture. The coordinated control architecture of the microgrid based on IEC 61850, proposed in this paper, can realize the digitization, standardization and intelligentization of information exchange in the control process of the microgrid, and in the decentralized arrangement and implementation of advanced application functions of the microgrid. It can also achieve the interoperability among the IEDs in the microgrid with maximum limit and realize the 'plug and play' of distributed generations and controllers and achieve the stable and economical operation of the microgrid.

The ultimate goal of the IEC61850 standard is to achieve excellent interoperability among IEDs from different manufacturers. At present, many electric power automation equipment suppliers have already mastered the related technology of IEC61850. But the suppliers of electrical and electronic equipment in the microgrid are not the traditional electric power automation equipment suppliers, relatively lagging behind the understanding and application of the IEC61850 standards. There are very few suppliers who can provide primary equipment for DGs and power electronic interface devices that comply with IEC 61850. The interoperability level in the field of microgrid control has to be improved, and this needs the unremitting efforts and cooperation of suppliers of primary equipment for DGs and power electronic interface equipment.

ACKNOWLEDGMENTS

This paper is supported by Science and Technology Project of Guangdong Power Grid Co. Ltd (K-GD2013-044) and the National High Technology Research and Development Program of China (863 Program) (2011AA05114).

REFERENCES

A. Colet-Subirachs, A. Ruiz-Alvarez, O. Gomis Bellmunt, et al. "Centralized and distributed active and reactive power control of a utility connected microgrid using IEC61850," IEEE Systems Journal, Vol. 6, No. 1, 2012, pp. 58–67.

A. P. Apostolov, "Modeling systems with distributed generators in IEC61850," Proceedings of Power Systems Conference, March 10–13, 2009.

B. G. Yoo, H.-S. Yang, S. Yang, et al. "CAN to IEC 61850 for microgrid system," Proceedings of 2011 International Conference on Advanced Power System Automation and Protection, October 16–20, 2011.

B. Zhao, X.-S. Zhang, P. Li, et al. "Optimal design and application of energy storage system in Dongfushan island stand-alone microgrid," Automation of Electric Power System, Vol. 37, No. 1, 2013, pp. 161–127.

C.-S. Wang, Z.-X. Xiao and S.-X. Wang, "Synthetical Control and Analysis of Microgrid," Automation of Electric Power Systems, Vol. 32, No. 7, 2008, pp. 98–103.

C.-S. Wang, Z.-X. Xiao and S.-X. Wang, "Synthetical Control and Analysis of Microgrid," Automation of Electric Power Systems, Vol. 32, No. 7, 2008, pp. 98–103.

C.-S. Wang, F. Gao, P. Li, et al. "Control strategy research on low voltage microgrid," Proceedings of the CSEE, 2012.

H. Frank, S. Mesentean and F.Kupzog, "Simplified. Application of the IEC 61850 for distributed energy resources," Proceedings of the First International Conference on Computational Intelligence, Communication Systems and Networks, July 23–25, 2009.

IEC. IEC61850-7-2 "Communication networks and systems in substations. Part 7-2: basic communication structure for substation and feeder equipment – abstract communication service interface," Geneva: IEC, 2003.

IEC. IEC61850-7-4 "Communication networks and systems in substations. Part 7-4: basic communication structure for substation and feeder equipment-compatible logical node classes and data classes," Geneva: IEC, 2003.

IEC. IEC61850-8-1 "Communication networks and systems in substations. Part 8-1: specific communication service mapping – mapping to MMS and to ISO/IEC 8802-3," Geneva: IEC, 2004.

IEC. IEC61850-7-420 "Communication networks and systems for power utility automation. Part 7-420: Basic communication structure – Distributed energy resources logical nodes," Geneva: IEC, 2009.

K.-L. Wang, Y.-G. You and Y.-Q. Zhang, "Energy management system of renewable stand-alone energy power generation system in an island," Automation of Electric Power System, Vol. 34, No. 14, 2010, pp. 13–17.

L. Guo, X.-P. Fu, X.-L. Li, et al. "Coordinated control of battery storage and diesel generators in isolated AC microgrid systems," Proceedings of the CSEE, 2012.

L.-H. Wang, T. Jiang, X.-H. Sheng, et al. "Analysis on protection function modelling based on IEC61850 standard," Automation of Electric Power Systems, Vol. 31, No. 2, 2007, pp. 55–59.

W. Deng, W. Peiand Z.-P. Qi, "Microgrid Information Exchange Based on IEC61850," Automation of Electric Power Systems, Vol. 37, No. 3, 2013, pp. 6–11.

W. Gu, Z. Wu and R. Wang, "Multi-objective optimization of combined heat and power microgrid considering pollutant emission," Automation of Electric Power Systems, 2012, 36(14): 177–185.

W.-J. Shi, S.-Q. Feng and Y.-D. Xia, "New intelligent power distribution automation terminal self-describing," Automation of Electric Power Systems, Vol. 36, No. 4, 2012, pp. 105–109.

W.-S. Ling, D. Liu, Y.-M. Liu, et al. "Model of intelligent distributed feeder automation based on IEC61850," Automation of Electric Power Systems, Vol. 36, No. 6, 2012, pp. 90–95.

W.-S. Tan, "An introduction to substation communication network and system – IEC61850," Power System Technology, Vol. 25, No. 9, 2001, pp. 8–9.

X.-C. Mou, D.-Q. Bi and X.-W. Ren, "Study on control strategies of a low voltage microgrid," Automation of Electric Power Systems, Vol. 34, No. 19, 2010, pp. 91–96.

X.-Y. Tong, Y.-C. Li, L. Zhang, et al. "Interaction model of protection functions based on IEC61850," Automation of Electric Power Systems, Vol. 32, No. 21, 2008, pp. 41–45.

X.-P. Liu, M. Ding, Y.-Y. Zhang, et al. "Dynamic economic dispatch for microgrids," Proceedings of the CSEE, 2011.

Y.-Z. Chen, B.-H. Zhang, J.-H. Wang, et al. "Active control strategy for microgrid energy storage system based on short-term load forecasting," Power System Technology, Vol. 32, No. 8, 2011, pp. 35–40.

Z.-X. Lu, C.-X. Wang, Y. Min, et al. "Overview on Microgrid Research," Automation of Electric Power Systems, Vol. 31, No. 19, 2007, pp. 100–107.

The design of an IED for a high voltage switch operating mechanism based on IEC 61850

Z.-Q. Liu & X.-R. Li
College of Electrical Engineering, Zhejiang University, China

ABSTRACT: A new type of Intelligent Electronic Device (IED) for a high voltage switch operating mechanism, which is based on IEC 61850 and is used at bay layer in an intelligent substation is discussed. Its design is based on the ARM-Linux embedded operating system, using the modular structure design method. The modelling of the IED, including Logical Node (LN) design, using the new LNs and data design according to the latest version of IEC 61850, is presented. The newly designed IED has the ability of fault alarm and failure diagnosis, which increases the reliability and intelligence of this equipment.

1 INTRODUCTION

As a key equipment in power systems, the running state of a high voltage (HV) switch is directly related to the security and stability of any power system. Statistic results show that in all failures of HV switch, the fault of the switch itself accounts for only about 30%, followed by the control circuit fault, which accounts for about 21%, while the malfunction of operating mechanism accounts for 43% of all cases. In addition, some problems, such as aging and loose contact, appeared in the operating mechanism as a result of long-term use, may not cause an accident immediately, but there is danger that accidents, such as an HV switch misoperation, caused by untimely maintenance of HV switch operating mechanism, will happen someday. Therefore, the real-time condition monitoring of HV switch operating mechanism is of great significance.

On-line monitoring technology for operating mechanism has existed for no more than 10 years, and most of the existing systems are based on custom data type which often result in poor interoperability. With the development of digital substations and the widespread of IEC 61850, an on-line monitoring model of operating mechanism for HV switches based on IEC 61850 needs to be established. In this paper, the information model of the IED for a HV switch operating mechanism is built according to the latest version of IEC 61850 and based on a functional analysis of HV switch operating mechanisms and the data requirement of on-line monitoring systems. In addition, the hardware and software design of the IED are discussed.

2 INFORMATION MODELLING

2.1 *IEC61850 modelling techniques*

IEC 61850 is an international standard on substation automation communication. According to IEC 61850, there are three physical layers in substation automation: the bay layer, the process layer, and the substation layer. A physical device is usually modelled as an IED according to IEC 61850 which uses object-oriented modelling technology. An IED, which is a device with network communication functions, contains a server object necessarily. Each server object contains at least one Logical Device (LD), which is composed of several LNs. Each LN is the smallest functional unit and contains Data Objects (DOs) and Data Attributes (DAs). DOs and DAs carry all the information of their parent LN.

2.2 *IEC 61850 IED model*

According to IEC 61850, IED for HV switch operating mechanism is at bay layer in intelligent substation. The main function of the IED is monitoring the running state of operating mechanism through the collection and analysis of vibration, current and stroke signals, so as to realize status early-warning and failure diagnosis. On the basis of the data requirements of the system, the function of the IED is decomposed and modelled respectively as logical node zero (LLN0), physical device information (LPHD), circuit switch supervision (SSWI), supervision of the operating mechanism (SOPM), vibration supervision (SVBR), disturbance recorder channel analogue (RADR) and disturbance recorder function (RDRE) according to the second edition of IEC 61850 protocol. Details are shown in Table 1.

In this model, LLN0 and LPHD are necessary for the modelling of an IED. SSWI, SOMP and SVBR, which are newly defined in the second edition of IEC 61850 protocol, are respectively used to monitor the HV switch, the operating mechanism and vibration. Use of the LNs above avoids extension of general input/output LN GGIO and makes the core data of the IED more concise and intuitive.

Table 1. A description of logical nodes related to the IED.

Name	Logical node	Explanation
LLN0	Logical node zero	Processing general data of logical device
LPHD	Physical device information	Simulate general data of physical device
SSWI	Circuit switch supervision	Supervising all switches
SOPM	Supervision of operating mechanism	Supervising operating mechanism for switches
SVBR	Vibration supervision	Supervising vibrations
RADR	Disturbance recorder channel analogue	Simulate analog disturbance recorder channel
RDRE	Disturbance recorder function	Simulate disturbance recorder

To meet the needs of actual monitoring, properties in addition to the necessary attributes of the corresponding LNs are listed below:

1) DO position (Pos) is added to SSWI to mark the state of the circuit breaker. Besides, the optional DOs in SSWI include operational time open (OpTmOpn), operational time close (OpTmCls), contact stroke (Stk), coil current (ColA), switch operating time exceeded (OpTmAlm), operation counter (OpCnt) and coil alarm (ColAlm).
2) The participating DOs of SVBR include vibration level (Vbr) and vibration alarm level reached (Alm). To indicate the amplitude of vibration signal of each phase, Vbr is extended. VbrA, VbrB and VbrC are added to SVBR and represent the vibration of the contact of A phase, B phase and C phase respectively. In addition, AlmA, AlmB and AlmC are added to indicate the vibration alarm of each phase.
3) SOMP contains operating time of the motor (MotTm) and motor current (MotA). MotAAlm is added to indicate the alarm of an abnormal energy storage motor current.

2.3 Data transmission

2.3.1 Report service

Warnings, alarms and measurement information are transferred by the IED through the IEC 61850 reporting mechanism. Two types of report control blocks are defined in IEC 61850, Buffered Control Block (BRCB) and Unbuffered Control Block (URCB). Buffered Report (BR) is a type of report with caching mechanism.BR is reliable and usually used to transfer warnings and alarms. BR is controlled by BRCB. Unbuffered Report (UR), which is controlled by URCB, is faster but less reliable than BR. Measurement information is transferred by UR. Data sets, whose values are to be reported, are referenced in URCB and BRCB. The reporting model in IEC 61850 is shown in Figure 1.

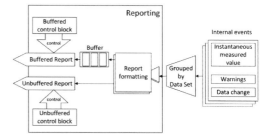

Figure 1. Reporting model in IEC 61850.

Table 2. Description of data sets related with the IED.

Data set	Description	Report type
dsAlm	Data set for warnings and alarms	Buffered
dsMeaA	Data set for Analog measurement	Unbuffered

Two Data Sets (DSs), which are shown in Table 2, are defined in the IED in this article. One is a DS for warnings and alarms and it contains all the related DOs, the other is defined for measuring with DOs related to measurement in it. All the DSs are defined in LLN0.

2.3.2 File service

The main function of an HV switch operating mechanism is completed by tripping and closing coil and iron core of electromagnetic module, so the coil current holds important information. Thus storage and analysis of the coil current are of great importance to the status monitoring and failure prediction of an HV switch operating mechanism. Meanwhile, the current waveform of the energy storage motor is also important. Hence, current waveform of tripping and closing coil and energy storage motor are stored in a transient data transformation standard COMTRADE format and the COMTRADE files can be accessed by a client through file services.

The exciting coil current and energy storage motor current are monitored when the HV switch operating mechanism operates and their waveform are recorded. This process is controlled by RDRE and RADR. When wave recording is done, the instance of records made (RcdMade) in RDRE is set true and client is informed that there is a new COMTRADE file in the server.

3 HARDWARE DESIGN

3.1 Hardware structure

As is shown in Figure 2, the IED for a HV switch operating mechanism is composed of a sensor module, a serial communication module, an ethernet communication module, a digital I/O module, and an ARM

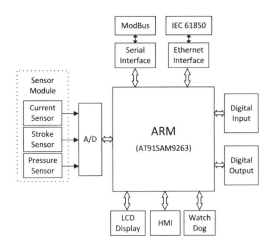

Figure 2. Hardware architecture of the IED.

processor, etc. Among them, the sensor module comprises the current sensor, the stroke sensor and the pressure sensor. The serial communication module supports RS485 serial communication based on ModBus protocol. Network communication based on IEC 61850 is supported by the ethernet communication module. The ARM processor is responsible for data processing and communication with the PC.

3.2 Signal detection

The tripping and closing coil current signal, energy storage motor current signal, the contact stroke signal and vibration signal are monitored by the IED, using a non-contact detection method.

The current in the tripping and closing coil and in the energy storage motor is direct-current and measured by a Hall current sensor. The output signal of the sensor is weak, so it is sent to the amplification circuit and is transformed into 4–20 mA current signal, and then it is sent to an A/D conversion module. The Hall sensor is a component that can convert an electromagnetic signal into a voltage signal based on Hall effect. When the material and size of the Hall component is determined, the output voltage is proportional to the control current magnitude I and magnetic induction intensity B, that is $U = KIB$, where K is the sensitivity factor.

The contact connection stroke is measured by measuring the angular displacement of the linkage, using an incremental photoelectric encoder, which is widely used in the position and angular measurement, due to its light weight, small size, and reliability. The output signal of the incremental photoelectric encoder is a pulse signal which is relevant to rotary displacement rather than its absolute position, therefore the measurement is precise. The encoder is mounted on the spindle of the breaker linkage.

Vibration diagnosis is an important means of HV switch mechanical condition monitoring, thus the vibration signal is measured by a piezoelectric

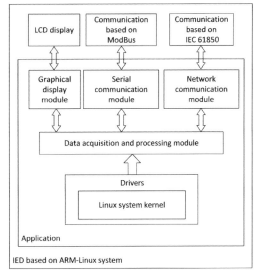

Figure 3. Software architecture of the IED.

acceleration sensor which has the character of wide frequency, large dynamic range, and high reliability.

3.3 Chip selection

The IED will be mounted in a mechanism case, thus a smaller volume, a higher level of integration and stronger anti-jamming capability are required. As the ARM+DSP model needs a large space with a weak anti-interference ability, we choose ARM as the single processor. AT91SAM9263 chip of ATMEL Company is selected because the chip uses high performance ARM9 32b RISC and supports multi-tasking applications where full memory management, high performance, and low power are all important.

TLC2543I chip of TI Company is chosen as the A/D converter. The TLC2543I is a 12-bit-resolution converter with 10-us conversion time over operating temperature and 11 analog input channels. The operating temperature of the TLC2543I ranges from −40°C to 85°C which enables the chip to work normally under the atrocious weather condition.

4 SOFTWARE DESIGN

4.1 Software architecture

The software part of the IED consists of data acquisition and processing module, graphical display module, serial communication module and Ethernet communication module. The architecture is shown in Figure 3.

The main program is based on Qt/Embedded and supports graphical display. Qt for Embedded Linux (Qt/Embedded) is a C++ framework for GUI and application development for embedded devices. The system supports serial communication and network communication at the same time because multithread programming is adopted. RS485 serial communication based on ModBus protocol facilitates testing and debugging while network communication based on IEC 61850 increases the interoperability of the IED.

4.2 Timing mode design

Data acquisition with high rate is a key problem for software design. The duration of the tripping and closing process is very short, so a high sampling rate is required so that the tripping and closing process can be reproduced by the waveform record to accomplish fault diagnosis. Timing is usually achieved by calling the timing service of the ARM-Linux system. But the sampling rate obtained in this way does not satisfy the requirements. Additionally, the operating system will be slow and not conducive to human-computer interaction when the sampling rate becomes very high.

To get a high sampling rate, the sampling interval is controlled by the timer/counter (TC) module of the AT91SAM9263 chip and data acquisition is controlled by the interrupt handler function. Here is how it works:

1) The system function *request_irq* () is called to request an interrupt for a TC module when the driver is initialized. At the same time, a low speed clock is selected as the clock source of the TC module.
2) Each channel of the TC module has three trigger modes. Set the trigger mode to the RC for a comparison trigger mode.
3) Write the interrupt interval to the RC register.
4) When the timer interrupts, the A/D converter is controlled to convert analog signals to digital signals by the interrupt handler function and data acquisition is completed.

5 CONCLUSION

The IED designed in this paper has the function of online monitoring, fault alarming, wave recording and file transmission, etc. Furthermore, the IED supports both serial communication and network communication, which facilitates field debugging and remote control. Support for IEC 61850 equips the IED with good interoperability. Through experimental verification, the IED is stable and reliable and can complete the online monitoring task well.

REFERENCES

[1] Chen, B.-L., Wen, Y.-N. 2010. Introduction of spring operating mechanism for CB. *High Voltage Apparatus* 45(10): 75–80.

[2] Chen, S.-Y., Song, S.-F., Li, L.-X., Shen, J. 2009. Survey on smart grid technology. *Power System Technology* 33(8): 1–7.

[3] Chang, G., Zhang, Z.-Q., Wang, Y. 2011. Review on mechanical fault diagnosis of high-voltage circuit breakers based on vibration diagnosis. *High Voltage Apparatus* 47(8): 85–90.

[4] Du, F.-Q., Sheng, G.-M., Xu, J., Liu, Y.-D., Y, J.-Q., Jiang, X.-C. 2013. Information modeling for GIS smart monitoring based on IEC 61850 and development of information interaction system. *Electric Power Automation Equipment* 33(6): 163–167.

[5] Guo, X.-S., Li, Z.-F., Chen, X.-S. 2002. Selection on on-line monitoring parameters of circuit breakers' operating mechanism. *High Voltage Apparatus* 38(1): 24–30.

[6] Guan, Y.-G., Huang, Y.-L., Qian, J.-L. 2000. An overview on mechanical failure diagnosis technique for high voltage circuit breaker based on vibration signal. *High Voltage Engineering* 26(3): 66–68.

[7] He, W., Tang, C.-H., Zhang, X.-W., Zhu, S.-Y., Liao, W.-G., Liu, S., Yuan, H., Sun, D., Li, J. 2007. Design of data structure for IED based on IEC 61850.*Automationof Electric Power Systems* 31(1): 57–61.

[8] Li, J., Jiao, S.-H. 2004. Online monitoring device for high voltage circuit breaker based on DSP. *Electric Power Automation Equipment* 24(8): 44–47.

[9] Qin, M., Liu, Z.-G., Liu, G., Zhang, R.-X. 2013. Research on high frequency vibration signal detection technology based on accelerometer. *Measurement & control technology* 32(5): 1–4.

[10] Yi, H., Yin, X.-G., Zheng, H. 2007. On-line monitoring device for high voltage circuit breaker based on DSP. *High Voltage Apparatus* 43(1): 35–39.

A profile of charging/discharging loads on the grid due to electric vehicles under different price mechanisms

Mingyang Li & Bin Zou
School of Mechatronic Engineering and Automation, Shanghai University, Shanghai, China

ABSTRACT: In this paper, a profit maximization model of electric vehicle charging/discharging is constructed and is aimed at the maximum operating profits, while being constrained by power batteries charging/discharging capacities and the travel needs of electric vehicles, which can express the charging/discharging decision of electric vehicles well. A calculation and analysis of the economic benefit and charge distribution of electric vehicle charging/discharging have been made by simulating user travel needs with Monte Carlo method, on the basis of the user travel rule derived from National Household Travel Survey (NHTS) in 2009. The results indicate that the economic benefits of the rational charging/discharging model can be significantly improved by responding to the time of use and real-time electricity price. Meanwhile, due to the relatively cheaper off-peak electricity price at night in contrast to the expensive on-peak electricity price during the day, electric vehicles tend to charge at low load time and discharge inversely at peak load time in the distribution system so as to achieve peak load shifting. The battery storage function of electric vehicles is worth further developing.

1 INTRODUCTION

Electric vehicle charging load forecasting is the basic issue of the electric vehicle charging management. Electric vehicle charging load distribution is defined as the distribution characteristics of the charging load in 24 hours a day. If the charging load of the electric vehicles can be arranged in the off-peak periods, it could reduce the construction investment of the distribution network and save plenty of power grid investment costs.

Recently, scholars around the world have done a lot in the electric vehicle charging load forecasting. According to the parking statistics, reference has further discussed the charging load on different spaces. This kind of research is based on the travel laws of the existing traditional cars and has well solved the forecasting of the charging load distribution of electric vehicles by the travel characteristics of electric vehicles. The concept of a vehicle to grid has been a focus since its proposition. According to the travel statistics of the existing traditional vehicles, a vehicle is moving more than 90% of the time, which makes it possible that its power batteries can provide energy to charge in the low load periods and provide peak load shifting, frequency modulation, and steady voltage by V2G in the peak load periods. Nowadays, the economic suitability, and the technical feature and management mode of V2G are widely researched. However, there is a lack of research into the synthesized impacts on the power load curve of the charging of electric vehicles and the discharging of V2G.

This paper has made a decision model to ensure the maximum operating income where charging or V2G discharging is needed. It has also discussed the effects of differently charged electricity prices on the charging load distribution.

2 THE TRAVEL DATA OF EVS

2.1 NHTS

The NHTS survey conducted by the Department Of Transportation (DOT) represents millions of vehicles in the United States. It reports on hundreds of thousands of vehicle trips made across the United States in 2009. The vehicle travel behaviour data used in this paper are derived from the 2009 National Household Travel Survey (NHTS). The NHTS data contains four major files, including a daily trip dataset and a vehicle dataset.

Several processing steps were required in order to prepare the data for input to the model. For each trip, the daily trip file contains the data on the distance travelled, the type of vehicle, the time of departure, and the type of departure. Vehicle-level information, such as the state of the vehicle and the vehicle's age, comes from the vehicle file. From the full survey, we only select trips taken by private vehicles, driver ID and vehicle ID. Table 1 shows a typical daily trip of a vehicle, and the start time, end time, duration, average velocity, and distance.

Table 1. A typical daily trip of a vehicle.

Trips	Start time	End time	Duration /h	Velocity /km	/(km/h)
1	8.80	9.05	0.25	7.2	28.8
2	12.47	12.72	0.25	8.0	32.0
3	15.00	15.17	0.17	6.4	39.4
4	17.50	17.75	0.25	6.4	25.6

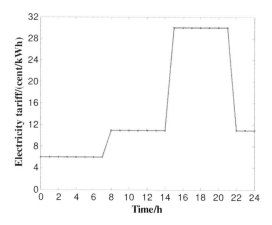

Figure 1. Time of use electricity rate.

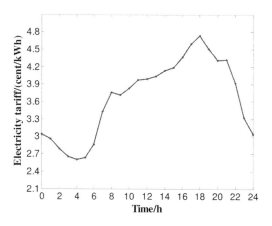

Figure 2. Real time electricity rate.

2.2 Electricity tariff structure

To some extent, the EV load demand can be dictated by the electricity tariff structure. Two types of typical electricity tariff structures are taken into consideration: Time-Of-Use (TOU) pricing and Real-Time Pricing (RTP). TOU pricing divides the tariff into three main blocks: peak, flat, and valley price, as shown in Fig. 1. The latter option provides a real-time rate which varies by season of the year, hour of the day, and by weekday and weekend, as shown in Fig. 2.

3 CHARGING AND DISCHARGING DECISION-MAKING MODEL OF EVS

Electric vehicles are associated with a power grid by the battery charge/discharge, whose key index is the battery state of charge. SOC is the ratio of current electric quantities and the full electric quantities of a battery, which describes the current available in the capacity of a battery by centesimal grade. Battery energy is in kWh, which is denoted by the product of the battery capacity and rated current.

3.1 Calculation of power battery SOC and loss cost

In this paper, the battery charge/discharge procedure involves the SOC calculation of the electric vehicles in driving, of the charging and of the V2G service. To avoid over-discharging of a battery, the SOC must be greater than the allowed minimum SOC (SOC_{min}), or else, the discharging must be terminated. Similarly, to avoid the over-charging of a battery, the SOC must be less than the allowed maximum SOC (SOC_{max}), or else, the charging must be terminated. We assume that the initial value of SOC is 0.90 (marked as SOC_S).

1) Under the condition that the travelling distance of an electric vehicle is given, its SOC_{trip} cost during the trip can be obtained by the following formula

$$SOC_{trip} = \frac{W_{100}}{100 * C_B} d \quad (1)$$

where W_{100} is the power consumption per hundred kilometers of the power batteries (/kWh/(100 km)), d is the travelling distance (/km), and C_B is the capacity (/kWh).

2) When charging electric vehicles, the way of fully-charging within six and four hours is adopted, during which time the battery is in the allowed SOC interval and the charging can be regarded as constant current charging. Then, the relationship between the charging time and SOC is

$$SOC_{Charge} = \frac{Q_C}{C_B} = \frac{\eta^{down} * p^{down}(t)}{C_B} t \quad (2)$$

where SOC_{charge} is the SOC after charging, Q_C is the charged electric energy (/kWh), $p^{down}(t)$ is the charge power (/kW), t is the charge time (/h) and η^{down} is the charge efficiency.

3) In the service that electric vehicles supply power to the power grid, we just discuss the peak load shifting, that is the selling power of the stored battery energy. Suppose that the electric energy of electric vehicles is sold by constant power, then the sold battery SOC is

$$SOC_{V2G} = \frac{Q_D}{C_B} = \frac{\eta^{up} * p^{up}(t)}{C_B} t \quad (3)$$

where SOC_{V2G} is the SOC of the sold batteries of the V2G in the discharging, Q_D is the discharged electric energy (/kWh), $p^{up}(t)$ is the discharge power (/kW), t is the discharge time (/h) and η^{up} is the charge efficiency.

4) The battery loss cost is the product of equivalent battery loss cost rate ω_d per kilowatt-hour and the battery electric energy of charging/discharging.

$$W_B = \omega_d(\frac{1}{2}\lambda^{up}Q^{up} + \frac{1}{2}\lambda^{down}Q^{down}) \quad (4)$$

where ω_d is the loss rate of power batteries (/\$/kWh), λ^{up} is the discharge power coefficient of V2G, λ^{down} is the charge power coefficient of power batteries, Q^{up} is the discharged electric quantities (/kWh) and Q^{down} is the charged electric quantities (/kWh).

3.2 Model

A. An objective Function

EV power batteries have charging and discharging energy processes throughout the course of one day. For example, a vehicle may make m trips during the course of 24 hours. The periods of battery SOC may decrease (i.e., electricity consumption) or increase during a period. In ith trip, a dwelling activity takes up a set of segments (i.e. $seg(i)$).

The objective of a charging/discharging problem is to maximize the operating income (minimize the operating cost) subject to the constraints of EV travel. The income function of EV can be represented as described in (5):

$$\max F_{profit} = \sum_{i=1}^{m}\sum_{j=1}^{seg(i)}[(g_{i,j}(t)\eta\eta^{up} - \frac{1}{2}\omega_d\lambda^{up})\int_0^{t_{i,j}^{up}} p^{up}(t)dt$$
$$-(f_{i,j}(t)\eta\eta^{down} + \frac{1}{2}\omega_d\lambda^{down})\int_0^{t_{i,j}^{down}} p^{down}(t)dt] \quad (5)$$

where $g_{i,j}(t)$ is the charging rate (\$/kWh), $f_{i,j}(t)$ is the discharging rate (\$/kWh), and η is the power grid convert efficiency. $t_{i,j}^{down}$, $t_{i,j}^{up}$ are the charging and discharging time of segment j for trip i, respectively.

B. Equality and Inequality Constraints:

1) The power balance constraints: in a period (usually 24h), the state of charge of the battery in the first stroke equals that in the last stroke, namely the difference of the charge electric quantities and the discharge electric quantities (vehicle discharge and V2G discharge) is zero. The power balancing constraint must be satisfied as follows:

$$\sum_{i=1}^{m}\sum_{j=1}^{seg(i)}[\eta\int_0^{t_{i,j}^{down}} p^{down}(t) - \eta^{up}\int_0^{t_{i,j}^{up}} p^{up}(t)] - \frac{w_{100}}{100}\sum_{i}^{m} d_i = 0 \quad (6)$$

2) Electric vehicles travel constrains: the discharge process must equal the battery SOC in the next trip. That is to say, the initial battery electric quantities of electric vehicles must satisfy the first trip.

$$SOC_{min}C_B \leq SOC_sC_B - \frac{W_{100}}{100}d_1 \leq SOC_{max}C_B \quad (7)$$

The electric vehicles start charging/discharging after the kth ($k = 1, 2, 3, \ldots, m$) stroke, but after charging and discharging, the electric quantities of electric vehicles must be satisfied in the $(k+1)$th stroke. Therefore, there are m constraints.

$$SOC_{min}C_B \leq SOC_sC_B + \sum_{i=1}^{k}\sum_{j=1}^{seg(i)}[\int_0^{t_{i,j}^{down}} p^{down}(t)$$
$$-\int_0^{t_{i,j}^{up}} p^{up}(t)] - \frac{w_{100}}{100}\sum_{i=1}^{k+1} d_i \leq SOC_{max}C_B \quad (8)$$

3) The battery SOC constraints: battery states of charge constraints. The summation of the charged electricity quantities and the minimum allowed electricity quantities is no more than the maximum allowed electricity quantities.

$$\int_0^{t_{i,j}^{down}} p^{down}(t) + SOC_{min}C_B \leq SOC_{max}C_B \quad (9)$$

The difference of maximum allowed electricity quantities and the total discharged electricity quantities is no more than the allowed minimum electricity quantities.

$$SOC_{min}C_B \leq SOC_{max}C_B - \int_0^{t_{i,j}^{up}} p^{up}(t) \quad (10)$$

The electricity quantities after charging/discharging in the lth ($l = 1, 2, 3, \ldots, seg(i)$) period of the kth($k = 1, 2, 3, \ldots, m$) stroke are not only no less than the minimum allowed electricity quantities but also no more than the maximum allowed electricity quantities. There are $\sum_{i=1}^{m} seg(i)$ such constraints.

$$SOC_{min}C_B \leq SOC_sC_B + \sum_{i=1}^{k}\sum_{j=1}^{l}[\int_0^{t_{i,j}^{down}} p^{down}(t)$$
$$-\int_0^{t_{i,j}^{up}} p^{up}(t)] - \frac{w_{100}}{100}\sum_{i=1}^{k} d_i \leq SOC_{max}C_B \quad (11)$$

4) Time constrains: After each stroke, the charging/discharging time of electric vehicles should be less than the time difference between the two periods and be non-negative. In the same period, charging and discharging can only be selected alternatively. Charging and V2G charging are prohibited during the running of electric vehicles.

$$0 \leq t_{i,j}^{down} \leq t_{i,j+1} - t_{i,j} \quad (12)$$

$$0 \leq t_{i,j}^{up} \leq t_{i,j+1} - t_{i,j} \quad (13)$$

$$t_{i,j}^{up} * t_{i,j}^{down} = 0 \quad (14)$$

$$t_{trip} = 0 \quad (15)$$

where t_{trip} is the electric vehicle travel time.

In this paper, the established orderly charging optimization model for electric vehicles is a mixed programming model and is solved by utilizing the YALMIP toolbox.

Table 2. EV batter parameters.

Types	C_B (kWh)	P (kWh)	ω_d (cent/kWh)	W_{100} (kWh/100km)
BYD	57	9.5	4.0	21.5
Nissan	24	4.0	4.0	14.9

Types	SOC_{min}	SOC_{max}	η^{up}/η^{down}	η
BYD	0.15	0.95	0.97	0.85
Nissan	0.15	0.95	0.97	0.85

4 THE RESULTS OF CALCULATION AND SIMULATION

4.1 The data of EVs

Lithium-ion, lithium-ion ferrous phosphate, and lead-acid have been the top three contending technologies for EV batteries due to a combination of performance, safety, life, capability, and cost. In this paper, two types of EVs—the BYDE6 and the Nissan—based on lithium-ion ferrous phosphate and lithium-ion batteries, respectively, have been chosen as examples to estimate the impact of battery charging/discharging load on the distribution system load profile, due to the well-documented data on these two EV batteries and their strong representative position in the market place.

Although the power demand, duration of charging process and capacity differ from each other, the charging/discharging characteristics, efficiency and related battery state-of-charge profiles of the two types of batteries have been approximately the same. The data of electric vehicles are listed in Table 2. To account for the growth in the number of lithium-ion ferrous phosphate battery powered EVs, the EV battery population is assumed to consist of 35% of lithium-ion and 65% lithium-ion ferrous phosphate.

4.2 Monte Carlo simulation considering on electricity rate and uncertainty schedule

Based on Part 2, Monte Carlo simulation is employed to sample the data, considering the change of EV trips, their duration, as well as the start time and the trip's distance. The charging/discharging load curve of all EVs is accumulated and a single EV charging/discharging load curve L is obtained. The entire calculation of charging and discharging decision-making modelling is carried out as described in Figure 3. The vehicle travel behaviours used in this paper are derived from Section 1.1, and the electricity tariff structures are derived from Section 1.2. The practical simulation results show that the average benefits remain invariant basically after simulating 5000 times. Considering the accuracy and the calculation speed, the simulation process is repeated for 5000 times.

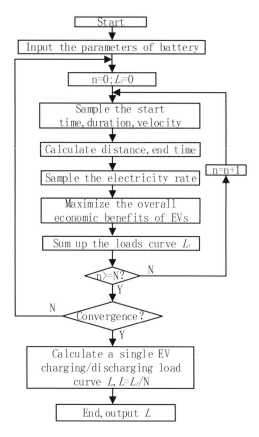

Figure 3. Flow chart of EV profits maximization algorithm based on Monte Carlo simulation.

4.3 Charging and discharging (V2G) profile base on a single EV

According to the algorithm of section 2.2, the time-of-use electricity rate is employed to determine the most economic start time of EV battery charging, while the real-time electricity price is adopted to discharge (V2G) during the peak time. In this paper, the peak load time is from 15.00 pm to 19.59 pm, while the valley load time is from 0.00 am to 6.59 am and 22.00 pm to 23.59 pm. At last the maximum operating profits of the average electric vehicle are 88.79 cents.

An obvious charging load peak appears from 0.00 am to 6.59 am and an obvious discharging (V2G) load peak appears from 16.00 am to 22.59 am. The charging peak and the regular load peak occur during different periods of time, while the discharging peak and the regular load peak almost occur simultaneously. Fig.4 shows the profile of charging/discharging load based on a single EV. The large scale development of private EVs will impose huge impacts on power grids. From 2021 to 2050, with the popularity of private EVs, the charging and discharging load of EVs increases dramatically. It is a controlled charging/ discharging strategy to smooth load the profile of a power grid. The main factor for this phenomenon is the charging and discharging electricity tariff structure. Based on

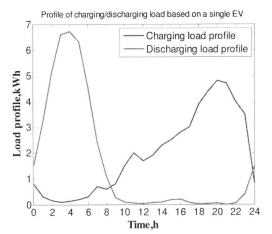

Figure 4. Profile of charging/discharging load based on a single EV.

the high level prediction, the large scale development of private EVs will pose huge impacts on power grids.

5 CONCLUSIONS

Electric vehicles are increasing their great potentials in the global automotive market share. EVs can be regarded as distributed energy storage units, if we can logically and reasonably control and operate them, they have great potentials to contribute to load shifting.

The charging/discharging load of EVs is influenced by many factors. Therefore it is difficult to build a deterministic model between the charging demand and these uncertain factors. In this paper, under the constraints of power battery charging/discharging capacity and customer travel needs, an electric vehicle charging/discharging model is proposed to maximize the electric vehicle economic benefits. The vehicle travel behaviour's data used are derived from the 2009 NHTS. A Monte Carlo simulation based model is proposed to forecast the charging load and discharging load of a single EV. On the top of that, several conclusions are reached below:

1) The uncertainties of EVs charging behaviour bring great challenges to power grid operation and peak-valley regulation. EVs charging load are mainly determined by EVs ownership, EV types, charging duration, charging modes, charging frequency, battery characteristics, and electricity tariff structure.
2) While a single EV charging load has spatial temporal uncertainties, a large number of EVs charging behaviour obeys certain probability models. Therefore, we can calculate and analyse an EV's charging load, based on Monte Carlo EVs charging load calculation model.
3) The results of the simulation indicate that with guiding and controlling EV owners' charging behaviour, the peak discharging load appears at the same time when the whole grid's load reaches peak value, which further narrows the gap between peak and valley.

REFERENCES

[1] 2009 National Household Travel Survey, User's Guide. BYDE6Data. http://www.bydauto.com.cn/.
[2] C. Madrid, J. Argueta, and J. Smith, "Performance characterization-1999 Nissan Altra-EV with lithium-ion battery," Southern California EDISON, Sep. 1999.
[3] D. Linden and T. B. Reddy, Handbook of Batteries, 3rd ed. New York: McGraw-Hill, Aug. 2001.
[4] Hu Zechun, Song Yonghua, Xu Zhiwei, et al. Reference to a book: Impacts and Utilization of Electric Vehicles Integration into Power Systems [J]. Proceedings of the CSEE, 2012, 32(4): 1–10.
[5] Löfberg J. Modeling and solving uncertain optimization problems in YALMIP[C]//Proceedings of the 17th IFAC World Congress. 2008: 1337–1341.
[6] Lu L, Han X, Li J, et al. A review on the key issues for lithium-ion battery management in electric vehicles [J]. Journal of power sources, 2013, 226: 272–288.
[7] Lu Lingrong, Wen Fuquan, Xue Yusheng, et al. Economicanaly of ancillary service provision by plug-in electric vehicles [J]. Automation of Electric Power Systems, 2013.
[8] Luo Zhuowei, Hu Zechun, Song Yonghua, et al. Study on Plug-in Electric Vehicles Charging Load Calculating [J]. Automation of Electric Power Systems, 2011, 35(14): 36–42.
[9] Ma Y, Houghton T, Cruden A, et al. Modeling the benefits of vehicle-to-grid technology to a power system [J]. Power Systems, IEEE Transactions on, 2012, 27(2): 1012–1020.
[10] Reference to a chapter in an edited book: Joao A. Pecas Lopes, Filipe Joel Soares, Pedro M. Rocha Almeida. Integration of Electric Vehicles in the Electric Power System. Proceedings of the IEEE, 2011, 99(1): 168–183.
[11] Yao Weifeng, Zhao Junhua, Wen Fushuan, Xue Yusheng, Chen Fengyun, Li Liang. Frequency Regulation Strategy for Electric Vehicles with Centralized Charging [J]. Automation of Electric Power Systems, 2014, 38(9): 69–76.
[12] Zhang Hongcai, Hu Zechun, Song Yonghua, Xu Zhiwei Jia Long. A Prediction Methodfor Electric Vehicle Charging Load Considering Spatial and Temporal Distribution [J]. Automation of Electric Power Systems, 2014, 38(1): 13–20.
[13] Zhang L, Brown T, Samuelsen S. Evaluation of charging infrastructure requirements and operating costs for plug-in electric vehicles [J]. Journal of Power Sources, 2013, 240: 515–524.37(14): 43–49.

Algorithm design of the routing and spectrum allocation in OFDM-based software defined optical networks

Shun Liu, Xingming Li & Dezhi Zhao
University of Electronic Science and Technology of China, Chengdu, Sichuan, China

ABSTRACT: An OpenFlow-based software defined optical network architecture (SDON) which introduces OFDM as optical modulation technology is designed in this paper. And to address the problem of routing in light path establishment and spectrum fragment waste caused by spectrum consistency, spectrum continuity constraints in OFDM-based SDON architecture, we propose a new routing algorithm which applies A* dynamic search to finding the shortest path satisfying the above spectrum restricted condition and a spectrum allocation algorithm which aims to spectrum fusion. The results prove that the routing and spectrum allocation (RSA) algorithm can effectively improve network resource utilization, reduce business blocking.

Keywords: SDN; OpenFlow; OFDM; A* algorithm; RSA

1 INTRODUCTION

At present, optical transmission network, to serve as the physical foundation of the next generation network, bears more than eighty-percent information transmission. However, the flexibility lack, expansibility insufficient and the separate divisions of operation, management and technology development between IP bearing layer and optical transmission layer in traditional optical transmission network make it difficult to adapt to the diversity and real-time performance of future business, as well as the inherent expansibility in multi-layer and multi-domain optical network system [1].

Software defined network (SDN), a new type of network architecture, provides a new guiding ideology for the next generation optical transmission network development. This network architecture decouples the network device control plane and the data forwarding plane, abstract the infrastructure, provide programmable network equipment and open the network capacity [2]. Through the flexible programmable hardware, SDN-based optical transmission network could dynamically adjust spectrum resource, more finely manage network and achieve reunification of the entire network resource planning and management in net global view.

On the other hand, in existing WDM-based optical transmission network, the granularity of bandwidth resource scheduling and signal transmission is wavelength channels and the channel spacing, signal rate and format parameters are fixed [3]. As a result, network flexibility is insufficient, bandwidth waste severely and it's hard to meet the future large capacity and high speed optical transmission requirements.

With the development of optical transmission technology, OFDM-based and bandwidth variable elastic optical switching technology have caused wide public concern. Compared with WDM, OFDM could realize the switching of finer granularity by introduce subcarrier and carry large capacity data transmission and exchange by gathering multi-subcarrier [4], [5], [6]. But the strict spectrum conditions in light path establishment, such as spectrum continuity and spectrum consistency, and the complicated spectrum state caused by fined granularity spectrum undoubtedly exacerbate the complexity of RSA problem and even make it be a NP-hard problem [7].

In view of the above analysis, this paper will discuss to apply OpenFlow, a suitable candidate for the realization of SDN [9], to OFDM-based optical transmission network and construct the software defined optical network (SDON) architecture, and then design traffic routing method and spectrum allocation scheme in it. According to the RSA problem, this paper puts forward an A* based multi-condition routing algorithm (AMCRA) which applies A* dynamic search to find the shortest path satisfying the above spectrum restricted condition and a spectrum fusion based-on spectrum allocation algorithm (SFBSA) which make the spectrum concentrated as far as possible.

2 SOFTWARE DEFINED OPTICAL NETWORK (SDON) ARCHITECTURE

Existing optical transmission network structure is based on IP over WDM or IP over OTN over WDM, and the characteristics of these traditional optical transport

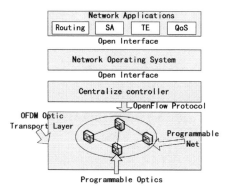

Figure 1. OFDM based SDON architecture.

networks is rigid bandwidth channel, fixed rate of the data flow and cooperation lack between each layer. Consequently, network's flexibility is not enough and has no ability to dynamically adjust resource to satisfy traffic demand in time. However, the programmability and openness of architecture in SDN can solve these problems coincidently [8], [10]. This paper apply SDN concept to optical network, which separate the optical network control plane and data forwarding plane, concentrate on logical control and use software to control entire network, and make use of the OpenFlow protocol to realize software defined traffic routing, spectrum resource planning and network management in logical centralized control plane. Meanwhile, finer granularity sub-carrier modulation in OFDM is more likely to achieve flexible spectrum allocation, so we take OFDM as the optical transmission technology in SDON. Figure 1 illustrates the proposed SDON architecture.

The SDON programmability means that the network can vary dynamically according to network environment; all kinds of resources have flexible programmable features which are also open to application layer. These features could make the whole optical transmission network equipped with software characteristics, improve the network overall performance and resource utilization.

The programmability of nodes, links and other optical components can realize the switching granularity programmability, light path resource programmability and spectrum efficiency programmability. And the entire optical network programmability is reflected in unifying the overall optical network resource, including all nodes, links and so on, planning and manage the entire resources in the controller in a global view and realizing the effective utilization of network resources and management convenience.

3 ROUTING AND SPECTRUM ALLOCATION ALGORITHM (RSA) IN SDON

To solve the RSA problem, we introduce a decomposition approach that divides the problem into two steps: search the shortest path and allocation spectrum.

3.1 A* based Multi-Condition Routing Algorithm (AMCRA)

Traditional Dijkstra algorithm can only be applied to the shortest route problem, but not to the OFDM optical network routing methods because of its spectrum consistency and continuity limitation in light path.

This paper proposes an A^* based multi-condition routing algorithm (AMCRA) which combines traditional Dijkstra algorithm and A^* dynamic searching method together to find the shortest path satisfying the spectrum consistency and continuity restriction. In the above SDON architecture, centralized controller could obtain the entire network switching nodes and optical links spectrum utilization by OpenFlow protocol, and then apply A^* search method to the entire network to find the shortest path satisfying spectrum restriction.

Assuming that finding the shortest path from source node S to destination node D, then for intermediate node N the formula of $A*$ algorithm could be expressed as:

$$f_s(n) = g_s(n) + h_d(n) \qquad (1)$$

In the equation, $f_s(n)$ represents the estimating total "weight" from source node S to destination node D going through the subpath $subPath_{s->n}$, $g_s(n)$ represents the actual "weight" of $subPath_{s->n}$ which satisfying spectrum consistency and continuity limitation and $h_d(n)$ represents the estimating "weight" from N to D on the next $A*$ searching. The selecting of evaluation function $h_d(n)$ and $g_s(n)$ are the key points to heuristic route searching in $A*$ algorithm. During each step to expand the subpath, AMCRA routing algorithm select the subpath $subPath_{s->n}$ which has the minimum $f_s(n)$ from the existing set of subpaths Θ_{sub_path} which containing all former valid subpaths. To guarantee the final path will be the optimal route and make the searching process be more efficient so as to $A*$ search completing soon, the algorithm sets $h_d(n)$ as the minimum "weight" from destination D to intermediate N calculated by Dijkstra. When expanding $subPath_{s->n}$, the algorithm chooses the next hop node n_{next} which satisfying spectrum consistency and continuity when adding to $subPath_{s->n}$ and make the "weight" from N to destination D will be as minimal as possible comparing to other adjacent nodes. On the basis of the above consideration, AMCRA conserves every valid subpath which is consisted of the iterating node and satisfies spectrum constraints into subpath pool Θ_{sub_path} and always preferred the subpath having the smallest $f_s(n)$ in the pool to expand until finding the destination D or Θ_{sub_path} being empty.

Description of AMCRA:

Step 1: Calculate the minimal "weight" value from destination node D to every other node N in the network by traditional Dijkstra algorithm and set it as $h_d(n)$ for every node N.

Step 2: Data initialization. For each node N, apart from source node S (because of setting $g_s(s)$ as 0), set $g_s(n)$ as *inf*, and then computing $f_s(n)$ according to formula (1); put single node path {s} into subpaths set Θ_{sub_path}.

Step 3: Dynamically search all nodes and fiber link in A* searching algorithm.

Step 3.1: Check if Θ_{sub_path} is empty, return false, otherwise popup $subPath_{s->n}$ whose endpoint n has minimal estimating "weight" value $f_s(n)$ among Θ_{sub_path}. If n is the destination, return $subPath_{s->n}$ as the final path, else go to the next step;

Step 3.2: Check if adjacent node set Θ_{adj_n} of node n is empty, go to *Step 3.1*, otherwise popup the node n_{next} which making $w_{n->n_{next}} + h_d(n_{next})$ be the least from Θ_{adj_n}. If n_{next} and $subPath_{s->n}$ satisfy the spectrum consistency and continuity when adding n_{next} into $subPath_{s->n}$, insert the new subpath $subPath_{s->next}$ into Θ_{sub_path} and go to Step 3.1; otherwise, go to *Step 3.2*.

3.2 Spectrum fusion based spectrum allocation algorithm

The carrying of large-grained business is realized through the combination of fine granular bandwidth in OFDM optical network. In this paper, we define the minimum bandwidth granularity of traffic distribution and spectrum resource scheduling and switching as frequency slot (FS). Accordingly, traffic requesting bandwidth is measured by the number of FS. In some existing research literature, the spectrum allocation method often adopts random selection which selecting one frequency block satisfying spectrum constraints from all available blocks randomly or first fit, namely selecting the first frequency block satisfying spectrum constraints as result when iterating all alternative frequency blocks. Those methods are easy to implement, but don't consider the optimizing configuration of FS combination. Consequently, under heavy load network, much follow-up traffic may route failure even left bandwidth resource in single fiber link is enough because of hardly to make one path which may consisted of several fiber links conform to spectrum consistency and continuity restriction. Consequently, make serious traffic blocking.

In this paper, under the SDON architecture we will make use of OpenFlow protocol to obtain the spectrum utilization in each link along one route from centralized controller device and then allocate spectrum for the route aiming to make spectrum be fusion as far as possible from link direction and path direction two dimensions. The goal in this paper's spectrum allocation is to decrease the disturbance to follow-up traffic may caused by current scheduling traffic. Through this way, we can increases the probability of follow-up traffic request succeed and optimize the whole spectrum utilization.

3.2.1 Spectrum fusion in link direction dimension

Spectrum fusion in link direction means that the frequency slots which have the same state (assume that each frequency slot has busy or available two states) are as concentrated as possible in each link of the route.

Assume there exists an available FS block $block_1$ (label range: $s1\#–e1\#$), and select FS block $block_2$ (label range: $s2\#–e2\#$ and s.t. $s1\# \leq s2\#$, $e1\# \geq e2\#$) in it, then we define link direction fusion factor c:

$$c = (e1 - e2) * (s2 - s1) \qquad (2)$$

That is to say, the more little of the c is, the more fusion of the spectrum will be in $block_1$ after allocating $block_2$. It is same to the spectrum of one link when choosing an available $block_1$ which can totally include the selecting FS block $block_2$ in the link to compute c as equation (2).

As for one route R, define the link direction fusion C_l as the average of each link direction fusion factor c in the route R. It signifies the more little the C_l is, the more fusion of the spectrum will be in route R on the whole.

3.2.2 Spectrum fusion in path direction dimension

Spectrum fusion in path direction means the uniformity of frequency slot state among adjacent links with route R.

Assume the route $R = \{l_1, l_2, \ldots, l_i, \ldots, l_n\}$, find the set of adjacent link set $A = \{l_{a1}, l_{a2}, \ldots, l_{aj}, \ldots, l_{am}\}$, each link is neighboring to one of links in R, then define the path direction spectrum fusion C_p of R when choosing the vacant FS block $block_i$ (label range: $s\# - e\#$) to be allocated.

$$C_p = \sum_{l \in A} N_l \qquad (3)$$

N_l means the number of FS which has the busy state and its label is between $s\#$ and $e\#$ in link l. The value of C_p reflects the distraction degree of $block_i$ to the adjacent links of route R and the larger the C_p is, the smaller the obstruction made by current traffic to follow-up traffics will be when allocating $block_i$.

3.2.3 Spectrum fusion based spectrum allocation algorithm (SFBSA)

Based on AMCRA to get the shortest route R and find the common available FS of all links FS_{com} in route R, SFBSA set C_p as the first planning target, and then consider C_l as the second planning target to find the FS block among FS_{com} to allocate spectrum for route R.

Description of SFBSA:

Step 1: Data preparation.

Acquire the adjacent links set A of route R and these links spectrum utilization data by using OpenFlow protocol from central controller device; select out all available continuous FS blocks FS_{com} that could carry traffic request and put them into alternative FS blocks set Θ.

Step 2: Iterate each FS block $block_i$ in Θ, for each $block_i$ compute corresponding C_p and C_l, and then choose the FS block having maximum C_p and the

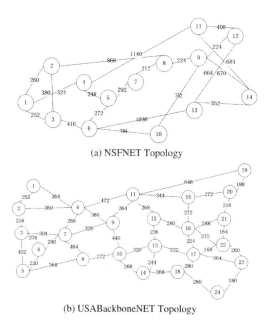

(a) NSFNET Topology

(b) USABackboneNET Topology

Figure 2. Simulating net topology.

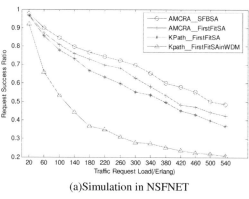

(a)Simulation in NSFNET

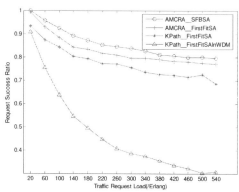

(b)Simulation in USABackboneNET

Figure 3. Traffic throughputs of different RSA.

minimum C_l (if there exists several FS block have maximum value of C_p) as the final spectrum allocation result. Output the FS block as result. If the set Θ is empty, the spectrum allocation failed.

4 SIMULATION

In this paper, we realize the SDON described in section 2 on software circumstance, apply proposing AMCRA and SFBSA to the optical layer, comparing with other routing and spectrum allocation algorithm, then simulate various types of traffic by using different RSA algorithms and optical transmission technology on network NSFNET (shown in Fig. 2(a)) and USABackboneNET (shown in Fig. 2(b)).

Assume the bandwidth of each bi-direction fiber link between each pair of nodes is 400G, practically set services include E-type, FE, GE, 40GE, STM-i (i = 1, 4, 16, 64), ODU-i (i = 0, 1, 2, 3, 4) and OCH-i (i = 1, 2, 3, 4), each kind of traffic arriving equably and dynamically. For each service, the reaching time is subject to λ Poisson distribution, and the serving time is subject to μ Exponential distribution. Other algorithms comparing with paper's RSA (recorded as AMCRA_FSBSA) are:

(1) AMCRA_FirstFitSA. Namely, combine AMCRA routing algorithm and existing FirstFit spectrum allocation. Firstly, find the shortest route by AMCRA, then select the spectrum fitting constraints firstly when searching spectrum from lowest label.
(2) Kpath_FirstFitSA. Namely, combine K-reserve route (k = 8 in NSFNET, and k = 20 in USABackboneNET, it is same to bellows) and FirstFit spectrum allocation. Firstly, compute K routes by KSP routing algorithm in advance as later route alternatives, choose the path having top priority when traffic request arriving, and then allocate spectrum by FirstFit strategy in the above selected route.
(3) Kpath_FirstFitWDM. Namely, Apply Kpath_FirstFit-SA in traditional WDM network.

Figure 3 illustrate network throughput performance of each algorithm. The vertical axis means request traffic success ratio, and the horizontal axis means traffic request load measured by the unit of Ireland. Analyzing Fig. 3 deeply (including Fig. 3(a) and Fig. 3(b)), it's no hard to find the next four verdicts: (a) Comparing AMCRA_SFBSA curve and AMCRA_FirstFitSA curve, we can know this paper's SFBSA spectrum allocation algorithm can more easily improve the success rate of traffic request; (b) The AMCRA_FirstFitSA and Kpath_FirstFitSA two curves indicate paper's AMCRA routing algorithm have better traffic throughput performance than K-Reserve route algorithm which can't acquire optimal route according to the real-time resource utilization of network; (c) Comparing Kpath_FirstFitWDM curve with the other three curves, it's obvious that OFDM-based optical network can greatly improve the traffic requesting success rate than traditional WDM-based network. OFDM has smaller spectrum granularity to scheduling, and has the ability to adapt to traffic bandwidth flexibility. Therefore, OFDM-base optical network can more easily make full use of spectrum resource and improve traffic throughput. (d)Through the comprehensive comparison of the AMCRA_SFBSA cure with the other three curves, it is obvious that this paper's routing and spectrum allocation integrated arithmetic have very good performance in improving traffic request success ration. Because AMCRA_SFBSA make full use of advantage of A* dynamically and heuristically searching, OpenFlow real-time acquiring network resource and spectrum fusion, it can find optimal route and rational spectrum allocation.

5 CONCLUSION

The OFDM-based software defined optical network (SDON) proposed in this paper, adopt IT and software thinking and architecture to innovate optical transmission network, sharply improve the transmission efficiency and ability of optical equipment and can plan the global network spectrum resources flexibility and reasonably. The strict spectrum constraints of routing and complex spectrum state pose a great challenge to RSA in OFDM, however, the proposed AMCRA and SFBRA algorithms in this paper can provide an effective solution. As the simulation results show, paper's RSA algorithm in SDON network can show good throughput performance and resource utilization.

REFERENCES

[1] N. Sambo, P. Castoldi, F. Cugini, *et al*. Toward high-rate and flexible optical networks [J]. IEEE Communications Magazine, 2012, 50(5): 66–72.

[2] Open Network Foundation (ONF).Software-Defined Networking: The New Norm for Networks [EB/OL]. http://www.opennetworking.org/images/stories/downloads/white-papers/wp-sdn-newnorm.pdf, 2012-04-23.

[3] Sang-Yuep Kim, Sang-Hoon Lee, Jae-Seung Lee. Upgrading WDM networks using ultra dense WDM channel groups [J]. Photonics Technology Letters, IEEE, 2004, 16(8): 1966–1968.

[4] W. Shieh, C. Athaudage. Coherent optical orthogonal frequency division multiplexing [J]. Electronics Letters, 2006, 42(10): 587–589.

[5] Guoying Zhang, De M. Leenheer, A. Morea. A Survey on OFDM-Based Elastic Core Optical Networking [J]. Communications Surveys & Tutorials, IEEE, 2013, 15(1): 65–87.

[6] M. Jinno, H. Takara, B. Kozichi, Y. Tsukishima, and Y. Sone. Spectrum-effient and scalable elastic optical path network: Architecture, benefits, and enabling technologies [J]. Communications Magazine, IEEE, 2009, 47(11): 66–73, 2009.

[7] K. Christodoulopoulos, I. Tomkos, A. Varvarigos. Elastic bandwidth allocation in flexible OFDM-based optical networks [J]. Light wave Technology, 2011, 29 (9): 1354–1366.

[8] D. Simeonidou, R. Nejabati, M. P. Channegowda. Software Defined Optical Networks Technology and Infrastructure: Enabling Software-Defined Optical Network Operations [J]. Optical Communications and Networking, IEEE/OSA, 2013, 5(10): A274–A282.

[9] N. McKeown, T. Anderson, H. Balakrishnan, *et al*. OpenFlow: enabling innovation in campus networks [J]. ACM SIGCOMM Computer Communication Review, 2008, 38(2): 69–75.

[10] Ankitumar, N. Patel, Philip, N. Ji, and Ting Wang. Qos-Aware Optical Burst Switching in OpenFlow Based Software-Defined Optical Networks [C]. Optical Network Design and Modeling (ONDM), Brest, 2013, 274–280.

The impact of Negative Bias Temperature Instability (NBTI) effect on D flip-flop

J.L. Yan, X.J. Li & Y.L. Shi
School of Information, East China Normal University, Shanghai, China

ABSTRACT: As semiconductor manufacturing has entered into the nanoscale era, performance degradation due to Negative Bias Temperature Instability (NBTI), has become one of the major threats to circuit reliability. This paper evaluates the severity of the NBTI-induced degradation in a D flip-flop based on a master-slave structure. The effectiveness of this framework is demonstrated by using a 40-nm technology model. First, the impact of the NBTI on a Positive Channel Metal Oxide Semiconductor (PMOS) device is investigated and the increase of a threshold voltage over time at different duty cycle values is presented. Using this model, the NBTI-induced degradation on the inverter and the TG are respectively discussed. Simulation results reveal that the NBTI can cause up 30% total delay both to the inverter and the TG over ten years' operation. In particular, this work includes a simple framework integrated with the NBTI effect on D flip-flop, analyzing the impact of the degradation of propagation time and setup time under different operational conditions.

Keywords: Duty cycle, negative bias temperature instability (NBTI), performance degradation, D flip-flop

1 INTRODUCTION

With the rapid development of Complementary Metal Oxide Semiconductor (CMOS) technology, reliability has become one of the most serious concerns. Besides the hot Carrier Injection (HCL), the negative bias temperature instability (NBTI) has been the main responsible of reliability with gate oxide becoming thinner [1], [2].

NBTI has a great effect on a PMOS device and results in the threshold voltage shifting (V_{th}). In digital circuit, when logic '0' is applied to the gate terminal (gate-to source voltage $V_{gs} = -V_{dd}$), the PMOS stays in the 'stress' state, and its V_{th} will be increased over time, thus resulting in a degradation of the drain current and slowing down the speed of circuit. On the contrary, when logic '1' is applied to the gate terminal, the NBTI stress is removed and the device is switched into a 'recovery' state [3], then the degradation due to the NBTI will be partially recovered. An experiment in [2] shows a threshold voltage difference up to 20mV between stress and recovery over 10^5 second.

Many researches focus on the analysis of NBTI and proposed various design techniques to reduce the effect of the NBTI [4], [2]. In general, compensation and mitigation techniques are the two main methods used in the NBTI-aware circuit optimization. Delay guard-banding of the former is proven to be less complexity and design effort compared with gate sizing and logic synthesis [5]. As for the mitigation technique, reducing stress time is the most efficient way rather than leakage reduction. Much progress has been made in the modeling and analysis of the basic logic gate such as NAND, NOR, INV and etc.

It is known that D flip-flop plays an important role in digital circuits. Unfortunately, precise NBTI analysis and model of D flip-flop are still desired. In this paper, the predictive models provided in [2] are used to integrate the NBTI effect into D flip-flop, analysing the impact of propagation time and setup time degradation under different operation conditions over operation time.

The remainder of this paper is organized as follows. In section 2, some related works and the NBTI degradation model adopted are reviewed, Section 3 analyses the NBTI degradation of the transmission gate and inverter. The effect due to the NBTI to the setup time and propagation delay of D flip-flop are described in section 4, and finally section 5 gives a summary.

2 NBTI PHYSICS AND MODEL

The primary impact of an NBTI at the device level can be interpreted to increase a transistor threshold voltage, which directly leads to the decrease of the drain current, resulting in an increase in the delay of digital gate, and further results in speed loss of logic circuits. In order to estimate the NBTI effect, the threshold voltage shifting of the PMOS under different stress and recovery conditions have been intensively studied [2], [3]. Based on those analyses, one of the possible physical explanations is the reaction-diffusion (R-D) model which is popularly used in many circuit analyses [4]. In this model, the NBTI is described as the accumulation of the Si-/Oxide surface charges. During the reaction phase, some holes in the inversion layer interact with the Si-H bonds in the interface. These Si-H bonds can be easily broken by hole capturing under the vertical electrical stress [2]. Consequently, interface charges are generated. Moreover, the dissociative hydrogen atoms (H^+) diffuse away from the interface and into the gate oxide, which leads to the increase of threshold voltage. On the contrary, when a PMOS device is switched to the recovery phase, some hydrogen atoms will be brought back and the interface states decreases as well.

The NBTI has a different threshold voltage shift (ΔV_{th}) under different conditions in stress and recovery states. To predict the long term threshold voltage degradation of the two states, a closed form for the ΔV_{th} has been proposed in [4].

$$\Delta V_{th} = \left(\frac{\sqrt{K_v^2 \alpha T_{clk}}}{1-\beta_m^{1/2n}} \right)^{2n}$$

$$\beta_m = 1 - \frac{2\xi_1 t_e + \sqrt{\xi_2 C(1-\alpha)T_{clk}}}{2t_{ox} + \sqrt{Ct}}$$

$$K_v = \left(\frac{qt_{ox}}{\varepsilon_{ox}}\right)^3 K_1^2 C_{ox}(V_{gs} - V_{th})\sqrt{C}\exp\left(\frac{2E_{ox}}{E_{01}}\right)$$

$$\varepsilon_{ox} = 3.453133 \times 10^{-11}, E_{ox} = (V_{gs} - V_{th})/t_{ox}$$

$$C = T_0^{-1} exp(-E_a/kT)$$

$$E_a(eV) = 0.49, K_1(C^{-0.5} nm^{-2.5}) = 3 \times 10^4$$

$$\xi_1 = 0.9, \xi_2 = 0.5,$$

$$T_o(s/nm^2) = 10^{-8}, E_{01}(V/nm) = 0.335$$

(1)

where t_{ox} is the oxide thickness, T_{clk} is the time period of one stress-recovery cycle, α is the duty factor (the ratio of time spent in stress to time period). $n = 1/6$ for H_2 diffusion model, while $n = 1/4$ is for H-based model. t_e equals either t_{ox} or the diffusion distance of hydrogen in the initial stage of recovery. From Equation (1), it can be seen that V_{th} strongly depends on the duty cycle. The R-D based long-term model in predicting aging of a PMOS transistor in static and a dynamic operation is shown in Figure 1. When the time is close to 0, ΔV_{th} changes rapidly with a slight change in α and the aging sensitivity to α becomes less at higher α values.

The above analysis is critical in predicting the total V_{th} shift of the CMOS flip-flops since they are widely used in the building systems such as portable systems, personal computers and etc. The rapid scaling of CMOS technology has result in new reliability concerns, and the NBTI-induced degradation of flip-flops has gradually become the dominant factor [2] for affecting circuit speed. In order to ensure reliable system operation, it is essential to develop design methods to understand, simulate and minimize the degradation of flip-flops performance in the presence of the

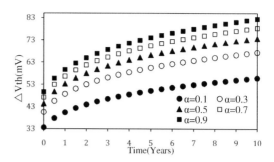

Figure 1. Results of (1): dependence of NBTI on the duty cycle. The model parameters are obtained from Equation (1), $t_{ox} = 1.85$ nm, $T_{clk} = 0.01$, $T = 378K$, $V_{th} = 0.29V$, $V_{gs} = 1.1V$, $n = 1/6$.

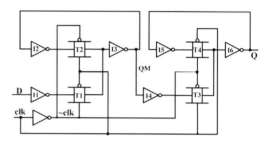

Figure 2. Circuit diagram of conventional D flip-flop.

NBTI. Along with addressing the challenge stated above, this work focuses on aged timing analysis of the conventional Master-Slave constructed D flip-flop. Which is widely used as a digital standard cell [6]. This framework is compatible with other new flip-flop implementations.

3 DEGRADATION ANALYSIS OF INVERTER AND TRANSMISSION GATE

In the following paragraphs, the TG-based D flip-flop will be discussed. Referring to Fig. 2, it illustrates the circuit of structure of a conventional rising edge D flip-flop. The operation of the flip-flop is described as follows: when the clock control signal CLK is logical "low", the input D is inversed by inverter I_1 and latched in master latch T_1 but cannot be transmitted to slave latch T_3; when the clock control signal CLK is changed from logical "low" to logical "high", the input D cannot be latched in T_1 and the data previously latched in master latch is finally transmitted to output Q of the flip-flop. Therefore, the conventional rising D flip-flop is capable of performing one bit access during a clock cycle [7].

As shown in Fig. 2, the D flip-flop consists of an inverter (NOT) and a transmission gate (TG), therefore, the NBTI-induced degradation in the NOT and TG needs to be considered first while performing aged timing analysis of D flip-flop. A detailed description of the NOT and TG degradation are demonstrated in this

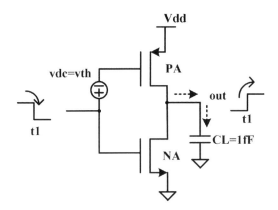

Figure 3. NBTI model of inverter.

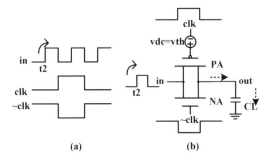

Figure 4. NBTI model of TG.

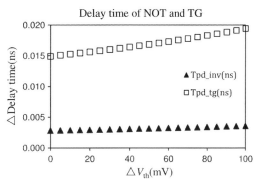

Figure 5. Intrinsic delay time extraction vs. ΔV_{th}.

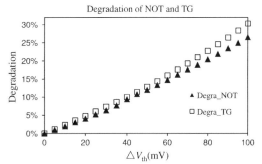

Figure 6. Delay model extraction for different ΔV_{th}.

section using SPICE simulation. A 40-nm technology model is used in the simulation.

Given a set of operational conditions, the (V_{dd}, environmental temperature T, duty cycle α and the input signal period t), ΔV_{th} can be calculated using Equation (1). With the known degradation, the propagation gate delay (t_p) in the condition of three parameters, including an input signal slew rate (t_i), output capacitance (C_L) and power supply voltage (V_{dd}). The nominal intrinsic delay for each gate is obtained from the simulation at $C_L = 1$ fF, $V_{dd} = 1.1$ V, $t_i = 10$ ps, $V_{th} = 0$ V [2]. According to these discrete values of t_p, the relationship between t_p and ΔV_{th} can be obtained, and further we can build up the connection between t_p and working time (years).

3.1 Inverter degradation model under NBTI effect

In this section, we provide detailed description on inverter degradation under the NBTI effect. Fig. 3 presents the experimental inverter aging model circuit. In this model, we select ΔV_{th} on the interval [0, 100 mV] since ΔV_{th} will reach the maximum value 90 mV at $\alpha = 1.0$ as shown in Fig. 1. The results are shown in Fig. 5 and Fig. 6.

3.2 TG degradation model under NBTI effect

Similarly, the TG experimental conditions are the same. Fig. 4 is the schematic of the TG NBTI-induced model, along with working conditions. NBTI-induced delay shifts t and the degradation of NOT and TG are respectively presented in Fig. 5 and Fig. 6.

From Fig. 5 and Fig. 6, we can see that the delay time of TG is much longer than that of the inverter, while the degradation is almost the same. Fig. 6 shows the ΔV_{th} degradation caused by NBTI will increase 30% over ten years stress.

4 D FLIP-FLOP NBTI DEGRADATION MODEL

The device and gate level aging models presented in the previous sections are crucial for the D flip-flop aging analysis. For the given set of operation conditions, the NBTI-induced delay in D flip-flop is condensed into setup time (T_{setup}) and propagation time (T_{c-q}) while hold time (T_{hold}) is zero [8]. The minimum period of the clock signal required can be defined as follows [9],

$$T_{clock} \geq T_{setup} + T_{c-q} + T_{plogic} + T_{skew} \quad (2)$$

where T_{plogic} and T_{skew} is the delay of combination logic and clock skew, respectively. In order to guarantee the circuit speed and lifetime, it is essential to analyse how T_{setup} and T_{c-q} work in D flip-flop aging circuit performance. In the following paragraphs, we analyse NBTI-induced delay shifts in T_{setup} and T_{c-q} under different operation conditions over operation time.

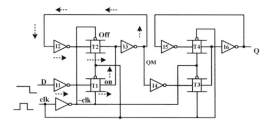

Figure 7. T_{setup} path delay shifts in D flip-flop.

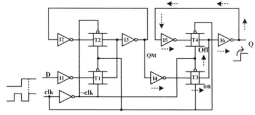

Figure 9. T_{c-q} path delay shifts in D flip-flop.

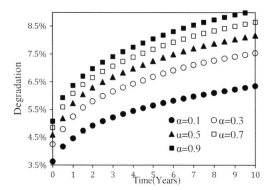

Figure 8. Delay degradation over time for various duty values of T_{setup}.

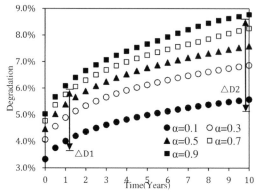

Figure 10. Delay degradation over time for various duty cycle values of T_{c-q}.

The increased threshold voltage caused by the NBTI effect strongly relies on the duty cycle and temperature [2]. In the D flip-flop, the TG and the NOT are respectively controlled by different signals. The TG is relating to the clock signal CLK where α is 0.5 while the NOT lies in the input signal D where α ranges from 0 to 1. Thus, the increase threshold voltage is different. To evaluate the timing degradation due to an NBTI, two parameters are required: 1) ΔV_{th1}, threshold voltage shift in TG at $\alpha = 0.5$; 2) Similarly, ΔV_{th2} is the threshold voltage shift in NOT at particular value of duty cycle which varies from 0 to 1. In order to analyse the impact of different α on the D flip-flop performance, five typical values: 0.1, 0.3, 0.5, 0.7, 0.9 are chosen in the experiment. We assume that the two variables are independent.

T_{setup} is the setup time of capturing data before the clock is ready. Figure 7 presents the experimental setup time circuit. The NBTI aging aware library is used to calculate the delay shift. Before the clock rising edge is ready, data D passes through I_1, T_1, I_3, I_2, thus,

$$T_{setup} = 3 \times T_{pd\_inv} + T_{pd\_ta} \quad (3)$$

where T_{pd_inv} is an inverter propagation time, T_{pd_ta} is a TG propagation time. Fig. 8 presents how the delay degradation of T_{setup} changes with t at different duty cycle values. T_{c-q} is the propagation delay through the combinational block. Fig. 9 illustrates the T_{c-q} path delay shifts in D flip-flop. When the clock rising edge is ready, Q_M has become $\sim Q_M$, now it includes the time of T_3, I_6, then

$$T_{c-q} = T_{pd\_inv} + T_{pd\_ta} \quad (4)$$

Figure 10 shows the relationship between the T_{c-q} delay degradation and the time caused by NBTI for different duty cycle values.

From these figures, we conclude that the following important observations for T_{setup} and T_{c-q} operation:

1) A different α value can result in very different timing degradation. For example in Figure 10, after one year stress, the delay degradation of T_{c-q} at $\alpha = 1.0$ is nearly 1.5× larger than that $\alpha = 0.1$. In addition, the gap in delay degradation increases with time. i.e., ΔD_2 is larger than D_1. Similarly, it is the same as the T_{setup}.
2) T_{setup} and T_{c-q} respectively has different contributions on the D flip-flop degradation at different α over time as presented in Fig. 8 and Fig. 10. When α is close to 0, NBTI-induced degradation is larger in T_{setup}, the D flip-flop delay mostly depends on it, while α is close to 1, T_{c-q} dominates the performance degradation.
3) The NBTI-induced degradation in TG and NOT nearly reaches 30% over ten years stress. However, the aging-induced degradation in D flip-flop is only up 10% over time stress. This shows that the NBTI induced degradation in combinational logic is considerably lower than that of a single transistor [5].

5 CONCLUSION

A simple framework to integrate the NBTI effect into D flip-flop *based on master-slave structure* is discussed in this paper. A simple analytical model to predict the NBTI-induced delay shift in propagation time and setup time under different operation conditions over operation time is analyzed. The analysis of the aging-induced delay, obtained through a Spice-based exploration framework, shows that, NBTI increases the D flip-flop delay shift which depends on the duty cycle and other working conditions. A different α value can result in very different timing degradation, and the gap in delay degradation increases with time. Moreover T_{setup} and T_{c-q} respectively has different contributions on the D flip-flop degradation at different α over time. The time delay caused by TG is much longer than NOT, therefore, TG-induced delay dominates in D flip-flop aging shift. Overall, this analysis provides a solid basis for further design exploration to improve logic gate reliability under NBTI effect.

ACKNOWLEDGMENTS

In this paper, the research was sponsored by the National Youth Science Fund (Grant 6120438) and the Youth Innovation Fund of East China Normal University (Project No. 78210090).

REFERENCES

[1] Andrea Calimera, Enrico Macii and Massimo Poncino. 2012. *Design Techniques for NBTI-Tolerant Power-Gating Architectures* in Transactions on circuits and systems. vol. 59, No. 4, pp. 249–253.

[2] Wenping Wang, Shengqi Yang, Sarvesh Bhardwaj, Sarma Vrudhula Frank Liu and Yu Cao. 2010. *The Impact of NBTI Effect on Combinational Circuit: Modeling, Simulation, and Analysis* in Transactions on Very Large Scale Integration (VLSI) Systems, vol. 18, pp. 173–183.

[3] Jyothi Bhaskarr Velamala, Ketul B. Sutaria, Venkatesa S. Ravi, and Yu Cao 2013. *Failure Analysis of Asymmetric Aging under NBTI* in Transactions on Device and Materials Reliability. Vol. 13, No. 2 pp. 340–350.

[4] X.M. Chen, Yu Wang, Huazhong Yang, Yu Cao, Yuan Xie. 2013. *Assessment of Circuit Optimization Techniques Under NBTI* in Design and Test. pp. 40–49.

[5] Sang Phill Park, Kunhyuk Kang** and Kaushik Roy. 2009. *Reliability Implications of Bias Temperature Instability in Digital ICs* in Digital Object Identifier 10.1109/ MDT.

[6] Hongli Gao Fei Qiao Dingli Wei Huazhong Yang. 2006. *A Novel Low-Power and High-Speed Master-Slave D Flip-Flop* in NSFC.

[7] Yu Chien-Cheng. 2007. *Design of Low-Power Double Edge-Triggered Flip-Flop Circuit* in National Science Council, Republic of China. pp. 2054–2058.

[8] J. M. Rabaey, Anantha Chandrakasan, Borivoje Nikolic. *Digital Integrated Circuits: A Design Perspective*.

[9] Ashutosh Chakraborty and David Z. Pan. 2013. *Skew Management of NBTI Impacted Gated Clock Trees* in transactions on computer-aided design of integrated circuits and systems, vol. 32, No. 6, pp. 918–927.

The electrical property of a three dimensional graphene composite for sensor applications

Min Sik Nam, Iman Shakery, Jae-hoon Ji & Chan Jun Seong
Department of Mechanical Engineering, Yonsei University, Republic of Korea

ABSTRACT: The gas sensing properties of graphene composite are already reported. Graphene has unique electrical and chemical properties. The transition metal oxides, such as ZnO, MoS$_2$ are a widely used material for gas sensing because of their optical reactivity. In this study, we fabricated a three dimensional graphene and its very high specific surface areas are used for sensor and sensing material platforms. Furthermore, the transition metal oxide improves the reactivity of a gas with an active carbon structure.

1 INTRODUCTION

1.1 A three dimensional graphene

Graphene is one of the most attractive materials due to it having a unique thermal and chemical stability, an extraordinarily high electrical conductivity, a large specific surface area, and a great mechanical strength. However, a strong π–π interaction between graphene sheets and defects in oxygen-containing chemical groups leads to a decrease of electrical properties. These shortcomings limit the performance of a graphene device. However, if we use a CVD-grown 3D graphene, it can provide a highly conductive network due to the large surface area of porous structures and the high conductivity of a defect-free graphene.

1.2 Transition metal oxide

Transition metal oxides, such as MnO$_2$, Co$_3$O$_4$, ZnO, and MoS$_2$ are widely used for gas sensing and photocatalyst materials because of their high power density and optical reactivity. However, these materials also have some demerits such as low conductivity and poor mechanical stability In this study, we use a zinc oxide nanorod, which has a large band gap and chemical stability, so it is appropriate for use in gas sensing.

2 EXPERIMENTAL DETAILS

2.1 Preparing the three dimensional graphene

A 3D graphene was synthesized by Chemical Vapour Deposition (CVD) using nickel foams for the substrate. The nickel foam, cleaned by nitric acid, was put in a furnace and heated to 1000°C for 60 min using 200 sccm of hydrogen gas and 500 sccm of argon gas. After 60 min, flowing methane gas under 30 sccm, and maintaining it for at least 25 min. In order to obtain nickel in 3D graphene, we performed ordinary etching process of graphene by using a PMMA as support material. After etching the nickel foam, the PMMA was removed by acetone vapour.

2.2 Synthesis of zinc nanorod

A zinc oxide nanorod was arrayed on the 3D graphene surface by the hydrothermal method using 0.02M of zinc nitrate hexahydrate (Zn(NO$_3$)$_2$.6H$_2$O) and 0.02M of hexamethylenetramine (HMT) mixture solution for 4h, which was then twice washed by deionized water.

2.3 Gas sensing

In order to measure the electrical property of a hybrid structure, we checked the I-V characteristics and compared them with each material before the synthesis progress. Furthermore, the same experiments are used in NH$_3$ and NO$_x$ gas condition to find the reactivity of the comfound. We used an in house gas sensor assembly with Mass flow controllers to control up to $\sim$100 ppm of gas flow; all of the study's measurements are at room temperature.

3 RESULT

3.1 The material synthesis

We can confirm the sensor materials by using a scanning electron microscope (SEM) and Raman spectroscopy. Figure 1(a) and (c) show large surface areas, which means that a sensing device can have high electrical sensitivity. The average porous size of the three dimensional graphene is $\sim$100 μm and its holes are coated by the ZnO nanorod whose size is 2 μm long.

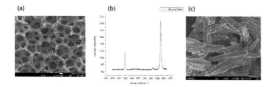

Figure 1. (a) SEM images of the CVD growth 3D graphene, (b) Raman spectra of a 3D graphene, (c) the SEM image of a ZnO nanorod.

Figure 2. A zinc oxide growth on the graphene at (a) 100°C (b) 105°C.

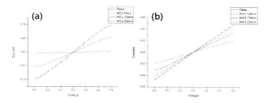

Figure 3. Response of the ZnO-3D graphene according to time (a) with NO_x (100 ppm) and (b) with NH_3 (100 ppm).

The zinc oxide nanorod growth is optimized at 105°C in order to make a large surface area.

3.2 Electrical properties

Figure 3 shows the I-V conductivity of the ZnO graphene composition at room temperature. We can show the increase and decrease reaction to NH_3 and NO_x gas after a specific time interval.

Figure 4 shows the comparison of the NO_x gas reactivity of a graphene and the ZnO deposited graphene. There is clearly a high reactivity from the ZnO-graphene, but also we can also demonstrate a poor recovery property. It is because NO_x gas is well attached and reacts to zinc oxide.

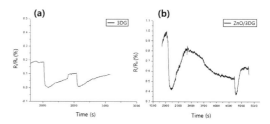

Figure 4. Response of pure 3D graphene and ZnO-3D graphene to NO_x (a) Pure 3D graphene, (b) ZnO-3D graphene composite.

4 CONCLUSIONS

In conclusion, the ZnO nanorods have been successfully synthesized at low temperature on a 3D graphene using the hydrothermal method which is a single step that is simple, easy, and efficient for large scale surfaces. In this study we investigated the possibility of ZnO-3D graphene to be used for sensing an application using NO_x and NH_3 gas responses. As a result we can show that the ZnO-3DG structure has more improved sensitivities than the active carbon structure, but it also has some limitation about recovery. The reason for this could be that that the gases are strongly attached to ZnO nanorods. For future works we will study the high performance sensing device with active carbon – oxide metal composites.

REFERENCES

[1] Gaurav Singh, Anshul Choudhary, D. Haranath, Amish G. Joshi Nahar Singh, Sukhvir Singh, Renu Pasricha*," ZnO decorated luminescent graphene as a potential gas sensor at room temperature" CARBON 50 (2012) 385–394.

[2] R. Arsat*, M. Breedon, M. Shafiei, P.G. Spizziri, S. Gilje, R.B. Kaner, K. Kalantar-zadeh, W. Wlodarski, "Graphene-like nano-sheets for surface acoustic wave gas sensor applications", Chemical Physics Letters 467 (2009) 344–347.

[3] Jiaqiang Xu a, Qingyi Pan, Yu'an Shun, Zhizhuang Tian, "Grain size control and gas sensing properties of ZnO gas sensor", Sensors and Actuators B 66 (2000) 277–279.

A method of automatically generating power flow data files of BPA software for a transmission expansion planning project

Baorong Zhou
Electric Power Research Institute, China Southern Power Grid, Guangzhou, China

Tao Wang, Lin Guan, Qi Zhao, Yaotang Lv & Lanfen Cheng
School of Electric Power, South China University of Technology, Guangzhou, China

ABSTRACT: Transmission expansion planners are completely dependent on a manual modification approach in order to generate different power flow calculation files which are a significant workload and a high error rate job. This paper proposes a method of automatically generating a BPA power flow data file for a transmission expansion planning project, based on the maximum common graph algorithm. The algorithm could identify the BPA data card which is corresponding to the transmission expansion planning projects and aggregate those BPA data card to generating line planning project groups and substation planning project groups. Furthermore, the power system planners could choose one or two groups and put them into the original network BPA data file in order to form a new planning network BPA data file. The whole process does not need manual intervention. Thus, it improves the automatic level of power system planning. Finally, this paper chooses a two years planning grid of the China Southern Grid to illustrate the validity of the method proposed.

1 INTRODUCTION

During the process of transmission expansion planning, a power flow checking of the different candidate projects proposed for a different planning period must be done. With the development of distributed generation and other fluctuated generation, the probabilistic reliability assessment is going to become an important link in the evaluation of transmission expansion planning. These calculations and evaluations depend on power flow analysis tools and models: BPA is one of the most popular tools which is widely used in China.

However, the transmission expansion planners are completely dependent on a manual modification approach used to generate different power flow calculation files, such as the dat extensions file in BPA software, which usually contains combinations of different candidate's planning projects. Because there are a large number of different years and different combination schemes, generating suitable power flow calculation files is a significant workload and a high error rate job. Actually, it is very common that a transmission expansion planning project appears in the wrong year power flow data file or it appears in some scheme that it should not be in. Different annual load levels and other operation conditions error setting and etc. are also common problems.

These problems not only hinder the improvement of a planning design level but also make it difficult for a power grid management department to evaluate and assess the necessity and reliability of the proposed transmission expansion plan. The automation and intelligent level of a transmission expansion planning assessment needs to be improved. The literature proposed a design method of urban power network planning, supporting decision systems, and improving the comprehensive decision-making level of the substation sites' capacity and layout of the network structure under uncertainty. The literature proposed a graphical-based automated supporting system for power system planning, integrating the common calculation tools, such as the flow calculation software, short circuit calculation software and a reliability calculation software for the unified platform, and solved the problem that requires switching software all the time during the process of planning. The literature applied the artificial intelligence algorithms to power system planning. All of the above methods only study different approaches used to assess the existing transmission expansion plans. However, a method of how to generate different ways of collating the power flow calculation files which contain different levels in the planning year and different combinations of proposed planning projects, has not yet been reported.

The aim of this paper is how to solve the problem of how to automatically generate power flow calculation files of BPA software used for the assessment of transmission expansion planning, in order to improve the automation level of power system planning assessment.

2 A BRIEF INTRODUCTION OF THE MAXIMUM COMMON SUB GRAPH ALGORITHM

The maximum common sub graph algorithm has been widely applied in many research fields, such as text similarity detection, three-dimensional CAD model retrieval, molecular structure similarity detection, and pattern recognition, etc. The relevant definitions are given below.

Definition 1: A connected graph

For a given undirected graph G, if the connecting paths exist between any two nodes in the graph, then the graph G is called a connected graph. Because the actual power grid is usually a connected graph, so all the graphs mentioned in the following text represent undirected connected graphs.

Definition 2: Sub graph

For the given graphs G_1 and G_2, it is defined that graph $G_1 = (V_1, B_1)$, $G_2 = (V_2, B_2)$, V_1 and V_2 indicate the vertex set of graphs G_1 and G_2, and B_1 and B_2 indicate the branch set of graphs G_1 and G_2. If $V_1 \subseteq V_2$ and $E_1 \subseteq E_2$, we define graph G_1 as the sub-graph of G_2, expressed as $G_1 \subseteq G_2$.

Definition 3: Common sub graph

For the given graphs G_0, G_1 and G_2, if $G_0 \subseteq G_1$ and $G_0 \subseteq G_2$, then G_0 is called the common sub graph. If G_0 is the common sub graph of G_1 and G_2, and there is no such a graph G_0', which is the common sub graph of G_1 and G_2 and its node number is bigger than G_0, then G_0 is called the maximum common sub graph of G_1 and G_2.

Because the search for the maximum common sub graph of two graphs belongs to a Non-deterministic Polynomial (NP) complete problem, the algorithm complexity is very high. With the increase in the number of nodes in the graph, the amount of calculation will show an explosive growth and the curse of dimensionality problem. At present, the main algorithms are: 1) the backtracking search strategy based on the depth-first, such as McGregor algorithm; 2) the detection based on maximum clique, a typical representative is the Durand-Pasari algorithm; 3) the neural network method. The algorithm used in this paper is a backtracking search strategy based on the depth-first, which the next section will describe in detail.

3 A PLANNING PROJECT DATA CARD RECOGNITION ALGORITHM BASED ON THE MAXIMUM COMMON SUB GRAPH

The premise of the planning project data card recognition algorithm based on the maximum common sub graph is that converting a card in BPA into a form which is convenient to generate the maximum common sub graph. Therefore, the first steps are: *a* pre-treatment of the data card in BPA power flow calculation files (dat extensions); removing the noted card and converting a valid card into the form of graph structure.

The node card in BPA (such as card type B, card type X, etc.) corresponds to the vertices in the graph; the branch card including a line (card type L) or a transformer (card type T), etc. corresponds to the edges in the graph. Therefore, the maximum common sub graph solves the process as follows:

Step 1: Convert the data of the two BPA files into the corresponding graphs G_1 and G_2; vertex sets are V_1, V_2, and edges sets are B_1 and B_2. For any one vertex in a graph G_1 $v_i \in V_1$, $i = 1, 2, \ldots, n$; if $v_i \in V_2$, then it is added into the vertex set V_c of a common sub graph G_c. After scanning all of the vertices in V_1, the vertex set V_c is the vertex set V_{mc} of the maximum common sub graph G_{mc}.

Step 2: Take any two vertices in V_{mc} to form the edge $b_0 = (v_1, v_2)$. If $b_0 \in B_1$ and $b_0 \in B_2$, then add $b_0 = (v_1, v_2)$ to the edge set B_c of the common sub graph G_c. If $b_0 \in B_1$ and $b_0 \notin B_2$, then add it to the edge set B_f, which indicates that the branch is deleted due to reforming; if $b_0 \notin B_1$ and $b_0 \in B_2$, then add it to the edge set B_p, which indicates that the branch is a planned added branch. After scanning all of the nodes in the vertex set V_{mc}, then the edge set B_c is the vertex set V_{mc} of the maximum common sub graph G_{mc}, which indicates that these branches did not change during the planning year. At this moment, the maximum common sub graph G_{mc} has been obtained.

Step 3: For each vertex in the graph G_1 ($v_i \in V_1$, $i = 1, 2, \ldots, n$), if $v_i \notin V_{mc}$, then add to a non-common vertex set V_f of the graph G_1, which represents deleted vertex cards in the planning year. For each node in the graph G_2 ($v_i \in V_2$, $i = 1, 2, \ldots, n$), if $v_i \notin V_{mc}$, then add to non-common vertex set V_p of the graph G_2, which represents new added vertex cards in the planning year.

Step 4: Find out nodes and branch cards corresponding to B_f, B_p, V_f, V_p, synthesizes them into a file and mark which one belongs to the new card or the deleted card.

4 AN AUTOMATICALLY GENERATING METHOD OF THE POWER FLOW CALCULATION FILES IN A POWER SYSTEM PLANNING PROJECT

4.1 Modules chart and introduction

Step S1: Recognition of the data card of a planning project. In terms of the given BPA power flow data files of two years, identify the planning projects and their corresponding BPA data rows through using the maximum common sub graph algorithm to achieve the automatic matching for data.

Step S2: Grouping of the planning project data cards. According to the identified planning projects and data rows, judge the topological relationship between the different planning projects to generate a planning project group;

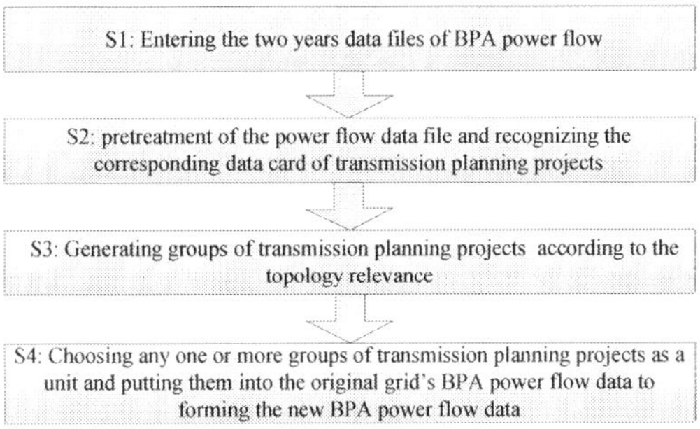

S1: Entering the two years data files of BPA power flow

S2: pretreatment of the power flow data file and recognizing the corresponding data card of transmission planning projects

S3: Generating groups of transmission planning projects according to the topology relevance

S4: Choosing any one or more groups of transmission planning projects as a unit and putting them into the original grid's BPA power flow data to forming the new BPA power flow data

Figure 1. The module diagram of automatically generated BPA power flow data file of a transmission expansion planning project.

Step S3: Selecting one or more planning project groups by experts and adding them to the BPA data file of the original grid to generate the new BPA data file which needs to be accessed.

4.2 Recognition of the planning project data card

Use the algorithm in Section 3 to obtain a different data card corresponding to the planning projects.

4.3 Grouping of the planning project data card

Group planning project data obtained in the previous section according to its topological relations. The following are the steps:

4.3.1 Grouping by project category and outputting two types including the substation planning project and the line planning project

The method divides the transmission network planning project into two types: a substation planning project and a line planning project. The substation planning project includes all substation related construction projects, such as the new construction, an extension and alteration of the substation, an extension of the main transformer capacity etc. The line planning project includes the new construction and alteration of the line.

The design ideas of a grouping method for the two types of projects are as follows:

a) The grouping method of a substation planning project.
For any vertex n_i corresponding to the BPA data card defined as datacard1 of the vertices set in V_p and V_f, find a BPA line data card or a transformer data card in B_p which has the same bus name with the n_i and put those line data or transformer data cards into a group. If those found line data or transformer data cards contain other vertex of the vertices set in V_p and V_f, put this new vertex to this group and repeat the previous step. In contrast, the searching process of vertex n_i ends. If a vertex in the vertices set in V_p and V_f has been added to a group, it should skip the search process

of this vertex. Finally, we could get the result of the substation's planning project groups: TGroup1, TGroup2, TGroup3, etc.

b) The grouping method of a line planning project.
For any branch b_i corresponding to the BPA data card is defined as datacard2 of the branch set in B_p and B_f, and if this branch does not belong to any substation planning project group, then search the other branch in the B_p and B_f to find those which do not belong to any group and have a common bus name with datacard2. Finally, we could obtain the line planning project groups: TGroup1, TGroup2, TGroup3, etc.

4.3.2 Aggregation of the associated substation and line project group

a) For a different line planning project group. If these are a common bus name between the different line planning project groups, aggregate them into one group.
b) For a substation planning project group. If there are two with same cards between different substation planning project groups, aggregate them into one group.

4.4 Generating new flow files of an assessment grid

The purpose of this step is to obtain the BPA power flower data files for reliability assessments.

a) The transmission expansion planner could choose one or two planning project groups in a planning year through the menu mode.
b) Modifying load and generator output parameters. Extracting load and generator output data from all of the B cards and BQ cards from the BPA data file of a planned terminal year and compare them with the beginning of the planned year. After the operating personnel set an annual load level, the annual load and generator output data of each node could be obtained through linear interpolating. In addition, it is allowed that a transmission expansion planner can make some adjustments.
c) Generate the BPA power flow calculation data files which are corresponding to the planning project group.
First, for each planned project group, add all the new projects' data from the project group to the original network data. Then, find the entire project data needed to be deleted and removed from the original network. Finally, on the basis of the load and generator output level from the previous step, a new mode of operation could be set and a new planning project power flow test file could be obtained.

5 CASE STUDIES

The two BPA power flow calculation data files in 2011 and 2012 (dat extensions) are used in this case. As shown in Figure 2, the process methods based on the

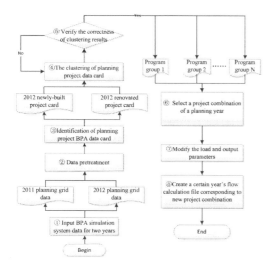

Figure 2. The flow chart of a case study.

Table 1. The 500kV grid planning project statistics of the year 2012.

Transmission Planning Project Group	Project Group Name
NB-21	HUADU0H
NB-27	MEIZH0H Transformer expansion
NB-34	SIHUI0H Power transmission and substation project
NB-36	TONGH0H Power transmission and substation project
NB-44	ZHANJ50H Transformer expansion

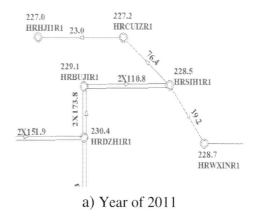

a) Year of 2011

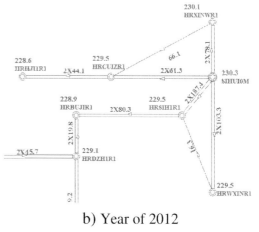

b) Year of 2012

Figure 3. The comparison of the BPA geography diagrams.

planned grid data in 2011 and 2012 can be divided into the following three stages:

The authors developed the software to achieve this algorithm using the Python programming language. In this case, the output files contain:

a) One output file consists of 54 project groups BPA data files, including 44 substations planning project groups, expressed as NB-1, NB-2, ..., NB-44; 10 line planning project groups, expressed as CL-1, CL-2, ..., CL-10;
b) 54 BPA data files which add each project group into the original network BPA data file to form the grids after construction of NB-1~NB-44 and the grids after construction of CL-1~CL-10

The analysis of the grouping result is shown as follows:

a) Examining the grouping count
The number of output project groups corresponding to the 500kV newly-built substation is 5, in accord with the statistics of 500kv planning project in the "Twelfth Five-Year Plan" report, demonstrating the correctness of the grouping method. The result is shown in Table 1.

b) Examining the network data
Run the output BPA power flow files to test the grouping rules and the correctness of the grid data. In addition to it, adjust the network data when it does not converge in the flow calculation. It is found that 50 project groups' results of power flow will converge and 4 project groups' results will not converge, the result can converge by adjusting the reactive power compensation.

c) Examining the grouping rule
Take an NB-34 power transmission and substation planning project group as an example for analysis. According to topological association between various data cards, after grouping the planning project data card, you can get the NB-34 power transmission and the substation project BPA data card.

Thus, it can be seen that the power transmission and substation planning project 500kV and 220kV supporting projects are all included in the same project group. The NB-34 includes alternated 500kV lines double circuit lines (ZHAOQ0H ~ HUADU0H double loops); four circuit of new 500kv lines (ZHAOQ0H ~ SIHUI0H double loops, and

ZHAOQ0H ~ HUADU0H double loops); nine circuit of new 220kv lines (SIHUI0M ~ HRCUIZR1 double loops, SIHUI0M ~ HRSIH1R1 double loops, SIHUI0M~ HRWXINR1 double loops, SIHUI0M ~ HRXINWR1 double loops and the HRCUIZR1 ~ HRHJI1R1 single loop), one loop of alternated 220kv line (HRCUIZR1 ~ HRSIH1R1 single loop). There are two new sets of three-winding transformers.

In addition, because of the two new substation connecting lines, the SIHUI0M ~ HRXINWR1 double loops, you can also put HRXINWR1 substation node, HRXINWR1 ~ HRCUIZR1 single loops and SIHUI0M ~ HRXINWR1 double loops into this project group in order to assure the integrity of flow calculation grid data. Experience confirms that the output of the NB-34 project group meets the requirements of identifying the planning project and its grouping rules etc.

d) The comparison of the NB-34 transmission and substation project geography wiring diagram.

The red lines represent the same project group. Among them, the chart a is for the year 2011 geographic wiring diagram, where the red lines indicate from where the project group was deleted. The chart b is for the 2012 geographic wiring diagram, in which the red lines indicate that the project group is newly-built. It can be seen from the graph, that the program grouping results meet the topological structure requirements.

6 CONCLUSIONS

This paper applies the maximum common sub-graph algorithm and uses the connection of the different planning grid data files sufficiently to achieve a group of planning projects. Transmission expansion planner could choose a different combination of the planning groups according to the request of the planning year, and the program developed by authors could automatically generate the BPA power flow data file. Meanwhile, it could avoid frequently generating the BPA power flow data files in different years.

The algorithm proposed in this paper could not only be applied in choosing the best planning scheme in many other planning schemes, but it could also be used in the fields of objective assessment of power system planning schemes for a power grid management department. In the future, it will become the basis of priority level assessment of planning programs and offer a guidance function for the decision making in planning power systems.

REFERENCES

[1] Sanfelieu and K. S. Fu. "A distance measure between attributed relational graphs for pattern recognition". IEEE transactions on systems, man, and cybernetics, Vol. SMC-13, No. 3, 1983, pp. 353–362.

[2] Bai Xiaoliang, Zhang Shusheng, Zhang Kaixing. "Algorithm for 3-Dimensional CAD model retrieval based on reuse". Journal of Xi'an Jiao Tong University, Vol. 45, No. 01, 2011, pp. 74–78.

[3] Chen Yuehui, Chen Jinfu, Duan Xianzhong. "Study on graphic and automatic system planning". Electric Power Automation Equipment, Vol. 22, No. 2, 2002, pp. 38–40.

[4] Durand P J, Pasari R, Baker J W, et al. "An efficient algorithm for similarity analysis of molecule". Internet Journal of Chemistry, Vol. 2, No. 17, 1999, pp. 1352–13589.

[5] E. Mjolsness and C. Garrett. "Algebraic transformation of objective functions". Neural Networks, Vol. 3, No. 6, 1990, pp. 651–669.

[6] J. R. Ullman. "Analgorithm for subgraph isomorphism". Journal of the ACM, Vol. 23, No. 1, 1976, pp. 31–42.

[7] J. I. McGregor. "Backtrack search algorithms and the maximal common subgraph problem". Software-Practice and Experience, Vol. 12, No. 1, 1982, pp. 23–34.

[8] Krotzky T, Fober T, Hullermeier E, et al. "Extended graph-based models for enhanced similarity search in cavbase". IEEE/ACM Transactions on Computational Biology and Bioinformatics, Vol. 11, No. 5, 2014, pp. 878–890.

[9] Levi G. "A note on the derivation of maximal common subgraphs of two directed or undirected graphs".Calcolo, Vol. 9, No. 4, 1972, pp. 341–354.

[10] Li Xiang, Cao Yuan-guo, Zhang Zhi-na, et al. "An automated planning-based conceptual design synthesis approach". Journal of Shanghai Jiao Tong University, Vol. 48, No. 8, 2014, pp. 1134–1141.

[11] Liu Y L, Yu Y X. "Probabilistic steady-state and dynamic security assessment of power transmission system". SCIENCE CHINA Technological Sciences, Vol. 56, No. 5, 2013, pp. 1198–1207.

[12] Liang Hao, Zhang Pei, Jia Hongjie, et al. "Economic assessment of future transmission line base on probabilistic power flow". Power System Technology, Vol. 38, No. 3, 2014, pp. 675–680.

[13] Shoukry A, Aboutabl M. "Neural network approach for solving the maximal common subgraph problem". IEEE transactions on systems, man, and cybernetics, Vol. 26, No. 5, 1996, pp. 785–790.

[14] Wu, Jiangning, Liu Qiaofeng. "Research on text similarity computing based on max common subgraphs". Journal of the China Society for Scientific and Technical Information, Vol. 29, No. 5, 2010, pp. 785–791.

[15] Xiao Jun, Luo Fengzhang, Wang Chengshan, et al. "Design and application of a multi-attribute decision-making system for power system planning". Power System Technology, Vol. 29, No. 2, 2005, pp. 9–13.

[16] Yang Xiaobin, Zhang Yan, Zhou Zhichao, et al. "Development and application of a software for quantitative reliability assessment of medium voltage distribution network". Automation of Electric Power Systems, Vol. 28, No. 18, 2004, pp. 83–85.

[17] Zhou Kunpeng, Fang Rengcun, Yan Jiong, et al. "Design and implementation of an intelligent decision-making system for power grid planning". Automation of Electric Power Systems, Vol. 37, No. 3, 2013, pp. 77–82.

The analysis of training schemes for new staff members from substation operation and maintenance departments

Yitao Jiang & Yubao Ren
State Grid of China Technology College, Jinan, Shandong, China

Xiuhe Zhou
State Grid Fujian Maintenance Company, Fuzhou, Fujian, China

Li Mu, Yang Jiang & Haike Liu
State Grid of China Technology College, Jinan, Shandong, China

ABSTRACT: This analysis has built an all-round training system which consists of multiple parts, e.g., training demands, goals, content, implementation, and assessments, etc. on the basis of summarized practical experience and analysed training measures in detail, while systematically discussing several features of new staff members training in terms of their characteristics.

1 INTRODUCTION

New staff members training, which entails fundamental content and methods enabling new staff members to understand their job from various perspectives of, for example, responsibilities, procedures, standards, etc. and primarily implant ideas expected by the company for them to have, i.e. attitudes, regulations, values, and behaviour patterns, can help them successfully adapt to the company's environment and action patterns and pick up the role as soon as possible. This thesis makes a discussion about how to improve new staff members training on empirical grounds of past training schemes and noticeable features of training schemes for new staff members from substation operation and maintenance department.

2 CHARACTERISTICS OF TRAINING SCHEMES FOR NEW STAFF MEMBERS FROM SUBSTATION OPERATION AND MAINTENANCE DEPARTMENT

2.1 *The characteristics of new staff members*

The new staff members of the company, most born in 1980s or 1990s and mostly graduates from universities or colleges, possess different qualities and characteristics compared with their senior colleagues. When communicating with them, the company should implement efficient measures for the purpose of embracing them into the company as soon as possible.

Currently, their characteristics are reflected by multi-facets as follows:

(1) A mix with positive and negative working attitudes. New staff members, born in the 1980s or 1990s are, on one hand, more devoted to creative activities, while showing a significant lack of loyalty to the company and satisfaction with the work.
(2) Capricious career values. Traditional career values emphasise the loyalty of staff to the company, while the company provides protection for the staff. Existing new staff members possess capricious career values with a desire to endeavour in different areas and focus on whether the company is capable of nurturing "transferable" competition abilities.
(3) A special definition of success. They long for many, including personal achievements, recognition by the public, better development opportunities, higher salary, flexible work hours, and lifelong learning opportunities. Compared to seniors in terms of demands, new staff members are similar, although they express their thoughts and demands to the company more explicitly.
(4) A personal view in relation to authority. They will not blindly respect their supervisors or seniors because of their positions, but rather, occasionally a contempt of authority is expressed. More respect is needed from them for the manners and leadership skills that their supervisors possess.
(5) Not attracted by daily work routine. A dislike for repeated work is commonly seen, while preferring challenging and interesting jobs.

(6) Keep to their own lifestyles instead of working. In the opinion of most of them, work is the foundation of life, which shall not be sacrificed by laborious work, rather than for its purpose.

2.2 Features of the training model

Under the existing training model of the "SANJI-WUDA (intensive management in human resources, financial resources and material resources, large-scale movements in programming, construction, operation, overhaul and marketing.)", for better realisation of cost-efficiency and of the positive efficacy of a human resources department, concentrated training schemes have been implemented, i.e. all new staff members are concentrated at the State Grid of China Technology College. Several attributions of considerable practical difficulties with relation to concentrated training scheme, when compared to its predecessors, are as follows:

(1) Provincial origins of new staff members are of considerable diversity, which include different provinces with different local subcultures and even cronyism in some circumstances, increasing the difficulties in communication;
(2) Training period is quite long, generally 3–5 months. Some new staff members will be slack and have their learning efficiency decrease;
(3) Backgrounds and professional capabilities with regard to each one are various when compared one to another, which poses a certain number of difficulties to the training scheme.
(4) There are some differences in understanding the extent of each supervisory department concentrated training for new employees, training of students learning performance requirements are not the same, affect the students learning in the life of the psychology and behaviour in the training performance.
(5) In their own right, the new staff members have different understanding of the importance of training and variant personal training goals.

2.3 The features of trainers

The State Grid of China Technology College has hired both contract and temporary trainers. The contracted lecturers have academic experience while lacking practical working experience. The temporary experts are from the production frontline of each department of the company with strong practical skills and abundant work experience; their academic experience and training skills, however, are limited in most circumstances.

2.4 The features of substation operation and maintenance department

The features of Substation Operation and Maintenance Department are tri-faceted. First, the department training curriculum includes safety regulations, operation regulations and substation management rules, the primary transformation equipment knowledge, the components protection and secondary transformation, the substation inspection and maintenance, the back brake operation, substation abnormalities and accidents, etc. which require full mastery of the electrical primary and secondary equipment knowledge and practical analysis. Second, repeated practice is required as a result of the high standard of operation safety, a strict compliance to substation rules and regulations demands, and operations in accordance with operation tickets and the requested proceedings. Third, wide-ranging abilities are required and judgment is expected for personnel on duty in the face of emergencies and accidents on the grounds of comprehensive analysis of a variety of signals.

3 PRACTICE OF SUBSTATION OPERATION AND MAINTENANCE DEPARTMENT TRAINING

In order to guarantee training quality, the substation operation and maintenance training research team has, in response to the above-mentioned characteristics, carefully arranged the training programmes in various aspects as follows:

3.1 A full mastery of training needs

Training needs analysis is the most important experience for new staff as well as the foundation for the guarantee of achieving good training effects. In order to meet the training needs of units who send new employees to The State Grid Technology College, the department makes practical and feasible training plans, ensures the effectiveness and rationality of training by means of sending training needs survey, exchanging with professionals face to face, receiving the feedback of graduates and so on.

3.2 Clear definition of training goals

The new staff training goals can be achieved by three measures. First of all, by establishing a sense of belonging for new staff members so that they can understand the company corporate culture, business process and conditions of production and management, while adapting to the new work environment, be consonant with the values of the company and the spirit of enterprise, accept the company's management idea and acting standards, and establishing a firm belief in order to make contribution to the company. Second, by clarifying the company's expectations and job requirements of each of them, learning relevant rules, regulations and laws, mastering the knowledge and skills the posts are in need of, and cultivating good working practices and capacities. Third, by helping new staff members to master communication skills with their seniors and the team, the training will build a harmonious work relationship and help manage the work faster.

3.3 Design of training contents

According to the new staff members training targets, training contents have been set in modules, including corporate culture, the company's main business awareness, professional skill modules, etc. in order to resolve new staff lack of knowledge, skills and attitudes.

The new employees have different knowledge bases, so on one hand strengthening the study of basic knowledge to maintaining their confidence of those whose knowledge base is weak. On the other hand, several extracurricular learning projects such as technology project and PPT making are designed to make some excellent new employees to have greater harvest.

In view of the major changes are, more content, professional and strong features, two aspects are mainly taken into consideration to design the training content: one is to grasp the direction and the importance of the company and the professional development direction in training, focusing on the key technology breakthrough in substation operations and maintenance of professional skills, clear the direction of learning for the new staff members. The second one is to focus on demand and course selection according to the field work, selection of core knowledge points and key skills, so that new staff members can enter into the practical work as soon as possible after graduation.

3.4 Implementation of training plans

The new staff members knowledge and skill training take "action" teaching mode, the implementation of the integrated teaching of situational teaching and field practice, seminars and discussions, for the "China State Grid Corp production skill personnel vocational ability training standards" level I ability requirements as the main content. They pay attention to training and simulation environment by using the established high fidelity production training, which by driving the specific implementation task or project, arouse the study interest and enthusiasm for new staff members.

In addition, in the process of carrying on training program, strengthening trainers' words and deeds guide and establishing good examples for new employees can help them to learn more knowledge, skills and necessary quality of State Grid employees, which can help them to establish the correct career outlook.

3.5 Assessment of training effects

According to the characteristics of new staff members training, training effects are assessed by using three level evaluation methods, eg response evaluation, and assessment of learning and behaviour evaluation.

(1) A response assessment: assessment of a response to the training is a recognized degree evaluation. Using a regular two-ways communication between the trainers and trainees will help to understand the comprehensive evaluation of new staff members' understanding of the content of training and training mode, so that the trainer can understand the training of the new staff's demands better.

(2) An assessment of learning: a learning evaluation is an important part of the training results, mostly through a combination of regular examinations and daily inspections, which carries out the appraisal of the new staff members' learning attitudes, learning achievements and learning effects, so that the training and teaching ways and methods can be further improved.

(3) A behavior assessment: It mainly means to evaluate the use of knowledge and skills learned during the training process and is carried out by means of questionnaires and face to face communication with new employees and their executive officers. The effectiveness of the training and new employee's performance in the training process are exchanged, which can help the department to find the defects of recent training and improve the quality of training.

4 CONCLUSION

In short, a substation operation and maintenance of professional in training has the following effects in full consideration of various factors and by designing the effective training solutions and management system and carefully organizing their implementation: it has made training not only a way to obtain qualification certificates, but also made the new staff members adapt to the company as soon as possible with a comprehensive improvement of the new staff members' knowledge, skills and professionalism, which finally contribute to becoming more qualified staff members and excellent assets for the development of the company.

REFERENCES

[1] Xu Yanyang. The scheme discussion on the Integration of operation and maintenance of substation [J]. China Electric Power Education, 2012, (33).
[2] Luo Yanjuan. Research on training, practice and development of the integration of operation and maintenance of substation [J]. China Electric Power Education, 2013, (20).
[3] Zhou Zhuojun, Cai Hongjuan, Wang Huifeng. Discussion on operation and maintenance work of substation [J]. Urban construction theory research. 2013, (32).
[4] Xiong Yanbin, Peng Yanchun. The analysis on problems during the implementation of the integration of operation and maintenance of substation [J]. Power Technology, 2013, (10).
[5] Wu Kexin. The discussion on problems during the implementation and strategy on the integration of operation and maintenance of substation [J]. Chinese E-Commerce, 2014, (6).

Research of source-grid-load coordinated operation and its evaluation indexes in ADN

Wei Liu, Mingxin Zhao & Hui Hui
Department of Power Distribution, China Electric Power Research Institute (CEPRI), Beijing, China

Chang Ye & Shihong Miao
School of Electrical and Electronic Engineering, Huazhong University of Science and Technology, Wuhan, China

ABSTRACT: With the large-scale access of Distributed Generations (DGs) and diversity loads, the so-called "Active Distribution Network" (ADN) is put forward. In this paper, coordinated operation mechanism in ADN is reviewed and proposed. Research approach and new technologies of source-grid-load interaction is analyzed, including its internal mechanism, realization means and fundamental principles. Implication of source-grid-load multiple target coordinated operation is put forward. A comprehensive index system is established in order to evaluate the coordinated operation level.

1 INTRODUCTION

The appearance of various distributed generations (DGs), controllable loads and new type energy storage systems has brought significant change to power system in the respect of both system structure and operation mode. Meanwhile, growing demand for electricity results in power supply shortage and environmental pollution. That leads to global energy-saving and emission-reduction. To solve these problems, the revolution that makes power system to be more efficient, flexible, intelligent and sustainable brooks no delay.

Contrast to transmission network, the large-scale access of DGs and diversity loads has greater influence on distribution networks. Although the development of micro-grid technology provides solutions for integrated control of DG accessing, capacity of micro-grid and different control target restrict its application. Aim at this problem, researchers put forward the concept called "Active Distribution Network" (ADN). ADN manages DGs in the whole distribution network level, concerning both autonomous control in partial area and optimal coordination in whole network. Source-grid-load coordinated operation mechanism research is based on ADN, studying overall interaction and coordinated balance of source, grid and load, thus it will provide holistic solution for optimal operation of ADN.

In this paper, the internal mechanism, realization means and fundamental principles of source-grid-load coordinated operation are further studied. A new comprehensive index system based on above principles is established, this comprehensive index system can evaluate the coordinated operation level effectively.

2 DEVELOPMENT OF ACTIVE DISTRIBUTION NETWORK

2.1 Features of active distribution network

As mentioned before, development of distribution network faces challenges such as energy demand, security, quality requirement and environmental protection. One solution of energy conservation and emission reduction is to make the distribution network active, enabling technologies provide solutions for the transition of distribution networks from the usual, passive fit-and-forget approach. The concept of active distribution network can be characterized as flexible, intelligent, integration and cooperation. The active network is flexible because it utilizes controllable resources throughout the network. According to CIGRE Working Group C6.11, Active Distribution Networks (ADNs) are defined as distribution networks that have systems in place to control a combination of distributed energy resources (generators, loads and storage). ADN is based on micro-grid coordination for DG control technology, focusing on the utility value of information, and using the design conception of top-down. ADN is an open architecture that can compatible with micro-grid and other new energy integrated technologies. It's the advanced stage for development of smart distribution grid.

The active control characteristics of ADN manifest mainly in the aspects such as absorption of intermittent energy, dispatching of DG, protection of DG and monitoring of DG. It needs new flexible network topology, protection, communication, and integration into existing systems. The most common applications of ADN are power flow congestion management, voltage

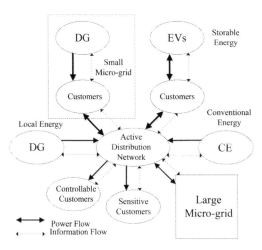

Figure 1. Structure of active distribution network.

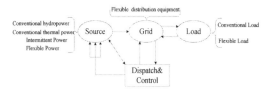

Figure 2. Model of source-grid-load coordinated operation

regulation, DG and load control, and fast reconfiguration. In ADN, the distribution system operator can interact with DGs and control the production of active and reactive power. In addition, load response programs are being stimulated as a way to provide electrical capacity during critical demand periods. ADN needs flexible and intelligent planning methodologies in order to properly exploit the integration of DG and load response, while still satisfying quality and reliability constraints. The structure of ADN is shown as Figure 1.

2.2 Research status of active distribution network

Because of the advantages that ADN can solve the problems such as grid compatibility and application of large-scale intermittent renewable energy, many countries actively develop correlation research of ADN. The most influential project is ADINE project led by "the sixth framework programme for research" (FP6) in EU, which is the EU demonstration project of active distribution network. The aim of the ADINE project is to develop new management methods for the electric distribution network including DG. The project develops and demonstrates not only protection and voltage control of distribution network including DG, but also new-generation medium voltage STATCOM. The ADINE project started in October 2007 and ended in November 2010, achieving gains including the interactions of the proposed ANM method and the functioning of enabling solutions like protection relays, fault location functions, de-centralised and centralised voltage controllers, and new generation medium-voltage STATCOM. In addition to the ADINE project, ATLANTIDE is a three years research project funded by the Italian Ministry of Economic Development under the framework of the Italian Research Fund for the Power System development. Its ambition is to realize a repository of reference models for passive and active Low Voltage (LV) and Medium Voltage (MV) distribution networks, specifically tailored to the Italian distribution system. In China, School of Electronic Information and Electrical Engineering in Shanghai Jiao Tong University has studied the optimal dispatching strategy of ADN, they choose the optimization objective of minimum operation cost in a complete dispatching period to obtain the result, considering correlation constraint condition and using the ISOP algorithm.

However, the lack of experience, the increased complexity, and the use of novel communication systems are perceived weaknesses of ADN and potential threats to the ADN development. The ADN research still rests on theoretical level, few examples of pilot installations of ADN are applied at present.

3 STUDY OF SOURCE-GRID-LOAD COORDINATED OPERATION MECHANISM

3.1 Outline of source-grid-load interaction

To obtain better economic objective and environmental objective, source, grid and load in ADN should work coordinately and interactively, which leads to the research of source-grid-load coordinated operation mechanism. In ADN, the situation in which traditional source side and grid side adjust passively according to the users' behaviors is broken. Except for demand side management, both source side and grid side are expected to participate in achieving the power balance. The model of source-grid-load coordinated operation is shown as Figure 2. Although the source-grid-load interactive operation and control method and its overall research framework are being discussed based on energy development researches, multivariate energy utilization researches and energy consumption structure, the study of source-grid-load coordinated operation mechanism is in its infancy.

Implication of source-grid-load multiple target coordinated operation in ADN can be described as "Under the aegis of corresponding monitor and control means, distribution networks take the dispatching institution for bridge, coordinate the schedulable resources in source side, grid side and load side integratively, and realize the aim that source-grid-load can work together to participate in the optimization of power system operation through various interactive means, thus to gain optimal objective of security, economy and environmental benefits, and to make distribution networks to operate safely, reliably, high-qualitatively and efficiently." Source-grid-load

Table 1. Comprehensive index system of source-grid-load coordinated operation.

Indexes of Source Side			Indexes of Grid Side		
Safety & Reliability	Reserve Capacity Coefficient of Generating Units		Safety & Reliability	Capacity-Load Ratio	
				N-1 check	
	Fluctuation of System Power			Heavy/Light Load Rate of Lines	
				Heavy/Light Load Rate of Transformers	
	Failure Rate of Generating Units			Proportion of Reactive Allocation	
				Mean Time to Failure of Power System	
Economy Benefits	Utilization Ratio of Generating Units			average failure probability of System	
	Start-up/Shut-down Costs		Economy Benefits	Line Loss Rate	
	Cost of Fuel				
	Maintenance Cost			Peak-valley Difference	
	Cost of Pollution Treatment				
Environmental Benefits	Generating Proportion of Clean Energy		Indexes of Load Side	Satisfaction	Power supply Availability / Power Supply Reliability Rate
					Satisfaction of Electricity Charge Expenditure
	Carbon Emissions				
Indexes of Interactivity Level	Proportion of Distributed Energy Resources and Energy Storage				Satisfaction of the Methods of Using Electricity
	Proportion of Incentive Electricity Consumption			Economy Benefits	Cost of the Peak-Load Regulation
	Maximum Absorptive Distributed Energy Resources				Social Benefits
	Interaction Potential				Average Interruption Hours

coordinated operation mechanism is the key technology of holistic solution for interaction including source, grid and load.

The modes of distribution coordinated operation include source-source complementarity, source-grid coordination, source-load interaction, grid-load interaction, etc. Source-source complementarity not only covers the shortage that single renewable energy is easily limited by region, environment and meteorological phenomena, but also the randomness and volatility of single source. It uses the correlation, wide-area complementary features and smoothing effect of multi-energy to increase utilizing efficiency of renewable energy and decrease the spinning reserve of system. Source-grid coordination utilizes combined application of various energy sources, improves the autonomy regulating capacity of grid operation, and reduces influence to grid secure and stable operation when renewable energies are con-nected to grid. Source-load interaction considers both flexible source and flexible load as schedulable resources. Through effective management mechanism, flexible loads like storages and electric vehicles participate in power regulation and supply-demand balance of ADN. Grid-load interaction makes resources in the load side cooperate with network regulation, strengthening the security and stability of power system by dispatching interruptible or flexible load.

3.2 *Key technologies of source-grid-load interaction*

Basic principle of source-grid-load coordinated operation is to gain better economic benefits and environmental benefits under the condition of satisfying the constraint conditions of system security. The key technologies of source-grid-load coordinated operation include modeling of complex interaction characteristic and behavior, identification of power system interaction capacity and analysis of security operation, and coordinative optimization control of source-grid-load in the environment of complex wide-area interaction. The technologies of Virtual Power Plant (VPP), power Demand Side Management (DSM) and Multi-Level Cluster Control Approach have been applied to the research of source-grid-load coordinated operation. However, to realize source-grid-load interaction comprehensively, more methods and technologies are imperative.

4 ESTABLISHMENT OF THE COMPERHENSIVE INDEX SYSTEM

To evaluate the source-grid-load coordinated operation level, a comprehensive index system is expected to be established. The basic principles include consistency, systematicness, independence, maneuverability, authenticity, and unity of qualitative analysis and quantitative analysis. The comprehensive index system should both consider technical indexes and economic indexes, as well as environmental indexes. Specific to the ADN, index system can be established in source side, in grid side and in load side. Considering the source-grid-load coordinated operation in ADN, the indexes of interactivity level are necessary.

In the source side, security of generator units, energy efficiency, cost of generator units and environmental treatment, and carbon emissions are considered. In the grid side, overloading rate of line and transformer, line loss rate, interruption duration are taken into account. While in the load side, indexes include power quality, satisfaction of using electricity, cost of peak regulation, etc. According to relevant national standards and considering the characteristics of source-grid-load coordinated operation in ADN, the comprehensive index system is shown as Table 1. The coordinated operation performance is scored by integrated weighting method, so the selecting of weighting coefficient is important. We use Delphi method to determine weighting coefficient. Detail meanings of indexes will not be discussed here.

5 CONCLUSION

As the requirement of safety, reliability and economy for power grid is increasing, ADN has become the focus of attention among the related practitioners in power system. Source-grid-load coordinated operation mechanism research in ADN is of crucial importance in adopting distributed generations and optimizing operation of distribution network. It results in greater security, higher economic benefits and better environmental benefits. This research will push further development of active distribution network technologies.

ACKNOWLEDGEMENT

This research is supported by the scientific research project of State Grid Corporation of China (SGCC) "The Research and Application on Coordinate Planning Method for Active Distribution Network with Large scale of Distributed Generation and Various types of Load" (Project No. EPRIPDKJ (2013) 4366).

REFERENCES

[1] Bracale, R. Caldon, M. Coppo, D. Dal Canto, R. Langella, etc. Active Management of Distribution Networks with the ATLANTIDE models, 8th Mediterranean Conference on Power Generation, Transmission, Distribution and Energy Conversion. 2012.

[2] F. Pilo, G. Pisano, G. G. Soma. Advanced DMS to Manage Active Distribution Networks, IEEE Bucharest Power Tech Conference. 2009.

[3] S. Chowdhury, S. P. Chowdhury, P. Crossley. *Microgrids and Active Distribution Networks*. London, U.K.: IET, 2009.

[4] S. Repo, K. Mäki, P. Järventausta, O. Samuelsson. Adine – EU Demonstration Project of Active Distribution Network, *CIRED Seminar: SmartGrids for Distribution*. 2008.

[5] V. F. Martins, C. L. T. Borges. Active Distribution Network Integrated Planning Incorporating Distributed Generation and Load Response Uncertainties, *IEEE Transactions on Power Systems*. 2011, 26(4).

[6] Y. Jianguo, Y. Shengchun, W. Ke, Y. Zhenglin, S. Xiaofang. Concept and research framwork of smart grid "Source Grid Load" intercative operation and control, *Automation of Electric Power Systems*. 2012, 36(21), 1–6.

[7] Z. Chen, X. Xiao, C. Luo. New Technologies of Active Distribution Network in Smart Grid, *IEEE Conference Publications*. 2013. 177–180.

[8] Z. Guoqing. Sketch of the future power system, *State Grid*. 2008. 44–45.

Progress on the applications of cascaded H-bridges with energy storage systems and wind power integrated into the grid

Shanying Li & Tao Wu
North China Electric Power Research Institute, Beijing, China

Yashuai Han, Wei Cao & Yonghai Xu
North China Electric Power University, Beijing, China

ABSTRACT: With the rapid development of new energy based on wind power, grids have to face a series of problems of large-scale wind power and other new energy sources being integrated into the power grid. Energy storage technology is one of the efficient methods of resolving the key problems. The development progress of different energy storage types and their application in wind power are concluded together for the first time in this paper. Power Conversion System (PCS) are an important part of energy storage systems. The topology of energy storage systems accessed to the DC sides of PCS is summarized. The topology of a multilevel cascade-type power conversion system and the control strategy of Cascaded Multilevel Converter with Energy Storage System (CMS/ESS) are concluded and analysed. According to the progress discussed above, the suggestions about the interesting research directions of the next research work, such as energy storage device selection, control strategy, and the cascaded multilevel topology are addressed. This paper will contribute to better research and application of the CMC/ESS intended for improving the operational characteristics of wind power in the future.

1 INTRODUCTION

Wind power and solar power have a natural volatility and intermittency, so there is greater uncertainty in their large-scale integration into the grid, and the response speed is slow. This means that new energy directly connected to the grid may lead to some influence on the safety and reliability of the power system (Chen 2009). In order to stabilize power fluctuations, we can allocate some energy storage devices with certain capacity close to new energy supplies or the new combination of distributed energy supplies. The devices can also play a role in load shifting, thereby providing a stable power output to the grid (Coppez 2010).

Large-capacity energy storage systems should access the grid at high voltage levels. The systems are integrated into the power grid through the Power Conversion System (PCS) in the usual practice. We can control the current amplitude and phase of the network side and the current of the DC side in changing the voltage amplitude and the phase of the AC ports of a PCS. Currently there are three basic forms of the topology of a PSC: converter modules in parallel topology, multi-level topology, and converter module cascaded in multi-level topology. The converter module cascaded multi-level topology has the following advantages: small voltage stress of switching devices, high output voltage, with low harmonics that are easy to expand, high reliability, and a high degree of modularity .The energy storage device with a cascade multi-level inverter emerged because of these advantages. This paper elaborates the energy storage system and the current research situations of the cascaded multi-level topology.

This article first analysed the role that energy storage technologies played in the current wind power system. Second, this article inductively analysed a variety of energy storage technologies and their applications in wind power system. Then we focused on the analysis of topologies, control strategies of CMC/ESS etc. Finally, we make recommendations on key issues for research into CMC/ESS.

2 ANALYSIS OF VARIOUS ENERGY STORAGE TECHNOLOGIES IN WIND POWER

Different energy storage technologies have different levels of application in wind power system because of different characteristics. Energy storage devices are divided into four categories as physical, electrochemical, electromagnetic, and phase change system (Wen-liang 2008). Physical storage refers to pumped storage, compressed air energy storage, and flywheel energy storage. Electrochemical energy storage mostly refers to the super capacitor energy storage and battery energy storage; battery energy storage contains lead-acid battery energy storage batteries, lithium ion batteries, vanadium redox flow batteries, and sodium sulphur batteries. Electromagnetic energy storage mostly refers to a superconducting magnetic energy storage (SMES) device. Phase change mostly refers to the ice storage tank, since development in the phase change material is very immature; there is no application in wind power system, so we do not do analysis on the phase change.

Jun (2010), Zhi-zhong (2008), Shuangqing (2002), Nganroo (2009), Hongxin (2009), Jinghong (2009), and DIAF (2008) did some analysis on applications of different energy storage devices utilized in the wind farms .Overall, the pumped storage and compressed air energy storage technology are bound by the construction period and geographical conditions, therefore the applications of the two technologies are not widespread in the wind power. As new energy storage technologies, the flywheel energy storage and superconducting magnetic energy storage technologies have many problems that need to be solved. Since the development of battery storage technology which started earlier, and of which there are numerous battery types, the battery storage technology is widely used in large-scale wind power. As the super capacitor energy storage has a longer life, the specific power and the specific energy can meet the requirement of smooth wind farm output. In terms of application in wind farms, the main constraint is still the relatively high cost.

3 A CMC/ESS TOPOLOGY STUDY

3.1 *Characteristics of the cascaded H-bridge*

A cascaded H-bridge inverter is composed of several single-phase inverters connected in series, and as each single-phase inverter has a DC power supply, the H-bridge cascades into a tandem structure as a basic power unit (Marchesoni 1990). A cascaded multi-level inverter is often used in several H-bridge inverter units connected in series and parallels, where each H-bridge inverter unit use an independent low-voltage DC power supply. A cascaded multi-level inverter has the following advantages (Villanueva 2009, Kouro 2009), such as small voltage stress of switching devices, high output voltage, small harmonics, it is easy to expand, high reliability and a degree of modularity.

3.2 *The research status of the cascaded H-bridge PCS topology*

One important part of the energy storage system is the voltage source inverter based on the Pulse Width Modulation (PWM) technology—Power Conversion System (PCS). The PCS can achieve a bi-directional power transfer between a battery energy storage system and an AC grid. When energy storage system is utilized in large-scale wind farms and a PV power plant, the rated power should achieve a KW or a MW, which requires a high-power PCS. A PCS is used to stabilize power fluctuations when the new energy connected to the grid, so that the grid stability and dynamic characteristics could be improved. There are two main topologies of cascaded multi-level energy storage systems:

(1) A traditional multi-level cascade topology combined with the energy storage device.

This PCS can combine decentralized low-voltage battery packs into high-power battery energy storage systems. The AC side can be directly incorporated into a high voltage level grid by cascading without transformers; this can reduce the cost and the size. The devices' equivalent switching frequency is increased after cascading, and the current quality is higher. The transmission efficiency is improved because each module's low carrier frequency can reduce switching losses. Feng (2013) proposed one model where a multilevel cascaded PCS is connected to the energy storage system through a DC/DC link. Qing (2014) proposed another one where at multi-level cascaded PCS is connected to the energy storage system through an isolated DC/DC link.

Feng (2013) proposed a unified power compensation system (UPC) by using a cascaded multi-level inverter plus battery energy storage system. The power fluctuations of wind turbines at the high voltage side may have an impact on the life of the battery system. In order to prevent this situation, we use a DC/DC boost converter between a battery energy storage system and a high voltage side. The system can balance DC voltages of H-bridge modules because of the two grade structures, and the cycle life of the battery would not be reduced when compensating for active and reactive power.

(Qing 2014) proposed a new topology of the hybrid cascaded megawatt PCS. The PCS consists of battery packs, an isolated half-bridge DC/DC converter and a cascaded H-bridge DC/AC converter. The coordinated control between the two kinds of a converter is emphatically studied on this topology. The phase-shift control at a DC/DC side and the dual-loop control at a DC/AC side are adopted to realize the bi-directional power exchange; the duty-cycle control at a DC/DC side and the global DC voltage control at a DC/AC side are adopted to maintain the DC capacitor voltage constant; the real-time power instruction of grid side is fed back to the DC/DC side in order to increase the response speed of a device and improve the quality of a DC capacitor voltage. A scheme of the soft start up control is applied to improve the start up characteristics of a device. The simulation results demonstrate that the device under the proposed control has a wide voltage matching capability, an excellent battery status adaptability, a fast response, a large energy-storage capacity and a bi-directional power regulation capability.

(2) A hybrid cascaded multilevel converter topology combined with the energy storage device.

Compared to the traditional multi-level cascade topology, the multi-level hybrid cascade topology not only improves the output voltage, but also can control

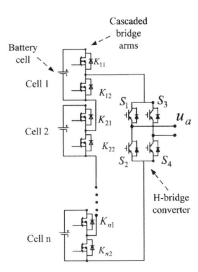

Figure 1. Hybrid cascaded multilevel converter.

each unit independently. In the charge and discharge control circuit, the balanced combination of the capacity of the energy storage cell is able to average voltage actively during the charging and discharging process, and can also output a multi-level voltage during the discharge process, reduce the harmonic voltage, reduce the harmonic current during parallel operation, and improve the power control performance (Fengqi 2013). Figure 1 shows a kind of topology of a hybrid cascaded multilevel converter.

Although the hybrid cascade multi-level converter can control each unit independently, it meanwhile increases the complexity of the control. In addition, the mixed multi-level is a mixture of a variety of topologies, which will increase the volume and footprint of the unit to some extent, and add the manufacturing cost.

4 RESEARCH OF CASCADED MULTI-LEVEL WITH STORAGE SYSTEM APPLICATION AND CONTROL STRATEGY CMC/ESS TOPOLOGY STUDY

The CMC/ESS is mostly used in solving the following problems of wind power: to increase the wind power low-voltage ride-through capability, stabilize power fluctuations, improve power quality, and improve the stability of power systems, including wind power and other aspects. Based on solving different problems the control strategies of storage system are not the same.

4.1 Increasing wind power low-voltage ride-through capability

Low voltage ride through (LVRT) have been one of the problems of wind power technology development that need to be addressed, and it is one of the key factors that affect the stability of power system. The literature (Hong 2013) compares the low voltage ride through capability of cascade multilevel STATCOM and modular multilevel STATCOM. The analysis results are that delta connection does not have the neutral point, so when the single-phase grounding fault occurred in the system, the device can still running.

However, only using the control strategies to solve LVDT problem, it is also difficult to achieve the desired effect.

4.2 Stabilizing power fluctuations

It is hard to control the fluctuation of wind power. In wind power integrated to the grid, by introducing the ESS, the corresponding control strategy is developed, and the effects of wind speed random variation on wind power can be reduced, which can control the wind power output fluctuation.

The traditional STATCOMs used in reactive power compensation are not capable of controlling active power flow. In the flexible AC transmission system (FACTS), (Qian 2002) the battery energy storage systems (BESS) are incorporated into a cascaded converter-based STATCOM to implement both active and reactive power flow control using a PQ-decoupled PI control strategy. The simulation model is established in the EMTDC/PSCAD; the simulation results are provided in order to verify the feasibility and practicality of these ideas.

4.3 Improve power quality

The CMC/ESS can efficiently control the active power and reactive power, rapidly balance the unbalanced power of the system, improve the power factor of the system, eliminate the system harmonic current, control the voltage sag and swell, reduce the impact of disturbance to the power grid, thereby achieving the purpose of improving the power quality.

For the STATCOM/BESS this type of power quality regulating device and the operation ability under a three-phase unbalanced condition is the key index to evaluate the performance of the grid connected power electronics device. (Yanjun 2012) proposes a control method for star connected cascaded STATCOMs under the unbalanced grid voltage. By applying an individual phase current tracking, each phase leg (cluster) of the cascaded STATCOM is treated as an individual converter. Each cluster has its own phase current tracking loop and a DC voltage feedback loop. This control method generates the necessary negative sequence current and the positive sequence voltage automatically in order to rebalance the unbalanced power, along with the unnecessary and harmful zero sequence current. An optimal zero sequence current reference separation algorithm is developed to remove the unwanted zero sequence current, by making a small variation of a reactive power command. The proposed algorithm can decouple the three-phase currents in the 3 wire system. Even in the star-connection, each cluster of the CMC can be controlled individually, such as in a delta-connection. Therefore, no extra means need to be taken when unbalance occurred.

4.4 Improve the stability of power systems including wind power

In a traditional power system, stability means that the system can restore from the original state or transition to a new state ability to maintain a stable operation in the large or small turbulence. The power system with wind power has the same stability problems. Furthermore, its stability shows some special characteristics, because the wind turbines and the synchronous generators are different. However, the fundamental reason for this stability problem is the system instantaneous power imbalance. The ESS has a fast power response ability, thus providing a new way for improving the stability of the power system with wind power.

(Lirong 2013) presented a hybrid power control strategy to control the STATCOM/BESS. The control strategy includes two closed-loop controls. The double closed loop controller consists of an inner-loop for current regulations and an outer-loop for active power and voltage control. The outer-loop control includes the hybrid power control composed of the battery charging and discharging control and active and reactive power coordinated control. The proposed control strategy is validated by simulation on the wind power system using the cascaded STATCOM/BESS as a compensation device. Simulation results show that the cascaded STATCOM/BESS can effectively improve the characteristics of a wind farm integration and provide a dynamic support to the grid.

At present, the majority view is that the influence of oscillation characteristics between the power system regions is not significant, when the wind power integrates to the power system. Therefore, the study of using the ESS to improve a small disturbance stability does not see more.

5 RESEARCH DIRECTION OF CMC/ESS

To sum up, the research on cascaded multilevel storage system should focus on the selection of an energy storage device, the study of control strategy, the topology structure of the cascaded multilevel, etc.

5.1 The selection of energy storage device

At present, people mostly use batteries to store energy. Among high-power batteries, the lithium battery, an important research in battery development, is more popular in the industry. A super capacitor is generally not used for a large-scale energy storage device because of its low energy density and small capacity. However, due to its rapid charge and discharge rate, a super capacitor can quickly provide active and reactive power for a wind farm, whose power fluctuates fast. A hybrid energy storage device provides a good idea for the development direction of the future energy storage device. A battery has low power density and high energy density, while a super capacitor has high power density and low energy density. Hybrid energy storage system combines the lithium battery and the super capacitor, taking advantage of two complementary techniques to simultaneously provide high power and high energy. It is foreseeable that the hybrid energy storage technology has a broad prospect for stabilizing wind power fluctuations.

5.2 The study of control strategy

Reasonable control strategy has been the focus of energy storage technology applied research, because it is not only the prerequisite for the ESS to work as expected, but also directly related to issues such as the energy storage capacity and economy. Along with the development of research, the function of the ESS in the wind power integration is more and more diverse; it can perform different functions at the same time or at time-sharing times, thus making the control strategy very complex. In addition, multiple coordinated control of the multiple composite energy storage systems, wind power/energy storage joint coordination and control problems, etc. are making the control strategy particularly important. Therefore, control strategy is still an important research topic in the future of energy storage technology in the application of wind power integration.

5.3 The study of the topology structure of cascaded multilevel

Because the topological structure of traditional cascade multilevel is too complex, many power semiconductor switches are needed, but their volume is too large. Therefore, it is necessary to deeply study the new proposed topology. Under the premise of power electronics and control technology, it is vital to optimize the traditional cascade multi-level circuits, in order to adapt them to the continuous development of new power electronic devices, and gradually reducing the size and cost of cascaded PCS to further promote their applications in areas such as energy storage in the new energy grid.

6 CONCLUSION

With the rapid development of technologies, new energies such as wind and solar energy will bring great instability to the grid network when they are connected to the grid on a large scale, due to their fluctuation and intermittency. However, energy storage technology has largely solved this problem. This paper mainly analysed and summarized research into energy storage systems and the cascaded PCS. The main work is summarized as follows:

(1) This paper summarizes the various energy storage technologies in the application of wind power integration, according to their characteristics.

(2) This paper summarizes the way that an energy storage device connects to the DC side of PCS. Currently the most widely used model is the one that connects to the grid through a non-isolated DC/DC converter.

(3) This paper summarizes the research status of the PCS energy storage system based on the multi-level cascade. Moreover, the research mainly focuses on the selection of energy storage devices, energy storage unit access ways, the study of control strategy and the topology structure of cascaded multilevel, etc.

(4) This paper puts forward some suggestions for the above problems, which will provide certain references to the study and use of the cascaded multilevel storage system in improving the working characteristics of wind power being connected to the grid.

ACKNOWLEDGEMENT

This paper is funded by the National High Technology Research and Development of China (863 Programme: 2011AA05A113).

REFERENCES

[1] Chang Qian; Crow, M.L., "A cascaded converter-based StatCom with energy storage," Power Engineering Society Winter Meeting, vol. 1, 2002, pp. 544–549.
[2] Coppez G., Chowdhury S., Chowdhury, S. P. "The importance of energy storage in Renewable Power Generation: A review," Universities Power Engineering Conference (UPEC), 2010 45th International, Aug. 31 2010–Sept. 3 2010.
[3] Diaf S, Belhamel M, Haddadi M. et al. "Technical and Economic Assessment of Hybrid Photovoltaic/wind System with Battery Storage in Corsica Island," Energy Policy, Vol. 36, No. 2, 2008, pp. 743–754.
[4] E. Villanueva, P. Correa, J. Rodriguez, and M. Pacas. Control of a single-phase cascaded H-bridge multilevel inverter for grid-connected photovoltaic systems [J]. IEEE Transactions on Industrial Electronics, Vol. 56, No. 11, 2009, pp. 4399–4406.
[5] Fengqi Chang, Yongdong Li, Zedong Zheng. "Research on grid connected strategy model of battery energy storage topology," Electric Drive, Vol. 1, 2013, pp. 98–102.
[6] Gao Feng, Zhang Lei, Zhao Yong, "A unified power compensation system for the large-scale grid-tied renewable energy generation system," Industrial Electronics Society, IECON 2013 – 39th Annual Conference of the IEEE, 2013, pp. 6692–6697.
[7] Hong Rao, Jun Chen, Shukai Xu. "Selection of multi-level main circuit schemes of STATCOM for transmission system." Automation of Electric Power System, Vol. 23, 2013, pp. 83–87.
[8] Hongxin Jia, Yu Zhang, Yufei Wang. "Energy storage for wind energy applications," Renewable Energy Resources, Vol. 27, No. 6, 2009, pp. 10–15.
[9] NGAMROO. "Power Oscillation Suppression by Robust SMES in Power System with Large Wind Power Penetration". Physical: Vol. 469, No. 1, 2009, pp. 44–51
[10] Jinghong Shang, Xu Cai, Liang Zhang. "Application and research of batteries power storage system of huge wind power generators," Applied Science and Technology, Vol. 36, No. 10, 2009, pp. 1–3.
[11] Jun Tian, Yongqiang Zhu, Caihong Chen. "Application of Energy Storage Technologies in Distributed Generation". Electrical Engineering. Vol. 8, 2010, pp. 28–32.
[12] Lirong Zhang; Yi Wang; Heming Li; Pin Sun, "Hybrid power control of cascaded STATCOM/BESS for wind farm integration," Industrial Electronics Society, IECON 2013 – 39th Annual Conference of the IEEE, Nov. 2013 pp. 5288–5293.
[13] Marchesoni M., M. Mazzucchelli, ATenconi. "A non-conventional power converter for plasma stabilization," IEEE Transactions on Power Electronics, Vol. 5, No. 2, 1990, pp. 212–219.
[14] Qing Miao, Junyong Wu, Hongke Ai. "Coordinated control of hybrid cascaded megawatt power regulation device," Electric Power Automation Equipment, Vol. 07, 2014, pp. 43–49.
[15] S. Kouro, A. Moya, E. Villanueva, P. Correa, B. Wu, and J. Rodriguez. "Control of a cascaded h-bridge multilevel converter for grid connection of photovoltaic systems," Proceedings of IEEE 35th Annual Conference of the Industrial Electronics Society, Porto, Portugal, 2009, pp: 3976–3982.
[16] Shuangqing Tang, Jiajun Yang, Daoxun Liao. "Summarization of Research on Fly wheel Energy Storage System," J of China Three Gorges Univ. (Natural Science), Vol. 25, No. 1, 2002, pp. 78–82.
[17] Wen-liang ZHANG, Ming QIU, Xiao-kang LAI. "Application of Energy Storage Technologies in Power Grids". Power System Technology, Vol. 32, No. 7, 2008, pp. 1–9.
[18] Yanjun Shi. "Key Control Techniques in Cascaded Multilevel STATCOM/BESS," WU Han: Huazhong University of Science & Technology, 2012.
[19] Yun Chen. "Researches on wind power and photovoltaic generation integration into power system," Shanghai: Shanghai Jiao Tong University, 2009, pp. 10–15.
[20] Zhi-zhong Tan, De-you Liu, Chuan-qi Ou. "Optimal operation model for wind-powered pumped storage system". Journal of Hohai University(Natural Sciences), Vol. 36, No. 1, 2008, pp. 58–62.

Aluminium alloy plate flaw sizing by multipath detection

Dubin Guo, Xizhong Shen & Lei Wang
Electrical and Automatic School, Shanghai Institute of Technology, Shanghai, China

ABSTRACT: This paper mostly explorers the detection of ultrasonic multipath in the application of aluminium alloy plate, using ANSYS for modelling simulation accumulated related parameters, and for detecting defects locating plate and its size.

Keywords: Ultrasonic, flaw, multipath, ANSYS, and sizing

1 INTRODUCTION

With the rapid development of science and technology, and manufacturing technology in, all walks of life, there is a growing demand for aluminium sheet metal which is being widely used in aerospace, high-speed trains, ships, rail transportation and other fields. Because of bad manufacturing technology, uncertain operating environments and especially long-term over-fatigue, resulting from use in harsh environments, metal plate exterior or internal surfaces will develop defects. The existence of defects will affect the safe operation of equipment, or even cause accidents.

In this paper, the main use of multipath ultrasonic testing tools is used to test the type 6061 aluminium alloy. The ultrasonic multipath is formed from the ultrasonic propagation path of a multipath and corresponding algorithm development, trying to solve these defects is a detection problem. Multiplicity refers to: encountered in the process of sound waves in the material space defect or discontinuous happens when the wave reflection, scattering, diffraction, and the phenomenon such as ultrasonic energy absorption. Ultrasonic testing is based on the phenomenon of material defect detection and evaluation. Using the ultrasonic multipath detection, we can detect the size and location of the defect.

2 MULTIPATH PRINCIPLE ANALYSIS

Figure 1 illustrates the standard of an ultrasonic transceiver measurement system. We receive the ultrasonic signal from the sensor for the modelling, $r(t)$, as follows:

$$r(t) = \sum_i \alpha_i s(t - \tau_i), \qquad (1)$$

where $s(t)$ is shown the ultrasonic echo pulse wave, α_i and τ_i are i-th path reflectivity and delay time respectively.

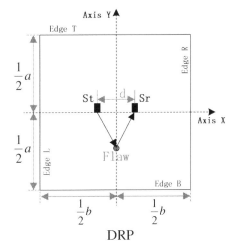

Figure 1. The standard directly reflects the scene in the ultrasonic testing system.

Figure 2 lists some of the second order of multiplicity.

A sized flaw is asuumed to be circle-like hole, whose centre at the position $f_p(x_{f,p}, y_{f,p}, z_{f,p})$. Assuming that the m-th transmitting sensor located in $S_{t,m}(x_{t,m}, y_{t,m}, z_{t,m})$, then the nth receiving sensor is located in $S_{r,n}(x_{r,n}, y_{r,n}, z_{r,n})$. The DRP Direct Reflection Path (DRP) corresponding to the nth receiving signal is represented as a sensor by

$$r_{o,mn}(t) = \alpha_{0,mn}(f_t)s(t - \tau_{0,mn}(f_t)), \qquad (2)$$

$\alpha_{0,mn}(f_t)$ are said defects of the DRP reflectivity, and $\tau_{0,mn}(f_t)$ is the delay time of the signal. Assuming a homogeneous medium of aluminium alloy sheet has a constant ultrasonic velocity v_p, the corresponding delay time to f_t can be calculated as:

$$\tau_{0,mn}(f_t) = (\|S_{t,m} - f_t\| + \|S_{r,n} - f_t\|)/v_p, \qquad (3)$$

where $\|\cdot\|$ denotes the Euclidean norm operation.

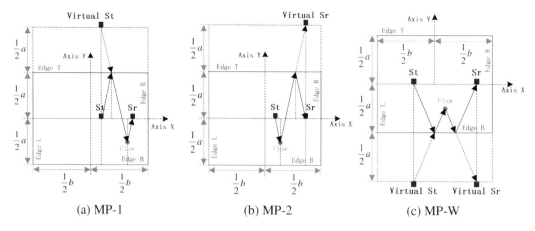

(a) MP-1 (b) MP-2 (c) MP-W

Figure 2. The second order reflection of multiplicity of scenarios.

Where a denotes the thin slab width, b is length, with aluminum plate surface center for the origin coordinate system is established. The virtual sensors are respectively located in $S_{t1,m} = S_{t,m}, S_{r1,n} = (x_{r,n}, y_{r,n} + a, z_{r,n})$ for MP-1, $S_{t2,m} = (x_{t2,m}, y_{t2,m} + a, z_{t2,m})$, $S_{r2,n} = S_{r,n}$ for MP-2, and $S_{t3,m} = (x_{t,m}, y_{t,m} - a, z_{t,m})$, $S_{r3,n} = (x_{r,n}, y_{r,n} - a, z_{r,n})$ for MP-W, as shown in Figure 2. Corresponding with the target size of MP-1, MP-2 and MP-W delay times are expressed by the following calculation respectively as:

MP-1: $\tau_{1,mn}(f_t) = (\|S_{t1,m} - f_t\| + \|S_{r1,n} - f_t\|)/v_p$ (4)

MP-2: $\tau_{2,mn}(f_t) = (\|S_{t2,m} - f_t\| + \|S_{r2,n} - f_t\|)/v_p$ (5)

MP-W: $\tau_{W,mn}(f_b) = (\|S_{t3,m} - f_b\| + \|S_{r3,n} - f_b\|)/v_p$ (6)

where $f_t = (x_{f,t}, y_{f,t}, z_{f,t})$ and $f_b = (x_{f,b}, y_{f,b}, z_{f,b})$ are respectively the flaw top point and the flaw bottom point.

When the flaw echoes and the bottom strong reflection wave distances are far enough to separate, using traditional technology (for example, defect detection methods) can easily recognize the flaw echoes of the DRP Direct Reflection Path (DRP). If the transmitting and receiving sensor distances are far enough, we can avoid the overlapping of MP-1 and MP-2.

3 ANSYS FINITE ELEMENT SIMULATION

Using ANSYS finite element simulation software for three-dimensional finite element modelling and simulation.Taking of ANSYS to establish a thin plate with hole and without hole defect model. At the same time meshing model and incentive. Through the finite element model of the same observation nodes received defective echoes with zero defect model. Using of extraction of defect echo information, can be calculated by multiplicity defects are obtained by the coordinate and size.

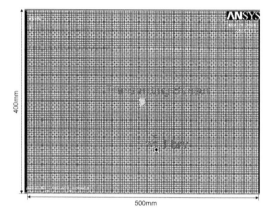

Figure 3. ANSYS model.

The finite element model of a plate test is shown in Figure 3. The 6061 type aluminium plate, whose size is 400 mm × 500 mm × 2 mm has been tested.

The Modulus of elasticity is $E = 68.9$ GPa, Poisson's ratio is $\upsilon = 0.330$, and the density of a plate is $\rho = 2.8 \times 10^3 \text{kg/m}^3$. Transmitting of PZT transducer is 4 mm radius of circular sensors, coordinates of the centre are $(0, 0, 0)$, and the radius of circular hole defects is $r = 2$ mm, whose centre coordinates are $(30, -100, 0)$. We use A0 Lamb wave mode as incentive mode, the simulation of the excitation signal is:

$$s(t) = \frac{A}{2}\left[H(t) - H(t - \frac{n}{f_c})\right]\left(1 - \cos\frac{2\pi f_c t}{n}\right)\sin\frac{2\pi f_c t}{n}$$

(7)

It is using narrow band excitation of the Lamb wave signal, with a single frequency sine signal and a window function (Hanning window) as excitation signal is shown in Figure 4.

$A = 10^{-5}$V denotes the signal amplitude, $f_c = 200$ kHz denotes the centre of the signal frequency, $n = 5$ is the signal periodicity, $H(t)$ is Heaviside function, which denotes when $t > 0$, then $H(t) = 1$, or when $t < 0$, then $H(t) = 0$, and when $t = 0$, then $H(t) = 0.5$.

We carried out on the ANSYS model 2 mm mesh generation and grid computing.

By using the ANSYS simulation calculation, we can obtain and extract the DRP and MP-W defect echo as shown in Figure 5 at the coordinates (60, 0).

As shown in Figure 5, DRP and MP-W defect echo time are 66.8 μs and 195 μs respectively. At the same time, we measured the speed of Lamb wave in A0 mode is $v_p = 3070.5$ m/s from the ANSYS simulation. Therefore, the simulation can be measured using the delay time and the propagation velocity generation in the formulas (3) and (6), where we can obtain the top and the bottom point coordinates with (30, −98.07, 0) and (30, −102.13, 0), respectively.

Again, we are in the node (182, 0, 0) and (36, 0, 0) with the MP-1 and the MP-2 receiving and extracting the defect echo is shown in Figure 6.

As shown in Figure 6, the MP-1 and the MP-2 defect echo times are 203 μs and, 194 μs respectively. Therefore, the simulation can be measured using the delay time and the propagation velocity generation in the formulas (4) and (5), where we can obtain the top of the point coordinates with (30, −98.07, 0) and, (30, −97.23, 0) respectively.

From what has been discussed above, the DRP and the MP-1 and MP-2 are used to detect defects of top positions, so we can take the average of the three and we can get the top positions as (30, −97.79, 0). Through the MP-W path measured the coordinates of the end point (30, −102.13, 0). Thus, we can obtain the diameter of the defect which was 4.36 mm at this time. The simulation results are very close to the actual size. Simulation of the hole size is greater than the original size, for the reason that the ANSYS mesh is not precise to select the nodes that caused a deviation of migration.

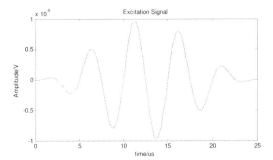

Figure 4. Excitation signal.

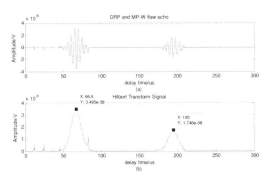

Figure 5. DRP and MP-W flaw echo and Hilbert Transform.

4 CONCLUSION

The ANSYS simulation results proved that the ultrasonic detection method of a multipath is feasible and that it can locate the defects and accurately calculate their size. The ultrasonic signals of a multipath broke through the sensor detection range, which can generate the virtual array aperture extension from the defects' coordinates for more accurate positioning. It

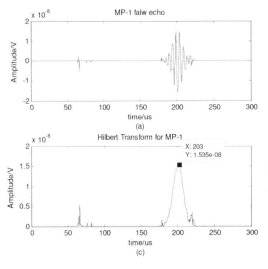
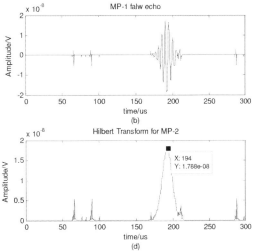

Figure 6. MP-1 and MP-2 flaw echo and Hilbert Transform.

is observed that the multipath of the signal provides additional data. Using this additional data information, which through a multiplicity ultrasonic testing can determine the position of the cracks, and using traditional technology can identify them with Direct Reflection Path (DRP). They can be determined through the DRP in the range estimate and the MP-C and the MP-W delay time, and then determine the scope of the DRP (directly reflecting the path delay). Detection of defects at the top of the DRP and the MP-C can be used, whilst at the same time the bottom can be used by MP-W testing. According to the multiple detection method can get extra sensor data, and further analysis of these additional data, can be achieved by a single measurement to determine the defects only coordinates and detect its size.

REFERENCES

[1] B. Poddar, A. Kumar, M. Mitra et al. Time reversibility of a Lamb wave for damage detection in a metallic plate [J]. Smart Materials and Structures, 2011, 20 (2): 25001–25010.

[2] Lester W. Schmerr Jr, Sung-Jin Song. Ultrasonic Nondestructive Evaluation Systems Models and Measurements [M]. Springer-Verlag, 2007.

[3] T.M. Meksen, B. Boudraa, R. Drai et al. Automatic crack detection and characterization during ultrasonic inspection [J]. Journal of Nondestructive Evaluation, 2010, 29 (3): 169–174.

[4] Xizhong Shen, Li Pan. Material Flaw Sizing by Ultrasonic Multipath Detection [J]. Periodical of Advanced Materials research vol. 712–715, pp: 1067–1070.

[5] X.Z. Shen, K. Yuan. Ultrasound Multipath Signature of Material Flaw Depth with Single Mono-bi-static Measurement [C]. Advanced Material Research, vols. 468-1-471, 2012, pp: 2505–2508.

… Electronics and Electrical Engineering – Zhao (ed.)

An equilibrium algorithm clarity for the network coverage and power control in wireless sensor networks

Li Zhu, Chunxiao Fan, Zhigang Wen, Yang Li & Zhengyuan Zhai
School of Electronic Engineering Beijing University of Post & Telecommunications, Beijing, China

ABSTRACT: An equilibrium algorithm for network coverage and power control is proposed for those nodes that differ in energy. The strategy, according to the different residual energy in nodes, combined with a network coverage rate, is to develop an adjustment to the sensing range of nodes, which aims to achieve a balance between network energy and network coverage, to reduce the premature death rate of nodes, and to try to balance the network's energy. Simulation results show that the strategy can effectively enhanced the network's coverage rate, prolong the network's life-time, improve the network's quality, and realize the optimization of the network's coverage control.

Keywords: Wireless sensor networks, the network coverage rate, energy awareness, Voronoi, coverage holes

1 INTRODUCTION

Wireless sensor networks (WSNs) are widely used in various fields [1, 2], How to utilize the limited network energy consumption and improve the network's coverage as much as possible, is one of the key issues to be solved [3–6]. In the study of existing studies about network coverage, Wu [7] proposed a node redundancy based on the probability calculation method. However, this method ignores the contribution of a neighbouring node to its sensing area coverage. The literature [8] proposed an adaptive node scheduling algorithm: an algorithm that is constantly sending messages between the nodes, making some nodes undertake a large amount of tasks, and appearing with a premature "death" phenomenon. Literature [9] proposed a fair override mechanism, but the mechanism does not take into account the energy factors, which cause the premature failure of nodes, and affect the network lifetime. Literature [10] only considers the coverage and network distribution of nodes, does not compare the node's remaining power, which reduces the network lifetime, all of which have some limitations.

This paper introduces an equilibrium algorithm for network coverage and power control in Wireless Sensor Networks, The algorithm seeks an equilibrium between network coverage and the power control, which, considering the node's residual energy, combined with its energy weight, makes a dynamic adjustment for the sensor node's range, and divides the effective sensor area, achieving an equilibrium on network coverage and power control, so as to prolong the lifetime of the WSN.

2 AN ENERGY-COVERAGE CONTROL STRATEGY

The coverage strategy needs to consider the optimization of network energy consumption under the condition of improving the network' coverage rate. If a node with low energy in the network still has the same work tasks as the other nodes, the node will be "killed" ahead of time, which may affect the entire network transmission performance and reliability [11–13]. In order to solve the optimization problem of coverage balance and the contradiction between individual nodes and the whole network's performance, this article proposed an equilibrium algorithm on network coverage and power control in the Wireless Sensor Networks which referred to as EACP. The strategy is divided into two main stages: an energy consumption perception phase, which determines the perception area of a sense node according to the residual energy of every sensor node in the monitoring area; and a network coverage adjustment phase, which combined with the coverage conditions of the whole network, adjusts the node's sensing radius, effectively reduces the phenomenon of network overlapping coverage, and reduces the network redundancy coverage, thus prolonging the network's lifetime.

In a wireless sensor network, as time goes on, the number of network survival nodes will be changed, due to various factors such as signal disturbances which lead to an unbalanced energy consumption in communications, and makes a difference in the node's residual energy. Considering the relationship between the node's residual energy and the perceived area, it is necessary to set a reasonable sensing range for each node, which can balance network energy consumption and prolong the network's lifetime.

In the energy consumption perception phase, when the node si in the sensing area A_i at work after the t time, the consumption electricity W_i and the sensing area R_i satisfy the following relations (k is a constant):

$$W_i = kR_i^2 \quad (1)$$

between any adjacent notes si and sj, its remaining power is Qi and Qj r, after a period of the t time, during which time the battery run out, the perception radius of the two nodes meet the formula (5):

$$R_i = d(i,j) \cdot \frac{\sqrt{Q_i}}{\sqrt{Q_i} + \sqrt{Q_j}} \quad (2)$$

d(i,j) is the Euclidean distance between nodes; according to the formula the relationships of the node si and sj between perception radius and residual energy can be seen:

$$R_i : R_j = \sqrt{Q_i} : \sqrt{Q_j} \quad (3)$$

Select any node si in the monitoring area, get on a stage of the energy perception scope, which sent some information to the neighbour node such as own perception radius, energy consumption and so on. According to the perceived intensity of the node, determine the scope of each node sensing and monitoring area. When coverage hole is produced in the monitoring area, the node monitoring area was divided again, Considering the overlap phenomenon of the node perception scope, and no new coverage hole, combined with nodes remaining power, which adjust perception radius, reduce redundant monitoring area, thereby reducing unnecessary energy consumption. Algorithm steps are as follows:

Step one: any node si was analysed through probability-aware, which determine node set s, meanwhile, some information about themselves were sent to subset of the node set s, such as perception radius, residual energy, at the same time, the node in the node set s can also get their information, each node according to the perception radius which determine their own monitoring area and establish the Voronoi polygon.

Step two: choose three adjacent Voronoi polygon areas; a shadow area was constituted by the midpoint of the edge of $\Delta S_1 S_2 S_3$; because the remaining power of the three sensor node is different, the shaded part is a coverage hole which was caused by a coverage area of the three nodes. Q1 is probability perceived strength of the node S1. Simultaneously, Q2 and Q3 are Probability perceived strength of node S2 and S3, the shaded part will exist a certain point ϑ and $Q1 = Q2 = Q3$, ϑ is critical point, shadow area will be divided to perceived range of node, point ϑ was just covered by three nodes.

Step three: node S1 get some information of neighbouring nodes such as node S2, S3, S4, node S1 will

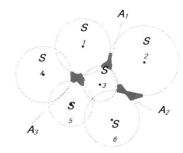

Figure 1. The coverage hole location diagram.

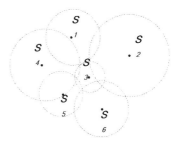

Figure 2. The coverage hole repair diagram.

forward relevant information to its neighbour node, find the coverage blind area the node A1 and A3, according to method of the second step to determine location of coverage blind point ϑ local (ϑ), combined with the formula (3) to adjust node sensing radius, as shown in Figure 1 and Figure 2.

Step four: Under no new coverage hole, adjust redundancy coverage area of network. Meanwhile, compare the remaining power of the node S1 and its neighbouring nodes S2, S3, S4. Assuming that S3 is the maximum remaining power, $S3 = \text{Max}\{S1, S2, S3, S4\}$, S1 is the minimum namely remaining power $S1 = \min\{S1, S2, S3, S4\}$, according to the relationship of formula (3), reset perception radius of the node S3 and S1, the coverage redundancy of adjacent nodes T is greater than the threshold a, if greater, repeat the above operation until $T < a$.

3 SIMULATION ANALYSIS

In this paper, the algorithm based on the network model will be simulated, meanwhile, the experiment environment was set up, which has a target area for $100 \times 100 \, \text{m}^2$. 100 sensor node were random uniform deployed. The biggest communication radius of node is Rc, perception radius Rs is variable. The EACP algorithm should be compared with the Random Distributed Algorithm (RMDA), in order to evaluate the performance of EACP algorithm performance. Some performances of algorithms were analysed in their following properties: (1) the network coverage rate (2) the network control overhead.

In this section, the experiment compared the network coverage rate of two algorithms, that is the

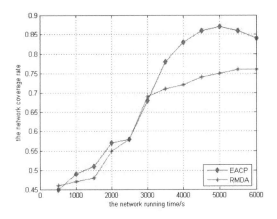

Figure 3. The network coverage rate.

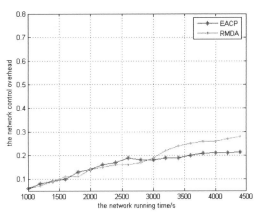

Figure 4. The network control overhead.

algorithm RMDA and the EACP. The network coverage rate is the ratio of the sum of all sensors coverage area and the whole monitoring area. As shown in Figure 3, when the number of initial nodes remain unchanged in the network. The network coverage rates of the EACP algorithm were higher than the RMDA algorithm. In the initial stage of the network, the network coverage rates of two algorithms are very close and there is no obvious change. With the change of network running time, coverage rate of RMDA algorithm was lower than EACP algorithm, EACP algorithm although has certain redundancy area, but it can satisfy the requirements of network coverage quality.

Figure 4 shows the network control overhead of two algorithms, when the network coverage rate reaches 70%. It can be seen from Figure 4, that the control overhead of the EACP algorithm is less than in the RMDA algorithm. In the EACP algorithm, based on the node perception range and coverage hold, the network redundancy area was adjusted and repaired. In the process of the area being repaired, the change tendency of the sending and receiving control overhead is relatively slow, and the whole network energy consumption cost less than the RMDA algorithm. Accordingly, the EACP algorithm, based on the coverage requirements being fulfilled, can play an important role in the optimization of the network's energy consumption.

4 CONCLUSION

This paper proposes an equilibrium algorithm on network coverage and power control in Wireless Sensor Networks, which according to the residual energy of the working node, the node sensing task was reassigned. At the same time using a probability perception adjusted probability radius, it improved the network coverage and reduced network redundancy coverage and network energy consumption. The experimental results show that compared with random method, the paper make an important indicators which is energy, to avoid the node with less energy to assume the heavy task of nodes, lead to premature "failure". The EACP algorithm effectively reduce the unnecessary energy consumption, and network coverage meet the actual demand.

ACKNOWLEDGMENTS

The work presented in this paper was supported by the National Natural Science Foundation of China (Grants No. NSFC- 61471067), the National Natural Science Foundation of China (Grants No. NSFC-61170176) and the Fund for the Doctoral Program of Higher Education of China (Grants No. 20120005110002). The work presented in this paper was supported by the National Great Science Specific Project (Grant No. 2012ZX03005008).

REFERENCES

[1] Hong Zhen, Yu Li, and Zhang Gui-jun. Efficient and dynamic clustering scheme for heterogeneous multi-level wireless sensor networks [J]. Acta Automatica Sinica, 2013, 39(4): 454–464.
[2] Naderan M, Dehghan M, and Pedram H. Sensing task assignment via sensor selection for maximum target coverage in WSNs [J]. Journal of Network and Computer Applications, 2013, 36 (1): 262–273.
[3] Asim M, Mokhtar H, and Merabti M. A fault management architecture for wireless sensor network[C]. International Wireless Communications and Mobile Computing Conference (IWCMC), Crete Island, 6–8 Aug. 2008: 779–785.
[4] Thai M T, Wang F, and Du D Z. Coverage problems in wireless sensor networks: designs and analysis [J].International Journal of Sensor Networks, 2008, 3(3): 191–200.
[5] Elad M. Optimized projections for compressed sensing [J]. IEEE Transactions on Signal Processing, 2007, 55(12): 5695–5702.
[6] Khedr A M and Osamy W. Minimum perimeter coverage of query regions in a heterogeneous wireless sensor network [J]. Information Sciences, 2011, 181(15): 3130–3142.

[7] Wu K, Gao Y, Li F, Xiao Y. Lightweight deployment aware scheduling for wireless sensor networks, J. ACM/Kluwer Mobile Networks and Applications (MONET). 6, 10 (2005).

[8] Ye F, Zhong G, Lu SW, Zhang LX. A robust energy conserving protocol for long-lived sensor networks. Proc. of the 10th IEEE Int'l Conf. on Network Protocols. IEEE Computer Society, 2001. 200–201.

[9] S. Meguerdichian, F. Koushanfar, M. Potkonjak, M.B. Srivastava, "Coverage Problems in Wireless Ad-hoc Sensor Networks," IEEE InfoCom, 2001.

[10] T. V. Chinh, and Y. Li, "Delaunay-triangulation based complete coverage in wireless sensor networks," IEEE Per Com, 2009.

[11] Zhang H. H. and Hou J. C. Maintaining sensing coverage and connectivity in large sensor networks [J]. Ad Hoc and Sensor Wireless Networks, 2005, 1: 89–124.

[12] Yang A Y, Gastpar M, Bajcsy R, et al. Distributed sensor perception via sparse representation [J]. Proceedings of the IEEE, 2010, 98(6): 1077–1088.

[13] Sengupta S, Das S, Nasir M D, et al. Multi-objective node deployment in WSNs: in search of an optimal trade-off among coverage, lifetime, energy consumption, and connectivity [J]. Engineering Applications of Artificial Intelligence, 2013, 26(1): 405–416.

Three-layer architecture based urban photovoltaic (PV) monitoring system for high-density, multipoint, and distributed PV generation

Hong Gang & Peng Qiu
Jinzhou Power Supply Company of Liaoning Province of State Grid, Jinzhou, Liaoning, China

Dacheng He
School of Electrical Engineering of Northeast Dianli University, Jilin, China

ABSTRACT: Traditional PV monitoring technology is limited in data acquisition and processing. This paper describes an urban PV monitoring system for high-density, multipoint, and distributed PV generation, which provides a platform for the concentrated processing of large data collection and storage, and simplifies the processing of the monitoring system's terminal data. The whole system is based on three-layer architecture and contains a PV monitoring unit (Layer 1), a regional PV management system (Layer 2), and a municipal PV monitoring centre (Layer 3). This paper also shows the process of communications and monitoring among three layers of architecture.

1 INTRODUCTION

In recent years, photovoltaic technology is developing rapidly in the world. As well as for the research of a photovoltaic power station's operational performance and the optimization design of a photovoltaic power station, the monitoring technology of photovoltaic power generation system is also beginning to development in photovoltaic technology. Before and after the 90s most research tended to be on monitoring systems overseas, and has formed a basically mature system. Several typical features of the research status are: most research focuses on the aspect of system control and on system maintenance, but the management of research is relatively limited; the research on grid types is more developed, but research projects on independent operational research are relatively few. In addition, the research is aimed at collecting and analysing data of the system to solve certain types of special problems.

Technique of photovoltaic power station data acquisition and monitoring work has been started inland. However, photovoltaic power station has the characteristic of the large area, and the traditional data acquisition is too scattered. Moreover, there is no centralized platform to observe and store data. All above factors lead to inconvenience and difficulties of the analysis and control. The traditional monitoring equipment is not flexible and expansible for the large capacity of photovoltaic power generation systems so that few technologies will be mature and put into use. What is more, there is very little research on urban power distribution network cluster monitoring technology that accesses solar power in high density.

In terms of the power grid, more and more serious effects that photovoltaic power causes on the power grid under a high permeability requires that we can not only examine the traditional indicators such as voltage, current, power factor, island, harmonic, flickering, short-circuit capacity etc. but also involve them in the whole power grid's power flow in order to examine them. Photovoltaic cluster monitoring technology provides relevant information to the dispatching departments of the municipal or provincial electric power companies for real-time management and control. In addition, it is also convenient for the distributed power users to trades with power companies and query their own real-time generated energy.

2 THREE-LAYER ARCHITECTURE BASED MONITORING SYSTEM

Photovoltaic cluster monitoring technique based on the design idea of hierarchical comprehensive and open system adopts modular, hierarchical and distributed structure to ensure the functions which include collection, metering, billing, statistics, analysis and decision-making, and the reliability of each link. Each module can be distributed to run on the every node of the network. The whole system in the form of three-layer architecture mostly contains photovoltaic monitoring unit (Layer 1), regional photovoltaic management system (Layer 2) and municipal photovoltaic monitoring centre (Layer 3) as Figure 1 shown. All these modules will be elaborated respectively later.

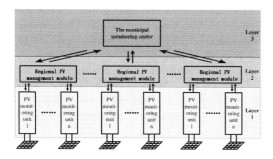

Figure 1. 'Photovoltaic monitoring unit + regional photovoltaic monitoring centre + provincial or municipal monitoring centre' of the three-layer architecture.

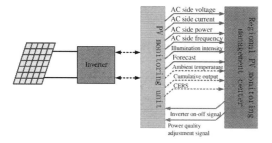

Figure 2. Communication structure of a PV monitoring unit.

3 PHOTOVOLTAIC MONITORING UNIT

As the basis of the whole cluster monitoring system, photovoltaic monitoring unit that is the core of the first layer architecture is responsible for collecting data, sending information to the regional photovoltaic monitoring centre, receiving instruction from regional monitoring centre and regional coordination control, etc. as shown in Figure 2.

3.1 Data collection

Cluster monitoring technique can not only examine traditional indicators such as voltage, current, power factor, island, harmonic, flickering, short-circuit capacity and so on, but also involve them in the whole power grid's power flow in order to examine and to obtain the electric statistics in different regions and the daily output forecast by the detection and processing of light intensity and environment temperature. Then, the analogue quantity that is collected from various types of sensors will be converted to digital quantity by the DSP which can process and compute the acquisition of signal to obtain more data acquisition indirectly and improve the function of monitoring unit.

3.2 Information feedback

A monitoring unit will sent data gained by monitoring and calculating, which contains a series of important data such as ac voltage, ac current, ac frequency, ac power and daily generation forecast of the regional monitoring centres through private network. At the same time, this unit can selectively transmit light intensity, environmental temperature, and CO_2 emissions in order to process data intensively and regulate and control photovoltaic distributed power supply at the Angle of macroscopic. The process of private network transmission is as follows: regional monitoring centres achieve two-way communication through the telecom department's special line between municipal monitoring centres or the monitoring unit and themselves. Each test control unit and higher level monitoring and control centre accesses the GPRS network through a GSM base station, then connects to a regional monitoring centre via the GGSN gateway, telecom operator's firewall, router and DDN special line. After completing regulation and control in a regional monitoring centre, this unit will give feedback switching signals of inverter, conditioning signal of power quality, etc. to each monitoring unit to achieve the high efficient and reasonable use of photovoltaic power generation and maximum benefit.

3.3 Monitoring

After the collected information was processed by a DSP, we can gain the data of harmonic, power factor, grid fault, and then control the inverter and MPPT link of photovoltaic panels effectively to ensure the efficiency of distributed photovoltaic power generation and achieve the monitoring of distributed photovoltaic power supply.

4 REGIONAL MONITORING CENTRE

The main body of a regional monitoring centre is the regional monitoring management system, whose functions contain data statistics, analysis and management. The regional monitoring centre will store the detailed running data that is received from each test control unit to automatically form a variety of reporting statistics that can be printed or displayed on screen in the form of a real-time curve, then upload the overall regional operation state to urban monitoring centre after statistics processing. Next, the regional monitoring centre will receive the scheduled information from the urban monitoring centre, and automatically adjust operation mode of each unit in the optimal way. The regional monitoring centre realizes the information interaction with the monitoring unit through Internet, and provides a friendly man-machine interface.

4.1 Communication mode

Regional monitoring centre achieves the two-way communication through the telecom department's special line between the municipal monitoring centre or the monitoring unit and itself. Each test control unit and a higher level monitoring and control centre connects to the regional monitoring centre through the GSM base station, accessing the GPRS network, collected by the GGSN gateway, passing telecom operator's firewall and router via the DDN special line of electric power and telecommunications operator.

4.2 Receive information

Regional monitoring system collects operation parameters of each photovoltaic power generation unit and information from higher level management centre. When data arrives, the system will analyse the data and store it to the background database.

4.3 User management

The system will offer a user interactive interface via the Internet, which contains a series of functions, such as parameters setting, real-time display, communication settings, real-time curve, historical data and print. It is more intuitional and convenient for user to quickly understand each unit's running situation. In addition, when operation fault appears, the software will issue a warning and save the record.

5 PROVINCIAL & MUNICIPAL MONITORING CENTRE

The provincial and municipal monitoring centre has complete function of statistics, analysis, and management as Figure 3 shows. Through regional monitoring centre we can obtain regional real-time electricity, forecast capacity and real-time generation cost, and store the historical data in the integrated communication management terminal. We will obtain the real-time and historical power cost in the integrated communication management terminal, and grasp the local forecast capacity and real-time capacity. On the electric power dispatching data network, the municipal monitoring centre will upload the real-time and historical data to provincial photovoltaic monitoring subsystem based on the D5000. At the same time, the provincial monitoring centre receives the weather forecast information, and uploads local power prediction results to photovoltaic monitoring subsystem according to the calculation, analysis, correction and checking the history and the running data, and afterwards accepts the results of the distributed power generation plan from system of master station which will be sent to the photovoltaic side to execute through the integrated communication management terminal. In the meantime, the provincial control centre can also order the municipal monitoring centre and regulate generating capacity.

5.1 Communication mode

In the link of data communications, information interaction between the master station and the substation will be realized through the electric power data network. There is a double set of redundant configuration, each one accordingly accesses network by 2x2M.

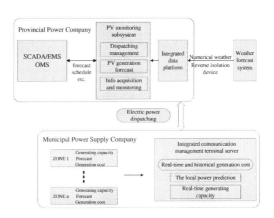

Figure 3. Provincial and municipal monitoring centres.

5.2 Upload information

The provincial and municipal monitoring centre gathers this area's running parameter information of each photovoltaic power generation unit, which covers the real-time and historical generation cost, the local generating capacity prediction, and the real-time generating capacity.

5.3 Receive information

The total regional photovoltaic generating capacity, the regional photovoltaic power generation trend prediction, and the regional power generation cost.

6 CONCLUSION

The results of this design will be applied to a distributed PV power generation and urban distribution network, in which the monitoring system will collect, analyse and integrate the data from the equipment that accesses the distributed PV distribution network. At the same time, online or offline energy will be settled accounts respectively. In addition, the results lay the foundation for the distributed PV system's smooth grid-connection, and promote the development of an upstream and downstream industrial chain of a distributed PV system.

REFERENCES

[1] A. Bagnasco, G. Allasia, M. Giannettoni and P. Pinceti. Innovative solutions for photovoltaic plants remote monitoring. Remote Engineering and Virtual Instrumentation (REV), 2012, 9: 1–5.
[2] A. Rivai, N. A. Rahim. A low-cost photovoltaic (PV) array monitoring system. IEEE Trans. on Clean Energy and Technology (CEAT), 2013: 169–174.
[3] A. Tsagaris, D.G. Triantafyllidis. Data monitoring system for supervising the performance assessment of a photovoltaic park. International Symposium on Computational Intelligence and Informatics (CINTI), 2012, 13: 385–389.

[4] B. Liu, Y.B. Che, L.H. Zhao. The design of photovoltaic monitoring system. Power Electronics Systems and Applications (PESA), 2011, 4: 1–4.

[5] C. Ranhotigamage, S.C. Mukhopadhyay. Field trials and performance monitoring of distributed solar panels using a low-cost wireless sensor network for domestic application. IEEE Sensors Journal, 2011, 11: 2583–2590.

[6] C. Ranhotigamage, S.C. Mukhopadhyay. Field trials and performance monitoring of distributed solar panels using a low-cost wireless sensor network for domestic application. IEEE Sensors Journal, 2011, 11: 2583–2590.

[7] D. Dustegor, N.M. Felemban. Wireless sensor network based monitoring system for photovoltaic panels in extreme GCC climate conditions: A literature review of current approaches. Proc. IEEE GCC Conference and Exhibition (GCC), 2013, 7: 326–329.

[8] J.H. Ye, W. Wen. Research and design of solar photovoltaic power generation monitoring system based on TinyOS. Computer Science & Education (ICCSE) on 9th International Conference, 2014: 1020–1023.

[9] P. Jonke, C. Eder, J. Stockl and M. Schwark. Development of a module integrated photovoltaic monitoring system. IEEE Trans. on Industrial Electronics, 2013, 39: 8080–8084.

[10] U.B. Mujumdar, D.R. Tutkane. Development of integrated hardware set up for solar photovoltaic system monitoring. Proc. IEEE India Conference (INDICON), 2013: 1–6.

[11] W. Kolodenny, Efficient Data Analysis with Modern Analytical System Developed for Use in Photovoltaic (PV) Monitoring System. Photonics and Microsystems on International Students and Young Scientists Workshop, 2006: 26–29.

[12] W. Kolodenny, M. Prorok, T. Zdanowicz, N. Pearsall, R. Gottschalg. Applying modern informatics technologies to monitoring photovoltaic (PV) modules and systems. Photovoltaic Specialists Conference (PVSC) on 33rd IEEE, 2008: 1–5.

[13] X.B. Duan, Z.R. Liang, X.M. Li, S.S. Niu, W. Zhou. Research of distributed photovoltaic power supply access simultaneous power quality monitoring system. 2010 China International Conference on Electricity Distribution (CICED), 2010: 1–4.

[14] X.W. Zhang, R. Chen, C. Wang, X.L. Zhang. Design of remote monitoring and control system center of photovoltaic power station. Innovative Smart Grid Technologies – Asia (ISGT Asia), 2012: 1–3.

[15] Y. Tsur, A. Zemel. Long-term perspective on the development of solar energy. Solar Energy, 2000, 68: 379–392.

Modelling and estimation of harmonic emissions for Distributed Generation (DG)

L.F. Li, N.H. Yu, J. Hu & X.P. Zhang
Electrical Power Research Institute of Guangdong Power Grid Ltd., Guangzhou, China

ABSTRACT: Novel harmonic problems that emerge along with more and more distributed generators are involved in the distribution system. This paper aims to develop a harmonic emission model of an IGBT-based DG. Parameters from both the inverter and the grid system, i.e. filter parameters, system X/R ratio, and short circuit capability are investigated in order to access their effects on harmonic emissions. Based on the proposed harmonic emission model, bode plots are analysed to explore the facts of harmonic emissions. Considering the vector summation theory, a DG capacity can be larger than the scalar summation considered. Simulation and experimental results are provided to validate theoretical analysis and conclusions.

Keywords: harmonics, distribution network, modelling, analysis, resonance

1 INTRODUCTION

One essential purpose of modern distribution network is to accommodate as much DG as possible. Consequently, power quality problems emerge along with the highly penetrated DGs. Harmonics are among the problems to which much attention must be paid. [1]–[6]. A line-commuted DG, such as a 6-pulse inverter or a 12-pulse inverter, used in early days, had a great harmonics emission level, i.e. the minimum THDi tested during one day was reported as 14% [2]. Thanks to the Pulse-Width-Modulation (PWM) technology, high-frequency switching, filtering techniques as well as advanced digital control, the IGBT-based DGs have much fewer harmonic emissions.

However, there are reports that the harmonic emission of the IGBT-based DGs exceeding IEEE 1547 standard [3]. Sometimes the DGs on site have more harmonics than tested in a lab. Possible reasons are listed as follows: 1) The output power is much less than the normal power [4]. 2) There are original voltage distortions of the grid, where a DG was even tripped in the worst case [5]. 3) Facilities of distribution network, e.g. a transformer, a transmission line, a compensate capacitor, etc. result in resonance along with the DG [6]. 4) A type of active anti-islanding techniques leads to a transient current distortion [4].

This paper aims to develop a harmonic emission model of the IGBT-based DG. Parameters from both the inverter and the grid system are investigated to access their effects on the harmonic emission. Based on the random vector summation theory, this paper further estimates the DG capacity allowed in one PCC according to harmonic emissions restriction in the GB/T 14549-93 [7]. Simulations and experiments are conducted to provide proof for the theoretical analysis and conclusions.

2 HARMONICS EMISSION MODELLING

Figure 1 shows a typical two-stage distributed generator that has been widely used as a PV inverter or a wind converter. Usually there are two control objects: a DC-DC converter that is used for the Maximum Power Point Tracking (MPPT) and the DC-AC converter that is used for the current control. The later one is the interfacing power grid which has a great relationship with

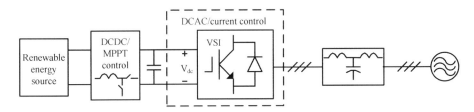

Figure 1. A typical IGBT-based DG structure for the grid-connected.

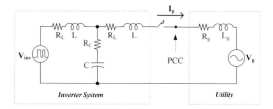

Figure 2. Harmonics emission model for IGBT-based DG.

power quality issues. Since there is a decouple capacitor directly before a DC/AC converter, a voltage source with a Pulse-Width-Modulation (PWM) waveform is taken as the interfacing inverter model. Figure 2 shows the equivalent single-phase IGBT-based DG model that is connected to the power grid, where LCL filter is employed. L and R_L are filter inductance and resistance, respectively; C and R_C are filter capacitance and ESR, respectively; L_g and R_g stand for a system impedance of the power grid; V_{inv} and V_g are voltage source referring to the inverter and its utility.

Based on this model, the grid voltage Ig(s) can be written as:

$$I_g(s) = V_{inv}(s) \cdot G_1(s) + V_g(s) \cdot G_2(s) \quad (1)$$

$$G_1(s) = \left.\frac{I_g(s)}{V_{inv}(s)}\right|_{V_g=0} = \frac{CR_C s + 1}{[(CL^2 + CLL_g)s^3 + (CR_C L_g + 2CR_C L + CR_L L_g \\ + CR_g L + 2CR_L L)s^2 + (L_g + 2L + CR_L^2 + CR_C R_g \\ + 2CR_C R_L + CR_g R_L)s + (R_g + 2R_L)]} \quad (2)$$

$$G_2(s) = \left.\frac{I_g(s)}{V_g(s)}\right|_{V_{inv}=0} = \frac{[(-CL^2 - CLL_g)s^3 - (CR_C L_g + CR_C L + CR_L L_g \\ + CR_g L + 2CR_L L)s^2 - (L_g + L + CR_L^2 + CR_C R_g \\ + CR_C R_L + CR_g R_L)s - (R_g + R_L)]}{[(CL^2 + CLL_g)s^3 + (CR_C L_g + 2CR_C L + CR_L L_g \\ + CR_g L + 2CR_L L)s^2 + (L_g + 2L + CR_L^2 + CR_C R_g \\ + 2CR_C R_L + CR_g R_L)s + (R_g + 2R_L)]} \quad (3)$$

All resistance has remained in this model. If some parameters are ignored in practice, this model can be simplified by making the related parameter to be zero.

3 HARMONICS EMISSION ANALYSIS FOR A SINGLE DG

3.1 Harmonics emission with input of Vinv

Equation (2) gives full transfer function of inverter voltage to output current, according to which one can draw bode plots as shown in Figure 3. In this figure, the quality factor of filter has been considered as $Q = \infty$, 50 and 5, respectively. It can be seen that there is a difference in the high frequency spectrum where the attenuation rate is low with a small Q. This implies a less harmonics suppress capability at high order with a small Q.

It is also observed that the magnitude attenuation at the low frequency spectrum is not so high in all cases, i. e. 9.74 dB at 250 Hz (5th order). This implies that the low order harmonics in V_{inv} will be remained in the output.

Based on the parameter sweep of L, C, R_L, R_C, the system short circuit level S_{sh}, and the system X/R ratio, harmonics emissions (both magnitude and phase) are observed. Figure 4–Figure 6 shows bode plots of the parameter sweep of L and C, X/R and S_{sh} respectively.

It can be seen from Figure 4 that with 50% difference of inductance and capacitance value, the gain at low frequency spectrum changes little. The only difference are harmonics suppression capabilities at a high frequency part. Considering there is ±5% tolerance of filter parameters from various devices, the harmonic emission at low harmonic order (under 25th) would not change too much among different devices. From Figure 5 and Figure 6, one can see that less X/R ratio, or less short circuit capacity leads to lower gains at a low frequency spectrum. This will be difficult to control and will sometimes result in instability and an additional harmonics emission.

3.2 Harmonics emission with grid voltage distortion

Equation (3) gives a grid voltage to the output current transfer function, according to which one can draw bode plots as shown in Figure 7. It is observed that there is no harmonic suppress effect for disturbance from the grid, especially for low order harmonics. This is the reason why voltage distortion will lead to further current distortion of the grid-connected DG.

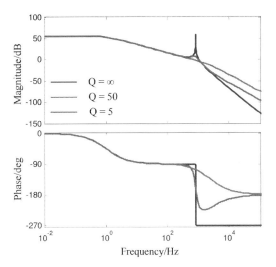

Figure 3. DG Bode plots of V_{inv} to I_g: full model.

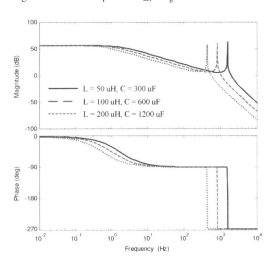

Figure 4. DG Bode plots of V_{inv} to I_g: L, C swept.

4 EXPERIMENTAL RESULTS

A product of the PV inverter has been used to test harmonics emissions with the grid voltage distortion, where a programmable power source served as a distorted grid.

The system setup is shown in Figure 8. There is a measurement right following the PV output. Since the programmable AC source cannot absorb real power, a three-phase load branch is connected between a PV inverter and an AC source. In this way the load consumes the power from both the PV inverter and the AC source.

Test data are recorded as a CSV file and then analysed by Matlab/Simulink. Figure 9–Figure 10 are experimental results of the PV output under different conditions.

Figure 9 shows a normal condition at 5 kW output power, where there is no distortion to the background

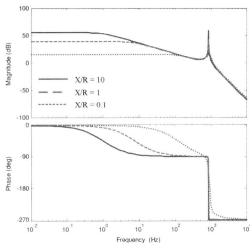

Figure 5. DG Bode plots of V_{inv} to I_g: X/R swept.

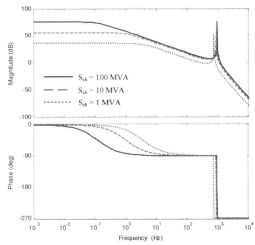

Figure 6. DG Bode plots of V_{inv} to I_g: S_{sh} swept.

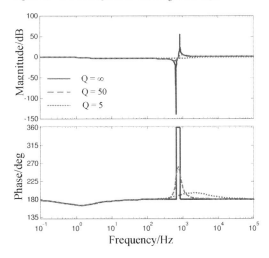

Figure 7. Bode plots of a grid voltage to the grid current.

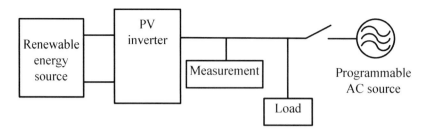

Figure 8. Experimental setup for a PV inverter testing.

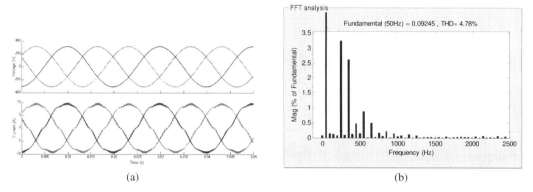

Figure 9. An experimental result at the 5 kW output, THDu = 0% (a) waveforms; (b) FFT analysis.

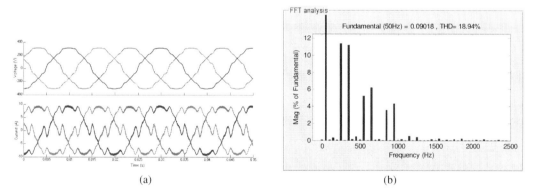

Figure 10. An experimental result at the 5 kW output, THDu = 5% (a) waveforms; (b) FFT analysis.

voltage. It is measured by a powergui/FFT tool that shows that the current total harmonic distortion is 4.78%, which satisfies the IEEE standard.

Figure 10 shows a condition at 5 kW output power with 5% of total harmonic distortion on a background voltage. One can see an obvious distortion of a current that is also injected into the grid. It is measured that the current total harmonic distortion is 18.94%, which again exceeds harmonic limitations quite a lot again. This result is consistent with the theoretical analysis that voltage distortion will lead to further current distortion of the grid-connected DG. It was further measured that a THD_i was 2.29%, 3.13%, 3.95%, 4.66% and 5.51% where the THD_u was increased from 1% to 5%, respectively. The current distortion suppression technique must be used to solve this problem.

5 HARMONICS EMISSION ESTIMATION

It is common that DGs are connected collectively to one PCC. Since it has been noticed that the phase of harmonics changes with inverter or system parameters, vector summation should be considered when estimate system distortion. [8] gives a statistical approach for the harmonics calculation, where (4) was presented in order to calculate the harmonics summation of several DGs.

$$I_h = k \sqrt[\alpha]{\sum_i I_{hi}^{\alpha}} \qquad (4)$$

In (5), the parameters k and α are decided by a magnitude and a phase distribution situation of harmonic vectors. Based on this, one can estimate that a DG capacity of one PCC could be 12 MW in order to reach the 5th order harmonic limit of 20 A, assuming that every DG is 1 MW with a 4% distortion of the 5th order harmonic emission. In comparison, the DG capacity could be only 8.66 MW if only a scalar quantity summation was considered.

6 CONCLUSION

This paper develops a harmonics emission model for the IGBT-based DG, which reveals that the following factors have the main effect on a magnitude and a phase of DG harmonics: 1) low order harmonics in V_{inv}, such as that initiated by a dead band of inverter arms; 2) the X/R ratio, filter parameters of L, C or the quality factor Q, and the system short circuit level S_{sh}; 3) low order harmonics in Vg. Further estimation of a DG capacity in one PCC is presented too. Considering the vector summation theory, a DG capacity can be larger than the only scalar summation considered.

ACKNOWLEDGEMENT

This research work is fully supported by the National High Technology Research and Development Program (863 Program) in China, No. 2012AA050212.

REFERENCES

[1] A. Sundaram, "Power Quality Impacts of Distributed Generation: Guidelines," Palo Alto Electric Power Research Institute 2000.
[2] S. John, "The Issue of Harmonic Injection from Utility Integrated Photovoltaic Systems Part 1: the Harmonic Source," IEEE Transactions on Energy Conversion, Vol. 3, No. 3, Sep. 1988, pp. 507–510.
[3] IEEE Standard for Interconnecting Distributed Resources with Electric Power Systems, IEEE Std. 1547, 2003.
[4] G. Chicco J. Schlabbach F. Spertinof, "Characterization and Assessment of the Harmonic Emission of Grid-Connected Photovoltaic Systems," IEEE Power Tech Petersburg Russia 2005.
[5] S. Filippo, D. L. Paolo, C. Fabio, et al, "Inverters for Grid Connection of Photovoltaic Systems and Power Quality: Case Studies," 3rd IEEE International Symposium on Power Electronics for Distributed Generation Systems (PEDG) 2012. pp. 564–569.
[6] J. H. R. Enslin P. J. M. Heskes, "Harmonic Interaction between a Large Number of Distributed Power Inverters and the Distribution Network," IEEE Trans on Power Electronics Vol. 19, No. 6, 2004 pp. 1586–1593.
[7] GB/T 14549-93, "Quality of Electric Energy Supply Harmonics in Public Supply Network," National Standard of P. R. China, 1993.
[8] J. M. Crucq, A. Robert, B. Laborelec, "Statistical Approach for Harmonics Measurements and Calculations," CIRED, 1989, pp. 91–96.

Mechanism and inhibiting methods for cogging torque ripples in Permanent Magnet Motors

H. Zhang, G.Y. Li & Z. Geng
School of Electrical Engineering, Shandong University, Jinan, Shandong, China

ABSTRACT: Cogging torque is an inevitable problem in Permanent Magnet Motors (PMM), which has a significant influence on control accuracy and motor performance. This paper gives a detailed and easy to understanding description of the mechanism of cogging torque. Different from the previous studies whose descriptions are based on the intensive analytical expression, this paper studied the basic cogging effect, and then added up the basic cogging elements to make up an integrative cogging torque. Then, based on the superposition method, this paper studied the inhibiting methods for cogging torque diminution, which were also verified by the Finite Element Method (FEM).

Keywords: Cogging torque, PM motor, torque ripple

1 INTRODUCTION

Accompanied with the continuous improvement of the high performance permanent magnetic materials, permanent magnet motors are more and more widely used in high performance velocity and position control systems. However, interaction between the permanent magnets and the armature core inevitably brings about torque ripples, which is called the cogging torque. Cogging torque would cause additional torque ripples, vibrations and acoustic noises, which also has a significant influence on the accuracy of the motion control [1].

Cogging torque is a unique phenomenon in the permanent magnet motor. According to the rotor structures, permanent magnet motors are usually classified as two types, e.g., the Surface-mounted Permanent Magnet (SPM) motors and the Interior-mounted Permanent Magnet (IPM) motors. Both the SPM and the IPM motors are inevitably influenced by the cogging torque; this paper mostly studies the SPM type. Different industrial and robotic applications utilize the SPM motor drives for the high torque-to-current and torque-to-volume ratios they exhibit. However, such applications require smooth motor running in order to avoid vibrations and noises [2]. Particularly, how to reduce the level of cogging torque in the SPM motors becomes a significant problem at the design stage.

Based on the analytical resolution and the FEM analysis for a prototype SPM motor, this paper describes the basic mechanisms and attributes of the cogging torque. Then, based on the understanding of the cogging torque mechanism, this paper performs some optimization measures in order to diminish the cogging torque.

2 COGGING TORQUE MECHANISM

When no currents go through the armature windings, the torque, generated by the interaction between the permanent magnets and the stator core, is defined as cogging torque. Furthermore, the cogging torque is also precisely a result of the tangential component's fluctuation of the force between the stator core teeth and the PMs [1].

In short, the cogging torque is caused by the slot openings which dynamically distort the permanent-magnet-excited air-gap magnetic field distribution. In addition, the motors with closed-slots or slotless stators are not affected by cogging torque. With regards to energy, the cogging torque is caused by a variation of the magnetic energy W_m of the field due to the permanent magnet (PM) with the mechanical angular position ϑ of the rotor. As seen [1], the expression of the cogging torque T_{cog} can be represented as

$$T_{cog} = -\frac{\partial W_m}{\partial \vartheta} \qquad (1)$$

However, the analytical resolution of the cogging torque based on a partial differential of the magnetic energy is complicated and abstract, as the magnetic energy W_m's expression is complex [1]. Many of the attributes and optimizations are all within the level of mathematics. In order to describe the cogging torque in an easier but more effective way, some papers propose a new understanding of the cogging torque based on the superposition method [3, 4].

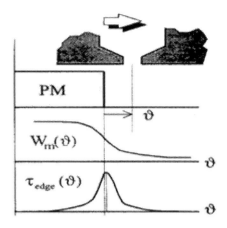

Figure 1. Elementary cogging torque of one pole edge and one slot.

Table 1. Original design parameters of the prototype motor.

Outer Diameter of Stator (mm)	Inner Diameter of Stator (mm)	Minimum Air Gap (mm)	Inner Diameter of Rotor (mm)
120	75	0.5	26
Length of Core (mm)	Max. Thickness of Magnet (mm)	Polar Arc Radius (mm)	Mechanical Pole Embrace
65	3.5	37	0.7

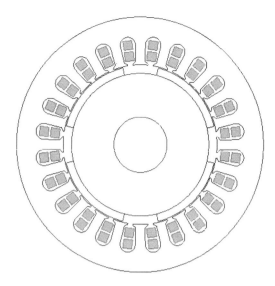

Figure 2. The original lamination of the prototype motor.

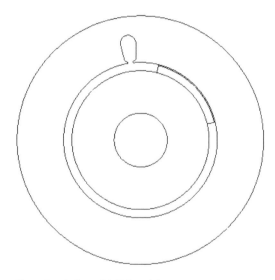

Figure 3. 1-slot and 1-PM model.

2.1 Cogging torque caused by one slot and one pole

The basic element of the cogging torque is generated by one slot and one pole edge, as is described in [2]. The integrative motor cogging torque may be considered as the sum of those cogging torque elements of all edges and slot openings, with each of them considered independent from the others.

Figure 1 shows an edge of a rotor PM pole and a single slot opening moving with respect to the PM edge. The angle ϑ indicates the position of the slot axis with respect to the PM edge. The magnetic energy $W_m(\vartheta)$, sum of the air and the PM energy contributions, is a function of the angular position ϑ. If the slot opening is in the middle of the PM, the variation of $W_m(\vartheta)$ with ϑ is null, while it is large if the slot opening is near to the PM edge [2].

According to (*1), the elementary cogging torque τ_{edge}, whose expression is $-\partial W_m(\vartheta)/\partial\vartheta$, is always a positive value as $W_m(\vartheta)$ is monotonously decreasing with ϑ.

Figure 1 shows the situation when the slot is moving apart from the PM edge. When the slot is approaching the PM edge, the variation tendency of $W_m(\vartheta)$ is opposite to that situation of moving apart. Thus, a reversed cogging torque element would occur.

Based on the FEM, the prototype of the SPM motor's cogging torque is studied. The original design parameters and lamination are respectively shown in Table 1 and Figure 2.

In order to study the elementary cogging torque of 1-slot and 1-PM, the lamination of Figure 2 is changed into Figure 3, which has only a single slot in the stator core and a single PM pole on the rotor surface.

The cogging torque of the situation in Figure 3 is then calculated by FEA software (Ansoft Maxwell 15), and the waveform of the elementary cogging torque is given in Figure 4. As shown in Figure 4, when the rotor in Figure 3 rotates constantly counter clockwise, a negative torque is encountered from about 15 degrees to 30 degrees as the slot is approaching the PM's anterior edge, and then a positive torque is encountered from about 75 degrees to 90 degrees as the slot is departing from the PM's posterior edge. This result is precisely

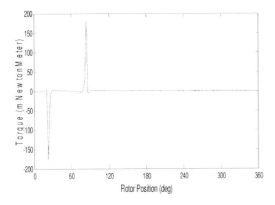

Figure 4. A waveform of the elementary cogging torque with 1-slot and 1-PM model.

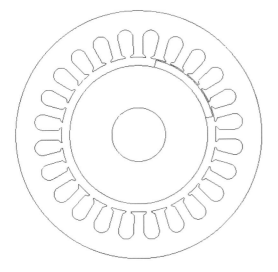

Figure 5. All slots and 1-PM model.

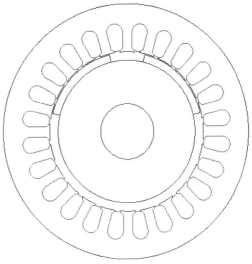

Figure 6. All slots and 1-pole-pair model.

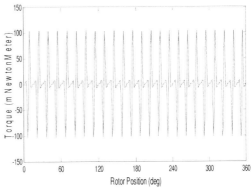

Figure 7. A waveform of the elementary cogging torque with all slots and 1-PM model.

consistent with the discussions above, which also indicates that the waveform in Figure 4 can be considered as a superposition of the cogging torque waveforms of the PM's bilateral edge-slot combinations as shown in Figure 1.

In the later part of this paper, the cogging torque in Figure 4 can be then summed up to make up the cogging torque generated by all the slots. Based on the superposition method, simple models in Figure 5 and Figure 6 are then studied with the assistance of the FEM.

2.2 Cogging torque caused by all slots and one pole

In Figure 4, it can be seen that the negative torque peak is leading the positive torque peak by 60 degrees, approximately to the pole arc's embrace in degrees. However, for a single PM, when all the slot openings are considered, negative torque peak would not lead the positive peak by 60 degrees. Actually, before the PM's anterior edge fully goes through the negative torque, the PM's posterior edge would be generating another positive torque with another slot, this case is shown in Figure 5, and the cogging torque given by FEA software is shown in Figure 7. It can be seen that the peak value is not as high as that in Figure 4, just because of the phased shifted superposition of the negative parts and the positive parts.

It should be noted that the sign of the cogging torque's value is determined merely by the relative positions of the slots and edges. Thus, the polarity of the PM poles does not affect the cogging torque waveform. Furthermore, an N pole and an S pole, or more couples of poles can work together to generate an integrative cogging torque.

2.3 Cogging torque caused by all slots and one pole pair

In a symmetric integral-slot motor (with respect to fractional-slot motors), all PM poles are posited in an

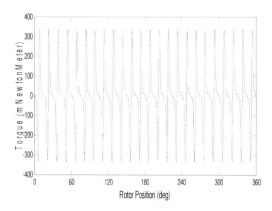

Figure 8. A waveform of the elementary cogging torque with all slots and one pole pair.

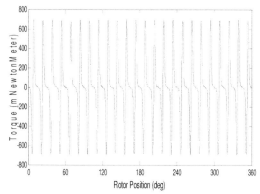

Figure 9. A waveform of the elementary cogging torque with all slots and all poles.

identical relative position with respect to the symmetric stator-slot structure. Thus, the elementary cogging torque of each pole can be composited at the same phase to generate an enlarged integrative cogging torque.

Moreover, when one or more pole pairs are applied, the magnetic energy variation per pole edge would increase to approximately a bit less than twice of the original value, which is also the cogging torque. Thus, for the situation in Figure 6, the cogging torque per pole should be slightly less than twice of that in Figure 5, and the cogging torque of a pole-pair should be slightly less than 4 times of that in Figure 5, which are both verified by the FEM as shown in Figure 8.

2.4 Cogging torque caused by all slots and all poles

Sections above have analysed the cogging torque elements of 1-slot and 1 PM edge, 1-slot and 1-pole, all slots and 1 pole, and all slots and 1-pole-pair. Based on the analyses above, the cogging torque of all slots with more pole pairs or all poles can be simply achieved in a symmetric motor structure by multiplication of the cogging torque waveform and the pole pair number.

When all slots and poles are considered as in Figure 2, the cogging torque is analysed by the FEM, and the waveform is shown in Figure 9. It can be seen that the cogging torque is approximately twice of that of all slots and one pole pair, just as predicted.

Thus, this paper fully achieves and describes the cogging torque of the prototype motor by analysing and summing up the elementary cogging torques of partial structures. Then the attributes of the cogging torque are obtained based on the analyses above.

2.5 Periodicity attributes of cogging torque

As the stator slots are all of self-symmetric structures, there is at least one cogging torque period per slot pitch. Possibly, there still can be more than one cogging torque period per slot pitch [6, 7]. According to equation (*1), the periodicity of the cogging torque is the same as the periodicity of M_m's variation, which is a reflection of the PM poles' MMF and the slot-modulated air-gap magnetic permeability. Thus, the cogging torque period is a function of the pole number $2p$ and the slot number z.

In short, the cogging effect is a process of the mathematical "product to sum" and then trigonometric functions' integration. The product is made up by the distribution of magnetic flux density distribution's square value and the slots' modulating function, both of which are dissolved in Fourier series. In addition, the integration of trigonometric functions makes an orthogonal screening.

The result of the screening is that, only when both N_p and $N_a ps/2p$ are integers, could the smallest N_p be the cogging torque period in one slot pitch. Definitely, for integral-slot motors, $N_a ps/2p$ is always an integer, thus, $N_p = 1$ and there is only one cogging torque period in one slot pitch; as for fractional-slot motors, such as $z = 12$ and $2p = 8$, $N_p = 2$, and there are two cogging torque periods in one slot pitch.

For this prototype motor, $N_p = 1$, and the period count in Figure 7 to Figure 9 are all 24, the same as the slot number.

3 INHIBITING METHODS FOR COGGING TORQUE

Method to inhibit the cogging torque ripples is a critical process in motor design. Also different from the measures basing on pure mathematics, this paper discusses the cogging torque diminutions basing on the superposition method.

3.1 Adjusting the pole embrace

From Figure 7 to Figure 9, it can be seen that the cogging torque waveforms are uniform in shape. Comparing Figure 7 and Figure 4, the cogging torques are all generated by one pole, except for the slot number difference. The peak difference in Figure 4 and Figure 7 reflects the difference in the superposition phases of the positive and negative peaks.

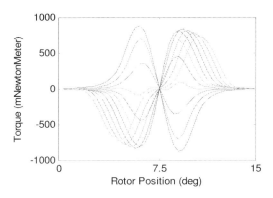

Figure 10. Cogging torque waveforms of different pole embraces in one slot pitch.

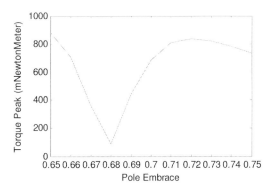

Figure 11. Cogging torque peaks according to different pole embraces.

Sometimes, if the pole embrace is well selected, the positive and negative peaks of the cogging torque generated by one pole and multiple slots can be neutralized to a relatively low cogging torque. Then the integrative cogging torque can be of a low amplitude. Figure 10 shows the cogging torque waveforms of different pole embraces in one slot pitch given by FEM. And Figure 11 maps the torque peaks in Figure 10 according to the pole embrace.

It can be inferred that, the cogging torque peak can be cut down to 80 m Newton meters if the pole embrace is changed from the original 0.7 to 0.68, which also does not make any significant changes in the output of the prototype motor.

3.2 Shifting the pole positions

As the integrative cogging torque can be treated as a superposition of the cogging torque generated by each pole, each pole can be shifted in a relatively small mechanic angle from the original position to perform a phase-shifted superposition of the cogging torque waveforms, which would neutralize the cogging torque. As discussed in [2], the shifting angles can be calculated by:

$$\varepsilon = \frac{(i-1)360°}{2\, pzN_p} \quad (*2)$$

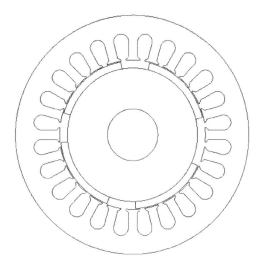

Figure 12. Modified motor structure with shifted poles.

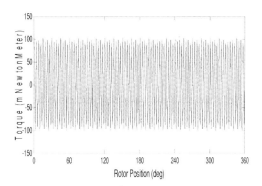

Figure 13. Cogging torque with shifted poles.

where, i stands for the serial number of each PM pole, $i = 1, 2, 3, 4$. This equation means that the shift angle of pole i is $(i-1)/2p$ of the cogging torque's period angle. It also means that, when the mutual deviations of cogging torque waveforms of all poles are $(i-1)/2p$ of the cogging torque's period angle, the fundamental element of the integral cogging torque would be diminished; only the harmonics are left or increased.

Based on the shifted angles calculated by (*2), the modified structure of the motor is shown in Figure 12 and the cogging torque waveform obtained by the FEM is shown in Figure 13. Actually, as the shifted angles are very small, minimum affects would take place in the performance of the prototype motor. Meanwhile, the vibrations and noises are significantly cut down. In Figure 13, the periodicity of the cogging torque is changed, and more periods appear in one slot, as a result of non-neutralized harmonics.

4 CONCLUSION

Cogging torque is a primary vibration source in the permanent magnet motors. Although its mechanism is complex, it can be described by the easy to understand superposition method. Based on the interpretation of the cogging torque superposition of the prototype motor, the characteristics of cogging torque are studied. Moreover, some inhibit methods are performed on the prototype motor, and the FEM results show good effects based these methods. Currently, many other studies are being conducted into the cogging torque analyses. Based on the magnetic analyses, further steps of mechanical motion analyses could be carried out on cogging torque and motor systems.

REFERENCES

[1] Wang, X.H. 2007. Permanent Magnet Electrical Machines. China Electric Power Press. (in Chinese).
[2] Bianchi, N & Bolognani, S.2002. Design Techniques for Reducing the Cogging Torque in Surface-Mounted PM Motors. IEEE Transactions on Industry Applications. IEEE Transactions on Industry Applications, volume: 38, issue: 5, pp. 1259–1265, 2002.
[3] Zhu, Z.Q. et al., 2006. Synthesis of cogging torque waveform from analysis of a single stator slot. IEEE Transactions on Industry applications. pp 125–130, 2005.
[4] Wu, L.J. et.al. 2011. Analytical Cogging Torque Prediction for Surface-Mounted PM Machines Accounting for Different Slot Sizes and Uneven Positions, IEEE International Electric Machines & Drives Conference, pp. 1322–1327, 2011.
[5] Zhang, H. et al. 2012. Stator Core Welding Slots' Influence on PM Motor's Cogging Torque. Small & Special Electrical Machines. volume 40, issue 10, 2012. (in Chinese).
[6] EL-Refaie et al. 2006. Analysis of surface permanent magnet machines with fractional-slot concentrated windings. IEEE Transactions on Energy Conversion. vol. 21, no. 1, pp. 34, 43, March 2006.
[7] Hua Zhang, et al. 2013. An FEM-Based Auxiliary Slot Design for an 8-Pole/12-Slot SPM Servo Motor Considering Saturation of Stator Tooth Coronas., Advanced Materials Research, Vols. 694–697, pp. 3155–3158, 2013.

Content-weighted and temporal pooling video quality assessment

Feng Pan
School of Internet of Things Engineering, Jiangnan University, Wuxi, China

Chaofeng Li
School of Internet of Things Engineering, Jiangnan University, Wuxi, China
Laboratory of computational geodynamics, Chinese Academy of Sciences University, Beijing, China

Xiaojun Wu
School of Internet of Things Engineering, Jiangnan University, Wuxi, China

Yiwen Ju
Laboratory of computational geodynamics, Chinese Academy of Sciences University, Beijing, China

ABSTRACT: Video Quality Assessment (VQA) plays an important role in video processing and communication applications. In this paper, we propose a Full Reference (FR) video quality metric using a content-weighted model in spatial domain and a temporal pooling strategy. The frame scores are got by a three-component weighted SSIM (3–SSIM) method, and then a temporal pooling strategy is used to get the adjusted frame–level scores, and finally these scores are averaged to get the overall video quality. The proposed metric is tested on the VQEG FR-TV Phase I database and on the EPFL-PoliMI video quality assessment database, which suggest it works well in matching subjective scores.

1 INTRODUCTION

With the development of network and the growth of the bandwidth, there are more and more video applications in our daily life. Generally, digital videos pass through several processing stages before they get to the end users. These stages, such as compression and transmission, reduce the quality of the videos. To measure the quality of video applications, we need a reliable video quality assessment metric.

It is customary to classify the objective VQA metrics into three types: full reference (FR), reduced reference (RR), and no reference (NR) metrics. A large body of study on VQA in the literature has focused on FR metrics. In FR video quality metrics, the reference sequence which is assumed to have the best quality is fully available. In these metrics, mean squared error (MSE) and the log–reciprocal peak signal–to–noise ratio (PSNR) have been widely used because they are simple to calculate, and have clear physical meanings, and are mathematically convenient. However, they are not correlated very well with perceptual quality. Assuming that visual perception is highly correlated to the structural information from the images, Zhou Wang et al. proposed a quality assessment method called structural similarity image index (SSIM). This assumption has been well recognized and applied in other image quality metrics. Meanwhile, most of the recently VQA metrics in the literature are based on image quality metrics.

The common way of a VQA method is operating frame-by-frame or several frames to get the local quality scores. These local quality scores are combined into a single score which is used to predict the score that get from the human observers. As we known, we should consider both spatial and temporal distortions in VQA metrics. Here, we propose a full-reference method that combines the three-component weighting method of C. Li (2010) with a temporal pooling strategy of S. Li (2012). In our metric, the pixels in an image frame are divided into three regions to get three region scores, and the frame scores are got by a weighted method. Finally, a temporal pooling strategy is used to get the adjusted frame-level scores and these scores are averaged to get the overall video quality. We test out proposed method on the distorted videos in the VQEG FR-TV Phase I database, and the EPFL-PoliMI video quality assessment database.

2 OUR PROPOSED METHOD

In our work, both the reference videos and distorted videos are transformed from the YUV color space to the Lab color space. Then a single frame is divided into three regions and these region scores are calculated separately. These region scores are weighted to get the frame score. Finally, we pooling the frame scores and get the final score of the video.

Figure 1. Illustration of quality maps: (a) reference image, (b) distorted image, (c) SSIM map, (d) gradient map of (a), (e) edge map, the black regions represent the edge regions, the gray regions represent the texture regions, and the white regions represent the smooth regions.

2.1 Color space

Lab color space is designed to approximate human vision. It is perceptually uniform which makes color distance linear. Lab color space consists of a lightness component and two opposite color components. Lab color space is usually represented as floating point value, which slows down the computation. YUV is a common color space that has been adopted in video coding. It consists of one luminance (Y) and two chrominance (U, V) components. As proved by N. Y. Chang (2008) that Y-only and L-only achieved better performance than using all 3 color spaces. The reason might be human observers focus on luminance or gray scale pattern more than color information to get the information from images. In our metric, the L component is used in the computation.

2.2 Component–weighted method

The SSIM method simply gets the SSIM value by calculating the mean of the entire SSIM map. However, human observers focus on some regions of an image, which is called salient image features, more than the other regions. So, divide the image into several regions will help to improve the IQA/VQA metrics. For example, intensity edges certainly contain considerable image information and are perceptually significant. Here we work on a three component SSIM method to make a full use of the edge regions.

In this method, the original image is divided into three regions: edges, textures, and smooth regions. According to these regions, the SSIM map of the reference and distorted images is divided into three regions as well. As shown in Figure 1, Figure 1(a) is the reference image, Figure 1(b) is the distorted image, Figure 1(c) is the SSIM map of Figure 1(a) and Figure 1(b).

According to the gradient image Figure 1(d), we can divide the SSIM map into three regions, as shown in Figure 1(e).

We can partition an image into three components using the computed gradient magnitude, as the way of C. Li (2010):

(1) Compute the gradient magnitudes using a Sobel operator on the reference and the distorted images.
(2) Determine thresholds $TH_1 = 0.12*g_{max}$ and $TH_2 = 0.06*g_{max}$, where g_{max} the maximum gradient magnitude value computed over the reference image.
(3) Denoting the gradient at coordinate (i, j) on the reference and distorted image by $p_o(i, j)$ and $p_d(i, j)$. While $p_o(i,j) > TH_1$ or $p_d(i, j) > TH_1$, the pixel in the SSIM map is considered as an edge pixel. Then, while $p_o(i, j) < TH_2$ and $p_d(i, j) \leq TH_1$, the pixel in the SSIM map is considered as a smooth pixel. The pixels left here are considered as texture ones.

Then, we can get the pooling scores of these components. Denoting E the score of the edge regions, F the score of the smooth regions, and T the score of texture regions, we get the score S_t of the frame:

$$S_t = w_1 \cdot E + w_2 \cdot F + w_3 \cdot T \qquad (1)$$

where the weighting factor w_1 is set for the edge regions, w_2 is set for the smooth regions and w_3 is set for the texture regions, which they can be determined by performance tuning, as explained in Section 3.

2.3 Temporal pooling

Different from still image quality metrics, there should be a temporal pooling method in a VQA metric. As proved, the human observers are more sensitive to the degradation of the quality than the improvement. we adopt the method of S. Li (2012) to get the adjusted frame score as equation (2). According to the frame-level scores S_t, we can get intermediate scores S'_t, $t \in \{1,\ldots,N\}$:

$$S'_t = \begin{cases} S'_{t-1} + a_- S_{\Delta t}, & S_{\Delta t} \leq 0 \\ S'_{t-1} + a_+ S_{\Delta t}, & S_{\Delta t} > 0 \end{cases} \quad (2)$$

where

$$S_{\Delta t} = S_t - S'_{t-1} \quad (3)$$

The parameters a_- and a_+ embody the asymmetric tracking human behavior. In S. Li's work (2012) the value of the two parameters are get by training ($a_- = 0.075$ and $a_+ = 0.431$), in our method the value of these two parameters are $a_- = 0.86$ and $a_+ = 2.42$, gained by performance tuning.

Finally, the overall video quality score S is calculated by averaging all the frame–level S'_t:

$$S = \frac{1}{N} \sum_{n=1}^{N} S'_t \quad (4)$$

3 EXPERIMENTAL RESULTS AND ANALYSES

To evaluate the performance of the proposed method, two metrics are used here. The first is the Spearman rank order correlation coefficient (SROCC), which is an indicator of the prediction monotonicity of the quality index. Another one is the linear correlation coefficient (LCC) between the DMOS or MOS and the algorithm scores following nonlinear regression. The nonlinearity chosen for regression was a five-parameter logistic function:

$$Q(x) = \beta_1 \times \left(0.5 - \frac{1}{1+\exp(\beta_2 \times (x - \beta_3))}\right) + \beta_4 \times x + \beta_5 \quad (5)$$

The fitting parameters $\{\beta_1, \beta_2, \beta_3, \beta_4, \beta_5\}$ are determined by minimizing the sum of squared differences between the mapped objective scores $Q(x)$ and the subjective scores.

We test on the VQEG FR-TV Phase I database and on the EPFL-PoliMI video quality assessment database.

To get the fitted weight for the three components of 3-SSIM, we varied the weight for edge regions from 0 to 1.0, in steps of 0.05. In each case the total weights summed to unity, and other equal weights were allocated to the texture and smooth regions. The plot of SROCC against the value of the weight for the edge regions are showed as Figure 2, which suggest the edge regions play the most important role, so in the following experiment, the weights of these regions are set as $w_1 = 0.90$, $w_2 = 0.05$ and $w_3 = 0.05$.

The two parameters in temporal pooling are also adjusted to get the maximum SROCC on the VQEG

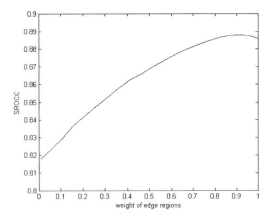

Figure 2. Plot of SROCC against the weight of edge region.

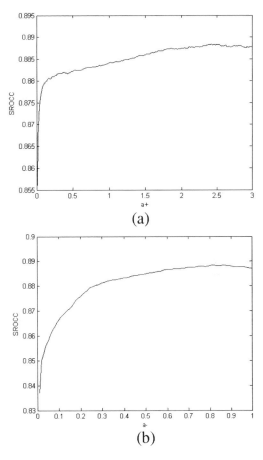

Figure 3. Plot of SROCC against parameter: (a) Plot of SROCC against the parameter a_+, (b) Plot of SROCC against the parameter a_-.

database. As shown in Figure 3, the values are set as $a_- = 0.86$ and $a_+ = 2.42$.

For comparison, the SROCC and LCC of the proposed metric and other metrics on the VQEG database is shown in Table 1.

Table 1. Performance comparison of video quali-ty assessment indices on VQEG database. M1 is the proposed method with the Y component in the YUV color space. M2 is the proposed method with the L component in the Lab color space.

VQA	SROCC	LCC
PSNR	0.780	0.781
SSIM	0.773	0.812
3-SSIM	0.870	0.865
Metric of J. You	0.803	0.817
MOVIE	0.860	0.858
M1	0.876	0.867
M2	**0.888**	**0.873**

Table 2. Database Performance comparison of video quality assessment indices on EPFL-PoliMI database.

VQA	SROCC	LCC
PSNR	0.795	0.800
SSIM	0.933	0.941
3-SSIM	0.962	0.963
Metric of J. You	0.945	0.947
MOVIE	0.920	0.930
M1	0.965	0.964
M2	**0.966**	**0.965**

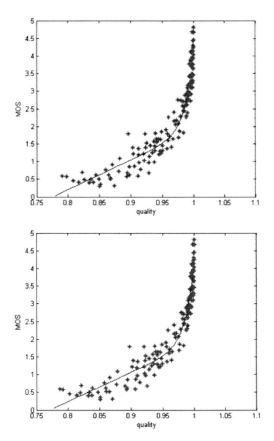

Figure 5. Scatter plots of predicted scores versus DMOS values of the VQEG database: (a) M1, (b) M2.

Figure 4. Scatter plots of scores calculated versus DMOS values of the VQEG database: (a) M1, (b) M2.

As shown in Table 1, the temporal pooling strategy can help improve the performance of the 3-SSIM method, and using the L component in the Lab color space can get the better performance than using the Y component in the YUV color space.

The scatter plots of predicted scores versus DMOS of the VQEG database are given in Figure 4, which also show the efficiency of our method.

Using two databases instead of one can remove any biases in analysis or interpretation that might be incurred by only using content from one database, so we evaluate our method on the EPFL-PoliMI video quality assessment database to verify the parameters. The results are shown in Table 2, and all the parameters used are the same with those used on the VQEG database.

From Table 2, we can find that our method also gets satisfying performance on the EPFL-PoliMI video quality database, which is verified by the scatter plots in Figure 5.

4 CONCLUSION

In this paper, we propose a content–weight and temporal pooling metric for video quality assessment. The main idea of the metric is to combine the 3-SSIM method with the temporal pooling strategy. According to the tests on the VQEG FR-TV Phase I database and on the EPFL-PoliMI video quality assessment database, our proposed method works well in matching subjective scores.

ACKNOWLEDGEMENTS

This research is supported by the National Natural Science Foundation of China (No.61170120), the program for New Century Excellent Talents in University (NCET-12-0881), and China Postdoctoral Science Foundation (No. 2011M500431).

REFERENCES

[1] C. Li, A. C. Bovik, 2010, "Content–weighted video quality assessment using a three–component image model," *J. Electron. Imaging*, vol. 19, no. 1, 011003-1 -011003-9.

[2] F. De Simone et al., 2010, "H.264/AVC video database for the evaluation of quality metrics," in Proc. IEEE Int. Conf. Acoustic Speech, Signal Processing, Dallas, TX, pp. 2430–2433.

[3] J. You, J. Korhonen, A. Perkis and T. Ebrahimi, 2011, "Balancing Attended and Global Stimuli in Perceived Video Quality Assessment," *IEEE Trans. Multimedia*, Vol. 13, No. 6, pp. 1269–1285.

[4] J. You, J. Korhonen, T. Ebrahimi and A. Perkis, 2014, "Attention Driven Foveated Video Quality Assessment," *IEEE Trans. Image Process.*, Vol. 23, No. 1, pp. 200–213.

[5] K. Seshadrinathan and A. C. Bovik, "Motion–tuned spatio–temporal quality assessment of natural videos, 2010," *IEEE Trans. Image Process.*, vol. 19, no. 2, pp. 335–350.

[6] N. Y. Chang, Y. Tseng, and T. S. Chang, 2008, "Analysis of color space and similarity measure impact on stereo block matching," Circuits Syst., IEEE Asia Pacific Conference on, 926–929.

[7] S. Li, L. Ma, and K. N. Ngan, 2012, "Full–Reference Video Quality Assessment by Decoupling Detail Losses and Additive Impairments," *IEEE Trans. Circuits Syst. Video Technol.*, Vol. 22, No. 7, pp. 1100–1112.

[8] VQEG, 2000, Final Report From the Video Quality Experts Group on the Validation of Objective Models of Video Quality Assessment, Phase I FR-TV I.

[9] Z. Wang, A. C. Bovik, H. R. Sheikh, and E. P. Simoncelli, 2004, "Image quality assessment: from error visibility to structural similarity," *IEEE Trans. Image Process.*, vol. 13, no. 4, pp. 600–612.

A filter structure designing for an EAS system

Min Lin & Jianliang Jiang
School of Optoelectronics, Beijing Institute of Technology, Beijing, China

ABSTRACT: Filter circuit is an important part of the tag identification and detection in the Electronic Article Surveillance (EAS) system and its performance has a direct impact on the tag detection sensitivity. In this paper, we propose a TL084-based filter circuit. By designing the proper parameter values combination of the peripheral capacitance and resistance, we obtain a filter structure with a centre frequency of 11 kHz, and a bandwidth of 20 kHz. Using simulation, we demonstrate that our design has much better filtering characteristics.

1 INTRODUCTION

EAS system is a special Radio Frequency Identification system, which prevents merchandise theft through the radio frequency identification tag. The main hardware design of an EAS system includes a transmitter, a receiver, and a tag-signal identification circuit. In between, filter circuit parameter design is one of the key tasks.

The filter circuit selects useful frequency signals while suppressing unwanted frequency signals. The traditional filter circuit uses a resistor, an inductor, and a capacitor. However, with the rapid development of the integrated operational amplifier (op-amp), the active filter circuit finds its application more widely.

In this paper, two TL084 and related RC networks constitute an active filter circuit (Yu, 2006). Through calculation and simulation, we determine the external circuit parameters and filter order to filter out low-frequency and high-frequency harmonics mixed in a tag signal.

2 CIRCUIT DESIGN

Through theoretical analysis, the tag signal frequency is about 8 kHz~14 kHz (M, 2004). In order to effectively filter out envelope signal around 300 Hz and frequency-sweeping signal around 8 MHz, and to attain better amplitude-frequency characteristics and rolling-down features of the filter circuit, the centre frequency of the filter circuit is set to 11 kHz, and pass-band range is set to 1 kHz~21 kHz.

2.1 Determination of integrated op-amp for filter circuit

Because the measured signal is relatively weak and very susceptible to interference, the proposed integrated op-amp is required to have a high input impedance, a strong anti-interference, a large open-loop gain, and a wide band. At present, the

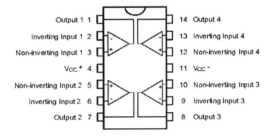

Figure 1. TL084 layout.

development of integrated circuits is very rapid; its input impedance can reach more than 10^{12} Ohms. Compared to the discrete components and the FET circuits, the integrated operational amplifies (op-amp) circuit has an absolute advantage in the conversion speed, open loop gain, and other properties. Therefore, a high-performance integrated operational amplifier (op-amp) is used for the filter circuit.

In the meanwhile the equivalent input noise voltage is an important indicator for detection sensitivity of the integrated operational amplifier (op-amp). In order to obtain a higher signal-noise ratio, the integrated operational amplifier (op-amp) should have a higher conversion rate and a lower equivalent input noise voltage. Based on the above considerations, we choose a TI's TL084 operational amplifier (op-amp) in this design. Its pin layout is shown in Figure 1. It should be noted that positive power supply and negative power supply must be equal in value in order to minimize offset voltage and offset current.

TL084 has four op-amps in one package, so each TL084 can be used for four-level filtering.

2.2 Determination of the topological structure of peripheral circuit

In 2008, Zhiwei Peng, put forward a type of filter circuit based on the TL084 (Peng, 2008). The circuit diagram is shown in Figure 2.

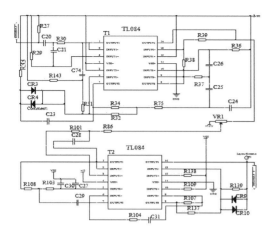

Figure 2. Filter circuit by Zhiwei Peng.

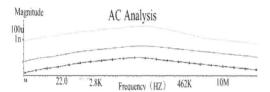

Figure 3. Amplitude-frequency characteristic of Figure 2.

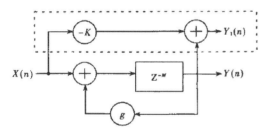

Figure 4. Comb filter block by Xihua University.

After having simulated the circuit performance with Multisim software, we obtained the amplitude-frequency characteristic curve as shown in Figure 3. It shows that the filtering feature is quite plain and rolling-down features are not so steep, and there are no obvious boundaries between the passband and the stopband.

Xihua University put forward a type of comb filter; the filter can process harmonic waves of the tag signal (Li, 2009). The idea is to adjust the comb gain and bandwidth of the spectral lines through the feedback loop, as shown in Figure 4.

Because it will come upon more trouble in filter implement, and it is susceptible to distortion in the process of time delay, and then causes the amplitude-frequency characteristic sag, it apparently does more harm than good, so this idea is not accepted in our design of the EAS system.

For simplicity and practicability of the integrated circuit, this paper proposes a voltage-controlled filter

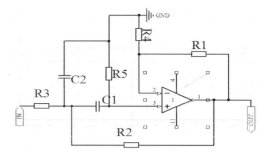

Figure 5. Basic filter circuit unit based on TL084 op-amp.

circuit based on the TL084 op-amp (Hou, 2014). The basic circuit unit is shown in Figure 5.

Since the capacitor within the pass-band can be regarded as "open", therefore C1 does not affect the filter pass-band voltage amplification factor, to simplify the calculation (Kang, 1998), set $R5 = R3 = R$, $C1 = C2 = C$, namely:

$$A_{up} = 1 + R_f / R_1 \quad (1)$$

According to the "virtual short" and "virtual-off" feature as well as the superposition theorem, we can derive the transfer function:

$$A_{us} = \frac{U_o(s)}{U_i(s)} = \frac{1}{1 + (3 - A_{up})sCR + (sCR)^2} A_{up} \quad (2)$$

To make the circuit work correctly, set:

$$Q = 1/(3 - A_{up}) \qquad A_0 = A_{up}/(3 - A_{up})$$

When $s = j\omega$, get the frequency response of the filter expression:

$$A_a = \frac{A_0}{1 + jQ(\omega/\omega_0 - \omega_0 - \omega)} \quad (3)$$

The filter passband width:

$$BW = \omega_0 / (2\pi Q) \quad (4)$$

2.3 *Determination of filter parameters and order number*

There is no ideal filter as to the actual filter amplitude-frequency characteristic, and there is no strict boundary between the pass-band and the stop-band; there is a transition zone between them. The composition within the transition zone will not be totally suppressed, it can only be reduced. The designed filter observed in this paper is set with the centre frequency of the band at 11 kHz, and at the pass-band range of 1 kHz–21 kHz. Within the pass-band, the gain is set to unity.

The order of the filter is referred to as the number of filtering harmonics. Generally speaking, with the

Table 1. Resistance values of four-order filter

Resistance	Order			
	1	2	3	4
R1	1.0K	1.0K	1.0K	1.0K
R2	4.7K	1.5K	10K	1.0K
R3	5.6K	1.8K	8.2K	820
R4	430	430	220	220
R5	2.5K	820	220	470

Table 2. Resistance values of eight-order filter

Resistance	Order							
	1	2	3	4	5	6	7	8
R1	1.0K	1.0K	1.0K	1.0K	1.0K	1.0K	1.0K	1.0K
R2	3.6K	1.5K	7.5K	1.1K	13K	930	20K	820
R3	5.6K	2.5K	7.5K	1.1K	9.1K	1.0K	12K	500
R4	430	430	270	270	200	200	150	150
R5	2.2K	1.0K	3.9K	560	5.6K	390	7.2K	300

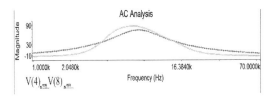

Figure 6. Amplitude-frequency characteristic of four-order and eight-order filtering circuits.

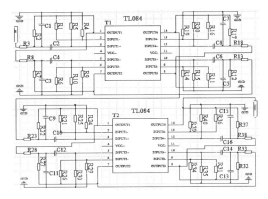

Figure 7. Eight-order filter circuit.

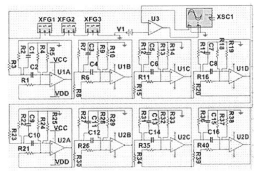

Figure 8. Simulation circuit topology of eight-order filter.

same filter, the higher the order number, the better the performance of the filter. However, the higher the order number, the higher the cost. Therefore, it is very important to choose the right order.

There are four integrated op-amps in one TL084 package. At first, according to the above formula, if the capacitance value is set to 10 nF, the resistance values of four-order or eight-order filtering circuits are shown in Table 1 and Table 2:

Figure 6 is the amplitude-frequency characteristic of the four-order (red curve) and the eight-order (green curve) filtering circuits in a Multisim simulation. We get the qualitative analysis diagram curves summarized together in one figure.

By comparison, the eight-order filter has better rolling-down and amplitude-frequency characteristics than the four-order one, so we choose the eight-order filter circuit, which adopts two TL084 op-amps as the main part of the circuit. A peripheral circuit uses the voltage-controlled filter circuit, as shown in Figure 7.

3 CIRCUIT SIMULATION

Multisim 10 is a well-known circuit design and simulation software, it does not require real circuit environmental intervention, and has many advantages such as high simulation speed, high precision, high accuracy, etc. The simulation analysis of the actual electronic circuit is of great significance in speeding up the design cycle, saving the cost of the design, and improving the quality of the design. Based on Multisim, the simulation circuit topology is shown in Figure 8.

3.1 *The signal waveform analysis*

In tags, signals from the antenna are linear superposition of an envelope signal, a tag signal, and a noise signal. The envelope signal is simulated by an 800 Hz sine wave, the tag signal is simulated by an 8 kHz sine wave, and the interference signal is simulated by an 8 MHz sine wave. The input of the filter is the superposition signal, and the output of the filter will be the tas signal. To make the filtering effect more obvious, the voltage comparator is used after the filter, and we will get a clear square wave signal, the tag signal. Figure 9 shows the waveform diagram before and after filtering. From the figure, we can see that the envelope signal and the interfering signal are filtered out.

3.2 *Ac small signal analysis*

Through an ac small signal analysis of the filter circuit, we can get the amplitude-frequency response one to

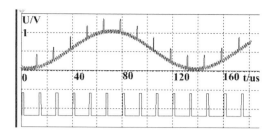

Figure 9. Waveform diagram before and after filtering.

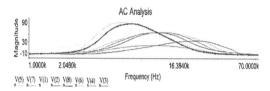

Figure 10. Amplitude-frequency response.

Table 3. Amplitude-frequency response and phase angle of different order filter.

Order	Amplitude-frequency gain	Phase angle gain
1	0.9997	−1.4962
2	0.9988	−2.9907
3	0.9972	−4.4817
4	0.9951	−5.9674
5	0.9923	−7.4463
6	0.9890	−8.9166
7	0.9852	−10.3768
8	0.9807	−11.8254

eight orders, as shown in Figure 10 and get the amplitude gain, as shown in Table 3. V(5) represents the amplitude-frequency response of the fifth order circuit and so on. We can see from the figure that V(8) has the best amplitude-frequency response (blue curve).

Transfer function of the band-pass filter circuit:

$$H(jw) = \frac{H_0}{1+jQ(\omega/\omega_0 - \omega_0/\omega)} = \frac{H_0}{1+jQ(\eta - 1/\eta)} \quad (5)$$

where $\eta = \omega/\omega_0 = f/f_0$

is the relative value of the frequency, namely the normalized frequency. Q is the quality factor of the circuit. To get the maximum amplitude-frequency response, set $\omega = \omega_0$, namely $\omega = 11$ kHz.

Phase-frequency characteristics of the bandpass filter circuit is:

$$\theta = -\arctan Q(\frac{\omega}{\omega_0} - \frac{\omega_0}{\omega}) = -\arctan Q(\eta - \frac{1}{\eta}) \quad (6)$$

This shows that no matter what the quality factor of the circuit is, the phase-frequency characteristic curve

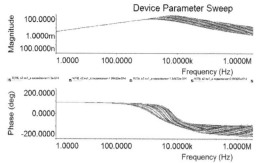

Figure 11. Obtained scanning curves.

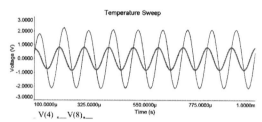

Figure 12. Temperature scanning diagrams.

of the band-pass filter circuit is always monotonically decreasing. The higher the frequency is, the lower the phase shift is. However, the relationship between the phase shift changes and the quality factor of circuits is larger.

Table 3 shows the corresponding amplitude-frequency response and the phase angle gain for each order filter circuit, the values show, the bigger the order, the amplitude-frequncy characteristics of the filter is more desirable.

3.3 Parameter sweep analysis

3.3.1 Capacitive scanning

The capacitances of the circuit are set between 1 nF and 21 nF, and the scanning step is set to 1nF. With the scanning curves obtained and shown in Figure 11 below, we can see that the capacitance has little effect on the amplitude-frequency characteristic and the phase-frequency characteristic below 10 kHz, and the large effect above 10 kHz. When capacitance = 10 nF, the characteristics are the best.

3.3.2 Temperature scanning

The temperature range is set between 1–150 degrees. The scanning analysis diagrams are shown in Figure 12. V(4) represents the fifth-order filter, V(8) represents the eighth-order filter. As seen from the two curves, the temperature has a little influence on the filter circuit. With the advancing time, the circuit voltage does not greatly change.

4 CONCLUSION

In this paper, we studied a type of filter circuit of the eight-order, based on a TL084 integrated op-amp. Through analysis and calculation, the peripheral circuit parameters are finally determined with the voltage-controlled filter circuit. Compared with the four-order filter circuit, it has better rolling-down features and filtering effects. Compared with a traditional filter circuit, it can improve the performance of the circuit integration and simplify debugging. In addition, simulation with Multisim software can intuitively describe the working process of the filter circuit, and is a great help for circuit designing.

REFERENCES

[1] Di Yu, "Study in the Optimization of Radio Frequency EAS system and Signal Identification Algorithm", Master thesis, Zhejiang University, 2006.
[2] Huaguang Kang, "Electronic Technology Foundation" (analog part) [M], 4th ed. Beijing: Higher Education Press Society, 1998.
[3] James Vanderpool and Olin S.Giles, "Managing EAS System and Medical Implant Interactions", IEEE International Symposium on Electromagnetic Compatibility, 2002, 2(4):925–930.
[4] Matthai Philipose and Kenneth P Fishkin, "Mapping and Local-ization with RFID Technology", Proc. 2004 IEEE Int.Conf. Robot. Automatic, pp.1015–1020.
[5] Weizhou Hou, "Design and simulation of two-order voltage source lowpass active filter circuit controlled by voltage", Experimental Technology and management, 2014, 31(10).
[6] Yunsheng Li, "The key technology of radio frequency EAS system", Application of Electronic Technique, 2009, 35(6).
[7] Zhiwei Peng, "Research and Implementation of EAS system based on DSP", Master thesis, Shanghai University, 2008.

A mini-system design based on MSP430F249

Mingming Yang, Yangmeng Tian & Hongwei Wang
Beijing Information Science and Technology University, Beijing, China

ABSTRACT: A mini-system, mini-application system, is made up of minimum components. With regard to MSP430 series microcontrollers, the mini-system generally includes MSP430F249, an oscillator circuit, and a reset circuit. In this paper, the characteristics of MSP430F249 are introduced. In addition, schematic diagrams of the circuit theory about certain modules such as the power module, the oscillator circuit module, the reset circuit module, the JTAG (Joint Test Action Group) interface module, the series module, and the memory module, are designed and presented. In addition, the functions of all modules are put forward.

1 SUMMARY OF MSP430

The hardware circuit design of an MCU(Micro Control Unit) application system consists of two parts: one is the system expansion, the internal MCU function unit, such as ROM, RAM, I/O and two is a timer/counter. An interrupt system cannot meet the requirements of an application system, therefore, we should extend off-chip, select appropriate chip, and design the corresponding circuit. The other is system configuration, to configure peripheral equipment according to the system functional requirements, such as power, and A/D and D/A converter, in order to design a suitable interface circuit. This paper will introduce the minimum system design process based on MSP430F249.

The Texas Instruments MSP430 is an ultralow-power microcontroller, with 60KB+256B Flash Memory, and 2 KB RAM. It contains a basic clock module, a watchdog timer, a 16-bit Timer_A With Three Capture/Compare Registers, a 16-bit Timer_B With Seven Capture/Compare-With-Shadow Registers, four 8-bit parallel ports, an analogue comparator, a 12-bit A/D Converter, a four serial communication interface and other modules.

The MSP430F249 has characteristics as follows:

- Ultralow-Power Consumption: Active Mode: 280 μA at 1 MHz, 2.2 V; Standby Mode: 1.6 μA; Off Mode (RAM Retention): 0.1 μA;
- High efficiency 16-bit RISC Architecture, 27 instructions, when clock frequency is 8 MHz, instruction cycle time is 125ns, most instructions complete in a single clock cycle; when clock frequency is 32 kHz, The 16-bit MSP430 microcontroller execution speed is higher than the typical 8-bit MCU 20 MHz clock frequency during the speed of execution;
- Low Supply-Voltage, wide operating voltage range: 1.8~3.6 V;
- Flexible clock system: two external clocks and an internal clock;
- Low clock frequency can realize high speed communication;
- Serial online programming ability;
- Powerful interrupt function;
- Wake up time is short, only 6s from a low power mode;
- ESD protection, strong interference resistance;
- Operating ambient temperature range of −40 to +85°, suitable for industrial environment.

In MSP430 series all the peripheral control modules are implemented by a special register, so their programming is relatively simple. Programming by special programmer, we can choose the assembly or C language, IAR company developed a special C430 language for MSP430 series microcontroller, which can direct compiler and debugging through the WORKBENCH and C-SPY, its flexible and easy to use.

2 MINIMUM SYSTEM DESIGN

The minimum system is composed of the basic circuit to ensure the reliable work of the required processor, and it mainly includes a power module, an oscillator circuit module, a reset circuit module, a JTAG interface module, a series module, and a memory module. A block diagram of the hardware system is shown in Figure 1.

2.1 *Power module*

This system requires a DC regulated power supply using +5 V and +3.3 V, MSP430F249 and some peripheral devices needed to +3.3 V power supply, the other part requires +5 V power supply. In this system, with +5 V DC voltage is the input voltage, +3.3 V directly obtained by +5 V linear Buck. The power supply module schematic diagram is shown in Figure 2.

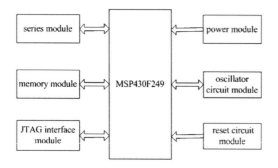

Figure 1. Minimum system hardware block diagram.

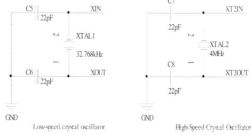

Figure 3. Oscillator circuit module schematic diagram.

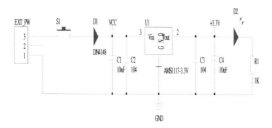

Figure 2. Power supply module schematic diagram.

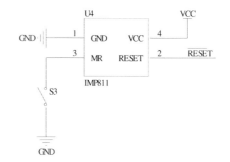

Figure 4. Reset circuit module schematic diagram.

2.2 Oscillator circuit module

The clock system of MSP430F249 is supported by a watch crystal oscillator, an internal Digitally-Controlled Oscillator (DCO) and a high frequency crystal oscillator. In order to solve the contradiction between the requirements of rapid processing data and low power requirements, a multiple clock source or a clock with kinds of different working modes are designed, which thus can solve some peripheral components of the real-time application clock requirements. Such as low frequency communication, LCD display, timer and counter, DCO has been integrated within the MSP430, high speed of crystal oscillator and low speed crystal oscillator circuit are designed in system.

A low speed crystal oscillator (LFXT1) meets the requirements of low power consumption and the use of a 32.768 kHz crystal; the default LFXT1 oscillator operates in a low frequency mode, namely 32.768 kHz. It can also work at a high frequency mode using the external 400 kHz–16 MHz high speed crystal oscillator or ceramic resonator. In this circuit, we use low frequency mode, which has two external 22 pF capacitors connect to the MCU through XIN and XOUT pin.

A high-speed oscillator is also known as a second oscillator XT2, which provides a clock for the MSP430F249 to work at a high frequency mode. In the system, the XT2, using a 4 MHz crystal, has two external 22pF capacitors which connect to the MCU through an XT2IN pin and XT2OUT pin. A schematic diagram is shown in Figure 3.

2.3 Reset circuit module

Reset circuit is set to return to the initial state of the circuit. Manual reset is commonly used in the minimum system; the system uses a dedicate reset chip IMP811 to achieve manual reset. Schematic diagram is shown in Figure 4.

2.4 JTAG interface module

2.4.1 JTAG debugger overview

JTAG is used for the initial testing of the chip, the basic principle is that define a TAP (Test Access Port) inside the device, through a dedicate JTAG test tools for testing internal nodes. The JTAG testing allows multiple devices together using the JTAG serial interface to form a JTAG chain, which can be tested separately. The JTAG interface is also often used to implement ISP (In-System Programmable online programming), for programming the FLASH and other devices.

With a JTAG interface chip, JTAG pin is defined as: TCK is the test clock input; TDI is the test data input; data input via the TDI pin to JTAG interface. TDO is the test data output; data output via the TDO pin JTAG interface. TMS test mode select, which is used to set JTAG interface a particular test pattern. TRST is the test reset input, active low.

MSP430F249 is a 60 kB+259 bytes FLASH memory MCU, and has a JTAG debug interface, therefore, compile program and download from the PC directly to the FLASH through JTAG debugger, then run the JTAG interface control program, read the on-chip CPU status, and memory contents and other information for designers to debug. The entire development

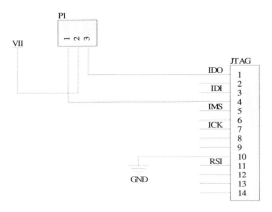

Figure 5. JTAG interface module schematic diagram.

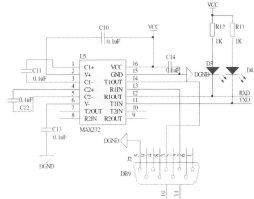

Figure 6. Serial communication module circuit design.

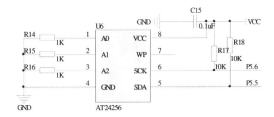

Figure 7. Memory module schematic diagram.

(compiling, debugging) can be integrated in the same software environment, and does not require a specialized emulator programmer. This kind of FLASH technology, JTAG debugging, and development methods combined with an integrated development environment, has a convenient, inexpensive, and practical advantages.

2.4.2 JTAG debug interface design

Because TI produce JTAG debug interface, we just need to lead the microcontroller interface according to the standard, when debugging connect to the JTAG debugger and it can debug programs online. A schematic diagram is shown in Figure 5.

2.5 Series module

Serial communication only needs a few ports to realize the communication between an MCU and a PC, it has the incomparable advantages. Serial communication in two ways: asynchronous and synchronous mode. The MSP430 series has the USART module to realize the serial communication. In this design, the USART module of MSP430F249 communicates with a PC via RS232 serial.

EIA-RS-232 standard is a serial data transfer bus standard developed by the American Electronics Industry Association (EIA). Earlier it was applied to a computer and a terminal for long-distance data transmission via telephone lines and modems. With the development of micro-computers and micro-controllers, not only the long-range communication, but also the short-range communication uses such communication module. In short-range communication systems, telephone lines and modems are no longer used; direct end-to-end connectivity is used instead. RS-232 standard uses negative logic, the standard logic "1" which corresponds to $-5V \sim -15V$ level, while the standard logic "0" corresponds to the $+5V \sim +15V$ level. Obviously, the communication between the two signal levels must be converted. This system uses the level conversion chip MAX3232.

The MAX3232 chip is the MAXIM company level conversion chip, which includes two receivers and a driver, and it a reliable performance. A schematic diagram is shown in Figure 6.

2.6 Memory module

Choose E2PROM CAT24WC256 as large-capacity data storage, it is a 256 K-bit serial CMOS E2PROM, the internal contains 32,768 bytes, each byte is 8-bit, The CATALYST's advanced CMOS technology reduces the power consumption of the device. The CAT24WC256 has a 64-byte pages write buffer to operate the device through the I2C bus interface. A schematic diagram is shown in Figure 7.

3 CONCLUSION

Minimum system can be directly applied as a core component engineering and scientific research, with good versatility and scalability. We can easily extend secondary development and function on the basis of the minimum system, it is possible to shorten the development cycle and reduce development costs. In this paper, we realized the basic functions of the minimum system and introduced the hardware circuit of each module and on this basis built a simple application platform. The minimum system can be used as learning, practice teaching experiment board, and after appropriate modification it can also be applied to electronics design, computer teaching, research, industrial control, and other fields. We can also load the appropriate module, and convert it into a useful product.

ACKNOWLEDGEMENTS

This work was supported by:

(1) Municipal key project of Natural Science Foundation of Beijing (B) (KZ201411232037).
(2) The development program of Beijing City Board of Education Science and Technology (KM201411232017) project.
(3) The development program of Shijiazhuang Heping Hospital (CBJ12C017) project.

REFERENCES

[1] Dake Hu. MSP430 series features 16 ultra-low FLASH MCU [M]. Beijing: Beihang University Press, 2001.
[2] Jianhua Shen. Ultra-low power MSP430 microcontroller series 16 Principles and Practice [M]. Beihang University Press, 2008:202–208.
[3] Lei Cao. MSP430 Microcontroller C Programming and Practice [M]. Beihang University Press, 2007:105–109.
[4] Ping Yang, Wei Wang. MSP430 family of ultra-low power microcontrollers and application [J]. Foreign Electronic Measurement Technology, 2008(12):48–50.
[5] Texas Instruments, MSP430x14x Family User's Guide[s]. 2003.
[6] (http://www.ti.com.cn).
[7] Xinghong Xie, Fanqiang Lin, Xiong ying Wu. MSP430 microcontroller-based and practice [M]. Beihang University Press, 2008:84–85.
[8] Bin Li, Chaoyang Wang, Tao Bu, Xuewei Yu. The minimum system design based on MSP430F149 [J]. Chinese core journals Science and Technology, 2009, (6):216–218.

A control system for the speed position of a DC motor

Violeta-Vali Ciucur
Department of Electrical Engineering, Constanta Maritime University, Constanta, Romania

ABSTRACT: The control system for the speed position of a DC motor, generally can be controlled by a microprocessor, or microcomputer, all these components which can determine the values for speed for each movement and passes. The system can contain a D/A converter. In order to obtain signals which can then be processed by a converter an optical encoder can be used, mounted on the shaft of a DC motor.

1 INTRODUCTION

First of all, DC machines have the power supply frequency or for restitution of energy in network, equal with zero, indifferent if the armature winding is crossed by alternative current. Therefore, frequency being null, gives opportunity to adjust the speed of machine in high limits.

There are indisputable advantages of DC motors compared to AC motors, with regards to speed control. Electromechanical characteristics of DC motors permitting adjustment of the angular velocity over a wide range, maintain a certain level of efficiency of the motor, independent of the speed adjustment method used.

2 THE ADJUSTING OF SPEED OF A DC MOTOR IN A CLOSED LOOP

In a mechatronic system the microcontrollers form the control logic or decision and engines form the execution part. The control part works with 5V voltage levels and low power while the execution part works with higher voltages and power. The interface between the two parts of the system is done by amplifying circuit, which frequently plays a role in the galvanic separation between the two systems. Due to the voltage and current very low values provided at the exit of microcontrollers are needed by the amplification circuits for driving DC motors. A simple and effective scheme to control these motors is the "H bridge."

The driven electrical scheme from above is realized based on the switches type transistor GTO. The bridge is built of four switches S1, S2, S3, and S4 driven to diagonal. When switches S1 and S4 are closed and switches S2 and S3 are opened a positive voltage will be applied to the motor. By opening the switches S3 and S4 and closing the switches S2 and S3, this voltage is inverted thus making it possible to rotate the motor in the opposite direction.

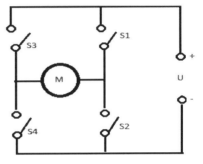

Figure 1. Diagram of a DC motor ordered by the H bridge.

The H bridges may be made of electronic components or are in a single integrated circuit. Their construction is based on the power required by the load. The H bridge method is generally used to reverse the polarity of the motor.

The same success, however, can be applied to engine braking. In this case the engine suddenly stops due to its short circuit of terminals. Also, it can be used to allow the engine to rotate freely until it stops. If the load power required is high will be used semiconductor devices execution like transistors, thyristors, GTO.

In most cases these devices cannot be linked directly to microcontroller outputs through resistances, like in the circuit below, but it needs by power amplifiers.

In case of power for a small load, the hundreds of A, specialized integrated circuits can be used whose outputs will be linked directly or through various interfaces at the motor [2], [6].

The integrated circuit L293D [6] can order motors supplied with a maximum voltage of 35 V and a specific maximum current by 600 mA. The two H bridges each have two input terminals: (INput) and an activation terminal (Enable).

– When the terminal EN is connected to 5 V the H bridge is active.

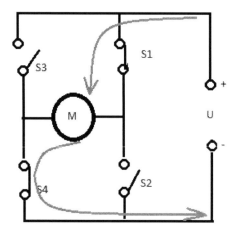

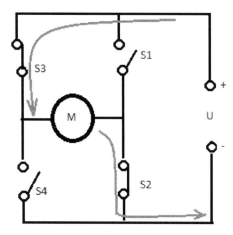

Figure 2. Diagrams of an electric drive for DC motors ordered.

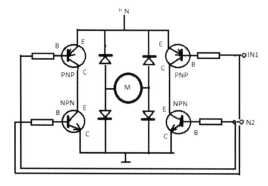

Figure 3. Diagram for a DC motor ordered by a closed loop.

- If entry IN1 is in 1 logic (+5 V) and entry IN2 is in 0 (0 V), the motor rotates.
- If the states of the two inputs are reversed, the motor will rotate in the opposite direction.
- When both inputs are in 0 logic, the engine stops.
- If both are in 1 logic, the motor shafts are braked.

The control and the adjusting in a closed loop where the duration of the PWM cycle can be adapted at torque, changes. It means that during the transition from a static to a dynamic regime, the system should adapt cycle duration PWM so as to provide more power per load.

If the electric diagram used consists of a power supply of 30 V for engine control and a power supply of 5 V to supply the microcontroller and other logic circuits, and a microcontroller with its associated circuitry was obtained, the proper operation of a DC motor for a change in speed between 0 and a maxim value, with a large number of intermediate states.

PWM motor control ensures motor speed control over a very large range with low energy consumption.

It optimizes the maximum energy consumption for motor control while ensuring increased flexibility and adaptability for that control scheme.

By modifying the algorithm written in microcontroller could engine control with variable speed according to various external factors or in both directions of movement [6].

By changing the algorithm and connecting the entry IN2 at microcontroller motor control in both directions at variable speed could [6].

By using external sensors in order to detect events or conditions, the engine speed could change according to the conditions detected by the sensors [6].

Through introducing the current sensors in circuit load (in series with the motor), the system could use that information to operate in a closed loop. The information could be used to change the cycle duration PWM depending on the power required by the load [6].

For a higher load, the current absorbed by the motor increased, as well as the cycle duration PWM.

For a lower load, the current absorbed by the motor decreased, and the cycle duration PWM could decrease also.

3 THE ADJUSTING SPEED OF A DC MOTOR IN AN OPEN LOOP

The circuit as shown in Figure achieves the speed control of a DC motor by varying the supply voltage

The operation of the engine is made in both directions through the potentiometer P (SET POINT) and the two power transistors which lead by turn in a one sense T1 and in other sense T2.

If supposed that the potentiometer P is in the middle and the voltage in a measuring point B1 is 0. In this case both transistors are blocked. As it moved the potentiometer left or right it changed the voltage positively or negatively at the base of transistor [2], [4].

The bipolar power transistor T1 is of the NPN type; when the BE junction is polarized directly and the current in the base increases, then the transistor enters the conduction and the engine starts. It increases the engine speed while the base current increases.

At this time, the transistor, T2 being of the PNP type, is blocked because it has reverse polarity.

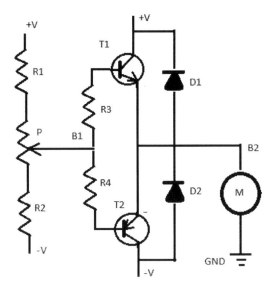

Figure 4. Diagram of a DC motor in an open loop linear adjustment.

However, if the potentiometer was turned the other way then the voltage at terminal B1 becomes negative. The transistor T1 is to block, and the T2 transistor enters the conduction because it is a PNP type. This makes the motor direction change. The diodes D1 and D2 serve to protect the transistors against excessive currents during their operation.

Under this scheme, the adjustment of the DC motor speed is linear due the potentiometer whose resistance changes linearly. From the point of view of the control system, this scheme is an open loop, where we have the prescribing block of reference (y*) and a proportional regulator (P type). Here, we have no negative reaction, so we have changes in the output of the system (y). Therefore, the measured speed of the engine may differ from a set point; the two measured values are compared and the difference between them is the setting error. If the value of this error is very small and tends to zero, we can say that the output signal follows the prescribed signal, so that the adjusting is accurate. If the error does not tend to zero have an output signal that is not respect the prescribed signal. Therefore, the adjusting is less accurate.

As such, this control system was used in applications that do not require high precision. This speed control system is easy to make, requiring fewer pieces and a reduced price.

Operating equations in steady state, underlying the adjustment speed of a DC motor are:
From equations in steady state operation of the DC motor with separate excitation:

$$U = E + R_a I_a + \Delta U_p \approx E + R_a I_a \qquad (1)$$

$$E = k\phi\Omega \qquad (2)$$

$$M = k\phi I \qquad (3)$$

It allows determining for expressions of the mechanical properties:

$$\Omega = \frac{U}{k\phi} - \frac{R_a}{k^2\phi^2}M = \frac{U}{k} - \frac{R_a}{k^2}M = \Omega_0(1 - \frac{M}{M_p}) \qquad (4)$$

Or

$$\Omega = \frac{U}{k\phi} - \frac{R_a}{k^2\phi^2}M = \frac{U}{k} - \frac{R_a}{k^2}M = \Omega_0(1 - \frac{I}{I_p}) \qquad (5)$$

At the constant, magnetic flux has the same configuration.

The artificial mechanical characteristics are constant rigidity:

$$\frac{d\Omega}{dM} = -\frac{R_a}{k^2\phi^2} = -\frac{\Omega_0}{M_{pN}} = ct \qquad (6)$$

and ideal no-load speed is dependent on rotor voltage linearly:

$$\Omega'_0 = \frac{U}{k\phi_N} \leq \frac{U_N}{k\phi_N} = \Omega_0 \qquad (7)$$

An adjustment method shows relatively high energy efficiency, but there is a disadvantage of complexity and there is also a significant cost associated with adjustable DC voltage sources.

The speed adjusting domain is constant, independent of torque resistant, and the torque control characteristics are linear.

In the area $\Omega_0 \leq \Omega \leq \Omega_M$ obtained from de-energization, torque limits are set in terms of satisfying the DC motor commutation, without touching the critical magnetic flux:

$$\phi_m = \frac{2R_a M}{kU} = \frac{M}{k\frac{I_k}{2}} \qquad (8)$$

$$\Omega_M = \frac{U^2}{4R_a M} = \frac{\frac{\Omega_0}{2} \cdot \frac{M_p}{2}}{M} \qquad (9)$$

which occurs at a current value of the induced equal to half of the short-circuit current.

By changing the inductor magnetic flux downwards $\phi \leq \phi_N$ with increasing speed stands decreasing mechanical rigidity of mechanical characteristic and of the load currents then the natural

$$\frac{d\Omega''}{dM} = -\frac{R_a}{k^2\phi^2} = -\frac{\Omega_0}{M_{pN}}(\frac{\phi_N}{\phi})^2 < -\frac{R_a}{k^2\phi_N^2} = -\frac{\Omega_0}{M_{pN}} \qquad (10)$$

$$\frac{d\Omega''}{dI} = -\frac{R_a}{k\phi} = -\frac{\Omega_0}{I_{pN}}(\frac{\phi_N}{\phi}) < -\frac{R_a}{k\phi_N} = -\frac{\Omega_0}{I_{pN}} \qquad (11)$$

and the increasing for the no-load ideal speed:

$$\Omega'_0 = \frac{U_N}{k\phi} > \frac{U_N}{k\phi_N} = \Omega_0 \qquad (12)$$

The efficiency method is relatively good, provided that that the losses in the inductor circuit are relatively low:

$$P_{je} = R_e I_e^2 + R_c I_e^2 \qquad (13)$$

because the excitation current is often negligible with respect to the rotor sizes.

The adjusting domain of the magnetic flux at variable speed control and torque depends on the latter, and tuning characteristics are nonlinear.

4 CONCLUSIONS

Most microcontrollers can generate PWM signals. They use timers in order to measure the time while the signal is in 1 logic and to measure time while the signal stays in 0.

To generate such a signal the numerator is incremented periodically and is reset at the end of each period at PWM. When the numerator value is greater than the reference value, PWM output transitions from state 1 logic in state 0 logic. It is possible to inverse this situation.

The advantages of PWM control are:

- less consumption and debited power
- the dissipated power on the electronic devices working in switching is much less than the power dissipation in the series regulating elements
- the consumption power is at its lowest since the power electronic device working in switching is close to the ideal performance of a power electric switch
- high efficiency
- low volume
- light possibility to include in the scheme of automation
- the PWM control circuit can provide output voltages from a large range controlling the fill factor
- the stabilized output voltages are isolated from the input voltage

REFERENCES

[1] Albu Mihai, Electronica de Putere, Ed. Venus Iasi, 2007.
[2] Haba C.G., Sisteme de comandă a maşinilor electrice, Ed. Gh.Asachi, Iaşi.
[3] John Iovine, PIC Microcontroller Project Book, Ed. McGraham Hill.
[4] M.P. Diaconescu, I. Graur, Convertoare Statice, Bazele te-oretice, elemente de proiectare, aplicatii. Ed. Asachi, Iasi, 1996. Niţulescu Mircea, Sisteme flexibile de fabricaţie, Note de curs.
[5] Vasile Surducan, Microcontrolere pt. toti, Editura Risoprint, Cluj, 2004.
[6] Viorel-Constantin Petre, Introducere in microcontrolere si automate programabile, Matrixrom.
[7] www.regieLive.ro
[8] www.aplicatii-automatizari.com
[9] Boţan, N.V., Comanda sistemelor de acţionare electrică, Editura tehnică, Bucureşti, 1977.
[10] Hilohi,S., Ghinea, D., Năstase, B., Elemente de comandă şi control pentru acţionări şi sisteme de reglare automată, Editura didactică şi pedagogică, Bucureşti, 2002.
[11] Catalogul Festo 2000 – Electromecanică aplicată, Bucureşti, 2000.
[12] www.robotics.ucv.ro/flexform/aplicatii/m1/Cerbulescu%20Claudia%20%20Comanda%20motoarelor%20de%20curent%20continuu/

Intelligent wireless image transmission system

Mingming Zhang, Jinyao Li & Mingfei Wang
School of Mechanical Engineering, Beijing Institute of Graphic Communication, Beijing, China

ABSTRACT: This system uses under-sampling methods, which are the basis for the requirements of an intelligent wireless image transmission system. Here we design the general framework of the system, then refine the Radio Frequency (RF) analogue front-end receiver, and establishes under-sampling down-conversion architecture. In this paper, the system architecture design of the broad receiver RF front-end is the principle line, then it researches and designs the LNA, the MIXER module, the AGC module, the BPF, and phase-locked local oscillator source module respectively. It designs every module with the latest design techniques and then debugs the circuits, in order to allow it at last makes it to meet the design requirements.

1 INTRODUCTION

In recent years, the rapid development of wireless communication technology has penetrated into all areas of life. 3G, CDMA, wireless LAN, and the wireless metropolitan area network are omnipresent. Furthermore, radio communication technology changes rapidly and further promotes the development of wireless data communication. It makes wireless monitoring, video telephone, wireless multimedia teaching, and wireless broadband Internet access become realities. All these have been widely used in social life. With ongoing economic development and greatly improved living standards, people are no longer satisfied with simple voice and image transmission. They now pay more attention to the quality of the image, to transmission bandwidth, transmission distance, and the requirements of the image transmission system with a focus on, moving target recognition and tracking, and other intelligent functions.

The purpose of this paper is to develop a broadband image transmission system under the condition of non-line of sight. The system with an intelligent function can track the target automatic recognition, and automatically focus on it. The system uses wireless communication mode to transmit images, realize the remote image monitoring, and is especially suitable for the layout of complex, seriously occluded terrains and buildings. It has significant meaning for high quality image transmission, intelligent image acquisition, emergency deployment, and the backup application area of the image signal wireless transmission.

2 SYSTEM FRAMEWORK DESIGN

2.1 *System framework*

The system design of the whole structure is shown in Figure 1. The whole system consists of 3 parts: Image acquisition and processing, wireless transceiver, and host.

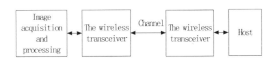

Figure 1. The overall structure of the system.

The image is processed through the wireless transceiver transmission, and eventually transferred to a PC through the USB interface. The image can be viewed, stored, and so on, and can be operated to set goals, adjusting the image acquisition range of the image acquisition part according to the demand.

2.2 *System detailed design*

A system framework of the image collection is shown in Figure 2. Using a DSP as the core of image acquisition and processing algorithms to complete the signal acquisition of the CCD and the image compression processing. The image format uses an MPEG2 or a JPEG. At the same time operation uses the intelligent image algorithm, including target identification, tracking, and automatic focusing in order to control the 3D PTZ according to the operation results of image processing algorithm. Using FPGA as the core to complete the OFDM modulation and demodulation, as well as the signal acquisition and processing. The function of a transmitter RF front-end module is frequency up convert for the image signals which output by the DAC, and then through the multi-stage power drive to transmit signal. The RF receiving module in image acquisition is mainly used to receive the control signal and the target image signal. The clock module provides clock signal which has a high frequency stability and low phase noise for ADC, DAC, and FPGA.

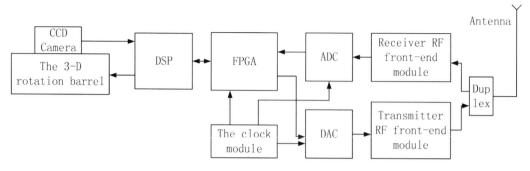

Figure 2. Image acquisition terminal system framework.

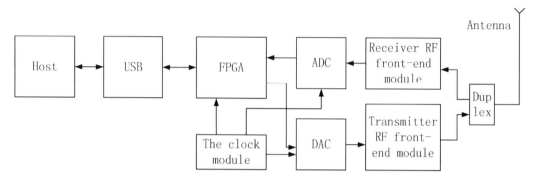

Figure 3. Image receiving terminal system framework.

The system overall framework of the image receiving terminal is shown in Figure 3. Using the FPGA as the core to complete the OFDM modulation and demodulation. The FPGA controls the ADC and DAC to complete signal conversion, and communicates with the host through USB interface to complete the transmission of image data. The receiver RF front-end module receives an image signal transmitted from antenna, and sends it to ADC after completing down conversion and other related processing. The transmitter RF module and the DAC module are used to transmit the control signal and the target image signal. The clock module provides a clock signal which has a high frequency stability and a low phase noise for ADC, DAC, and FPGA.

2.3 *Image processing algorithm*

The image processing algorithm is the core function of the system, which mostly includes the image compression algorithm, the automatic focusing algorithm, and the automatic detection and tracking algorithm for the moving target. The purpose of motion detection is to segment the change region from the background image in the sequence image. The effective motion object segmentation for tracking processing is very important later onHowever, due to the dynamic change of background images, motion detection is a very difficult job. At present, the motion detection algorithm is commonly used with a difference method, an optical flow method, and a background subtraction method. The difference method can adapt to the dynamic background, but generally cannot extract all feature pixels. The calculation time of the optical flow method is long, and the anti-noise performance is poor. This paper intends to select the background subtraction method. Due to errors in the process of image acquisition, changes in the background light and other interference factors in the environment, the background subtraction is also effected. Therefore, it needs to be some improvement in background subtraction method, so that method can adapt to the change of background conditions.

The essence of object tracking is to find the best match in the continuous frame. It can be achieved through the target character matching. The non-rigid object shape has changed over time, and with a simple template matching method, it is certainly not feasible to track the target. Because the probability of occurrence of the target colour, which is recorded by the target histogram, is not affected by the target shape change, this paper uses a colour histogram as a tracking mode, which has good stability. The mean shift algorithm is a non-parametric density gradient evaluation algorithm, which can obtain the local optimal solution. The mean shift has fast and effective characteristics, capable of real-time tracking of non-rigid objects.

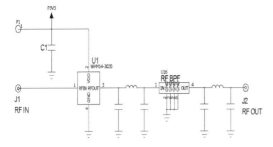

Figure 4. Schematic diagram of LNA module.

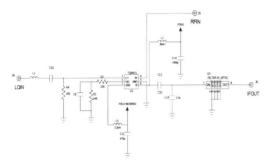

Figure 5. Mixer module circuit diagram.

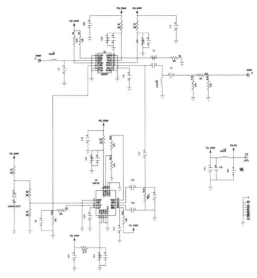

Figure 6. AGC module circuit.

3 WIDEBAND RECEIVER RF FRONT-END CIRCUIT DESIGN

RF analog front-end is one of the key technologies used in order to realize high dynamic, wideband receiver, which largely restricts the several key performance indicators of the receiver, such as sensitivity, bandwidth, and dynamic range. It consists of a low noise amplifier module, a mixer module, a local oscillator module, and an AGC module. It also designed the clock distribution system module, which gives the corresponding circuit design results and the corresponding index, respectively.

3.1 *LNA module design*

A Low Noise Amplifier (LNA) has the main function of weak signal amplification received by antennae from the air; it reduces the noise interference, and demodulates the data required by the system. LNA using WanTcom Company's WHM14-3020AE, is a super low noise wide band amplifier. It has a very low Voltage Standing-wave Ratio (VSWR). Taking into account the isolation of the RF module, the circuit is designed in the form of the module as shown in Figure 4 (LNA module principle diagram). The RF IN terminal is connected with the antenna or the superior LNA module. The output of the U1 has a frequency selective filter, U16 is RF BPF, and both sides of the filter have PI type matching circuit. The low noise amplifier gain is 31 dB, the loss of the filter is 7 dB, and the theoretical output gain of the centre is 24 dB. In theory, the total gain of two levels cascaded LNA module is 48 dB. The actual gain is about 45 dB, taking into account the actual circuit mismatch and connect the cable attenuation. The final system can consider increasing the number of cascade according to the transmission distance and transmission power. The module selected the integrated low noise amplifier with high gain. This method reduced the design and the debugging difficulty. Circuit debugging results and theoretical calculation have little difference, meet the design requirements.

3.2 *Mixer module design*

The mixer is an indispensable component of the RF and the microwave circuit system. According to the conditions, such as signal frequency, chip noise coefficient, conversion gain, and so on, we select TriQuint Company's TQ5M31. The chip is a general down conversion mixer. The mix's IF output is 140 MHz, and the bandwidth is 40 MHz. We need to filter the signal after the mix. In addition to filtering the mirror signal produced by the mixer, needed to filter the other noise signals need to be filtered, for example the system LO leakage. In this paper, we chose 9 order elliptic filter designed by discrete element. Mixer module circuit is shown in Figure 5.

3.3 *AGC module design*

In order to guarantee the dynamic range and bandwidth of A/D conversion, it requires the intermediate frequency circuit can provide amplification function which has a large dynamic range, high sensitivity, and large bandwidth. Usually, in order to meet the dynamic range requirements, circuit needs to design AGC in analogue front-end receiver. Traditionally, because the AGC circuit is controlled by factors such as the impact of the performance of the device, it is difficult to satisfy the requirements of bandwidth, dynamic range,

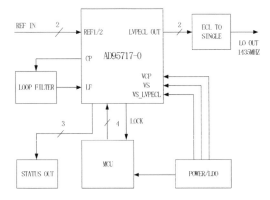

Figure 7. Local oscillator circuit diagram.

linearity and other technical indicators. In this paper, we chose broadband VGA (variable gain amplifier) AD8367 and logarithmic amplifier AD8318 developed by the ADI Company, and designed the AGC circuit which has the advantages of simple structure, small volume, and good design performance. The module circuit is shown in Figure 6.

3.4 *Phase-locked local oscillator design*

The local oscillator is an important component of the transceiver; its performance directly affects the whole index. Phase noise is a very important index to measure the ability of the local oscillator. This paper designs phase-locked local oscillator of low phase noise. According to the system of wireless transmission frequency and intermediate frequency sampling frequency, the local oscillator frequency of the mix is 1.435 GHz. According to the frequency, the performance of the phase noise, and the jitter requirements, the paper selects ADI Company's AD9517-0, the chip is a low phase noise clock generator. The local oscillator circuit diagram is shown in Figure 7.

After frequency doubling the differential input of 10MHz high precision reference clock, the 1.435 GHz vibration signal is outputted in the LVPECL differential form. The vibration signal test results are slightly worse than the simulation results. The spectrum analyser of phase-locked loop in a typical "shoulder" appeared in the vicinity of 100 KHz, at this point the phase noise was about −93 dBc/Hz, and the system could not satisfy the basic requirements.

4 CONCLUSION

This paper carried out the analysis of the intelligent broadband image transmission system, decompensated the functions, designed the overall framework of the system, and refined the framework. It also contributed to a further refinement of the RF analogue front-end receiver, and established a down conversion under sampling structure. We achieved the simulation, design and debugging of the RF analog front end, and designed the low noise amplifier module, the mixer module, the local oscillator module, and the AGC module. Finally, given the corresponding circuit design results and indicators, the debugging results basically satisfy the requirements of the system.

ACKNOWLEDGEMENTS

The paper is a sustentation fund from the Intramural Research Programme of the Beijing Institute of Graphic Communication (No. E-b-2014-19, 23190114036). It was also supported by the Curriculum Construction Project of the Beijing Institute of Graphic Communication (No. 22150113078), the General Program of Science and Technology Development Project of Beijing Municipal Education Commission of China (Grant No. KM201110015007), and the Beijing Institute of Graphic and Communication Talent Program (BYYC201316-001).

REFERENCES

[1] Ahmed Hafiz Zainaldin. Low bit-rate Video Transmission over Wireless ZigBee Networks [D].Canada: Department of Systems and Computer Engineering Caeleton University. 2007.
[2] Aleksandar Tasic', Su-Tarn Lim, Wouter A. Serdijn, John R. Long. Design of Adaptive Multimode RF Front-End Circuits [J]. IEEE Journal of Solid-state Circuits, Vol. 42, No. 2, 2007.
[3] Cao Peng, Wang Mingfei, Qi WeiFei, Yuanchun, Tian Haiyan, Meng Fanjun. The Design of Clock Distribution in the Parallel/interleaved Sampling System [J]. The 9th International Conference on Electronic Measurement & Instruments, 2009, Vol. 1, 544–546.
[4] Stockhammer T, Hannuksela M, Wiegand T. H. 264/AVC in wireless environments [J]. IEEE Transactions on Circuits and Systems for Video Technology, 2003, 13(7): 657–673.

Model predictive control for a class of nonlinear systems via parameter adaptation

C.X. Zhang & W. Zhang
Hebei University of Science and Technology Graduate School, Shijiazhuang, Hebei, China

D.W. Zhang
Hebei University of Science and Technology School of Information Science and Engineering, Shijiazhuang, Hebei, China

ABSTRACT: A number of methods have previously been proposed to deal with the predictive control of nonlinear models. In this paper it is assumed that state of the system is measurable. The research addresses the problem of adaptive models' predictive control and incorporates robust features to guarantee closed loop stability and constraint satisfaction. The parameter update law constructed is only associated with parameter error, and it guarantees non-increasing of the estimation error vector. Using this estimation to update the parametric uncertain set at every time step results in a sequence of gradually reduced sets. The last simulation results are exploited to illustrate the applicability of the proposed method.

Keywords: Model predictive control; Parameter adaptation; Nonlinear systems

1 INTRODUCTION

Model Predictive Control (MPC) is a very popular controller design method in the process industries, which aims to derive the sequence of control actions over a finite time horizon, by seeking to optimize the control performance based on the predicted states of the system under control. An important advantage of MPC is that the use of a finite horizon allows the inclusion of additional constraints. Furthermore, MPC can handle structural changes, such as sensor or actuator failures and changes in system parameters or system structure. Due to the advantage of control performance and robustness, MPC is able to deal with constraints effectively [1].

In general, most physical systems are nonlinear and possess parametric uncertainties or unmeasurable parameters. A lot of research has been done in this area. Pascal Gahinet solved linear systems with parametric uncertainty in [2]. Reference [3] proposed an adaptive MPC algorithm for a class of constrained linear systems, which estimates system parameters online and produces the control input satisfying input/state constraints for possible parameter estimation errors. While some results are available for linear systems with parameters, the design of adaptive nonlinear MPC schemes is very challenging still. In order to improve the design and performance of nonlinear systems with parameters, parameter adaptation MPC has attracted increasing interest in recent years [4,5]. Clarke proposed generalized predictive control which combined adaptive control with predictive control. The algorithm not only improved the adaptability of predictive control for uncertainty, but also enhanced the robustness of adaptive control. A more attractive method is to apply adaptive extensions of MPC in which parameter estimation and control are performed online [6,7,8].

In order to estimate the unknown parameters in systems and obtain more accurate solutions, a lot of adaptive laws have been applied [9,10]. Veronica Adetola presented a method for the adaptive MPC of constrained nonlinear systems [11]. The developed approach considered the effect of the parametric estimation error by combining a parameter adjustment mechanism with robust MPC algorithms. The adaptive parameter mechanism reduces the conservativeness of the solutions through the adaptation of a set including the true value of parameter which is directly adapted online.

This paper presents an algorithm to solve the problem of an adaptive MPC for a class of nonlinear systems. The developed approach combines parameter adaptation which considers the effects of the parameter estimation error with a robust MPC algorithm. In this paper a novel parameter update law and a receding horizon control law are proposed. The parameter adjustment mechanism guarantees that the parameter estimation has been close to the true value and the parametric set is smaller with the process of parameter estimation. First, the problem statement is given in section 2. The adaptation of parameters and the error bound of parameters are discussed in section 3. In section 4 a proposed algorithm is outlined. Lastly the simulation and conclusions are given.

2 PROBLEM STATEMENT

The paper focuses on the following nonlinear parameter system:

$$\dot{x} = f(x,u) + g(x,u)\theta \qquad (1)$$

where $x \in \Re^n$ and $u \in \Re^n$ are the state vector and control input vector of the system respectively; $\theta \in \Re^n$ is the unknown parameter vector. While it is supposed that θ is in the set $\Omega^0 = B(\theta^0, z_\theta^0)$, θ^0 denotes an initial nominal estimator and $z_\theta^0 = \sup_{s \in \Omega^0} \|s - \theta^0\|$ means an associated error bound. In this paper, the objective is to estimate θ and z_θ by adaptive MPC, and then present the robust MPC algorithms.

3 ESTIMATION OF UNCERTAINTY

3.1 Parameter adaptation

In this subsection, a parameter update law and adaptation algorithm are formulated. The procedure assumes the state of the system (1) is accessible for measurement, but does not require the measurement or computation of the velocity state vector $\dot{x}$.

Let $\hat{x}$ and $\hat{\theta}$ denote the state predictor for (1) and the parameter estimator as follows

$$\dot{\hat{x}} = f(\hat{x},u) + g(\hat{x},u)\hat{\theta} \qquad (2)$$

The parameter update law used in this work is as follows

$$\dot{\hat{\theta}} = h\tilde{\theta} \qquad (3)$$

where h is an appropriate constant vector and $\tilde{\theta}$ is the parametric estimation error. $\tilde{\theta} = \theta - \hat{\theta}$. The constant vector h is adjusted with the changing about $\tilde{\theta}$. Such update law is different from the traditional laws which depend on the state prediction error. The performance of the update mechanism is outlined in the following theorem.

Theorem 1. Let $\delta \in [0,1]$ as an appropriate constant, and define a Lyapunov function $V = \frac{1}{2}\tilde{\theta}\tilde{\theta}^T$. The parameter adjustment mechanism (3) is such that the estimation error $\|\tilde{\theta}\| = \|\theta - \hat{\theta}\|$ is non-increasing and:

$$V(t) \leq \delta V(t_0), \text{ for all } t \geq t_0 \qquad (4)$$

Proof. The non-increasing of $\|\tilde{\theta}\|$ is shown from (3):

$$\dot{V} = \dot{\tilde{\theta}}\tilde{\theta}^T = -\dot{\hat{\theta}}\tilde{\theta}^T = -h\tilde{\theta}\tilde{\theta}^T \leq 0 \qquad (5)$$

hence $V(t)$ is a non-increasing function, and $\|\tilde{\theta}\|$ is non-increasing.

Then $V(t) \leq V(t_0)$, $\delta \in [0,1]$ could be chosen to guarantee $V(t) \leq \delta V(t_0)$, for all $t \geq t_0$.

3.2 Set adaptation

In this paper, the set, $\Omega = B(\hat{\theta}, z_\theta)$, which includes the true value of parameter θ, is the adaptive object. The parameter estimator $\hat{\theta}$ is updated by the parameter update law. In the following, its associated error bound $z_\theta^0 = \sup_{s \in \Omega^0} \|s - \theta^0\|$ will be adjusted. Moreover the uncertain set $\Omega = B(\hat{\theta}, z_\theta)$ is obtained online:

$$z_\theta = \min(z_\theta^{\tilde{\theta}}, z_\theta^\delta) \qquad (6)$$

with $z_\theta^{\tilde{\theta}}$ and z_θ^δ generated from

$$z_\theta^{\tilde{\theta}} = \sqrt{V_{\tilde{\theta}}} \qquad (7a)$$

$$\dot{V}_{\tilde{\theta}} = -h\tilde{\theta}\tilde{\theta}^T, \quad V_{\tilde{\theta}}(t_0) = \frac{1}{2}(z_\theta^0)^2 \qquad (7b)$$

$$z_\theta^\delta = \sqrt{V_\delta} \qquad (8a)$$

$$V_\delta = \delta V_\delta(t_0), \quad V_\delta(t_0) = \frac{1}{2}(z_\theta^0)^2 \qquad (8b)$$

respectively.

Algorithm 1. The adjustment of the uncertain set Ω

(i) $\Omega(t_2) \subseteq \Omega(t_1) \subseteq \Omega(t_0), t_0 \leq t_1 \leq t_2$
(ii) $\theta \in \Omega(t_0) \Rightarrow \theta \in \Omega(t), \forall t \geq t_0$

Output $\Omega(t_{i+1})$

(1) Let $i = 1$
(2) For $(i = 1; ++i)$
(3) Let $\dot{\hat{\theta}}(t_i) = h\tilde{\theta}(t_{i-1})$
(4) Let $z_\theta(t_i) = \min[z_\theta^{\tilde{\theta}}(t_i), z_\theta^\delta(t_i)]$, where $z_\theta^{\tilde{\theta}}(t_i)$ and $z_\theta^\delta(t_i)$ are generated from (7) and (8) respectively.
(5) Let $\dot{\hat{\theta}}(t_{i+1}) = h\tilde{\theta}(t_i)$
(6) Let $z_\theta(t_{i+1}) = \min[z_\theta^{\tilde{\theta}}(t_{i+1}), z_\theta^\delta(t_{i+1})]$
(7) If $z_\theta(t_{i+1}) \leq z_\theta(t_i) - \|\hat{\theta}(t_{i+1}) - \hat{\theta}(t_i)\|$ do $\Omega(t_{i+1}) = B(\hat{\theta}(t_{i+1}), z_\theta(t_{i+1}))$
(8) Else let $\Omega(t_{i+1}) = B(\hat{\theta}(t_i), z_\theta(t_i))$.

In the following algorithm it is shown that the uncertainty set Ω is non-increasing during implement the Algorithm 1, and the true value about parameter θ is still in the ball Ω.

Algorithm 2. The evolution of $\Omega = B(\hat{\theta}, z_\theta)$ under (3), (6)–(8), and Algorithm 1 is such that:

(i) $\Omega(t_2) \subseteq \Omega(t_1) \subseteq \Omega(t_0), t_0 \leq t_1 \leq t_2$
(ii) $\theta \in \Omega(t_0) \Rightarrow \theta \in \Omega(t), \forall t \geq t_0$

4 ROBUST ADAPTIVE MPC

Predictive control is a kind of receding optimization control based on the objective function. In this subsection, a robust adaptive MPC algorithm is proposed.

The discrete system is considered as follows:

$$\hat{x}(k+1) = f(\hat{x}(k), u(k)) + g(\hat{x}(k), u(k))\hat{\theta}(k) \qquad (9a)$$

where $\hat{x}(k) = (\hat{x}_1, \hat{x}_2, \ldots, \hat{x}_n)^T \in \Re^n$ is the state of the system, $u(k) = (u_1, u_2, \ldots, u_n)^T \in \Re^n$ is the input of the system, and $\hat{\theta}(k) = (\hat{\theta}_1, \hat{\theta}_2, \ldots, \hat{\theta}_n)^T \in \Re^n$ is the uncertain parameter of the system.

Moreover, let F be the feedback control policies, and the receding horizon control law is defined by:

$$F = \arg\min J_{MPC}(\hat{x}(k), u(k), \hat{\theta}(k)) \quad (9b)$$

Then:

$$u(k/k) = F(\hat{x}(k/k), \hat{\theta}(k/k)) \quad (9c)$$

In order to implement the MPC algorithm a cost function is defined for the system as:

$$J_{MPC}(k) = \sum_{j=0}^{N_x} \hat{x}(k+j/k)^T Q\hat{x}(k+j/k) +$$

$$\sum_{j=0}^{N_u} u(k+j/k)^T Ru(k+j/k)$$

$$+ \sum_{j=0}^{N_\theta} \hat{\theta}(k+j/k)^T P\hat{\theta}(k+j/k) \quad (9d)$$

where N_x is the prediction horizon, N_u is the control horizon and N_θ is the uncertain horizon. $\hat{x}(k+j/k)$, $u(k+j/k)$ and $\hat{\theta}(k+j/k)$ denote the predicted state, control action, and parameter estimator at step k. $Q > 0$, $R > 0$ and $P > 0$ are positive definite states, control, and parameter weights respectively.

The robust MPC problem is defined as follows:

$$\min J(\hat{x}, u, \hat{\theta}) \quad (10a)$$

subject to:

$$\hat{x}(k+1) = f(\hat{x}(k), u(k)) + g(\hat{x}(k), u(k))\hat{\theta}(k) \quad (10b)$$

$$\hat{x}(k+j/k) \in \Re^n, j = 1, \cdots, N_x$$
$$u(k+j/k) \in \Re^n, j = 1, \cdots, N_u \quad (10c)$$
$$\hat{\theta}(k+j/k) \in \Re^n, j = 1, \cdots, N_\theta$$

The establishment of the robust MPC problem accounts for the effect of future parameter adaptation. The algorithm reduces the parameter error in the process of optimizing the cost function J.

Algorithm 3. The performance of the MPC
Input $\hat{x}_i(k)$ (could be measured) and $\hat{\theta}_i(k)$ (generated from (3))
Output $u_i(k)$

(1) Let $i = 1$
(2) For $(i = 1; ++i)$
(3) Input $\hat{x}_i(k)$ and $\hat{\theta}_i(k)$
(4) Solve the optimization problem (9) and (10).
(5) Apply the resulting feedback control law to manufacturing plant until the next-sampling instant.

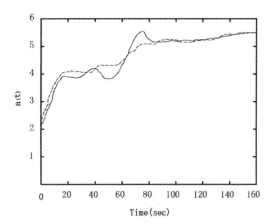

Figure 1. Closed-loop parameter a estimates profile for states at (0.5,323) is the solid line, and (0.8,323) is the dotted line.

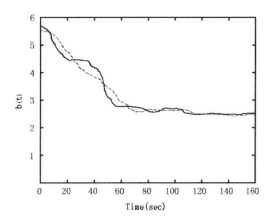

Figure 2. Closed-loop parameter b estimates profile for states at (0.5,323) is the solid line and (0.8,323) is the dotted line.

5 SIMULATION EXAMPLE

This section illustrates the effectiveness of the techniques which combining adaptive mechanism with robust MPC by a simple example with two parameters. Consider the following systems:

$$\dot{X}_1 = F(X_{si} - X_1) - a\exp(-\frac{G}{X_2})X_1$$

$$\dot{X}_2 = F(X_{cd} - X_2) + \frac{b}{C}\exp(-\frac{G}{X_2})X_1 + D(E - X_2) \quad (11)$$

X_1 and X_2 are states for the systems which could be measured. a and b are parameters satisfying $0.1 \leq a \leq 10$ and $0.1 \leq b \leq 10$. The states constraints are $0 \leq X_1 \leq 1$, $280 \leq X_2 \leq 370$, and $280 \leq E \leq 370$. The nominal operating conditions are given from [11], $F = 1$, $G = 8750$, $D = 2$, $X_{si} = 1$, $X_{cd} = 350$.

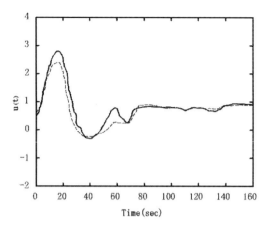

Figure 3. Closed-loop input profiles for states at different initial conditions: (0.5,323) is the solid line and (0.8,323) is the dotted line.

Defining $x = [\frac{X_1 - 0.5}{0.5}, \frac{X_2 - 350}{20}]$, $u = \frac{E - 300}{20}$. The cost function J was selected as:

$$J = x^T Q x + u^T R u + \hat{\theta}^T P \hat{\theta} \qquad (12)$$

where $Q = \begin{bmatrix} 0.6 & 0 \\ 0 & 1.134 \end{bmatrix}$, $R = \begin{bmatrix} 1.233 & 0 \\ 0 & 0.5 \end{bmatrix}$, $P = \begin{bmatrix} 1.414 & 0 \\ 0 & 2.3 \end{bmatrix}$.

The system was simulated from two different initial states $(X_1(0), X_2(0)) = (0.5, 323)$ and $(0.8, 323)$. The true values of the unknown parameters were chosen as $a = 5.5$ and $b = 2.4$.

The following are the results using Matlab control toolbox.

6 CONCLUSIONS

In this paper an approach is provided to deal with MPC for nonlinear systems with parametric uncertainties. With the update of the parameters, the performance of the system is improved. This work combined adaptive parameter and robust MPC. It guaranteed the non-increasing of the estimation error and the stability of the system. Particularly in adaptive mechanisms the parameter is placed in an uncertain set, and it reduced the conservatism of the solution. Lastly the simulation proved the effectiveness of the novel technique.

REFERENCES

[1] XI Yu-Geng. & LI De-Wei. 2008. Fundamental philosophy and status of qualitative synthesis of model predictive control. *Acta Automatica Sinica*: Vol.34 No.10.

[2] P. Gahinet., P. Apkarian., & M. Chilali. 1996. Affine parameter-dependent Lyapunov functions and real parameter uncertainty. *IEEE Transactions on Automatic Control* 41(3): 436–442.

[3] Hiroaki Fukushima., Tae-ahyoung Kim., & Toshiharu Sugie. 2007. Adaptive model predictive control for a class of constrained linear systems based on the comparison model. *Automatica* 43: 301–308.

[4] Alexandra Grancharova. & Tor A. Johansen. 2009. Computation, approximation and stability of explicit feedback min-max nonlinear model predictive control. *Automatica* 45:1134–1143.

[5] Ivan Y. Tyukin, Erik Steur, Henk Nijmeijer, & Cees van Leeuwen. 2013. Adaptive observers and parameter estimation for a class of systems nonlinear in the parameters. *Automatica* 49: 2409–2423.

[6] Guido Makransky & Cee A.W. Glas. 2013. Modeling differential item functioning with group-specific item parameters: A computerized adaptive testing application. *Measurement* 46: 3228–3237.

[7] Lian Lian Jiang., Douglas L. Maskell., Jagdish C. Patra. 2013. Parameter estimation of solar cells and modules using an improved adaptive differential evolution algorithm. *Applied Energy* 112: 185–193.

[8] L. Magni, G. De Nicalao, L. Magnani, & R. Scattolini. 2001. A stabilizing model-based predictive control algorithm for nonlinear systems. *Automatica* 37: 1351–1362.

[9] V. Naumova, S.V. Pereverzyev, & S. Sivananthan. 2012. Adaptive parameter choice for one-sided finite difference schemes and its application in diabetes technology. *Journal of Complexity* 28: 524–538.

[10] Wenyin Gong & Zhihua Cai. 2014. Parameter optimization of PEMFC model with improved multi-strategy adaptive differential evolution. *Engineering Applications of Artificial Intelligence* 27: 28–40.

[11] Veronica Adetola, Darryl Dehaan, & Martin Guay. 2009. Adaptive model predictive control for constrained nonlinear systems. *Systems & Control Letters* 58: 320–326.

Control strategy of BESS for wind power dispatch based on variable control interval

Ting Lei
Electric Power Research Institute of Shanghai Municipal Electric Power Company, Shanghai, China

Wei L. Chai, Weiying Chen & Xu Cai
Wind Power Research Center, Department of Electrical Engineering, Shanghai Jiaotong University, Shanghai, China

ABSTRACT: A novel control strategy of BESS for wind power dispatch based von variable control interval (for charge or discharge) is proposed to prolong the battery lifetime and smooth out the power fluctuation of wind farms. The fluctuation index and the capacity loss index based on BESS lifetime prediction model are introduced to evaluate the performance of this control strategy. Based on that, the relations between the control interval and the two indexes are investigated and a fuzzy controller is designed to determine the best control interval for optimal performance in terms of battery lifetime and damping power fluctuation of the wind farm, the effectiveness of this control strategy is verified by simulation results.

1 INTRODUCTION

To compensate the intermittent power output of wind farms, a BESS is connected to the output bus of the wind farm and is charged or discharged through a power conversion system to smooth the net power injected into the grid, which is called Hybrid Wind-power/Battery-energy-storage System (HWBS). The HWBS can smooth out the power fluctuation of the wind farm and makes wind power more dispatchable.

In order to utilize the wind energy and make wind power dispatch as easy as fossil fuel power plant or hydro power plant, related researches have been done across the world. In [1], two case studies to balance renewable sources and user demands in grids using BESS are presented, but the operation characteristics and battery lifetime is not considered. In [2], a multi-time scale operation model of hybrid wind energy storage system with constraints of battery operation taken into account is established, but the impact of randomness and fluctuation of wind energy on the grid is ignored. [3] Proposes short term control models of BESS for dispatching wind power generation, but the prediction of future wind power output is not included. In [4], Model Predictive Control method (MPC) is used for real-time control of the BESS to smooth out wind power fluctuation, but the impact on battery lifetime is not taken into account.

As can be seen, most of the researches on HWBS focus on minimization of object function under practical operating constrains. In spite of increased dispatchability, the impact on battery lifetime is rarely taken into account [5–7]. In allusion to this problem, this paper establishes a joint operation model

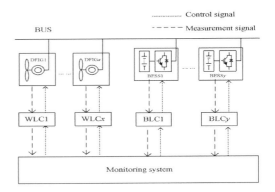

Figure 1. The system structure.

of wind power and BESS and battery lifetime model to investigate the relations between control interval (T_C) and two performance indexes (Fluctuation index δ and Capacity loss index γ_m), based on the relations, a fuzzy controller is designed to determine the best control interval for optimal performance in terms of battery lifetime and damping power fluctuation of the wind farm, the effectiveness of this control strategy is verified by the simulation results in case study.

2 PERFORMANCE INDEXES OF BESS FOR DISPATCH APPLICATIONS

2.1 *Dispatching model of HWBS and the fuctuation index*

As shown in Figure 1, each wind turbine is allocated with a Wind-turbine Local Controller (WLC) and each

BESS is allocated with a BESS local controller (BLC), the monitoring system sends dispatch set point P_r to WLC and BLC, P_r is set to be the average power output of the wind farms $P_r = P_a$. The power output of BESS can be set as P_b (positive sign for discharge), the power output of the wind farm is P_w, which means the equation below:

$$P_b = P_a - P_w \qquad (1)$$

To protect the battery from damage, the charge-discharge current and the State of Charge (SOC) should be kept within proper limits. Pb_MAX and Pb_MIN are the maximum and minimum charge-discharge powers, respectively. SOC_{max} and SOC_{min} are the maximum and minimum SOC, respectively. P_b should then meet the following conditions:

$$\begin{cases} P_b \geq (SOC - SOC_{max})SU/T_{CD}, P_r \leq P_w \\ P_b \leq (SOC - SOC_{min})SU/T_{CD}, P_r > P_w \\ P_{b\_MIN} \leq P_b \leq P_{b\_MAX} \end{cases} \qquad (2)$$

In equation (2), S is the rated capacity of BESS, U is the output voltage of BESS, and T_{CD} is the control interval, SOC can be calculated as follow:

$$SOC(t) = SOC(0) + \int_0^t I_{bat}(t)dt/S \qquad (3)$$

In equation (3), SOC(0) is the initial state of charge, which can be obtained using open loop voltage method.

In this dispatch application, BESS is used for damping power fluctuation of the wind farm, which enables P_g (net power injected into the grid) to track dispatch set points, the fluctuation index δ is defined to evaluate the performance of BESS.

P_g is sum of output power of BESS and wind farm:

$$P_g = P_b + P_w \qquad (4)$$

T_s is the sampling period of wind power output, t_0 is the initial time, the fluctuation index at $t = j \cdot T_s$ is defined as δ_j:

$$\delta_j = \frac{\left|P_g(t_0 + jT_s) - P_r(t_0 + jT_s)\right|}{P_r(t_0 + jT_s)} \qquad (5)$$

In time interval of $(t_0 + mT_s)$, the fluctuation index of HWBS is shown below:

$$\delta = \frac{1}{m}\sum_{j=1}^{m}\delta_j \qquad (6)$$

Funded by Science and Technology Commission of Shanghai Municipality (11dz1210300).

2.2 Lifetime prediction model of BESS and the capacity loss index

LiFePO4 battery is one of the most widely used batteries for BESS. The lifetime of LiFePO4 battery is influenced by many factors, such as temperature (T), charge-discharge rate (CDR), depth of discharge (DOD), state of charge (SOC) and so on. Establishing the lifetime prediction model requires not only research on the chemical reaction mechanism, but also a large amount of experiments for verification. With a mass of experiment results John Wang et al. has established a semi-empirical lifetime prediction model of LiFePO4 battery to predict capacity loss index, as shown below:

$$\gamma_{Si} = B_i \exp\left(\frac{-a_1 + a_2 C_i}{RT}\right) Ah^z \qquad i = 1, 2, 3, 4 \qquad (7)$$

This model takes four influence factors into account, that is cycle times, temperature, current rate and depth of discharge. In equation (7), R represents ideal gas constant, Ah is the Total throughput capacity, B_i is the corresponding parameter with current rate C_i, $C_1 = 0.5$, $C_2 = 2$, $C_3 = 6$, $C_4 = 10$. The capacity loss rate under these four different current rates is recorded in experiments, and the fitting curves are found with corresponding parameters B_i, a_1 and a_2.

The model in equation (7) cannot yet be used directly as the lifetime prediction model for BESS, because it is only the estimation for single cell battery, while BESS is a large energy storage system with much more capacity. It consists of a large amount of single cells, its cycle life is normally shorter than a single cell. The lifetime prediction model for BESS is shown below:

$$\gamma_{Mi} = B_i \exp\left(\frac{-a_1 + a_2 C_i}{RT}\right) \cdot \frac{N_1 S_1}{N_2 S_2} \cdot Ah^z \qquad i = 1, 2, 3, 4 \qquad (8)$$

In this model, N_1, S_1 and N_2, S_2 stand for cycle life and rated capacity of a single cell and BESS, respectively, these four parameters can be obtained from the manufacturer. In equation (8), N_1/N_2 represents the capacity loss increase due to Series-parallel imbalance of the batteries and S_1/S_2 represents equal sharing of the total throughput capacity on each single cell.

The capacity loss index in equation (8) comes from circulating charge and discharge experiments under the same test conditions. However, in the joint operation of HWBS the working conditions can be totally different, the BESS charges and discharges with variable current rate and TCD. Therefore, a revised lifetime prediction model considering cooperation with wind energy integration is proposed. The capacity loss index

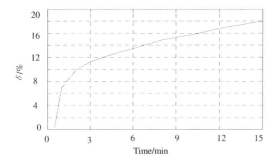

Figure 2. Relation between T_C and fluctuation index.

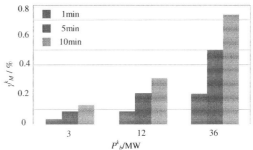

Figure 3. Relation between capacitor loss index and P_b, T_{CD}.

during each charge-discharge interval is defined as γ_m^k (kth interval):

$$\gamma_M^k = \begin{cases} B_1 \exp(\frac{-a_1+a_2C_1}{RT}) \times \left(\frac{P_b^k T_{CD}^k}{u_k}\right)^z \frac{L_N}{L}, 0 < \frac{P_b^k}{u_k I_{1C}} \leq 0.5 \\ B_2 \exp(\frac{-a_1+a_2C_2}{RT}) \times \left(\frac{P_b^k T_{CD}^k}{u_k}\right)^z \frac{L_N}{L}, 0.5 < \frac{P_b^k}{u_k I_{1C}} \leq 2 \\ B_3 \exp(\frac{-a_1+a_2C_3}{RT}) \times \left(\frac{P_b^k T_{CD}^k}{u_k}\right)^z \frac{L_N}{L}, 2 < \frac{P_b^k}{u_k I_{1C}} \leq 6 \end{cases} \quad (9)$$

In equation (9), P_M^k and T_{CD}^k represents charge-discharge power and time interval, respectively. u_k Represents the output voltage of BESS and I_{1C} represents 1C current rate. High current rate is not recommended because of battery aging. Therefore, the current rate is limited under 5C in this model. By accumulating the capacity loss index during each interval leads to the overall capacity loss index for BESS:

$$\gamma_M = \sum_{k=1}^{n} \gamma_M^k \quad (10)$$

3 VARIABLE TCD CONTROL TO DISPATCH WIND POWER

3.1 Discussion of T_C

As control interval of BESS, T_C affects directly on the performance of BESS, it can also be called as the control time of BESS. In this section the relations between T_C and the two performance indexes introduced in chapter 2 will be investigated, the results come from the studied system in section 4.1.

The fluctuation index δ is given in equation (6). On one hand, δ is affected by the limited capacity and energy storage of BESS, because sometimes BESS cannot track the dispatch set point close enough, on the other hand, since BESS can only deliver constant power output during T_C, it is affected by the real-time quick changes of the wind power generation. Therefore, the selection of T_C is directly related to the fluctuation index δ. As shown in Figure 2, the fluctuation index decreases with shorter T_C.

As can be seen in equation (9) and (10), the capacity loss index γ_M^k is directly related to the charge-discharge

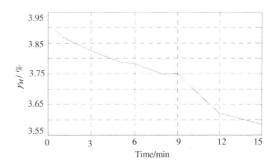

Figure 4. Relation between T_C and capacitor loss index

power P_b and control interval T_C, which is proved by simulation results in Figure 3 and Figure 4.

As shown in Figure 3, the capacity loss index increases with larger charge-discharge power during the same T_C, and with the same charge-discharge power the capacity loss index increase when T_C is extended.

As shown in Figure 4, the capacity loss index increases with shorter T_C. That means shorter control interval causes more damages on battery than longer T_C, which leads to larger capacity loss index.

3.2 Design of a fuzzy controller to determine T_C^k

According to the discussion in section 3.1, a fuzzy controller is designed to determine a proper control interval to delay battery aging while smoothing out wind power fluctuation.

T_C^k is the output of the fuzzy controller, its discourse domain is defined as {0.5, 1, 2, 3, ..., 15} (min), the linguistic value is defined as {ZE, VVS, VS, MS, MB, VB, VVB}. Parameter T_C^k depends on the fluctuation situation and the battery aging status. As fluctuation index, δ_j is selected as one input of the fuzzy controller, its discourse domain is defined as $(0, +\infty)$, its linguistic value is defined as {ZE, VS, MS, MB, VB}. As shown in equation (10), P_b^k, as the charge-discharge power of BESS, is also an important factor for battery aging, it is selected as the other input of fuzzy controller. Its discourse domain is defined as $(-24, 24)$ and its linguistic value is defined as {NB, NS, ZE, PS, PB}.

Table 1. Control rule table for fuzzy control.

δ_j P_b^k	ZE	VS	MS	MB	VB
NB	VVB	VB	MB	MS	VS
NS	VB	MB	MS	VS	VVS
ZE	MB	MS	VS	VVS	ZE
PS	VB	MB	MS	VS	VVS
PB	VVB	VB	MB	MS	VS

Table 2. Parameters of BESS.

Parameter	Value
SOC_{min}/%	20
SOC_{max}/%	80
P_{b_MAX}/MW	24
P_{b_MIN}/MW	−24
SOC(0)/%	50
η_c/%	90
η_d/%	90

The fuzzy control rules are defined as follows:

Rule 1: When δ_j is too large, T_c^k should be decreased to smooth out the fluctuation and increase the power quality.

Rule 2: When δ_j is small enough. Parameter T_c^k should be increased to delay battery aging.

Rule 3: When P_b^k is too large, T_c^k should be increased to delay battery aging.

Rule 4: When P_b^k and δ_j are both too large, the priority should be given to power quality, which means a shorter T_C^k.

According to the rules above, a rule table for fuzzy control is shown below:

4 CASE STUDY

4.1 Configuration of the studied system

A wind farm in eastern China is selected as the studied system, and the configuration of this HBWS is shown in Figure 1. There are altogether 30 wind turbines with installed capacity of 60 MW in this system. The actual wind farm data from the midnight on Oct. 1 2012 to the midnight on Nov. 1 2012 is used for simulation.

This wind farm is equipped with 12 BESSs, each has energy storage of 500 kWh, and the energy altogether is 6 MWh. These BESSs are parallel connected to the output bus of the wind farm through transformers. Each BESS can be charged and discharged for more than 6000 times, and with a capacity of 756Ah, the maximum charging and discharge rate is 6C and 10C, respectively. The parameters of BESS and the lifetime prediction model in section 2.2 are shown below.

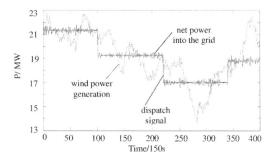

Figure 5. Comparison of power output with dispatch set points.

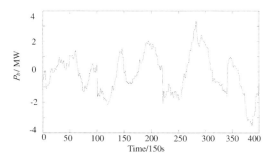

Figure 6. Power output of BESS.

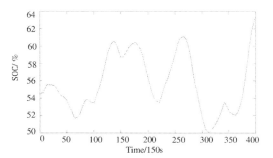

Figure 7. SOC of BESS.

4.2 Simulation results

Figure 5–7 are the simulation results of the HBWS using the proposed control strategy and the actual wind farm data from 18 o'clock on Oct. 20 2012 to 18 o'clock on Oct. 21 2012. Figure 5 shows the comparison of three curves: dispatch set points, wind power generation and net power injected into the grid. Figure 6 and Figure 7 shows the power output and SOC of BESS, respectively. As shown in these figures, with the proposed control strategy the net power of HBWS tracks the dispatch set points quite closely while keeping the SOC of the BESS within desired limits. According to the control rule 1 of the fuzzy controller, when δ_j is too large, T_C^k should be decreased to smooth out the fluctuation, which increases the power quality.

The overall simulation with the actual wind farm data from the midnight on Oct. 1 to the midnight on

Nov. 1 in 2012 shows a satisfactory result. The fluctuation index for the whole month is 12.93% and the capacity loss index is 3.77%, which is much lower than the control strategy using constant control interval T_C.

5 CONCLUSION

This study proposes a novel control strategy of BESS for wind power dispatch based on variable control interval. The performance of this new control strategy is verified by the simulations results using actual wind farm data. The following results and conclusions have been achieved: 1) The fluctuation index and the capacity loss index are introduced to evaluate the performance of HBWS, 2) Shorter control interval leads to smaller power fluctuation of the HBWS and larger capacity loss, 3) A fuzzy controller is designed to determine the best T_C to delay battery aging while keeping the fluctuation within proper limits.

REFERENCES

[1] Bragard M, Soltau N, Thomas S, et al. The Balance of Renewable Sources and User Demands in Grids: Power Electronics for Modular Battery Energy Storage Systems [J]. IEEE Trans on Power Electronics, 2010, 25(12): 3049–3055.

[2] Wu Xiong, Wang Xiuli, Li Jun, et al. A Joint Operation Model and Solution for Hybrid Wind Energy Storage Systems [J]. Proceedings of the CSEE, 2013, 33(13): 10–17 (in Chinese).

[3] Teleke S, Baran M E, Bhattacharya S, et al. Rule-Based Control of Battery Energy Storage for Dispatching Intermittent Renewable Sources [J]. IEEE Trans on Sustainable Energy, 2010, 1(3): 117–124.

[4] Teleke S, Baran M E, Bhattacharya S, et al. Optimal Control of Battery Energy Storage for Wind Farm Dispatching [J]. IEEE Trans on Energy Conversion, 2010, 25(3): 787–794.

[5] Yao D L, Choi S S, Tsend K J, et al. A Statistical Approach to the Design of a Dispatchable Wind Power-Battery Energy Storage Syste [J]. IEEE Trans on Energy Conversion, 2009, 24(4): 916–925.

[6] Li Q, Yuan Y, Yao D L. On the Determination of Battery Energy Storage Capacity and Short Term Power Dispatch of a Wind Farm [J]. IEEE Trans on Sustainable Energy, 2011, 2 (2): 148–158.

[7] Wang J, Liu P, Hick-Garner J, et al. Cycle-life model for graphite-LiFePO$_4$ cells [J]. Journal of Power Sources. 2011, 196(8): 3942–3948.

A study of the maintenance of transformer using a cost-effectiveness optimization model

Li-Juan Guo & Song-Mei Tao
Electric Power Research Institute of Guangxi Power Grid Corp., Nanning, China

ABSTRACT: This paper describes research aimed at the optimization of the maintenance decision of a power transformer, selecting economy and reliability as the analysis index, and considering the effect of different maintenance methods on transformer failure rate and maintenance cost. By analysing the relationship between different maintenance strategies and reliability, maintenance costs, risk return, a life cycle cost-effectiveness of maintenance optimization model is established. The example analysis shows that the model can effectively avoid disrepair and improve the economy and reliability of the transformer.

Keywords: reliability, economy, maintenance effectiveness, risk return, life cycle

1 INTRODUCTION

A power transformer is one of the key pieces of equipment in an electrical power system. It is also an expensive and high weight component. It is therefore essential that the power transformer be appropriately methodically maintained to ensure its availability and reliability during operation. In order to improve the reliability, many maintenance activities and models are proposed [1–3].

However, the existing methods do not consider maintenance strategies on the life cycle level. By analyzing the relationship between maintenance cost and failure rate, establishing maintenance strategies optimization model based on cost-effectiveness.

2 INDICATOR

Effectiveness cannot be measured in monetary terms, it is a measure of the effect of the cost achieved. Effectiveness factors include: the inherent capability, reliability, maintainability, advanced technology, durability, security, and survivability, etc [4–6]. Typically decision criteria are:

① the same expenses criteria: meet a given cost constraints, select the largest effectiveness of the program;
② the same benefits criteria: meet a given performance constraint, select the least cost program;
③ benefit-cost ratio criteria: select the biggest effectiveness and cost ratio program;.

In this paper, we select the effectiveness as the key optimization indicator, failure rate and the risk-benefit as constrains in the model.

3 FAILURE RATE AND EFFECTIVENESS IN DIFFERENT MAINTENANCE STRATEGIES

3.1 *Analysis of failure rate and maintenance cost*

Different maintenance strategies will lead to different transformer failure rates. In order to describe the effect of different maintenance, we introduce the concept of age reduction [7], in which t_n refers to the total number of the equipment operation, t_a refers to the effective age.

As shown in Figure 1, if the transformer has undergone running, the failure rate follows the trend of continuing to rise along curve 1. If the transformer has undergone minor repair pattern, the failure rate goes back to the equivalent age t_2 and follows the trend of continuing to rise along curve 2. If the transformer has undergone major repairs, the failure rate goes back to the equivalent age t_1 and follows the trend of continuing to rise along curve 3. If the transformer has been replaced, the failure rate goes back to the equivalent age t_{n-1} and follows the trend of continuing to rise along curve 4.

Considering cases reported in engineering practice, most of failures happen in non-core parts, and only need minor repairs to restore the transformer's functionality. If there are obstacles in the core parts of the transformer, a timely major repair is needed.

At the early stage of the operation, a transformer failure rate is low, fault risks tend to be smaller, and more easily repaired, therefore adopting the method of minor repair can better affect the transformer's health status.

But with the increase of maintenance times, the effects minor repairs worsen over time, the transformer failure rate is bigger, the minor repair method gradually cannot meet the demand of the reliability of

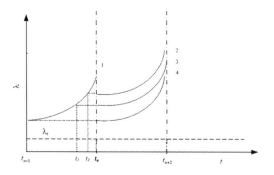

Figure 1. Change of failure rate under different maintenance strategies.

the transformer operation. Major repair can make the transformer's health return to the ideal, the failure rate go back to a low level, the gains are higher, so relatively major repair is more reasonable.

By the end of a transformer life, its mechanical properties and electrical performance degradation is serious, poor maintainability, and major repairs contribute little to health recovery and improvement of the transformer's reliability. At this time a transformer replacement is in accordance with practices of the power grid enterprise. The analysis of failure rate and effectiveness under different maintenance strategies are as follows:

① Continue running mode

Transformer in a continue running mode, no maintenance costs incurred $M_0 = 0$, equipment failure rate and recursion will not be affected. In this maintenance mode, the transformer failure rate and the actual recursion degradation will be in accordance with the current trend to continue running the transformer.

$$P_{f(t-)} < P_{f(t+)} \quad (1)$$

$$t_{a(t-)} < t_{a(t+)} \quad (2)$$

② Minor repair model

Based on reference [8], the transformer minor repair costs are roughly unchanged, a single minor repair cost value is up to 3% of the cost of purchasing transformer. A 110 kV 220 kV, 500 kV, single minor repair will cost roughly 240000 RMB, 510000 RMB, and 1800000 RMB.

Assume that before the minor repair the transformer failure rate is $P_{f(t)}$, after the minor repair the transformer equivalent recursion will be back, this will reduce the failure rate to $P_{f(t+)}$.

When the transformer's equivalent age is t_a, the function of age recursion $T_{\Delta 1}$ as shown in Equation (3), the failure rate variation ΔP_{f1} as shown in Equation (4).

$$T_{\Delta 1} = \begin{cases} 1, & 1 \leq t_a < 5 \\ 0.3 t_a, & 5 \leq t_a < 15 \\ 0.2 t_a & t_a \geq 15 \end{cases} \quad (3)$$

$$\Delta P_{f1} = P_f(t_-) - P_f(t_+) = \alpha \cdot e^{\beta \cdot t_-} - \alpha \cdot e^{\beta \cdot t_+}$$
$$= \alpha \cdot (e^{\beta \cdot t_a} - e^{\beta \cdot t_a - T_{\Delta 1}}) \quad (4)$$

③ Major repair model spending

Major repairs will cost 6% of the spending the of the transformer's purchase. A 110 kV, 220 kV, 500 kV, single major repair will cost roughly 480000 RMB, 1020000 RMB, and 3600000 RMB. In this model, reasonable assumptions of transformer equivalent age at the moment of t_a, the function of age recursion $T_{\Delta 1}$ as shown in Equation (5), the failure rate variation ΔP_{f1} as shown in Equation (6).

$$T_{\Delta 2} = \begin{cases} 0.85 t_a, & 1 \leq t_a < 10 \\ 0.75 t_a, & 10 \leq t_a < 25 \\ 0.5 t_a & t_a \geq 25 \end{cases} \quad (5)$$

$$\Delta P_{f2} = P_f(t_-) - P_f(t_+) = \alpha \cdot e^{\beta \cdot t_-} - \alpha \cdot e^{\beta \cdot t_+}$$
$$= \alpha \cdot (e^{\beta \cdot t_a} - e^{\beta \cdot t_a - T_{\Delta 2}}) \quad (6)$$

④ Replacement model

A transformer's replacement costs is roughly 110% of the equipment acquisition cost. A 110 kV, 220 kV, 500 kV transformer single replacement cost roughly 8800000 RMB, 18700000 RMB, and 66000000 RMB.

In replace mode, a new transformer is equal to the initial state with initial failure rate and age recursion. When equivalent age is t_a, the function of age recursion $T_{\Delta 1}$ as shown in Equation (7), the failure rate variation ΔP_{f1} as shown in Equation (8).

$$T_{\Delta 2} = t_a \quad (7)$$

$$\Delta P_{f3} = P_f(t_-) - P_f(t_+) = P_f(t_-) - P_f(0)$$
$$= \alpha \cdot (e^{\beta \cdot t_a} - 1) \quad (8)$$

3.2 *Analysis of risk return*

Let a transformer in the risk model i ($i = 1$, abnormal state; $i = 2$, Partial outage; $i = 3$, Complete outage; $i = 4$, damaged), maintenance level j (minor repairs, $j = 1$; major repairs $j = 2$; replacement $j = 3$), our risk return analysis is as follows:

① Abnormal condition

For a transformer in the abnormal risk condition, a minor repairs strategy will usually be taken. The transformer's maintenance risk return is show in Equation (9).

$$C_{p1} = R_{earn1} = \frac{(P_f(t_-) - P_f(t_+))}{P_f(t_-)} R_t = \frac{\Delta P_{f1}}{P_f(t_-)} R_t \quad (9)$$

ΔP_{f1}: variation of the. failure rate
$T_{\Delta 1}$: age reduction time.

② Partial outage condition

For a transformer in the partial outage risk condition, a minor repairs strategy usually be taken.

The transformer's maintenance risk return is show in Equation (10).

$$C_{p2} = R_{earn2} = \frac{(P_f(t_-) - P_f(t_+))}{P_f(t_-)} R_t = \frac{\Delta P_{f2}}{P_f(t_-)} R_t \quad (10)$$

③ Complete outage risk condition

For a transformer in a complete outage risk condition, a major repairs strategy usually be taken. The transformer's maintenance risk return is show in Equation (11).

$$C_{p3} = R_{earn3} = \frac{(P_f(t_-) - P_f(t_+))}{P_f(t_-)} R_t = \frac{\Delta P_{f3}}{P_f(t_-)} R_t \quad (11)$$

④ Damaged condition

For a transformer in a damaged condition, a replacement strategy should be taken.

$$C_{p4} = R_{earn4} + D_{earn} = R_{earn4} - CD \quad (12)$$

$$R_{earn4} = \frac{(P_f(t_-) - P_f(t_+))}{P_f(t_-)} R_t = \frac{(P_f(t_-) - P_f(0))}{P_f(t_-)} R_t \quad (13)$$

C_{p4}: replacement maintenance risk return
R_{earn4}: risk return due to failure rate due decline
D_{earn}: the revenue generated by the equipment retirement
CD: retirement cost

3.3 Maintenance strategy optimization model

① Based on the cost-risk return ratio optimization maintenance decision model

$$\min \; f_2 = \frac{C_m}{R_{earn}} = \frac{C_m}{\frac{(P_{f(t_-)} - P_{f(t_+)})}{P_{f(t_-)}} R_t}$$

$$= \frac{C_m}{(P_{f(t_-)} - P_{f(t_+)})(S \cdot M_i + \xi \cdot v \cdot R_{EENS} \cdot L_{ref} + D_{earn,i})} \quad (14)$$

C_m: Maintenance cost
R_{earn}: risk return

② Maintenance constraints

1) Reliability constraints

(1) If the transformer is in abnormal risk or partial outage mode, minor repairs or major repairs are needed. If the transformer is in complete outage risk mode, major repairs or replacement are needed. If the transformer is in the risk of damage mode, major repairs or replacement are needed.

(2) After repair, the failure rate must not exceed a defined threshold $P_{f,s}$.

$$P_{f(t+)} < P_{f,s} \quad (15)$$

$P_{f,s}$ is a threshold, which is related to voltage level and the importance of the substation.

2) Economic constraints

The maintenance cost should be less than the risk return:

$$C_M < R_{earn} \quad (16)$$

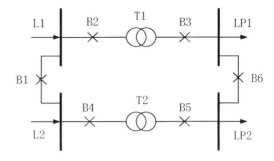

Figure 2. 500 kV substation main electrical equipment wiring structure simplified diagram.

Table 1. Online monitoring data of transformer's oil dissolved gas (μL/L).

Data	CO	H_2	CH_4	C_2H_2	C_2H_4	C_2H_6	total
2013/7/1	12.66	1.04	4.06	0.00	13.37	2.23	19.66
2013/9/14	32.38	3.27	3.95	0.00	15.06	2.16	21.17
2013/12/21	62.86	3.16	7.54	0.00	23.55	2.77	33.86
2013/12/6	76.33	4.47	25.64	0.00	26.09	3.31	55.04
2014/1/19	102.35	5.82	28.87	0.00	27.21	3.42	59.5
2014/3/3	137.11	7.79	37.21	0.00	33.74	5.08	76.03

3) Maintenance time constraints

Minor repair should be implemented within half a year after the maintenance decision, major repairs should be undertaken within one year after the maintenance decision.

$$\begin{cases} t_{m1} < t + 0.5 \\ t_{m2} < t + 1 \end{cases} \quad (17)$$

4 EXAMPLE ANALYSIS

In this paper, we take a 500 kV substation in Guangxi as an example to analyse and verify the cost-effectiveness optimization model. The substation main electrical equipment wiring show in figure 2. The #1 main transformer was put into the substation in June 2000, chromatography online monitoring data is shown in Table 1.

By analysis of the transformer fault diagnosis, the state of the transformer changes from "abnormal state – 3" into "fault state – 4", and the failure rate is $P_{fi} = 0.00917$, the equivalent age recursion is $t_a = 17.11$ years.

Assume that the transformer's designed operational life is $T_{age} = 35$ years, A maintenance strategy is as follows: (1) minor repairs: At the time of Fourth years in the second quarter, Seventh years in the fourth quarter, Sixteenth in the first quarter. (2) major repairs: At the time of Tenth years in the second quarter, Eighteenth years in the third quarter. In this strategy, the age recursion time is 4.92 years, and the minimum cost of operations and maintenance is 2317000 yuan.

For the main transformer (T1) #1, according to analysis of preventive test data, the partial discharge exceed threshold of partial discharge magnitude, the failure positions may be located in the casing.

For a transformer in the partial outage risk condition, the failure rate is $P_f = 0.00917$, the losses of comprehensive risk are 45571000 yuan. For current state, the value of the failure rate, risk value, risk return and the risk benefit/cost ratio when taking different maintenance strategies are shown in Table 4.

Adopting the minor repair method can make the failure rate fall to a low level, the reliability of the equipment is guaranteed, and the risk return is the highest. Therefore, for the main transformer we will take minor repairs based on optimization model.

5 CONCLUSION

1) Considering the influence of different maintenance strategies on failure rate, we use age recursion back function to reflect the difference, the transformer failure rate is more reasonable and accurate.
2) In the whole life cycle of a transformer, it is essential to have a fix maintenance strategies, and to establish a maintenance optimization model based on cost – effectiveness, which can improve the transformer's reliability and economy of the transformer operation.
3) The example analysis proves that the maintenance strategy based on cost-effectiveness model in this paper are reasonable.

REFERENCES

[1] Brophy James M., Erickson Lonny J. Cost-effectiveness of Drug-eluting Coronary Stents in Quebec, Canada [J]. International Journal of Technology Assessment in Health Care, 2005, 21(3): 326–333.

[2] Gao Zhe, Liao Xiao-zhong. Improved approximation of fractional-order operators using adaptive chaotic particle swarm optimization [J]. Journal of Systems Engineering and Electronics, 2012, 23(1): 145–153.

[3] Hardwick, G.M. Reliability centered maintenance at Florida Power Corporation [C]. IEEE Power Engineering Society Summer Meeting, 1999, 6(2): 1169–1170.

[4] Pan Lezhen, LU Guoqi; Zhang Yan. Decision-making Optimization of Equipment Condition-based Maintenance According to Risk Comprehensive Evaluation [J]. Automation of Electric Power Systems, 2010, 34(11): 28–33.

[5] Walker W.E. Overview and Cost-effectiveness Analysis [J]. Safety Science, 200035(1): 105–121.

[6] Wang Yi, Wang Hui-fang. Research on maintenance type selection with CBM technology based on effectiveness and costanalysis [J]. Power System Protection and Control, 2010, 38(19): 39–45.

[7] Zhao Dengfu, Duan Xiaofeng. Maintenance Scheduling of Power Transmission Equipment Considering Equipment Condition and System Risk [J]. Journal of Xi'an. Jiaotong University, 2012, 46(3): 84–89.

[8] Zhang Huai-yu, Zhu Song-lin, Research and Implementation of Condition-Based Maintenance Technology System for Power Transmission and Distribution Equipment [J]. Power System Technology, 2009, 33(13): 70–73.

[9] Zeng Qinghu, Qiu Jing, Liu Guanjun. On Equipment Degradation State Recognition Using Hidden Semi-Markov Models [J]. Mechanical Science and Technology for Aerospace Engineering, 2008, 27(44): 429–434.

A new decision support system for a power grid enterprise overseas investment

Liu Tuo
State Grid Energy Research Institute, Beijing, China

Zhang Yi
State Grid Corporation of China, Beijing, China

ABSTRACT: This paper constructs an optimization model for overseas strategic investments of power grid enterprise basing on the Monte Carlo method, and then develops a decision support system which can analyse the possibility and sensitivity of the power grid enterprise's overseas investment revenue.

1 INTRODUCTION

A power grid is made up of the strategic industries and infrastructure services industries on which the economic development of a country relies. Power grid enterprises are the lifeblood of the national economy. Most of the power grid enterprises focus on transnational investment in the power grid, power development, electrical equipment exports (electrical engineering contracting), and resources mergers and acquisitions.

At present they use the SWOT model (with a focus on the market and the investment opportunity analysis) to set their strategy target, and then calculate the overseas assets and profit share index.

In practice, there are mainly two problems. On the one hand, due to the sample size and data accumulation and information level, some advanced software tools, such as PIMS, IBP, Vanguard, cannot always be directly applied, which affects the quantitative analysis of the strategic planning ability. On the other hand, a financial sensitivity analysis can only analyze the impact of the change of one or two factors; this directly impacts the planning accuracy.

According to the feature of grid operating overseas business, such as weak predictability, profit model difference, and complex capital circulation, this article builds an optimization model for international business of power grid enterprise based on the Monte Carlo method, which is designed to fit more international business characteristics, improving the ability of the quantitative analysis.

2 MODEL AND DESIGN

2.1 *Mathematical model construction*

This paper divides the international business into investment acquisitions and international trade. The investment business category includes transnational mergers and acquisitions, power development, and resource acquisition. These can be divided into the Mergers and Acquisitions, and green field projects. The international trade category includes electrical equipment export and electricity engineering contracting businesses. Their profits are calculated by the main market scale change, electric enterprises' possible market share, and business sales calculating profit margins, etc. All kinds of business' overseas profits after taxes are converted into RMB(CNY).

In practice, almost all these factors are accidental uncertainty factors. Nonetheless, their probability distribution and range can be judged. This belongs to the uncertainty optimization problem with stochastic parameters. The Monte Carlo method is the main solution to these problems. This article adopts the Monte Carlo method to build a new model for power grid enterprises' international business strategy, which can calculate the probability of realizing the goal of international business benefits.

At present, the international business goals of Chinese power grid enterprise are proportion of overseas profits or assets. Meanwhile the overseas profits proportion are expecting to maximize. As shown in the following formulas:

$$\max y = \frac{\left[\sum_k \sum_i \sum_j s_{ijk} z_{ijk} g_{ijk} + \sum_p \sum_i \sum_q s'_{piq} z'_{piq} g'_{piq} + \sum_t \sum_m \sum_n r_{mn}(1+\gamma_{mn})^t b_{mnt} u_{mnt}\right] \kappa(1-\lambda)}{O_t} \quad (1)$$

$$E\left\{\frac{\left[\sum_i\sum_j z_{ij}g_{ij} + \sum_i\sum_j z'_{ij}g'_{ij}\right]\kappa}{P_t}\right\} \geq \eta_t \quad (2)$$

$$E(y) \geq \varepsilon_t \quad (3)$$

$$a_{ijk} \leq s_{ijk} \leq b_{ijk} \quad (4)$$

$$a_{ijk} \leq g_{ijk} \leq b_{ijk} \quad (5)$$

$$\xi_{ik} \leq \sum_j\sum_k z_{ijk} \leq \psi_{ik} \quad (6)$$

$$\pi \leq \kappa \leq \theta \quad (7)$$

$$\Re \leq \lambda \leq \hbar \quad (8)$$

$$\alpha_{mn} \leq \gamma_{mn} \leq \beta_{mn} \quad (9)$$

$$\chi_{mn} \leq b_{mn} \leq \varphi_{mn} \quad (10)$$

$$\Im_{mn} \leq u_{mn} \leq \ell_{mn} \quad (11)$$

$$0 \leq \sum_k\sum_i\sum_j z_{ijk}g_{ijk}(1-\lambda_{ijk}) \leq T_t \quad (12)$$

All variable meanings are shown in the table below:

Table 1. Variable meaning.

Name	Meanings
y	The proportion of overseas profits
$E(y)$	Expectations of Overseas profits share
s_{ijk}	ROE of No. j projects using No. k operations in the No. i investment business
z_{ijk}	the new assets of No. j projects using No. k operations in the No. i investment business
g_{ijk}	the company equity share of No. j projects using No. k operations in the No. i investment business
s'_{piq}	ROE of No. q projects using No. p operations in the No. i investment business
z'_{piq}	existing assets of No. q projects using No. p operations in the No. i investment business
g'_{iq}	the company equity share of No. q existing projects using No. p operations in the No. i investment business
r_{mn}	the No. m trade services business in the No. n target market capacity in the basic year
γ_{mn}	the average annual growth rate of No. n target market capacity in No. m trade services business in the No. t year
b_{mnt}	the share of No. n target market in No. m trade services business in the No. t year
u_{mnt}	the profit ratio of sales of No. n target market in No. m trade services business in the No. t year
λ_{ijk}	asset-liability ratio of No. j projects using No. k operations in the No. i investment business
ε_t	the planning target of the overseas profits proportion
η_t	the planning target of the overseas assets proportion
κ	exchange rate's changes
λ	the tax burden of shafting profits to homeland
T_t	the investment volume of international business in the No. t year
O_t	the planning target of total profit in the No. t year
P_t	the planning target of assets in the No. t year

To calculate the influence of the different allocation proportions of overseas profits of the company's overseas profits proportion target, the above model is adjusted as follows: $T_i = \{i = 1, 2, 3, 4\}$ means the total investment, including the M&A and greenfield investment on the grid assets, power development, and resources mergers and acquisitions. So, $z_{ijk} = \frac{T_{ijk}}{g_{ijk}(1-\lambda_{ijk})}$. T_{ijk} is the investment volume of No. j projects using No. k operations in the No. i investment business, and then plugged into equation 1, getting the following models:

$$\max y = \frac{\left[\sum_k\sum_i\sum_j \frac{T_{ijk}}{g_{ijk}(1-\lambda_{ijk})} s_{ijk}g_{ijk} + \sum_p\sum_i\sum_q s'_{piq} z'_{piq} g'_{piq} + \sum_t\sum_m\sum_n r_{mn}(1+\gamma_{mn})^t b_{mnt} u_{mnt}\right]\kappa(1-\lambda)}{O_t} \quad (13)$$

$$E\left\{\frac{\left[\sum_i\sum_j \frac{T_{ijk}}{g_{ij}(1-\lambda_{ij})}g_{ij} + \sum_i\sum_j z'_{ij}g'_{ij}\right]\kappa}{P_t}\right\} \geq \eta_t \quad (14)$$

$$E(y) \geq \varepsilon_t \quad (15)$$

$$\sum_k\sum_i\sum_j T_{ijk} = T_t, \quad \forall T_{ijk} \geq 0 \quad (16)$$

$$a_{ijk} \leq s_{ijk} \leq b_{ijk} \quad (17)$$

$$a_{ijk} \leq g_{ijk} \leq b_{ijk} \quad (18)$$

$$\pi \leq \kappa \leq \theta \quad (19)$$

$$\Re \leq \lambda \leq \hbar \quad (20)$$

All the parameters are not negative. In practice, almost all these factors are accidental uncertainty factors. Nonetheless, their probability distribution and range can be judged. This belongs to the uncertainty optimization problem with stochastic parameters.

2.2 Monte Carlo model construction

This article adopts the Monte Carlo method to build a new model for power grid enterprises' international business strategy, which can calculate the probability of realizing objective of international business benefits.

This paper is based mainly on the following considerations: Firstly, international businesses are affected by many factors, which can give full play to the advantages of the Monte Carlo method for simulating

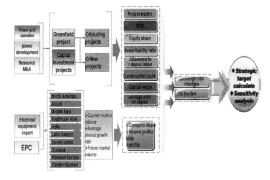

Figure 1. The new model for a Power Grid Enterprise's overseas investment.

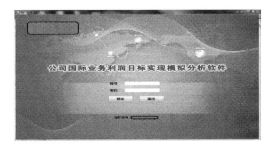

Figure 2. The login interface for the Decision Support System.

multivariate factors' changes at the same time. Secondly, the investment opportunities of the power grid operation are difficult to forecast. The global grid asset M&A, the greenfield investment history data, and the overall size of the Chinese grid enterprises who can participate need to be estimated, which is suitable for the Monte Carlo method. Thirdly, the ROE of the power grid operation is relatively stable; the probability distribution types and the range of the upper and lower bounds are apparent, which can be estimated by expert judgment on the basis of case analysis. Fourth, the electrical equipment export and electricity engineering contracting business overseas faces fierce market competition, while the type of business yields distribution probability and its range can be calculated according to the historical data. Fifth, using the Monte Carlo method with financial analysis models is closer to the characteristics of the power grid enterprise and its capital circulation process.

3 DECISION SUPPORT SYSTEM DEVELOPMENT

Based on the above model, this paper has developed a new type of decision support system. The Login interface as in Figure 2:

After entering the user name and password, it will enter the business profile display module, with an interface as in Figure 3:

Figure 3. The basic situation of International Business.

Figure 4. The background data module.

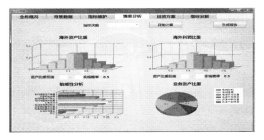

Figure 5. The scenario analysis module.

The business overview display module cannot only show the geographical distribution of projects, but also shows the proportion of global business conditions. However, these data can be maintained through the background data module.

In the indicator maintenance module, it can select the type and range of the distribution of all indicators. These data can be imported from existing data, they can also be assessed based on expert opinion.

After the basic work above, we can enter the scenario analysis module. This module can analyze the probability of achieving the specific gravity and the foreign asset share of the profits overseas, the sensitivity factors can also be analysed and the results can be visually displayed.

4 CONCLUSION

This new decision support system can carry out scenario analysis for power grid enterprise's overseas investment strategy, by analyzing the possibility and influence factors of overseas profits and assets proportion. It can help power grid companies easily analyze their overseas investment strategy, and set strategic planning objectives. It can then better meet the needs of a power grid enterprise's international business strategy planning analysis.

REFERENCES

[1] Liutuo, Chengcheng. Study on the Optimization Model for International Business Strategic of Power Grid Enterprise. The Seventh International Conference on Computational Sciences and Optimization.
[2] Liutuo, "On The International Management Strategy of Chinese Power Enterprise", East China economic management. 2012. 4: 67–72.
[3] Liutuo, "Construction Design of Grid Enterprise Strategic Plan's Performance Evaluation Model". 2011 2nd International Conference on Management Science and Engineering (MSE 2011). 2011. 11: 396–401.
[4] Liutuo, "Enlightenment from Transnational Enterprise Experience to Power Grid Enterprise", Economic research reference. 2013. 19: 75–79.
[5] Liutuo, "Enlightenment from the Central Enterprise Internationalization Experience". Foreign Investment of China. 2011. 12: 70–71.
[6] Liutuo, "Multinational enterprise internationalization strategy analysis". Energy Comments. 2011. 10: 100–107.
[7] Liutuo, "Power grid enterprise international business sector classification research." China's Urban Economy. 2011. 12: 157–15.

A coal mine video surveillance system based on the Nios II soft-core processor

Pingjun Wei
Zhengzhou University of Industry Technology, XinZheng, Henan, China
Zhongyuan University of Technology, Zhengzhou, Henan, China

Leilei Shi
Zhongyuan University of Technology, Zhengzhou, Henan, China

ABSTRACT: The design of the mine in digital video surveillance systems based on Field-Programmable Gate Array (FPGA) embedded Nios II soft-core processor. We use the System-on-a-Programmable Chip (SOPC) design concept, to give the system function and overall structural design, and then use a combination of software and hardware to achieve the design of the mine's video surveillance system. This design improves system performance to guarantee its stability. It has high real-time performance, reduces the development cycles, and reduces the development costs. This is of high significance for the improvement of the openness and reliability of mine video surveillance systems.

1 INTRODUCTION

In recent years, a series of mine accidents have had a significant impact on society. Research on coal mine video surveillance systems is very relevant and significant for the national economy and people's livelihood.

With the progress of large scale integrated circuit design technology and the improvement of manufacturing technology levels, the function and processing power of FPGA is growing stronger and stronger. Compared with the traditional design method, Single Chip Microcomputer (SCM), the design of coal mine video surveillance systems based on the Nios II soft-core processor has incomparable advantages.

This paper focuses on Altera's System-on-a-Programmable Chip (SOPC) design concept. It uses embedded Nios II soft-core processor and puts forward the implementation scheme of a mine's video surveillance system based on Nios II. A reconfigurable system-on-a-chip (SOC) based on an FPGA integrates the module processor, memory, and I/O ports needed by the system into an FPGA; it completes the logic function of the whole system. The system hardware is designed to be flexible, scalable, extendible, upgradeable and programmable, which is a flexible and efficient solution.

2 SYSTEM OVERALL DESIGN

Nowadays, dozens of kinds of monitoring systems are used by China's coal mine, but there are problems in terms of flexibility and interoperability:

(1) Most of the traditional master control parts use SCM, Advanced RISC Machine (ARM), which do not have the superior performance of flexibility and development that the Nios II soft-core processor has.
(2) Currently, the real-time transmission rate of the field bus and industrial Ethernet technology used at the beginning is relatively low, which is difficult to meet the demand for video transmission.
(3) In the industrial process monitoring and other automation fields, the application of $4 \sim 20\,mA$ analogue transmission is the norm. Field bus, industrial Ethernet, and multimedia technology are widely used in coal mine monitoring, enhancing the level of coal mine surveillance technology. However, each system is self-contained and incompatible with the other, which is not conducive to the development of an integrated coal mine monitoring system.

In order to promote openness and interoperability between systems, you can select the embedded system and FPGA technology to achieve the "triple play."

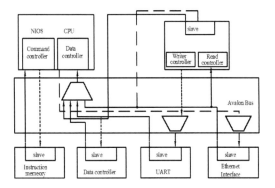

Figure 1. Nios II core structure.

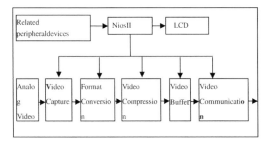

Figure 2. System scheme.

SOPC technology is the result of the integration of embedded systems and FPGA technologies, to achieve an embedded processor in the FPGA chip.

Nios II is a second generation configurable soft-core processor launched by Altera company for programmable logic devices. The biggest feature is that it can be tailored according to the needs of the actual applications. Nios II processor has a 32-bit instruction set, 32 data channels, and configurable instruction and data buffering. In particular, it has been optimized for programmable logic designs.

The Nios II core structure is shown in Figure 1. The system overall design is shown in Figure 2.

(1) The entire master control system essentially consists of core Nios II FPGA to complete, from the capture of the analogue, video transmission and communication, with the host computer.
(2) The Analogue video module consists mainly of the CMOS camera, and captures real-time video data in the mine.
(3) The video capture module converts analogue data from the camera into digital video data.
   The video Analog/Digital chip is selected as ADV7181B, and decode, sample the input analog video signal is sampled and simplify the process.
(4) Due to the large amount of data of video signals, it is necessary to change the format and compress the stored video data for processing and transmission
(5) The video display module designs VGA timing and display video captured by the camera via VGA interface on the local CRT monitor.

Figure 3. Nios II system components.

(6) The video communication partis controlled by the Nios II processor and completes to communicate with the host computer.

3 DESIGN AND IMPLEMENTATION OF MAIN PARTS

3.1 Nios II system components

The Nios II system consists of the Nios II processor, Universal Asynchronous Receiver/Transmitter (UART) components, RAM module, Erasable programmable configurable serial (EPCS) component, and the display components. Its internal components are automatically connected with an SOPC builder software. According to actual needs, users can add or remove components flexibly, which embodies the advantages of flexibility and adaptability of the Nios II soft-core processor. The system components are shown in Figure 3.

3.2 Video capture

Video capture is a prerequisite for video image, you can add 0.5 at the time during operation. The reason why the colour space conversion module is set up after the output buffer module mainly to reduce FIFO processing, transmission and the digital video images will directly affect the outcome of video processing. Figure 4 shows the structure of the video capture of an image pro cessing system. A multi-standard video decoder chip ADV7181B conducts analog-to-digital conversion for the video image captured.ADV7181B can automatically detect base band video signals such as NTSC, PAL and SECOM, and convert them based on 4:2:2 samples of 16/8 compatible digital video signal CCIR601/CCIR656 formats, with a six-channel analog video input port.

ADV7181B simultaneously inputs the luminance signal,the chrominance signal (TD_DAT), and horizontal and vertical sync signals (TD_HS/VS) of the real-time digital video image converted, into the FPGA chip. It extracts the digital image information by the image acquisition module, and moves it to the SRAM for buffering the pending image frames.

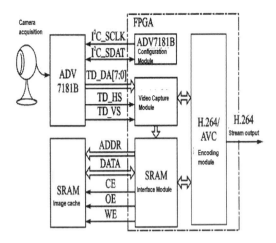

Figure 4. Video capture configuration of the image processing system.

3.3 Video buffer

The module is an asynchronous FIFO (First In First Out) device with a depth of 1024 and a width of 16 bits, as well as a write clock and read clock of 65 MHz. After the end of the SDRA Minitialization, the module firstly reads out 1024 of the data from the other SDRAM then it makes the timing generator module begin to work. When the enable reading signal is active, in synchronization with the read clock, the read address counter counts from 0 to 1023 cycles; when the counter reaches to 100 or 700, the module sends the writerequest command to the main controller module, so that each can read out 512 M in a row from the SDRAM. Since the period of output lines, and the vertical blanking, the enable signal is interactive, so there is none of reading void phenomenon.

3.4 Colour space converter module

The output buffer module outputs YC_bC_r (4:2:2) format signal, the output is in the following order:

$Y_0C_{b0}\ Y_1C_{r0}\ Y_2C_{b2}\ Y_3C_{r2}\ Y_4C_{b4}\ Y_5C_{r4}$ ......

This design uses then interpolation, the interpolated image format

$Y_2C_{b2}C_{r2}\ Y_3C_{b4}C_{r2}\ Y_4C_{b4}C_{r2}$ ......

The YC_bC_r colour space to RGB colour space conversion formula is:

$R = 1.164(Y- 16) + 1.596(C_r- 128)$
$G = 1.164(Y- 16) - 0.813(C_r- 128) - 0.392(C_b- 128)$
$B = 1.164(Y- 16) + 1.017(C_b- 128) + (C_b- 128)$

where Y is in the range (16,235), C_bC_r in the range (16,240), in order to prevent the subtraction overflow occurs, the deformationof theconversion formula is obtained by:

$R= (1.164Y+1.596C_r) -222.912$

$G=(135.616+1.164Y)-(0.392C_b+1.813C_r)$

$B=(1.164Y+2.017C_b)-276.8$

When describing the decimal multiplication by Very-High-Speed Integrated Circuit Hardware Description Language (VHDL) language, after the decimal expansion of integer power of 2 times, it rounds and multiplies; then the result is shifted operations. When the result of the operation needs to be rounded, the SDRAM memory in the input buffer and output buffer module will be done.For example, in the same precision, one frame of the image stored in SDRAM will occupy 1024 × 768 × 24 bits storage space, colour space conversion module after the output buffer module will reduce one-third of the storage space, which greatly improves the performance of the system.

4 CONCLUSION

Most FPGA integrated Nios II soft-core processor co-designs hardware and software on an FPGA and provides a powerful support to achieve programmable system-on-programmable-chip (SOPC). With the improvement of the semiconductor manufacturing process, FPGA integration will continue to increase; the cost will continue to decrease; FPGA is suitable.

This paper uses the Nios II embedded processor as the CPU to achieve the function of algorithms operation. In the system design process, it uses a GW48 type SOPC development platform and Cyclone II series EP2C35 chip to complete the basic functions. FPGA technology cannot only deal with the image data, but also voice data, which will help achieve integrated services in the coal mine; it can further develop the digital video surveillance system, form embedded wireless video transmission system; it may also increase appropriate language functions; Embedded systems port to this hardware platform could provide effective support for more advanced applications and also increase staff attendance, temperature display, gas concentration display. Therefore, it has great flexibility, practicality, and scalability, with better economic prospects.

REFERENCES

[1] Liang Haijun, Zhao Jianji in NIOSII high-resolution imageacquisition system [J]. China Measurement & Testing Technology, September 2008, Volume 34, No. 5.
[2] Pinson, EDA technology and VHDI [M]. Beijing: Tsinghua University Press, 2005.
[3] Zhang Yu, Qing Lin Bo, Su Yi Video capture systems based on FPGA [J], Chengdu University of Information Technology in April 2008 Volume 23, No. 2.

[4] Liya Chun, Huang Qian, Zhuhui Bin, Liu. SOPC-based digital video surveillance system design [J], Security Technology in October 2008.
[5] Wang Gang, Zhang Lian. SOPC FPGA-based embedded system design and typical examples [M], Electronic Industry Publishing House 2009.
[6] Yang Wang, Song Mouping, Chen Peng. Embedded network real-time video monitoring system [J]. Jiangnan University (Natural Science Edition) October 2008 Volume 7.

Research on a mechanism for measuring force in material moulding and the features of its measuring circuits

H.P. An
Institute for Mechanical Testing and Fault Diagnosis, Lanzhou City University, Lanzhou, China
College of Mechano-Electronic Engineering, Lanzhou University of Technology, Lanzhou, China

Z.Y. Rui
College of Mechano-Electronic Engineering, Lanzhou University of Technology, Lanzhou, China

R.F. Wang
LanZhou Jiao Tong University, Lanzhou, China

ABSTRACT: For detecting forces in material deformation, this paper systematically analyses the working mechanism of a strain transducer and the measuring principle of an electrical bridge. Strain gauges are based on the strain effect of metal and the piezoresistive effect of semiconductors, respectively. Factors affecting resistance-rate and sensitivity of a transducer involve its physical dimension and the specific-resistance rate after loading. Balancing conditions of several electrical bridges with transducers are offered for the design of detecting the force system in material formation. Performance features of direct and alternate current bridges are analysed.

Keywords: resistance sensor; strain effect; piezoresistive effect; measuring bridge

1 INTRODUCTION

Detection and controlling of forces in material moulding and processing are key links of process-system design and part-quality assurance [1, 2]. Using a strain gauge type transducer we can convert the applied force caused by material deformation to an electrical signal. It has less composing links and is easy to use for measuring forces, moment, pressure, acceleration, and weight [3, 4]. The peculiarity of a resistance sensor has great influence on its working performance, so there is practical significance with regard to giving full play to the resistance sensor's efficiency and broadening its applied range for rational design and correct selection and use of a sense element. Therefore, studying the working mechanism of a strain gauge type transducer and character of a detection circuit in this paper may supply a reliable basis for the design of a measuring force system and the effective use of a sense element.

2 WORKING MECHANISM OF A STRAIN-GAUGE TYPE TRANSDUCER

There are two kinds of strain-gauge-type transducers according to the mechanisms of electric resistance variation of materials. One is a metallic-resistance strain sensor based on strain effect switched force-strain-resistance, and the other is a

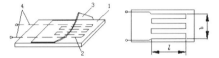

a) strain gage composite and structure b) length and width of sensitive grid

Figure 1. Structure diagram of resistance strain gage.

semiconductor-resistance strain sensor switched force-silicon piezoresistance based on piezoresistive effect.

2.1 Sense mechanism of a metallic-resistance strain

The alteration of length and sectional area for a metal conductor results in a variation of its resistance value, which is called strain effect because of the variation of the resistance value caused by the generation of strain. From electricity we know that resistance of a conductor is, $R = \rho l/A$, which relates to length (l) and sectional area (A). Applying force to it changes the two parameters and causes the resistance value to change, while the specific resistance remains nearly unchanged. Therefore, an applied load can be determined by means of measuring the resistance value and using suitable calibration.

Structure and composition of resistance strain gage. The resistance strain gage shown in Figure 1a consists of an alloy resistance wire (2) which is a mechanical coil, pasted to a substrate (1) made of organic materials. It is covered by a piece of thin film (3) and draws lead wires (4) from resistance, which forms an entirety. Figure 1b is the schematic diagram of a sensitive grid with length *l* and width *b*.

The initial resistance of a metallic conductor is given by:

$$R = \rho \frac{l}{A} \quad (1)$$

where l = length of resistance wire; A = cross sectional area; ρ = specific resistance.

When the resistance wire receives force along the length direction, changes of ρ, l and A cause changes of the resistance value. Make a complete differential to both ends of Equation 1, that is:

$$dR = \frac{\rho}{A} dl + \frac{l}{A} d\rho - \frac{\rho l}{A^2} dA \quad (2)$$

We obtain a total differential Equation 3 from Equation 1 and Equation 2 as:

$$\frac{dR}{R} = \frac{dl}{l} + \frac{d\rho}{\rho} - \frac{dA}{A} \quad (3)$$

Equation 3 shows that a relative change in resistance increases with the relative length and specific resistance, whereas it decreases with the cross sectional area add.

If the resistance wire is circular, and its section area is $A = \pi r^2$, taking the differential to the radius obtains $dA = 2\pi r dr$ or $dA/A = 2dr/r$. Assume $dl/l = \varepsilon_x$, which is an axial strain. Similarly, let $dr/r = \varepsilon_r$ is a radial strain. In an elastic range, a relation of axial and radial strains is $\varepsilon r = -\mu\varepsilon_x$, where μ is Poisson's coefficient of the crosswise efficient of the materials (between 0.25~0.5). Then we obtain:

$$\frac{dR}{R} = \frac{d\rho}{\rho} + (1+2\mu)\varepsilon_x \quad (4)$$

Make $K_s = dR/(R\varepsilon_x)$, then we have:

$$K_s = \frac{d\rho/\rho}{\varepsilon_x} + (1+2\mu) \quad (5)$$

where K_s = sensitivity coefficient, showing the relative change of resistance caused by per unit axial strain. From Equation 5 we know that the sensitivity of the resistance strain gage is affected by two factors, one is the change of physical dimension, i.e. $1+2\mu$, the other is the change of specific resistance after being forced, i.e. $d\rho/(\rho\varepsilon_x)$. For certain metal materials, the two are a constant, and $d\rho/(\rho\varepsilon_x)$ is very small. So, the former $(1+2\mu)$ plays a leading role. The value of which is between 1.5~2.

Stress-strain gauging. As a strain gage is fixed firmly on a measured material, the same deformation in strain gage and measured object occurs. By measuring the change (ΔR) in resistance, the strain value (ε) can be worked out. The relation of stress and strain is as follows:

$$\sigma = E\varepsilon \quad (6)$$

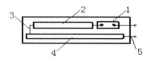

Figure 2. Semiconductor strain gage.

where σ and E are stress and elasticity modulus respectively, measured in (MPa).

Stress (σ) is proportional to strain (ε), so is to the relative change of resistance (dR/R), which is the principle underlying the use of a strain gage to measure stress. Strain gages must be bonded fast to object to prevent error.

2.2 *Sense mechanism of semiconductor resistance strain*

A semiconductor-strain gage is based on the piezoresistive effect of semiconductor materials. The effect is a phenomenon of resistance variation, which is caused by the change of specific resistance as an axial force is applied to materials. Figure 2 shows the structure of a strain gage composed of a sensitivity bar (2), intraconnection track (3), pin connection strap (4) and outer lead (5) on the substrate (1). The relative change of resistance of the strain gage is:

$$\frac{\Delta R}{R} = \frac{\Delta \rho}{\rho} + (1+2\mu)\varepsilon_x \quad (7)$$

where $\Delta\rho/\rho$ = relative change of specific resistance of materials. Its ratio to the axial strain from the semiconductor sensitivity bar is a constant, that is:

$$\frac{\Delta \rho}{\rho} = \pi E \varepsilon_x \quad (8)$$

where π = piezoresistance coefficient of the semiconductor material; E = elasticity modulus; ε_x = axial strain.

By substituting Equation 8 into Equation 7, we obtain:

$$\frac{\Delta R}{R} = \pi E \varepsilon_x + (1+2\mu)\varepsilon_x \quad (9)$$

where item $(1+2\mu)$ varies with the material's geometrical shape, πE (is the piezoresistive effect) relates to the specific resistance. Test shows that πE is 100 times bigger than $(1+2\mu)$, therefore $(1+2\mu)$ may be ignored. Then the sensitivity coefficient of the strain gage can be expressed as:

$$K_B = \frac{\Delta R/R}{\varepsilon_x} = \pi E \quad (10)$$

Big elastic modulus of semiconductor material makes strain gauge has such strong points, as small volume, sensitive, wide range of frequency response and big output amplitude that it can be used to connect with recorder directly. Its defects involve a big temperature coefficient, serious nonlinearity, and specific resistance relying on crystal orientation.

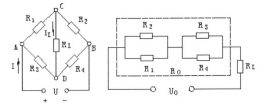

a) Bridge circuit   b) Equivalent bridge circuit

Figure 3.   Direct bridge circuit.

3 ELECTRICAL BRIDGE OF RESISTANCE-STRAIN SENSOR

The output signal of the sensor is so feeble that it cannot be detected. There is therefore a need to transform the signal for displaying or recording by an instrument. The change of value in the resistance or inductance or capacitance in an electrical bridge will cause that of voltage or current. By measuring the change and calibrating the parameters, we can determine the cutting force. This is the working principle of Direct Current (DC) and Alternating Current (AC) bridges.

3.1 Method of DC bridges

(1) Working principle. Figure 3(a) shows the fundamental form of an electrical bridge using DC power supply. The four arms of the bridge are composed of resistance R_1, R_2, R_3, and R_4. There is a DC main (U) between the A and B ends while there is a measuring meter between C and D with internal resistance (R_L) and electric current (I_L). The equivalent circuit of the electric bridge is shown in Figure 3(b). The floating voltage of the circuit is as follows:

$$U_o = U \frac{R_1}{R_1 + R_2} + U \frac{R_4}{R_3 + R_4} \qquad (11)$$

The equivalent resistance is:

$$R_o = \frac{R_1 R_2}{R_1 + R_2} + \frac{R_3 R_4}{R_3 + R_4} \qquad (12)$$

The current flowing internal resistance is:

$$I_L = \frac{U_o}{R_L + R_o}$$
$$= \frac{U(R_1 R_3 - R_2 R_4)}{R_L(R_1+R_2)(R_3+R_4) + R_1 R_2 (R_3+R_4) + R_3 R_4 (R_1+R_2)} \qquad (13)$$

The voltage of both ends of R_L is $U_L = I_L R_L$. When an instrument or amplifier with an extreme value of internal resistance is connected between C and D, i.e. $R_L \to \infty$, we can obtain Equation 14 from Equation 13:

$$U_L = \frac{U(R_1 R_3 - R_2 R_4)}{(R_1+R_2)(R_3+R_4)} \qquad (14)$$

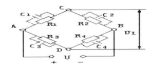

Figure 4.   Distributed capacitance circuit.

The above narration shows that the measuring of strain (ε), stress (σ) and force (F) by calibrating voltage U_L forms the principal for making a strain gauge, stress gauge, and measuring cell, respectively.

(2) Outline and difference of a bridge's properties. If all resistances of bridge arms vary, i.e. $R_1 \to R_1 + \Delta R_1$; $R_2 \to R_2 + \Delta R_2$; $R_3 \to R_3 + \Delta R_3$; $R_4 \to R_4 + \Delta R_4$, the bridge will output voltage. Suppose that resistance (ΔR_i) is far smaller than R_i, it satisfies $\Delta R_n \to 0$ ($n \geq 2$) and $R + \Delta R \approx R$. The high-order term of R in Equation 14 may be ignored. Then it may be simplified as:

$$U_L = \frac{U}{4R}(\Delta R_1 - \Delta R_2 + \Delta R_3 - \Delta R_4) \qquad (15)$$

① One-arm work bridge. If one arm (Assume R_1 works while others are fixed) works, Equation 15 becomes:

$$U_L = \frac{U}{4R} \cdot \Delta R \qquad (16)$$

② Neighbour two-arm work bridge. If both R_1 and R_2 bridge arms work, their increments are ΔR_1 and ΔR_2, respectively, Equation 15 becomes:

$$U_L = \frac{U}{4R}(\Delta R_1 - \Delta R_2) \qquad (17)$$

As it has $\Delta R_1 = \Delta R_2$, we obtain $U_L = 0$. Also it has $\Delta R_1 = -\Delta R_2 = \Delta R$, $U_L = \Delta U_R/(2R)$, the sensitivity of which is two times as big as the single-arm bridge.

③ Opposite two-arm work bridge. Assume R_1 and R_3 as the work-arm, their increments are ΔR_1 and ΔR_3, respectively. Equation 15 becomes:

$$U_L = \frac{U}{4R}(\Delta R_1 + \Delta R_3) \qquad (18)$$

Similarly, suppose $R_1 = R_3 = R$ in Equation 18, then $U_L = UR/(2R)$. Because $R_1 = -R_3 = R$, exist $U_L = 0$.

④ Four-arm differential bridge. Assume R_1, R_2, R_3, and R_4 as the work arm, their increments them are $\Delta R_1 = \Delta R_3 = \Delta R$, and $\Delta R_2 = \Delta R_4 = -\Delta R$, respectively. Equation 15 becomes:

$$U_L = \frac{U}{R} \cdot \Delta R \qquad (19)$$

Equation 19 indicates that the output of the differential bridge is five times as big as the one-arm electrical bridge. Its sensitivity has increased by four times.

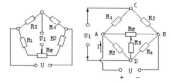
a) Resistance balance in series b) Resistance balance in parallel

Figure 5. Resistance balance circuit.

3.2 Alternating-Current (AC) bridges

An AC bridge can overcome the zero drift of a DC bridge. It includes distributed capacitance and inductance among wires. The influence of the capacitance is much bigger than that of the inductance, the circuit of which is shown in Figure 4. Let the voltage supplying bridge be:

$$U = U_m \sin \omega t \quad (20)$$

where U_m = maximum amplitude of voltage; ω = angular frequency of voltage; t = time. Impedance of j bridge arm is:

$$Z_j = \frac{1}{1/R_j + i\omega C_j} \quad (j=1,2,3,4) \quad (21)$$

where R_j = resistance of j bridge arm; C_j = capacitance of j bridge arm; i = an imaginary unit. The output voltage of the AC bridge is:

$$U_L = \frac{Z_1 Z_3 - Z_2 Z_4}{(Z_1 + Z_2)(Z_3 + Z_4)} U_m \sin \omega t \quad (22)$$

So the equilibrium condition of the bridge in Figure 4 is $Z_1 Z_3 = Z_2 Z_4$.

3.3 Methods of bridge balance

The equilibrium condition of a bridge is $U_L = 0$. A DC bridge differs from an AC bridge.

(1) Balance of DC bridge. A DC bridge is considered resistance-balance only. It may be a series balance or a parallel one. The series bridge balanced by adjusting a variable resistance (R_w) between R_1 and R_2 in series is shown in Figure 5(a). The parallel one in Figure 5(b) can also be balanced by adjusting R_w with a fixed resistance R_5.

(2) Balance of AC bridge. An AC bridge involves capacitance and inductance. A test proves that the influence of distributed capacitance is much bigger than that of the inductance. There are two methods of fixed and variable capacitance for a capacitive-balance circuit. The fixed one is shown in Figure 6(a); it consists of capacitance (C) and potentiometer (R_H). The bridge balance is performed by sliding the contact position of the potentiometer to change the values of the resistance and capacitance. The variable capacitance (C_V) in Figure 6(b) changes the capacitance

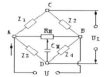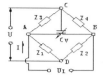
a) Fixed-capactance balance b) Variable capacitance balance

Figure 6. Capacitance balance circuit.

connected to Z_1 and Z_2 by increasing one part and decreasing another of the capacitance.

4 CONCLUSIONS

(1) Metallic-resistance sensors work based on a strain effect relating to the strain theory of elastic mechanics. Strain concerning dimension change of the strain gauge caused by a deformation of the object plays a leading role.

(2) Semiconductor-strain sensors work based on the piezoresistive effect of materials. This sensor has such merits as small volume, high sensitivity, wide range of frequency response, and big amplitude.

(3) A DC bridge is simple in structure and convenient to regulate and control, whereas an AC bridge is complex in structure, but it can overcome the zero drift problem of a DC bridge. A twin bridge is higher by one times than a wheat stone bridge in sensitivity. A differential bridge is higher by four times than a one-arm in sensitivity.

(4) Of the four methods of bridge balance, the resistance circuit is used for balancing a DC bridge and a capacitive circuit for balancing an AC bridge. The former is simpler than the later in operation.

ACKNOWLEDGEMENTS

The support of this study by the National Nature Science Foundation of China under Grant No. 51065014 and Natural Science Foundation of Gan Su Province No. 145RJZA134 is greatly appreciated.

REFERENCES

[1] Yu Yan. Engineering materials and processing basis [M]. Beijing: Press of Beijing Institute of Technology, 2012.7.
[2] Wang Xu ewen, zhang zhi yong. Fundamentals of sensors and application [M]. Beijing: Press of Beijing University of Aeronaut and Astronaut, 2004.
[3] Li Chen xi, Qu Ying dong, Hang Zheng xiang et al. Material molding detecting techniques [M]. Beijing: Chemical Industry Press, 2007.4
[4] H. P. An, Z. Y. Rui, R. F. Wang et.al. Research on cutting-temperature field and distribution of heat dates among a workpiece, cutter, and chip for high-speed cutting based on analytical and numerical methods [J]. Strength of Materials, Vol. 46, No. 2, March, 2014.

Author index

An, H.P. 351
An, J.S. 211

Cai, X. 333
Cao, Q.Z. 95
Cao, W. 275
Chai, W.L. 333
Chen, B. 195
Chen, H. 57
Chen, P.Z. 13
Chen, W.Y. 333
Chen, X.C. 185
Chen, Y. 135
Chen, Y.H. 71
Cheng, H.M. 113
Cheng, L.F. 185, 261
Ciucur, V.V. 321
Coanda, H.G. 167

Dang, J. 7
Darwis, F. 177
Di, P.Y. 185

Enescu, D. 167
Eskikurt, H.I. 63

Fan, C.X. 285
Fan, T. 129
Fang, Q.S. 113
Fang, Y. 39
Fu, X. 7

Gang, H. 289
Geng, Z. 299
Gu, W.S. 181
Guan, L. 185, 261
Guan, Y.P. 225
Guo, D.B. 281
Guo, L.J. 339
Guo, Y.G. 57
Guo, Y.J. 75

Han, S.M. 201
Han, Y.S. 275
He, D.C. 289
Hsieh, M.C. 161
Hu, J. 293
Hu, J.X. 123
Hu, X.G. 53
Huang, B. 135
Huang, H.Y. 231
Huang, L.J. 39

Huang, W. 75
Huang, X.Y. 231
Hui, H. 271

Iwai, H. 23

Jang, H. 191
Jeong, S. 151, 191
Ji, J.H. 143, 259
Ji, Y.K. 143
Jiang, H.T. 75
Jiang, J.L. 147, 311
Jiang, Y. 267
Jiang, Y.T. 267
Jin, H. 23
Jing, G. 57
Jong, G.J. 13
Ju, Y.W. 305
Jun, S.C. 87, 143

Kakushima, K. 23
Kang, J. 151
Karabacak, M. 63
Kim, Y. 151, 191

Lee, M.C. 161
Lee, S.C. 87
Lei, J.Y. 231
Lei, T. 333
Li, B.Z. 71
Li, C.F. 305
Li, F. 195
Li, G.S. 201
Li, G.Y. 299
Li, H.X. 33
Li, J.S. 215
Li, J.Y. 325
Li, L.F. 293
Li, M.Y. 241
Li, R.R. 185
Li, S. 211
Li, S.Y. 219, 275
Li, X.J. 253
Li, X.M. 247
Li, X.R. 237
Li, Y. 285
Li, Y.S. 33
Lim, J. 191
Lin, B.H. 13
Lin, M. 311
Liu, C. 219
Liu, H.K. 267

Liu, P.Y. 27, 147
Liu, S. 247
Liu, W. 271
Liu, X.S. 101
Liu, Y. 95
Liu, Z. 95
Liu, Z.Q. 237
Lu, Y.Y. 71
Luo, J. 155
Lv, X. 53
Lv, Y.T. 261

Miao, S.H. 271
Mu, L. 267

Nam, M.S. 259
Nedelcu, O. 167
Ning, C. 1
Ning, J. 7

Pan, C.Y. 13
Pan, F. 305
Patil, U.M. 87
Peng, F.J. 231
Peng, P.L. 13

Qin, K.Y. 195
Qiu, J.H. 83
Qiu, P. 289

Raza, A. 45
Ren, B. 219
Ren, Y.B. 267
Ruhiyat, D. 177
Rui, Z.Y. 351

Salisteanu, C.I. 167
Saputera, Y.P. 177
Seong, C.J. 259
Shakery, I. 259
Shen, X.Z. 281
Shen, Z. 231
Shi, L.L. 347
Shi, Y.L. 253
Shin, K.S. 143
Song, M.C. 7
Song, Q. 211
Song, X.F. 17, 205
Sun, P. 39

Tan, Y. 117
Tang, Y. 7
Tao, J.H. 57
Tao, S.M. 339

Tian, Y.M. 317
Tsai, T.I. 161
Tuo, L. 343

Ulansky, V. 45

Virjoghe, E.O. 167

Wahab, M. 177
Wang, C.F. 147
Wang, D.L. 89
Wang, F.Z. 155
Wang, G.Z. 129
Wang, H. 135, 181
Wang, H.W. 317
Wang, L. 281
Wang, M.F. 325
Wang, N.N. 83
Wang, R.F. 351
Wang, S. 185
Wang, S.T. 107
Wang, T. 261
Wang, W.L. 113
Wang, Y.F. 89
Wei, H. 57
Wei, P.J. 347
Wen, B.J. 155
Wen, Z.G. 101, 285
Wijaya, I. 177
Wong, H. 23
Wu, T. 219, 275
Wu, X.J. 305

Xian, X.H. 201
Xiao, L.Y. 83
Xiao, S.W. 201
Xie, Y. 39
Xu, A.D. 231
Xu, Y.H. 219, 275
Xue, F. 17, 205

Yan, J.L. 253
Yan, T. 123
Yan, X. 195
Yan, X.Q. 205
Yang, E. 195
Yang, M.M. 317
Yang, Y. 75
Yao, X.H. 123
Yao, Y. 39
Ye, C. 271
Yi, Z. 343
Yu, D. 117
Yu, L. 231
Yu, N.H. 293
Yu, X.D. 201
Yuan, C. 219

Zhai, Z.Y. 285
Zhang, C.Q. 225
Zhang, C.X. 329
Zhang, D.W. 329
Zhang, H. 27, 299
Zhang, J. 23
Zhang, J.G. 27

Zhang, J.M. 117
Zhang, J.Q. 23
Zhang, M.M. 325
Zhang, P.Y. 83
Zhang, W. 329
Zhang, X. 107
Zhang, X.H. 53
Zhang, X.P. 293
Zhang, Y. 83, 117
Zhang, Y. 83
Zhang, Z.L. 95, 117
Zhang, Z.W. 17
Zhang, Z.Y. 113
Zhao, D.Z. 247
Zhao, J.J. 53
Zhao, M.X. 271
Zhao, Q. 261
Zhao, Y. 117
Zhao, Z.Q. 215
Zhou, B. 261
Zhou, B.R. 185
Zhou, C.W. 155
Zhou, L. 17
Zhou, X.H. 267
Zhou, Y. 17, 135, 205
Zhu, L. 285
Zhu, Y. 211
Zhu, Y.H. 215
Zong, H. 83
Zou, B. 241